भारतस्य महान् गणितज्ञाः
(१९ शताब्द्याः आरम्भपर्यन्तम्)

# The Great Mathematicians of Bharat

## (Till Early 19th Century)

Compiled By:
**Partha Pratim Ray**

2023

**Rajmangal Prakashan**

An Imprint of **Rajmangal Publishers**

**ISBN : 978-8119251995**

*Published by :*

**Rajmangal Publishers**

Rajmangal Prakashan Building,
1st Street, Ozone, Quarsi, Ramghat Road
Aligarh-202001, (UP) INDIA
**Cont. No. +91- 7017993445**
www.rajmangalpublishers.com
rajmangalpublishers@gmail.com
sampadak@rajmangalpublishers.in

---

**प्रथम संस्करण :** नवम्बर 2023 – पेपरबैक

**प्रकाशक : राजमंगल प्रकाशन**

राजमंगल प्रकाशन बिल्डिंग, 1st स्ट्रीट,

ओजोन, क्वार्सी, रामघाट रोड,

अलीगढ़, उप्र. – 202001, भारत

**फ़ोन :** +91 - 7017993445

---

**First Published :** Nov. 2023 - Paperback
**Printed by :** Thomson Press India Ltd, Repro India Ltd & Manipal Tech Ltd.
**eBook by :** Rajmangal ePublishers (Digital Publishing Division)

तदेतत्त्रयँ शिक्षेद् दमं दानं दयामिति

Learn three cardinal virtues – temperance, charity and compassion for all life.

— *Brihadaranyaka Upanishad, V.ii.3*

# भारतस्य महान् गणितज्ञाः
## (१९ शताब्द्याः आरम्भपर्यन्तम्)

## The Great Mathematicians of Bharat
## (Till Early 19th Century)


Compiled
By
Partha Pratim Ray

2023

ॐ गं गणपतये नमः

Dedicated

To

The *Great*

*Munis* (मुनि), *Rishis* (ऋषी), *Darshaniks* (दार्शनिकाः), *Pandits* (पण्डिताः),
*Mathematicians* (गणितज्ञः) *and Bidwans* (विद्वाणाः)

of

Bharat

# Preface

In the enchanting tapestry of Bharat's heritage, woven with diverse threads of culture, philosophy, arts, and sciences, there resides a brilliant and intricate motif - the saga of our mathematical maestros. '*The Great Mathematicians of Bharat*' is not merely a chronological documentation of the luminaries who illuminated the realms of mathematics, but it is also a tribute to Bharat's ancient wisdom and its profound mathematical traditions ingrained in the very fabric of Bhartiya life and Hindu culture. Throughout this book, I have used Bharat to specify "India" and Bhartiya for "Indian" in almost of all places where I could best fit these words to signify its deep meanings. Also, I use used Sanskrit and Devnagari translations to many English words to give the readers the actual feel of the content and related words. I use data source majorly from Wikipedia pages to get information about the mathematicians while compiling this book.

The essence of this book emanates from the integral relationship between the Bhartiya culture and mathematical practices. Our mathematical odyssey begins with revered sages like Lagadha and Baudhayana and journeys through time, celebrating the ingenious minds such as Aryabhata, Bhaskara, and the remarkable Srinivasa Ramanujan, among others. Each page of this book unveils the contributions of these sagacious souls who deciphered the mysteries of numbers, geometry, astronomy, and metrics, which were seamlessly overlapped with various aspects of daily life, religious practices, and philosophical constructs in Bharat.

The pivotal notion that this compiled book upholds is the splendid symbiosis of science and spirituality. It delineates how mathematical principles were harmoniously intertwined with Hindu culture, manifesting in architectural marvels, astrological computations, rhythmic musical patterns, and even in the philosophical realms of existential inquiry. Concepts like zero, the decimal system, and sophisticated astronomical models emerged from the fertile lands of Bharat, reflecting a civilization deeply engrossed in mathematical explorations and scientific curiosity.

As you traverse through the pages of this book, you'll encounter the elegance of Hindu numerals, the meticulous precision of Hindu time and metrics, and the astronomical brilliance encapsulated in the Hindu calendar. These are not just cold calculations or rigid formulas, but they echo the wisdom, the devotion, and the celestial connection of our great mathematicians who viewed mathematics as a divine language, a sacred tool to unravel the universe's myriad mysteries.

*'The Great Mathematicians of Bharat'* is an ode to our mathematical heritage, an invitation to dive deep into the ocean of numbers and theories enriched by the vibrant currents of our culture and traditions. It is an effort to reignite the ancient pride, to revisit and revere the extraordinary mathematical minds and the incredible legacy they bequeathed to the world.
With a heart swelling with pride and a mind inspired by our ancient seers' intellectual grandeur, I invite you to embark on this enlightening journey through the mathematical marvels of Bharat.

This exhaustive compendium, *'The Great Mathematicians of Bharat'*, unfurls the glorious journey of Indian mathematics from the ancient to the modern era, encapsulating the essence and evolution of mathematical wisdom inherent in the cultural and spiritual ethos of Bharat. Our endeavor is to meticulously document and celebrate the luminaries of mathematical Bharat, who have intricately woven the fabric of mathematics with threads of our profound culture, spirituality, and daily existence.

Chapter 1 revisits the luminaries of the Ancient Era, where sagacious minds like Lagadha, Baudhayana, and Yajnavalkya graced the mathematical arena with foundational concepts. These profound thinkers not only pioneered various mathematical concepts but also interlaced them with the realms of astronomy, linguistics, and philosophy, showcasing the multifaceted brilliance of Bhartiya intellect. Their contributions resonate with the vibrancy of Hindu culture, revealing a universe where mathematics wasn't merely a scientific discipline but a holistic approach to understanding the cosmos and its divine principles.

In the Classical era, as illustrated in Chapter 2, luminaries like Aryabhata I and Varahamihira emerged, illuminating the path of astronomical and mathematical advancements. Aryabhata's astounding revelations in astronomy and mathematics echo the harmonic convergence of scientific rigor and cultural richness in Bharatiya tradition. This period signifies a golden era where the mathematical renaissance flourished with remarkable theories and practices, embodying the synthesis of spiritual inquiry and scientific curiosity.

Chapter 3 explores the Early Medieval Period, marked by mathematical giants like Brahmagupta and Bhaskara. This era was adorned with extraordinary contributions, especially in algebra, trigonometry, and astronomy. Their legacies, encapsulated in brilliant manuscripts, unfold the mysteries of zero, infinity, and planetary motions, intertwined with the spiritual and cultural aspects of daily life, resonating with the divine rhythm of the universe.

In the Late Medieval Period, delineated in Chapter 4, the tradition of mathematical brilliance was carried forward by scholars like Madhava and Nilakantha Somayaji. Their unparalleled contributions laid the foundation for mathematical analysis, calculus, and a refined understanding of astronomical phenomena. Here, we witness a spectacular confluence of tradition and innovation, where ancient wisdom blossomed with new insights and explorations.

Chapter 5 navigates through the Early Modern to Late Modern Period, embodying the spirit of continuous evolution and adaptation. It celebrates the era of Srinivasa Ramanujan, whose intuitive genius brought a refreshing vibrancy and depth to mathematical thought. This phase exemplifies the endurance and resurgence of Bhartiya mathematical traditions, continuing to resonate with the essence of our profound cultural and spiritual heritage.

The subsequent chapters delve into the specialized realms, emphasizing the Kerala School of Astronomy and Mathematics, exploring the elegance and global influence of Hindu Numerals, and illuminating the scientific brilliance encapsulated in Hindu Time, Metrics, and Calendar systems. These chapters embody the intricate architecture of Bhartiya mathematical traditions, showcasing the seamless integration of mathematical sciences in our cultural and cosmological explorations.

*'The Great Mathematicians of Bharat'* is a tribute to the timeless legacy of our mathematical traditions. It invites you on a transformative journey through epochs, revealing the symphony of mathematics and culture in the diverse and divine landscape of Bharat. Embark on this enlightening exploration to discover the treasures of wisdom bestowed by the great mathematicians, and experience the radiant confluence of science, spirituality, and cultural brilliance that signifies the mathematical heritage of Bharat.

The target audience for "*The Great Mathematicians of Bharat*" is quite diverse and expansive, reflecting the universality and profundity of the book's content. Here's a breakdown of the primary audiences:

- *Academicians and Scholars*

  Professors and educators involved in mathematics, history of science, Indology, and cultural studies. Researchers and scholars who specialize in the history of mathematics, especially those with a focus on Indian contributions.

- *Students*

  Students of mathematics who wish to delve deep into the historical and cultural roots of various mathematical concepts and discoveries. History students with a special interest in the history of science and technology in ancient civilizations, particularly India. As per NEP2020, this book can be considered as a special compiled book under the Indian Knowledge System (IKS) for both UGC and AICTE approved institutions.

- *Cultural Enthusiasts*

  Individuals interested in Indian culture, heritage, and historical contributions to global knowledge. Those who are curious about the synthesis of science and spirituality in ancient Indian traditions.

- *Educational Institutions*

  Libraries of schools and universities, to serve as a reference book for various disciplines. Cultural and historical societies within educational institutions.

- *Mathematical Societies and Organizations*

  Members of mathematical societies who appreciate a historical perspective on mathematical evolutions. Organizations promoting the rich tradition of mathematical research and development.

- *General Readers*

  Those with a curiosity about the evolution of mathematical ideas and their interplay with cultural and spiritual aspects. Readers who appreciate interdisciplinary reading that combines history, culture, spirituality, and mathematics.

- *Spiritual Seekers*

  Individuals exploring the intersection of spirituality and science, particularly in ancient wisdom traditions. Seekers interested in understanding the universal truths expressed through mathematical language in ancient Indian sciences.

- *Policy Makers and Educational Planners*

  Those involved in curriculum development who wish to integrate diverse historical and cultural perspectives in modern education. Policy makers aiming to promote the richness of indigenous knowledge and its relevance to contemporary education and research.

*"The Great Mathematicians of Bharat"* is not just a historical documentation but a confluence of the river of mathematical wisdom with the vast ocean of cultural and spiritual enlightenment. It appeals to a wide spectrum of readers, encouraging them to explore the multifaceted gems of ancient Indian mathematical brilliance and its profound connection with the essence of Bharat's cultural heritage.

Partha Pratim Ray
31 October, 2023

## Reviving the Radiance: Bharat's Mathematical Brilliance in the Global Landscape

In a world that increasingly leans towards Western paradigms of knowledge and innovation, *"The Great Mathematicians of Bharat"* is not merely a book, but a luminous vessel carrying the timeless flames of Bharat's mathematical genius. It is a monumental endeavor that aims to resurrect the forgotten or obscured treasures of Bhartiya mathematical heritage in the midst of the fast-paced advancements of the modern world.

- **Acknowledging the Roots**

  Western accomplishments in mathematics and science, undoubtedly profound, have often been spotlighted at the forefront of intellectual progress. However, the roots of many monumental discoveries trace back to the ancient scholars of Bharat, who nurtured the tree of mathematical wisdom with their profound insights and explorations. This book acts as a formidable bastion against the erosion of historical truth, safeguarding the acknowledgment and appreciation of Bharat's unparalleled contributions that have subtly flowed into the vast oceans of global mathematical thought.

- **Empowering Indigenous Wisdom**

  In an era where the shadows of external influences loom large, this magnificent volume emerges as a potent catalyst to empower the indigenous wisdom of Bharat. It passionately advocates for the recognition, respect, and revival of the native intellectual traditions, encouraging societies to stand steadfast in their unique identities and contributions amidst the currents of global change.

- **Cultural and Spiritual Harmony**

  One of the profound attributes of *"The Great Mathematicians of Bharat"* is its sublime portrayal of the intrinsic harmony between mathematics, culture, and spirituality. It reveals the divine tapestry where numbers, patterns, and geometries intertwine with philosophy, consciousness, and the rhythms of life, offering a multifaceted perspective that transcends the conventional boundaries of mathematical exploration.

- **Educational Enrichment**

As a beacon of knowledge, the book illuminates educational realms with the radiant lights of historical depth and cultural wisdom. It fosters a nurturing environment where learners can blossom in the rich soils of heritage, cultivating a profound appreciation for the roots of mathematical knowledge and its universal expressions.

- **Navigating Global Conversations**

In global conversations, where diverse voices converge to weave the narrative of humanity's journey, this book ensures that the vibrant threads of Bharat's mathematical brilliance are woven with dignity and grace. It inspires a dialogical ethos, fostering open-hearted exchanges and mutual enrichment in the global intellectual symphony.

- **Future Inspirations**

As we navigate towards future horizons, *"The Great Mathematicians of Bharat"* stands as an inspiring lighthouse, guiding minds and hearts towards the shores of holistic understanding and innovative possibilities. It encourages the blossoming of a future where mathematical exploration is not just a technical endeavor but a beautiful journey of the soul through the realms of existence and creativity. Through its pages, the book breathes life into the ancient wisdom of Bharat, allowing its timeless radiance to illuminate minds and hearts across the horizons of time and space.

Partha Pratim Ray
31 October, 2023

## Elevating the Tapestry of Bharatiya Knowledge System

*"The Great Mathematicians of Bharat"* unfolds as an epic canvas, intricately woven with the threads of mathematical genius, ancient wisdom, and the evolving tapestry of Bharatiya knowledge systems. Enriched by the legacy of *Jnan* (ज्ञानं), *Vignan* (विग्नानं), and *Jeevan Darshan* (जीवन दर्शनं), the manuscript brings forth a symphonic harmony that resonates with the ideals and objectives of the *National Education Policy (NEP) 2020*. This work becomes the confluence of rivers of divine wisdom, gushing from the spiritual mountains of ancient sages, mingling with the streams of contemporary thought, thus nurturing the terrains of modern education, culture, and holistic human advancement.

- **Sustainable Foundations and Universal Harmony**

  In the heart of this masterpiece lies the intrinsic sustainability and harmony encapsulated within the Bharatiya traditions. The book extends an invitation to embark upon a transformative journey where mathematics emerges not merely as numeric or algorithmic formulations but flourishes as a philosophical ethos guiding the realms of universal harmony and balanced prosperity. It nurtures an ecosystem of thoughts where mathematical concepts and their applications become the roots that absorb the timeless wisdom of sustainable practices and the branches that reach out towards the welfare of global humanity.

- **Nurturing the Blossoms of Scholarly Excellence**

  Embarking upon a voyage through the pages of this tome, readers are embraced by the nurturing warmth of Bharatiya intellectual heritage. The scholarship embedded within the chronicles aligns harmoniously with the essence of NEP 2020, fostering an educational ambiance where the vibrant blossoms of curiosity and creativity flourish amidst the gardens of traditional wisdom and contemporary insights. The rich soil of this manuscript encourages the seeds of young minds to sprout into robust trees of intellectual mastery and compassionate beings contributing to global knowledge and societal well-being.

- **Integrative Exploration and Evolutionary Wisdom**

  Every chapter of this book reverberates with the evolutionary vibrance and exploratory spirit that characterizes Bharatiya knowledge systems. It is a harmonious orchestra where the symphonies of ancient wisdom and contemporary innovations resonate together, echoing the rhythms of continuous learning, integration, and holistic enlightenment. This text is a tapestry where the threads of time-honored wisdom interweave with the colors of modern innovations, manifesting a fabric of knowledge that adorns the halls of educational excellence and lifelong learning.

- **Illuminating Influences and Societal Integration**

  This volume artistically unveils the multifaceted influences of mathematical brilliance across an expansive spectrum of societal arenas such as governance, arts, culture, and overarching human life dimensions. It illuminates the landscapes of societal norms and functionalities with the radiant light of mathematical intricacies and logical acumen, portraying a vivid canvas where numbers and equations breathe life into administrative sagacity, cultural vibrancy, and harmonious human existence.

- **Linguistic Richness and Diverse Transmission**

  Celebrating the linguistic diversity and multifarious transmission modes inherent to Bharatiya knowledge systems, this book blossoms as a vibrant garden where varied linguistic petals bloom, and the fragrances of oral, textual, and artistic expressions pervade the intellectual atmospheres. It fosters an enriched educational ecology where the nectars of knowledge flow through the diversified conduits of linguistic expressions and traditional pedagogical pathways, thus nurturing minds and souls in the delightful gardens of linguistic and cultural diversity.

- **Future Vision: Aspirations and Realizations**

This meticulous work of scholarship is not just a historical chronicle but a futuristic vision that aligns seamlessly with the aspirations and directions carved out by NEP 2020. It bridges historical grandeur with contemporary potentials and future aspirations, crafting a seamless pathway that winds through the evolutionary milestones of Bharatiya mathematics towards the illuminating horizons of global educational frontiers and innovative societal landscapes.

- **Conclusion: A Radiant Confluence**

In conclusion, *"The Great Mathematicians of Bharat"* emerges as a radiant beacon where the divine luminosity of ancient wisdom, the vibrant colors of contemporary insights, and the aspiring rays of future possibilities converge to illuminate the skies of educational renaissance and intellectual evolvement. Each page becomes a pathway lit with the profound knowledge of great minds, guiding humanity through the journeys of educational exploration, societal advancement, and holistic human blossoming in the divine lights of Bharatiya wisdom.

Partha Pratim Ray
31 October, 2023

## Key Takeaways

*"The Great Mathematicians of Bharat"* unravels a treasure trove of insights, knowledge, and inspiration. Here are some key takeaways from this magnum opus:

- *Rich Historical Legacy*

  Understanding the rich historical tapestry of Bhartiya mathematics. Appreciating the chronological development and evolution of mathematical concepts and techniques across different eras.

- *Confluence of Culture and Mathematics*

  Exploring the deep interconnections between the Bhartiya culture and mathematical practices. Gaining insights into how mathematics was seamlessly integrated into various aspects of daily life and spiritual practices.

- *Diversity of Contributions*

  Discovering the wide array of contributions made by Bhartiya mathematicians from various regions and times. Recognizing the multifaceted genius of these mathematicians in areas like algebra, trigonometry, calculus, astronomy, and geometry.

- *Innovation and Creativity*

  Witnessing the innovative approaches and creative solutions put forth by Bhartiya mathematicians. Learning about pioneering concepts such as zero, the decimal system, and sophisticated astronomical models originated from Bharat.

- *5. Integration of Science and Spirituality*

  Realizing the synthesis of science and spirituality in the pursuit of mathematical knowledge. Understanding mathematics as a divine language and a sacred tool to explore universal truths.

- *Global Impact*

  Appreciating the global impact and dissemination of Bhartiya mathematical knowledge. Acknowledging the influence of Bhartiya numerals and mathematical concepts on global mathematical developments.

- *Inspiration and Wisdom*

  Drawing inspiration from the lives, dedication, and wisdom of the great mathematicians. Recognizing the perseverance and passion of these mathematicians in overcoming challenges and contributing to global knowledge.

- *Philosophical Depth*

  Delving into the philosophical underpinnings and existential inquiries that often-accompanied mathematical explorations. Exploring mathematics as a pathway to understanding the deeper realities of existence and the cosmos.

- *9. Educational Enlightenment*

  Serving as an educational resource to enrich the learning experience with historical and cultural perspectives. Offering a broader view of mathematics beyond mere numbers and formulas, portraying it as a living discipline interwoven with culture and life.

- *Celebration of Heritage*

  Celebrating the remarkable heritage and ancient pride associated with the mathematical advancements in Bharat. Cultivating a sense of appreciation and respect for the indigenous knowledge systems and their timeless relevance.

*"The Great Mathematicians of Bharat"* is more than a book; it is a journey through time, a celebration of intellectual spirit, and a tribute to the incredible legacy of mathematical wisdom in the cultural heart of India. It allows readers to traverse through epochs, exploring the magnificence of mathematical contributions that have emanated from the Indian subcontinent, illuminating the world with the light of knowledge and wisdom.

Partha Pratim Ray
31 October, 2023

## Acknowledgments

Before embarking on this journey through the mathematical landscapes of Bharat, I would like to extend my deepest gratitude to those who have made this book possible. This work is not just a product of solitary research but a symphony of collective wisdom, support, and inspiration.

First and foremost, I express my sincere appreciation to the scholarly community, whose extensive works and tireless dedication to uncovering the historical and cultural depths of Bharatiya mathematics have been invaluable. Their rigorous research and insightful publications have provided a robust foundation upon which this book is built.

I am immensely thankful to my mentors and teachers, both past and present, who have guided my intellectual pursuits and instilled in me a profound respect for our ancient knowledge systems. Their wisdom and teachings have been the guiding stars in this endeavor. My gratitude extends to the various institutions and libraries that have been generous with their resources, allowing me access to rare manuscripts and texts that have been pivotal in compiling this comprehensive account. The staff and curators of these institutions deserve special mention for their assistance and enthusiasm.

I must also acknowledge the role of the digital age and technological advancements, particularly the vast repository of information available through platforms like Wikipedia. These resources have been instrumental in cross-verifying facts and expanding the scope of this book.

A special word of thanks goes to the community of Indology and mathematics enthusiasts. Their keen interest, insightful feedback, and encouraging words have been a source of constant motivation. Their passion for rediscovering and celebrating our heritage is a true reflection of the spirit this book endeavors to capture.

I am grateful to my family and friends for their unwavering support and patience. Their understanding and encouragement have been my strength throughout the journey of writing this book.

Lastly, I extend my thanks to you, the reader. Your interest in exploring the rich tapestry of Bharatiya mathematical legacy is what brings this book to life.

May this journey enlighten, inspire, and reconnect us with the glorious intellectual traditions of our ancestors.

As we turn the pages of history and mathematics together, let us remember that this book is more than a mere academic exploration. It is a celebration of our heritage, a homage to the giants upon whose shoulders we stand, and a testament to the timeless wisdom that Bharat has bestowed upon the world.

Partha Pratim Ray
31 October, 2023

## Contents

# Chapter 1

# Ancient Era
(Before 320 CE)

## Lagadha (लगाध)
## (c. 1300 BCE)

Lagadha (लगाध) was the earliest mathematician to whom definite teaching can be ascribed to, and who used geometry and elementary trigonometry for his astronomy. He remains a pivotal figure in the annals of mathematical and astronomical history due to his association with the "*Vedanga Jyotisha*" (वेदङ्ग ज्योतिष). As mentioned, the *Vedanga Jyotisha* is the earliest known Indian astronomical text. Let's delve deeper into its significance and its content. He developed oral methods for calculations of ellipse, and also for elementary arithmetic. He was the first to suggest the concept of *yuga* - युग (era), a period of cycle of repeating astronomical events.

### 卐 Key Contributions in Mathematics

### 卐 Vedanga Jyotisha

The *Vedanga Jyotisha,* though commonly credited to *Lagadha,* is likely an amalgamation of astronomical knowledge passed down through generations. The text exists in two versions: one associated with the *Rigveda* (ऋग्वेद) and the other with the *Yajurveda* (यजुर्वेद). It's essential to understand that the Vedanga Jyotisha is not purely an astronomical document in the modern sense. It focuses on timekeeping for ritualistic purposes, ensuring that Vedic rituals were conducted at the most auspicious times.
*Vedanga Jyotisha* has two parts: (i) *Rik* (रिक्) and (ii) *Yajur* (यजुर्) recensions. The former part contains *36 verses* in Sanskrit. They deal with eclipses, a *lunar calendar* (चन्द्रपञ्चाङ्गम्) and units in time calculation.

*Vedanga Jyotisha* deals with the calculations of solar and lunar eclipses. According to Vedanga Jyotisha, the lunar month begins on a new moon day; one lunar month is 29.5 lunar days approximately; one lunar year is 12 lunar months, called lunation, and is of about 354 lunar days.

A lunar month is divided into 30 equal parts called *tithi* (तिथि).

A period of 19 lunar years is called a *Lagadha era.*

*Lagadha Yuga* (लगाध युग) begins on a new moon day which coincides with winter solstice, i.e., December 22.

### 卐 Sophisticated Time Measurement

Timekeeping was crucial in the *Vedic* (वैदिक) era. Lagadha's contributions in this realm went beyond simple day-night cycles. The *Vedanga Jyotisha* presents a luni-solar calendar, wherein time intercalation was applied to bring the lunar and solar calendars in sync. This method involves adding an extra month (*Adhikamasa* - अधिकामस) periodically to ensure synchronization. The text mentions a cycle of 5 years (युग) comprising 1,830 days, 62 lunar months, and 67 solar months.

A cycle of 5 years (Yuga) comprising 1,830 days, 62 lunar months, and 67 solar months.

### 卐 Mathematical Foundations in Vedanga Jyotisha

The text lays the foundation for several mathematical concepts. For instance, it provides methods to determine the diagonals of a rectangle and the properties of right triangles, introducing rudimentary geometric concepts. In terms of arithmetic, it touches upon basic operations and the division of a number into parts. This might sound rudimentary, but it was the precursor to more advanced mathematical developments in India.

## y Astronomical Observations

gadha's observations, as captured in the text, revolved around the correlation between terrestrial and celestial events. For example, he marked the solstices, recognizing their importance in the annual solar cycle. The *Vedanga Jyotisha* also provides methods to determine the north direction during night-time using the Pole Star, highlighting the observational prowess of the time.

## 卐 Trigonometry and Celestial Observations

While the *Vedanga Jyotisha* does not contain trigonometry as we understand it today, the principles it introduces (like the aforementioned geometric principles) would later lay the groundwork for more advanced trigonometric studies in Indian mathematics. The text hints at the use of basic angular measurements and shadows, indicating a primitive form of trigonometric understanding essential for astronomical observations.

## 卐 Further Reading


1. Chakraverti 2007, p. 33.
2. Debiprasad Chattopadhyaya, History of Science and Technology in Ancient India, Firma K.L Mukhopadhyaya (1986), pp. 486–494
3. Hart Defouw (1996). Light on Life: An Introduction to the Astrology of India. Penguin. p. 3. ISBN 978-0-14-019507-1.
4. Helaine Selin, Encyclopaedia of the History of Science, Technology, and Medicine in Non-Western Cultures, Kluwer Academic Publishers (1997), p. 977
5. Klostermaier 2010, p. 977.
6. Lagadha (2005). "Vedāṅgajyotiṣam: Yajurvedināṃ paramparayāgatam vistr̥tasaṃskr̥tabhūmikayā Somākarabhāṣyeṇa Kauṇḍinnyāyanavyākhyānena ca sahitam: Saṅkṣiptahindībhūmikā-Hindyānuvādādiyutaṃ vividhapariśiṣṭavibhūṣitaṃ ca".
7. Pingree, David (1973), "The Mesopotamian Origin of Early Indian Mathematical Astronomy", Journal for the History of Astronomy, 4: 1–12, Bibcode:1973JHA.....4....1P, doi:10.1177/002182867300400102, S2CID 125228353
8. Subbarayappa 1989, p. 29.

9. Subbarayappa 1989, pp. 25–40.
10. Subbarayappa 1989, pp. 47–48.
11. Witzel 2001.
12. Chakraverti, Santanu (2007), Science in History. In: Jyoti Bhusan Das Gupta (ed.), "Science, Technology, Imperialism, and War", Pearson Education India
13. Indian Astronomy - An Introduction (2000) published by Hyderabad University Press by Balachandra S Rao.
14. Klostermaier, Klaus (2010), A Survey of Hinduism (Third ed.), SUNY, ISBN 978-0-7914-8011-3
15. Lagadha Maharishi – Hindu Astronomer – Mathematician – Lived During 1200 BC By Abhilash Rajendran, February 02, 2019. Available Online https://www.hindu-blog.com/2019/02/lagadha-maharishi-hindu-astronomer.html, Accessed online 13 October, 2023.
16. N. P. Subramania Iyer (1991). Kalaprakasika. Asian Educational Services. p. 3.
17. Subbarayappa, B.V. (1989). "Indian astronomy: a historical perspective". In Biswas, S. K.; Mallik, D. C. V.; Vishveshwara, C. V. (eds.). Cosmic Perspectives. Cambridge University Press. ISBN 978-0-521-34354-1.
18. Vedāṅga jyotiṣa of Lagadha, translated by TSK Sastry, edited by KV Sarma, https://web.archive.org/web/20110501084408/http://www.new.dli.ernet.in/rawdataupload/upload/insa/INSA_1/20005abd_s1.pdf
19. Vedanga_Jyotisha, Available Online https://en.wikipedia.org/wiki/Vedanga_Jyotisha, Accessed on 23 October, 2023.
20. Witzel, Michael (25 May 2001). "Autochthonous Aryans? The Evidence from Old Indian and Iranian Texts". Electronic Journal of Vedic Studies. 7 (3).

**Baudhayana (बौधायन)**
**(fl. c. 900 BCE)**

Figure 1.1. Baudhayana (बौधायन). (Image Credit: https://twitter.com/DialogueSamvaad/status/1219491949599608832)

To write a biography of Baudhayana (बौधायन) (Figure 1.1) is essentially impossible since nothing is known of him except that he was the author of one of the earliest *Sulbasutras (सुल्बासूत्रा)*. We do not know his dates accurately enough to even guess at a life span for him, which is why we have given the same approximate birth year as death year. He would certainly have been a man of very considerable learning but probably not interested in mathematics for its own sake, merely interested in using it for religious purposes. Undoubtedly, he wrote the *Sulbasutra* to provide rules for religious rites (धार्मिक संस्कार) and it would appear an almost certainty that *Baudhayana* himself would be a *Vedic* priest (पुरोहित). The *Baudhāyana sūtras* (बौधायन सूत्रम्)

are a group of *Vedic Sanskrit* (वैदिक संस्कृत) texts which cover *dharma* (धर्म), daily ritual, mathematics and is one of the oldest *Dharma*-related texts of *Hinduism* that have survived into the modern age from the 1st-millennium BCE. They belong to the *Taittiriya* (तैत्तिरीय) branch of the *Krishna Yajurveda* (कृष्ण यजुर्वेद) school and are among the earliest texts of the genre.

The *Baudhayana sūtras* consist of six texts as shown in Figure 2. The mathematics given in the *Sulbasutras* is there to enable the accurate construction of altars needed for sacrifices. It is clear from the writing that *Baudhayana*, as well as being a priest, must have been a skilled craftsman. He must have been himself skilled in the practical use of the mathematics he described as a craftsman who himself constructed sacrificial altars of the highest quality. The *Baudhāyana Śulbasûtra* is noted for containing several early mathematical results, including an approximation of the *square root of* 2 and the statement of the *Pythagorean theorem*. Figure 1.2 shows the various branches of *Baudhayana* sutras.

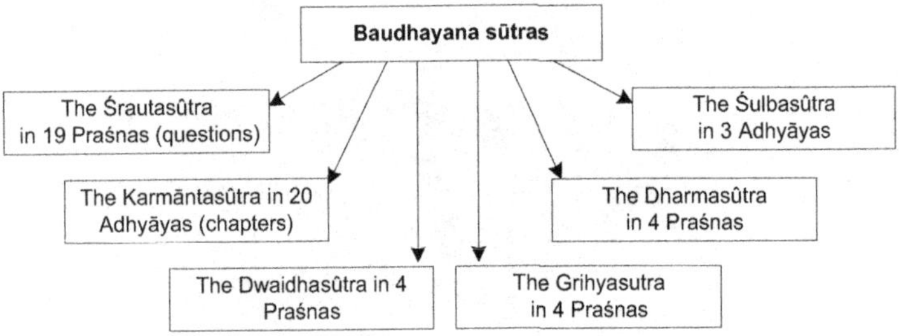


Figure 1.2. Texts of *Baudhayana* sutras.

## 卐 Key Contributions in Mathematics

### 卐 Inventor of Pythagorean theorem

The *Baudhāyana Śulvasūtra* states the rule referred to today in most of the world as the *Pythagorean Theorem*. The rule was known to a number of ancient civilizations, including also the Greek and the Chinese, and was recorded in Mesopotamia as far back as 1800 BCE. For the most part, the Śulvasūtras do not contain proofs of the rules which they describe. The rule stated in the *Baudhāyana Śulvasūtra* is as below.

| दीर्घचतुरस्रस्याक्षणया रज्जुः पार्श्वमानी तिर्यग् मानी<br>च यत् पृथग् भूते कुरूतस्तदुभयं करोति ॥ |
| --- |
| *dīrghachatursrasyākṣaṇayā rajjuḥ pārśvamānī, tiryagmānī,*<br>*cha yatpṛthagbhūte kurutastadubhayāṅ karoti.* |
| The diagonal of an oblong produces by itself both the areas which the two sides of the oblong produce separately.<br><br>Translation |

The diagonal and sides referred to are those of a rectangle (oblong), and the areas are those of the squares having these line segments as their sides. Since the diagonal of a rectangle is the hypotenuse of the right triangle formed by two adjacent sides, the statement is seen to be equivalent to the *Pythagorean theorem*.

*Baudhāyana* also provides a statement using a rope measure of the reduced form of the *Pythagorean theorem* for an isosceles right triangle:

> The cord which is stretched across a square produces an area double the size of the original square.

## 卐 Circling the Square

Another problem tackled by *Baudhāyana* is that of finding a circle whose area is the same as that of a square (the reverse of squaring the circle). His *sūtra* i.58 gives this construction:

> Draw half its diagonal about the centre towards the East–West line; then describe a circle together with a third part of that which lies outside the square.
>
> Translation - *sūtra i.58*

**Explanation:**

1. Draw the half-diagonal of the square, which is larger than the half-side by
$$x = \frac{r}{2}\sqrt{2} - \frac{r}{2}.$$
2. Then draw a circle with radius $\frac{r}{2} + \frac{x}{3} = \frac{r}{2} + \frac{r}{6}(\sqrt{2} - 1) = \frac{r}{6}(2 + \sqrt{2})$.
3. $(2 + \sqrt{2})^2 \approx 11.66 \approx \frac{36.6}{\pi}$.
4. So, the area is $\pi a^2 \approx \pi \times \frac{r^2}{6^2} \times \frac{36.6}{\pi} \approx r^2$.

## Values of π

Several values of $\pi$ occur in Baudhayana's Sulbasutra since when giving different constructions Baudhayana uses different approximations for constructing circular shapes. Constructions are given which are equivalent to below.

$$\pi = \frac{676}{225} = 3.004$$
$$\text{or} = \frac{900}{289} = 3.114$$
$$\text{or} = \frac{1156}{361} = 3.202$$

None of these is particularly accurate but, in the context of constructing altars they would not lead to noticeable errors.

## Square Root

Baudhāyana i.61-2 (elaborated in Āpastamba Sulbasūtra i.6) gives the length of the diagonal of a square in terms of its sides, which is equivalent to a formula for the square root of 2:

*samasya dvikaraṇī. pramāṇaṃ tṛtīyena vardhayet*
*tac caturthenātmacatustriṃśonena saviśeṣaḥ*

The diagonal [lit. "doubler"] of a square. The measure is to be increased by a third and by a fourth decreased by the 34th. That is its diagonal approximately. (Translation)

Thus,

$$\sqrt{2} \approx 1 + \frac{1}{3} + \frac{1}{3 \times 4} - \frac{1}{3 \times 4 \times 34} = \frac{577}{408} = 1.414215686$$

which is correct to five decimals i.e., 1.414216.

Out of other theorems, we present a few below:

| |
|---|
| *Diagonals of rectangle bisect each other* |
| *Diagonals of rhombus bisect at right angles* |
| *Area of a square formed by joining the middle points of a square is half of original* |
| *The midpoints of a rectangle joined forms a rhombus whose area is half the rectangle* |

It noteworthy that the emphasis on rectangles and squares; this arises from the need to specify *yajña bhūmikās* (यज्ञ भूमिका) - i.e., the altar on which rituals were conducted, including fire offerings (*yajna* - यज्ञ).

卐 **Further Reading**


1. Plofker, Kim (2007). Mathematics in India. p. 17. ISBN 978-0691120676. In relative chronology, they predate Āpastamba, which is dated by Robert Lingat to the sutra period proper, between c. 500 to 200 BCE. Robert Lingat, The Classical Law of India, (Munshiram Manoharlal Publishers Pvt Ltd, 1993), p. 20
2. Høyrup, Jens (1998). "Pythagorean 'Rule' and 'Theorem' – Mirror of the Relation Between Babylonian and Greek Mathematics". In Renger, Johannes (ed.). Babylon: Focus mesopotamischer Geschichte, Wiege früher Gelehrsamkeit, Mythos in der Moderne. 2. Internationales Colloquium der Deutschen Orient-Gesellschaft 24.–26. März 1998 in Berlin (PDF). Berlin: Deutsche Orient-Gesellschaft / Saarbrücken: SDV Saarbrücker Druckerei und Verlag. pp. 393–407.
3. Nanda, Meera (16 September 2016). "Hindutva's science envy". Frontline. Archived from the original on 17 July 2017. Retrieved 14 October 2016.
4. O'Connor, "Baudhayana".

5. Patrick Olivelle, Dharmasūtras: The Law Codes of Ancient India, (Oxford World Classics, 1999), p. 127
6. Patrick Olivelle, Dharmasūtras: The Law Codes of Ancient India, (Oxford World Classics, 1999), p. xxxi
7. Patrick Olivelle, Dharmasūtras: The Law Codes of Ancient India, (Oxford World Classics, 1999), pp. 128–131
8. Sacred Books of the East, vol.14 – Introduction to Baudhayana
9. "The Śulvasútra of Baudháyana, with the commentary by Dvárakánáthayajvan", translated by George Thibaut, was published in a series of issues of The Pandit. A Monthly Journal, of the Benares College, devoted to Sanskrit Literature:
10. (1875) 9 (108): 292–298
11. (1875–1876) 10 (109): 17–22, (110): 44–50, (111): 72–74, (114): 139–146, (115): 166–170, (116): 186–194, (117): 209–218
12. (new series) (1876–1877) 1 (5): 316–322, (9): 556–578, (10): 626–642, (11): 692–706, (12): 761–770
13. B.B. Dutta."The Science of the Shulba".
14. Baudhayana sutras, Available online, https://en.wikipedia.org/wiki/Baudhayana_sutras, Accessed on 23 October, 2023.
15. Baudhayana, Available online, https://mathshistory.st-andrews.ac.uk/Biographies/Baudhayana/, Accessed on 23 October, 2023.
16. G G Joseph, The crest of the peacock (London, 1991).
17. G Kumari, Some significant results of algebra of pre-Aryabhata era, Math. Ed. (Siwan) 14 (1) (1980), B5-B13.
18. George Gheverghese Joseph. The Crest of the Peacock: Non-European Roots of Mathematics, 2nd Edition. Penguin Books, 2000. ISBN 0-14-027778-1.
19. Ian G. Pearce. Sulba Sutras at the MacTutor archive. St Andrews University, 2002.
20. O'Connor, John J.; Robertson, Edmund F., "Baudhayana sutras", MacTutor History of Mathematics Archive, University of St Andrews St Andrews University, 2000.
21. O'Connor, John J.; Robertson, Edmund F., "The Indian Sulbasutras", MacTutor History of Mathematics Archive, University of St Andrews St Andrews University, 2000.
22. R C Gupta, Baudhayana's value of $\sqrt{2}$, Math. Education 6 (1972), B77-B79.

23. S C Kak, Three old Indian values of $\pi$, Indian J. Hist. Sci. 32 (4) (1997), 307-314.
24. S. Balachandra Rao, Indian Mathematics and Astronomy: Some Landmarks. Jnana Deep Publications, Bangalore, 1998. ISBN 81-900962-0-6
25. The English translation comes from George Thibaut's series of articles in The Pandit. (See References.) The translated passage is on page 298, volume 9. Thibaut remarks, "We should of course say 'rectangular triangles' instead of 'oblongs'. The length of the diagonals of these oblongs or of the hypotenuses of these rectangular triangles is not explicitly mentioned by Baudháyana. Ápastamba states it, when describing the different manners of constructing the vedi."
26. Vincent J. Katz. A History of Mathematics: An Introduction, 2nd Edition. Addison-Wesley, 1998. ISBN 0-321-01618-1

**Yajnavalkya (याज्ञवल्क्य)**
**(700 BCE)**

Figure 1.3. Yajnavalkya (याज्ञवल्क्य) teaches the *Brahma Vidya* to King Janaka. (Image Credit: https://en.wikipedia.org/)

Yajnavalkya (याज्ञवल्क्य) was the great *Vedic sage* (वैदिक ऋषि) figuring in the *Brihadaranyaka Upanishad* (बृहदारण्यक उपनिषत्). *Yajnavalkya* proposed and debated *metaphysical questions* (आध्यात्मिक प्रश्न) about the nature of existence, consciousness and impermanence, and expounds the epistemic doctrine of *neti neti* (नेति नेति) ("*not this, not this*") to discover the universal Self and *Ātman* (आतमन्). Texts attributed to him include the *Yajnavalkya Smriti* (याज्ञवल्क्य स्मृति), *Yoga Yajnavalkya* (योग याज्ञवल्क्य), various *Brahmanas* (ब्राह्मणा), *Aranyakas* (आरण्यक) and some texts of the *Vedanta* (वेदान्त) *school*. The *Brihadaranyaka Upanishad* is dated at c. 700 BCE. *Yajnavalkya* is derived from *yajna* (यज्ञ), which connotes ritual, *Yajnavalkya* is referred to as "*a thinker, not a ritualist*". Figure 1.3. depicts *Yajnavalkya* teaching *the Brahma Vidya* (ब्रह्म विद्या) to *King Janaka* (राजा जनक).

In the *Brihadaranyaka Upanishad*, a set of dialogues portrays *Yajnavalkya* as having two wives, one *Maitreyi* (मैत्रेयी) who challenges *Yajnavalkya* with philosophical questions like a scholarly wife; the other *Katyayani* (कात्यायनी)

who is silent but mentioned as a housewife. While *Yajnavalkya* and *Katyayani* lived in contented domesticity, *Maitreyi* studied metaphysics and engaged in theological dialogues with her husband in addition to "*making self-inquiries of introspection* (आत्मनिरीक्षणस्य आत्मजिज्ञासां कृत्वा)".

He concluded his dialogue on the "*inner self*", or *soul* (आत्मा), *Yajnavalkaya* tells *Maitreyi* as below.

> One should indeed see, hear, understand and meditate over the Self, O Maitreyi;
> indeed, he who has seen, heard, reflected and understood the Self – by him alone the whole world comes to be known.
>
> — Brihadaranyaka Upanishad 2.4.5b

## 卐 Key Contributions in Mathematics

### 卐 Refined Approximation to Pi (π)

*Yajnavalkya*'s approximation of pi, given as

$$\pi = 3.1428571$$

It stands out as a remarkably accurate representation for its time. The significance of this is amplified when one realizes that the accurate determination of $\pi$ has roots in circular geometry, which is fundamental to various scientific applications. This approximation suggests a sophisticated observational capability and hints at an advanced knowledge of geometry among the *Vedic scholars* (वैदिक विद्वांस).

### 卐 Geometry in the Context of Rituals

The construction of *ritualistic altars* (संस्कारात्मका: वेदी) demanded precision. *Vedic altars* came in various shapes: *squares, rectangles,* and even *complex bird-shaped structures.* Each had symbolic significance, and *Yajnavalkya* provided meticulous geometric rules for their construction. The need for perfect geometric constructions in rituals meant that the underlying mathematics was non-trivial. This activity indirectly furthered the understanding of geometry, laying foundational knowledge for future generations.

### 卐 Astronomical Contributions

*Yajnavalkya*'s association with astronomy was more than mere *star-gazing*. He demonstrated a keen understanding of the *celestial cycles* (आकाशचक्रम्), including the *synodic period of planets*, which refers to the time it takes for a planet to return to the same position relative to the sun as observed from *Earth* (पृथ्वी). His works provide methods to calculate specific celestial events, underlining the close relationship between mathematics and astronomy in ancient Bhartiya thought.

### 卐 Advanced Calendrical Systems

*Yajnavalkya*'s contributions to the calendar system demonstrate a profound understanding of both lunar and solar cycles. Recognizing the misalignment between these two cycles, he advocated for the addition of an intercalary month (*Adhikamasa* - अधिकामस) to reconcile the difference, ensuring *festivals* (उत्सवा) *and rituals* (संस्कारा) were observed at the appropriate season.

### 卐 Philosophical Underpinnings of Infinity

*Yajnavalkya*'s discussions on infinity in the "*Brihadaranyaka Upanishad*" are philosophical masterpieces. While the immediate context was *spiritual* (आध्यात्मिक), these discussions have mathematical implications. The very idea of discussing *infinity* (अनंतं) shows a deep conceptual understanding, which would later play a pivotal role in mathematical advancements in Bharat.

### 卐 Understanding of Large Number Systems

*Yajnavalkya*'s mention of large numbers, such as "*koti* - कोटि " (10 million), in the "*Shatapatha Brahmana* - शतपथ ब्राह्मण " is significant. It indicates an early number classification system and suggests that the ancients had a need to discuss, understand, and perhaps even compute with such large numbers.

**Further Reading**


1. "Yajnavalkya's Marriages and His Later Life". Shukla Yajurveda. Shuklayajurveda Organization. Archived from the original on 2 April 2015. Retrieved 2 April 2015.
2. Ben-Ami Scharfstein (1998), A comparative history of world philosophy: from the Upanishads to Kant, Albany: State University of New York Press, pp. 9-11 Olivelle 1998, p. xxxvi with footnote 20
3. Ben-Ami Scharfstein (1998). A Comparative History of World Philosophy: From the Upanishads to Kant. State University of New York Press. pp. 56–57. ISBN 978-0-7914-3683-7.
4. Brereton 2006, pp. 323–345.
5. Brereton 2006, pp. 323–45.
6. Brihadaranyaka Upanishad 4.4.5-6 Berkley Center for Religion Peace & World Affairs, Georgetown University (2012).
7. Deussen 2010, p. 435.
8. Hajime Nakamura (1968), Yajnavalkya and other Upanishadic thinkers in a Jain tradition, The Adyar Library Bulletin, Volume 31–32, pages 214–228.
9. Hino 1991, p. 5.
10. Hino 1991, pp. 54–59, 94–95, 145–149.
11. Hino 1991, pp. 5–6, 94.
12. Hino 1991, pp. 94–95.
13. Hock, Hans Henrich (2002). "The Yajnavalkya Cycle in the Brhad Aranyaka Upanisad". Journal of the American Oriental Society. American Oriental Society. 122 (2): 278–286. doi:10.2307/3087621. JSTOR 3087621.
14. Hume 1967, pp. 98–102, 146–48.
15. I Fisher (1984), Yajnavalkya in the Sruti traditions of the Veda, Acta Orientalia, Volume 45, pages 55–87.
16. Ian Whicher (1999), The Integrity of the Yoga Darsana: A Reconsideration of Classical Yoga, State University of New York Press, ISBN 978-0791438152, pages 27, 315–316 with notes.
17. John Muir, Metrical Translations from Sanskrit Writers, p. 251, at Google Books, page 246–251.
18. Jonardon Ganeri (2007). The Concealed Art of the Soul: Theories of Self and Practices of Truth in Indian Ethics and Epistemology. Oxford University Press. pp. 27–28, 33–35. ISBN 978-0-19-920241-6.
19. Larson & Bhattacharya 2008, pp. 476–477.

20. Macdonell, Arthur Anthony; Keith, Arthur Berriedale (1912). Vedic Index of Names and Subjects. Vol. II. John Murray. p. 189. Pechilis 2004, pp. 11–15.
21. Marvelly 2011, p. 43.
22. Max Muller, Brihadaranyaka Upanishad 4.3-4, Oxford University Press, pages 161-181 with footnotes.
23. Müller, F. Max (Friedrich Max) (1879–84). The Upanishads. Princeton Theological Seminary Library. Oxford, The Clarendon press. pp. 162–163.
24. Olivelle 1998, p. 28.
25. Patrick Olivelle (1992). The Samnyasa Upanisads: Hindu Scriptures on Asceticism and Renunciation. Oxford University Press. pp. 92, 140–146. ISBN 978-0-19-536137-7.
26. Patrick Olivelle (1993). The Asrama System: The History and Hermeneutics of a Religious Institution. Oxford University Press. pp. 92 with footnote 63, 144, 163. ISBN 978-0-19-534478-3.
27. Paul Deussen (2015). The System of the Vedanta: According to Badarayana's Brahma-Sutras and Shankara's Commentary thereon. KB Classics Reprint. pp. 173–174. ISBN 978-1-5191-1778-6.
28. Paul Deussen, Sixty Upanishads of the Veda, Volume 1, Motilal Banarsidass, ISBN 978-8120814684, page 482.
29. Paul Deussen, Sixty Upanishads of the Veda, Volume 1, Motilal Banarsidass, ISBN 978-8120814684, pages 475-507.
30. Scharfstein, Ben-Ami (1 January 1998). A Comparative History of World Philosophy: From the Upanishads to Kant. SUNY Press. p. 63. ISBN 978-0-7914-3683-7.
31. Scharfstein, Ben-Ami (1 January 1998). A Comparative History of World Philosophy: From the Upanishads to Kant. SUNY Press. p. 64. ISBN 978-0-7914-3683-7.
32. Vishwanath Narayan Mandlik, The Vyavahára Mayúkha, in Original, with an English Translation at Google Books, pages lvi, xlviii–lix.
33. White 2014, pp. xiii, xvi.
34. Brereton, Joel P. (2006). "The Composition of the Maitreyī Dialogue in the Bṛhadāraṇyaka Upaniṣad". Journal of the American Oriental Society. 126 (3). JSTOR 20064512.
35. Deussen, Paul (2010). Sixty Upanishads of the Veda. Motilal Banarsidass. ISBN 978-81-208-1468-4.
36. Frits Staal (2008). Discovering the Vedas: Origins, Mantras, Rituals, Insights. Penguin Books. p. 3. ISBN 978-0-14-309986-4., Quote: "Yajnavalkya, a Vedic sage, taught..." Olivelle 1998, pp. 3, 52–71.

37. Frits Staal (2008). Discovering the Vedas: Origins, Mantras, Rituals, Insights. Penguin Books. p. 365 note 159. ISBN 978-0-14-309986-4.
38. Hino, Shoun (1991). Suresvara's Vartika On Yajnavalkya'S-Maitreyi Dialogue (2nd ed.). Motilal Banarsidass. ISBN 978-81-208-0729-7.
39. Hume, Robert (1967). Brihadaranyaka Upanishad. (Translator), Oxford University Press.
40. Joseph, George G. (2000). The Crest of the Peacock: Non-European Roots of Mathematics (2nd ed.). London: Penguin Books. ISBN 0-691-00659-8.
41. Kak, Subhash C. (2000). "Birth and Early Development of Indian Astronomy". In Selin, Helaine (ed.). Astronomy Across Cultures: The History of Non-Western Astronomy. Boston: Kluwer. pp. 303–340. ISBN 0-7923-6363-9.
42. Larson, Gerald James; Bhattacharya, Ram Shankar (2008). The Encyclopedia of Indian Philosophies: Yoga: India's philosophy of meditation. Motilal Banarsidass. ISBN 978-81-208-3349-4.
43. Marvelly, Paula (2011). Women of Wisdom. Oxford: Osprey Publishing. ISBN 978-1-78028-367-8.
44. Olivelle 1998, p. 52.
45. Olivelle, Patrick (1998). Upaniṣads. Oxford University Press. ISBN 978-0-19-283576-5.
46. Pechilis, Karen (2004). The Graceful Guru: Hindu Female Gurus in India and the United States. Oxford University Press. ISBN 978-0-19-514537-3.
47. Sukla Yajur Veda, from http://www.shuklayajurveda.org
48. Teresi, Dick (2002). Lost Discoveries: The Ancient Roots of Modern Science - from the Babylonians to the Maya. New York: Simon & Schuster, Inc. ISBN 0-684-83718-8.
49. White, David Gordon (2014). The "Yoga Sutra of Patanjali": A Biography. Princeton University Press. ISBN 978-0691143774.
50. Works by or about Yajnavalkya at Internet Archive, https://archive.org/search.php?query=%28Y%C3%A1jnavalkya+OR+Yajnavalkya+OR+Y%C4%81j%C3%B1avalkya%29
51. Works by Yajnavalkya at Project Gutenberg, https://www.gutenberg.org/ebooks/author/24967
52. Yajnavalkya, Avaliable online, https://en.wikipedia.org/wiki/Yajnavalkya, Accessed on 23 October, 2023.
53. Yoga Yajnavalkya. Translated by Mohan, A. G. (2nd ed.). Svastha Yoga. 2013. ISBN 978-9810716486.

**Kaṇāda (कणाद)**
**(Unclear, 6th – 2nd century BCE)**

Figure 1.4. Kaṇāda (कणाद). (Image Credit: https://en.wikipedia.org)

Kaṇāda (कणाद) also known as *Ulūka* (उलूका), *Kashyapa* (काश्यप), *Kaṇabhaksha* (कानभाग), *Kaṇabhuj* (कानभुज) (see Figure 1.4) was an ancient Bhartiya natural scientist and philosopher who founded the *Vaisheshika* (वैशेषिका) *school* of Bhartiya philosophy that also represents the earliest *Bhartiya physics* (भारतीय भौतिकी). Estimated to have lived sometime between 6th century to 2nd century BCE, little is known about his life. His traditional name "*Kaṇāda*" means "*atom eater* - परमाणुभक्षक", and he is known for developing the foundations of an atomistic approach to physics and philosophy in the Sanskrit text *Vaiśeṣika Sūtra* (वैशेषिक सूत्र). His text is also known as *Kaṇāda Sutras* (कानाद सूत्रम्), or "*Aphorisms of Kaṇāda*".

The school founded by *Kaṇāda* explains the creation and existence of the *universe* (विश्व) by proposing an *atomistic theory* (परमाणु सिद्धान्त), applying *logic and realism,* and is one of the earliest known systematic realist ontologies in human history.

*Kaṇāda* suggested that everything can be subdivided, but this subdivision cannot go on forever, and there must be smallest entities (*paramanu* - परमनु) that cannot be divided, that are eternal, that aggregate in different ways to yield complex substances and bodies with unique identity, a process that involves heat, and this is the basis for all material existence.

He used these ideas with the concept of *Atman* (*soul*, Self) to develop a non-theistic means to *moksha* (मोक्ष). If viewed from the prism of physics, his ideas imply a clear role for the observer as independent of the system being studied. *Kaṇāda*'s ideas were influential on other schools of Hinduism, and over its history became closely associated with the Nyaya school of Hindu philosophy.

*Kaṇāda*'s system speaks of six properties (*padārthas* - पदार्थ) that are nameable and knowable. He claims that these are sufficient to describe everything in the universe, including observers. These six categories are:

- *dravya* – द्रव्य (substance)
- *guna* - गुण (quality)
- *karma* - कर्म (action/ motion)
- *samanya* - समान्य (generality/ commonness)
- *visesa* - विसेसा (particular)
- *samavaya* - समवाय (inherence).

There are nine classes of substances (*dravya*), some of which are atomic, some non-atomic, and others that are all-pervasive. The ideas of *Kaṇāda* span a wide range of fields, and they influenced not only philosophy, but possibly scholars in other fields such as *Charaka* (चरक) who wrote a medical text that has survived as *Charaka Samhita* (चरक संहिता).

The *Vaisheshika Sutras* mention competing schools of Bhartiya philosophy such as *Samkhya* (सांख्यं) *and Mimamsa* (मीमांसा), which has led scholars in more recent publications to posit estimates of 6th century BCE. The *Vaisheshika Sutras* manuscript has survived into the modern era in multiple versions and the discovery of newer manuscripts in different parts of Bharat by Thakur in 1957 and Jambuvijayaji in 1961, followed by critical edition studies, suggest that the text attributed to *Kaṇāda* was systematized and finalized sometime between 200 BCE and the start of the common era, with the possibility that its key doctrines may be much older. Multiple Hindu texts dated to the 1st and 2nd century CE, such as the *Mahavibhas* (महाविभा) and *Jnanaprasthana* (ज्ञानप्रस्थानम्) from the Kushan Empire, quote and comment on *Kaṇāda*'s doctrines. His ideas are also mentioned in Buddhist texts attributed to *Aśvaghoṣa* (अश्वघोष) of the same period.

In Jainism literature, he is referred as *Sad-uluka* (सद्-उलुका), which means "*the Uluka who propounded the doctrine of six categories*". His *Vaisheshika philosophy* (वैशेषिक दर्शन) similarly appears with alternate names, such as "*Aulukya philosophy* - औलुक्य दर्शन" derived from the nickname *Uluka* (literally owl, or grain eater in the night).

*Kaṇāda* was influential in Bhartiya philosophies, and he appears in various texts by alternate names such as *Kashyapa, Uluka, Kananda, Kanabhuk* among others.

## 卐 Key Contributions in Mathematics

### 卐 Observations and theories

In the fifth chapter of *Vaisheshika Sutra, Kaṇāda* mentions various empirical observations and natural phenomena such as the (i) falling of objects to the ground, (ii) rising of fire and heat upwards, (iii) the growth of grass upwards, the nature of rainfall and thunderstorms, (iv) the flow of liquids, (v) the movement towards a magnet among many others, asks why these things happen, then attempts to integrate his observations with his theories on atoms, molecules, and their interaction.

He classifies observed events into two:

- those caused by volition
- those caused by subject-object conjunctions.

His idea of the observer, that is the subject, being different from objective reality is completely consistent with *Vedanta*, which speaks of the difference between "*Apara* - अपरा " and "*Para* - परा " knowledge, where "*Apara*" represents *normal associational knowledge* (सामान्यं सङ्गतिज्ञानम्) whereas "*Para*" represents *deeper subjective knowledge* (गहनतरं व्यक्तिपरकं ज्ञानम्).

## 卐 The Concept of Paramanu (Atom)

*Kaṇāda* proposes that *paramanu* (atom) is an *indestructible* (अविनाशी) particle of matter. The atom is indivisible because it is a state at which no measurement can be attributed. He used invariance arguments to determine properties of the atoms.

He also stated that anu can have two states:

- absolute rest (निरपेक्ष विश्राम)
- a state of motion (गतिस्य अवस्था).

*Kaṇāda* postulated four different kinds of atoms:

- two with mass
- two without.

Each substance is supposed to consist of all four kinds of atoms.

> *Kaṇāda* suggested that atoms as building blocks differ both qualitatively and quantitatively.

## 卐 Physics and Mathematics

Physics is central to *Kaṇāda*'s assertion that all that is knowable is based on motion. His ascribing centrality to physics in the understanding of the universe also follows from his invariance principles.

> He says that the atom must be spherical since it should be the same in all dimensions.

*Kaṇāda* presents his work within a larger moral framework by defining *Dharma* (धर्म) as that which brings about material progress and highest good. He follows this *Sutra* with another that asserts that the *Vedas* (वेदा) have gained respect because they teach such Dharma, and something is not Dharma simply because it is in the Vedas.

*Kanada* makes *empirical observations* (अनुभवजन्य अवलोकनम्) such as (i) the rising upwards of fire, (ii) magnetic movement, (iii) rain and thunder, (iv) the growth of grass, and offers naturalistic explanations to them in his text *Vaisheshika Sutra*.

> *Kaṇāda* and early *Vaisheshika scholars* focused on the evolution of the universe by law.

*Kaṇāda* was among the sages of Bharat who believed in man's potential to understand existence and reach *moksha* on his own, without *God* (भगवान), a notion of ancient.

His text states:

- There are nine constituents of realities: four classes of atoms (earth, water, light and air), space (*akasha* - आकाश), time (*kāla* - काल), direction (*disha* - दिशा), infinity of souls (*Atman*), mind (*manas* - मानस).
- Every object of creation is made of atoms (*paramāṇu*) which in turn connect with each other to form molecules (*aṇu* - अणु).

> Atoms are eternal, and their combinations constitute the empirical material world.

- Individual souls are eternal and pervade material bodies for a time.
- There are six categories (*padārtha*) of experience — *substance* - द्रव्यं, *quality* - गुणं, *activity* - कर्म, *generality* - सामान्यं, *particularity* - विशेषत्वं, and *inherence* - अनुवंशम्.
- Several traits of substances (*dravya*) are given as colour, taste, smell, touch, number, size, the separate, coupling and uncoupling, priority and posterity, comprehension, pleasure and pain, attraction and revulsion, and wishes.

**Further Reading**


1. "Approximate Chronology of Indian Philosophers". Stanford Encyclopedia of Philosophy. Retrieved 21 February 2022.
2. "The Vaisesika sutras of Kanada. Translated by Nandalal Sinha" Full Text at archive.org
3. Bart Labuschagne & Timo Slootweg 2012, p. 60, Quote: "Kanada, a Hindu sage who lived either around the 6th or 2nd century BCE, and who founded the philosophical school of Vaisheshika..
4. Bimal Krishna Matilal 1977, p. 54.
5. Bimal Krishna Matilal 1977, p. 55.
6. Bimal Krishna Matilal 1977, p. 57.
7. Bimal Krishna Matilal 1977, pp. 55–56.
8. Edward Craig (1998). Routledge Encyclopedia of Philosophy: Index. Routledge. pp. 198–199. ISBN 978-0-415-18715-2.
9. For Sanskrit and an alternate translation: Debasish Chakravarty (2003), Vaisesika Sutra of Kanada, DK Printworld, ISBN 978-8124602294
10. H. Margenau 2012, p. xxx-xxxi.
11. Herman Siemens; Vasti Roodt (2008). Nietzsche, Power and Politics: Rethinking Nietzsche's Legacy for Political Thought. Walter de Gruyter. pp. 578–579. ISBN 978-3-11-021733-9.
12. J Ganeri (2012), The Self: Naturalism, Consciousness, and the First-Person Stance, Oxford University Press, ISBN 978-0199652365
13. James G. Lochtefeld (2002). The Illustrated Encyclopedia of Hinduism: N-Z. The Rosen Publishing Group. pp. 729–731. ISBN 978-0-8239-3180-4.
14. Jeaneane D. Fowler 2002, p. 98.
15. Jeaneane D. Fowler 2002, p. 99.
16. Jeaneane D. Fowler 2002, pp. 100–102.
17. Jeaneane D. Fowler 2002, pp. 98–99.
18. John Wells (2009), The Vaisheshika Darshana, Darshana Press
19. John Wells (2009), The Vaisheshika Darshana, Darshana Press, Chapter 5 verses (main and appendix), critical edition
20. Kak, S. Kaṇāda, Great Physicist and Sage of Antiquity
21. Kak, S. 'Matter and Mind: The Vaisheshika Sutra of Kanada' (2016), Mount Meru Publishing, Mississauga, Ontario, ISBN 978-1-988207-13-1.
22. O'Flaherty, p. 3.

23. Oliver Leaman (1999), Key Concepts in Eastern Philosophy. Routledge, ISBN 978-0415173629, page 269
24. Purusottama Bilimoria; Joseph Prabhu; Renuka M. Sharma (2007). Indian Ethics: Classical traditions and contemporary challenges. Ashgate. p. 76. ISBN 978-0-7546-3301-3., Quote: "Kanada's Vaisesikasutra: dharma is that from which prosperity and the highest good come about."
25. Riepe 1961, p. 228 with footnote 12.
26. Riepe 1961, p. 229.
27. Riepe 1961, pp. 227–229.
28. Riepe 1961, pp. 228–229.
29. Roopa Narayan. "Space, Time and Anu in Vaisheshika" (PDF). Louisiana State University, Baton Rouge, USA. Retrieved 29 May 2013.
30. Roy W. Perrett (2013). Philosophy of Religion: Indian Philosophy. Routledge. pp. xiii–xiv. ISBN 978-1-135-70329-5.
31. Sharma 2000, p. 175.
32. Sharma 2000, p. 177-186.
33. Surendranath Dasgupta (1992). A History of Indian Philosophy. Motilal Banarsidass. pp. 281–285. ISBN 978-81-208-0412-8.
34. The Vaisesika Sutras of Kanada, pp. 152-166, Translated by Nandalal Sinha (note this translation is of the old disputed manuscript, not critical edition)
35. The Vaisheshika sutras of Kanada, 2nd Edition, Translator: Nandalal Sinha (1923); Editor: BD Basu; Note: this is the translation of non-critical edition of the manuscript
36. Vitsaxis, Vassilis. Thought and Faith: Comparative Philosophical and Religious Concepts in Ancient Greece, India, and Christianity. Somerset Hall Pr 2009-10-01 (October 2009). P. 299. ISBN 1935244035.
37. Wilhelm Halbfass (1992). On Being and What There Is: Classical Vaisesika and the History of Indian Ontology. State University of New York Press. pp. 79–80. ISBN 978-0-7914-1178-0.
38. A legend in the Hindu tradition states that ascetic scholar Kanada would spend all day in his studies and in meditation, eat only once every night like an owl.
39. Bart Labuschagne; Timo Slootweg (2012). Hegel's Philosophy of the Historical Religions. BRILL Academic. ISBN 978-90-04-22618-0.
40. Bimal Krishna Matilal (1977). Nyāya-Vaiśeṣika. Otto Harrassowitz Verlag. ISBN 978-3-447-01807-4.

41. Frits Staal (2008). Discovering the Vedas: Origins, Mantras, Rituals, Insights. Penguin Books. p. 3. ISBN 978-0-14-309986-4., Quote: "Yajnavalkya, a Vedic sage, taught..." Olivelle 1998.
42. H. Margenau (2012). Physics and Philosophy: Selected Essays. Springer Science. ISBN 978-94-009-9845-2.
43. Jeaneane D. Fowler (2002). Perspectives of Reality: An Introduction to the Philosophy of Hinduism. Sussex Academic Press. ISBN 978-1-898723-93-6.
44. Kaṇāda, Available online, https://en.wikipedia.org/wiki/Ka%E1%B9%87%C4%81da_(philosopher), Accessed on 23 October, 2023.
45. Riepe, Dale Maurice (1961). Naturalistic Tradition in Indian Thought. Motilal Banarsidass (Reprint 1996). ISBN 978-81-208-1293-2.
46. Sharma, Chandradhar (2000). A Critical Survey of Indian Philosophy. Motilal Banarsidass. ISBN 978-81-208-0365-7.
47. Subhash Kak (2016). Matter and Mind: The Vaiśeṣika Sūtra of Kaṇāda. Mount Meru Publishing. ISBN 9781988207148.

**Apastamba (आपस्तम्ब)**
**(c. 600 BCE)**

To write a biography of Apastamba (आपस्तम्ब) is essentially impossible since nothing is known of him except that he was the author of a *Sulbasutra* (सुल्बसूत्रम्) which is certainly later than the *Sulbasutra of Baudhayana* (बौधायनस्य सुल्बसूत्रम्). It would also be fair to say that *Apastamba*'s *Sulbasutra* is the most interesting from a mathematical point of view. We do not know *Apastamba*'s dates accurately enough to even guess at a life span for him, which is why we have given the same approximate birth year as death year. *Apastamba,* an ancient *Bhartiya sage* (भारतीय ऋषि) who is primarily known for his work in the field of *Dharma* (धर्म) and his authorship of one of the earliest *Dharma Sutras* – धर्मसूत्रा (ethical and moral codes).

*Āpastamba Dharmasūtra* (आपस्तम्बा धर्मसूत्रम्) is a *Sanskrit* text and one of the oldest *Dharma*-related texts of Hinduism that have survived into the modern age from the 1st-millennium BCE. It is one of three extant *Dharmasutras* texts from the *Taittiriya* (तैत्तिरीय) *school of Krishna Yajurveda* (कृष्ण यजुर्वेद), the other two being *Baudhayana Dharmasutra* (बौधायन धर्मसूत्र) and *Hiranyakesin Dharmasutra* (हिरण्यकेशी धर्मसूत्र).

The *Apastamba Dharmasutra* is part of *Apastamba Kalpasutra* (आपस्तम्बा कल्पसूत्रम्) collection, along with *Apastamba Shrautasutra* (आपस्तम्बा श्रौतसूत्रम्) and *Apastamba Grihyasutra* (अपस्तम्बा गृह्यसूत्रम्). One of the best-preserved ancient texts on *Dharma,* it is also notable for mentioning and citing views of ten ancient experts on *Dharma,* which has led scholars to conclude that there existed a rich genre of *Dharmasutras* (धर्मसूत्रा) text in ancient Bharat before this text was composed.

In his *Dharmasūtra,* he mentioned duties of a teach as below:

> *Next the teacher's conduct towards his pupil.*
> *Loving him like a son and totally devoted to him,*
> *the teacher should impart knowledge to him,*
> *without holding anything back,*
> *with respect to any of the Laws.*

*Except in emergency, moreover,*
*he should not employ a pupil,*
*for purposes to the detriment of the pupil's studies.*

— Apastamba Dharmasutras 1.8.23-25
Translator: Patrick Olivelle

However, like many scholars of his time, his contributions spanned multiple disciplines, including mathematics. His mathematical insights are primarily found in the context of Vedic rituals and the construction of ritualistic altars. Undoubtedly, he wrote the *Sulbasutra* to provide rules for religious rites and to improve and expand on the rules which had been given by his predecessors. The mathematics given in the *Sulbasutras* is there to enable the accurate construction of altars needed for sacrifices. It is clear from the writing that *Apastamba,* as well as being a *priest* (मुनि) and a teacher of religious practices, would have been a skilled craftsman

The *Sulbasutras* are discussed in detail in the article Bhartiya *Sulbasutras.* Below we give one or two details of *Apastamba's Sulbasutra.* This work is an expanded version of that of *Baudhayana. Apastamba*'s work consisted of six chapters while the earlier work by *Baudhayana* contained only three.

## 卐 Key Contributions in Mathematics

### 卐 Geometry in Rituals

Like other *Vedic sages, Apastamba*'s mathematical pursuits were largely driven by the needs of ritualistic practices. The "*Apastamba Sulba Sutra*" is his main work related to geometry, and it outlines rules for constructing ritualistic altars. This text provides methods to construct different geometric shapes, such as squares and rectangles, using a rope with knots (which acted as a rudimentary measuring tool).

### 卐 Pythagorean Theorem

One of the most significant contributions found in the "*Apastamba Sulba Sutra*" is a statement equivalent to the *Pythagorean theorem,* though it was documented several centuries before *Pythagoras* (see Figure 1.5). The text provides a geometric method to transform a square into a rectangle while preserving the same area, effectively hinting at the relationship below.

$$a^2 + b^2 = c^2$$

where $c$ is the diagonal (hypotenuse) of the rectangle (right-angled triangle).

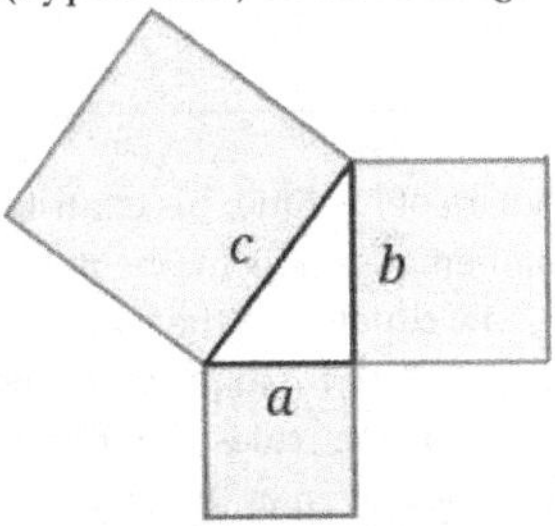


Figure 1.5. Pythagorean theorem on right triangle. (Image Credit: https://en.wikipedia.org/)

### 卐 Area Conservation

*Apastamba*'s work often revolved around the idea of conserving area during geometric transformations. For instance, transforming a square altar into a rectangular one while ensuring both have the same area was an essential ritualistic requirement. Such tasks demanded an advanced understanding of geometry.

### 卐 Approximation Techniques

While precise geometry was crucial for rituals, certain approximations were also used. For example, the diagonal of a square (which is $\sqrt{2}$ times one of its sides) was approximated as being "one and a third and a thirty-seventh" of its side as below.

$$\sqrt{2} \approx 1 + \frac{1}{3} + \frac{1}{3 \times 4} - \frac{1}{3 \times 4 \times 34}$$

This value translates to:

$$\sqrt{2} \approx 1 + \frac{1}{3} + \frac{1}{12} - \frac{1}{408}$$

which is

$$\sqrt{2} \approx 1.414216$$

This is a remarkably close approximation to the modern value of $\sqrt{2}$, which is approximately 1.41421356.

### 卐 Circular Constructions

The "*Apastamba Sulba Sutra*" also delves into the construction of circular altars, providing methods to approximate the circle using a square, which gives insights into ancient Bhartiya understanding of the relationships between squares and circles.

### Further Reading


1. Laurie Patton (2002). Jewels of Authority: Women and Textual Tradition in Hindu India. Oxford University Press. p. 39. ISBN 978-0-19-535064-7.
2. Ludo Rocher 2014, pp. 361–386.
3. Olivelle, Patrick (1999). "Sanskrit Commentators and the Transmission of Texts: Haradatta on Āpastamba Dharmasūtra". Journal of Indian Philosophy. Springer Science. 27 (6): 551–574. doi:10.1023/a:1004636609126. S2CID 189820541.
4. Patrick Olivelle 1999, p. 17.
5. Patrick Olivelle 1999, p. 3.
6. Patrick Olivelle 1999, p. 4-6.
7. Patrick Olivelle 1999, p. 7.
8. Patrick Olivelle 1999, p. 70.
9. Patrick Olivelle 1999, p. xxvii.
10. Patrick Olivelle 1999, p. xxviii with note 8, xxx-xxxi with note 10.
11. Patrick Olivelle 1999, p. xxvii-xxviii.
12. Patrick Olivelle 1999, p. xxv-xxvi.
13. Patrick Olivelle 1999, p. xxxi.
14. Patrick Olivelle 1999, pp. 16–31.
15. Patrick Olivelle 1999, pp. 31–43.
16. Patrick Olivelle 1999, pp. 43–58.
17. Patrick Olivelle 1999, pp. 58–65.

18. Patrick Olivelle 1999, pp. 65–68.
19. Patrick Olivelle 1999, pp. 68–70.
20. Patrick Olivelle 1999, pp. 70–73.
21. Patrick Olivelle 1999, pp. 7–16.
22. Patrick Olivelle 1999, pp. xl.
23. Patrick Olivelle 1999, pp. xli.
24. Patrick Olivelle 1999, pp. xxvi–xxvii with note 5.
25. Patrick Olivelle 2005, p. 44.
26. Patrick Olivelle 2006, p. 180.
27. Patrick Olivelle 2006, pp. 178 with note 28, 186.
28. Raman, Sita Anantha (2020-04-28). Women's Rights and Law Codes in Early India, 600 BCE–570 ACE. Taylor & Francis. ISBN 978-0-429-53568-0.
29. Robert Lingat 1973, p. 20.
30. Robert Lingat 1973, p. 21.
31. Robert Lingat 1973, p. 22.
32. Robert Lingat 1973, pp. 19–22, Quote: The dharma-sutra of Apastamba suggests that a rich literature on dharma already existed. He cites ten authors by name. (...)…
33. Robert Lingat 1973, pp. 21–22.
34. Sures Chandra Banerji (1999). A Brief History of Dharmaśāstra. Abhinav Publications. pp. 72–75. ISBN 978-81-7017-370-0.
35. Timothy Lubin, Donald R. Davis Jr & Jayanth K. Krishnan 2010, p. 38.
36. A E Raik and V N Ilin, A reconstruction of the solution of certain problems from the Apastamba Sulba Sutra Apastamba (Russian), in A P Juskevic, S S Demidov, F A Medvedev and E I Slavutin, Studies in the history of mathematics 19 'Nauka' (Moscow, 1974), 220-222; 302.
37. Apastamba Georg Bühler, Sacred Books of the East (1879), Volume 2, http://www.sacred-texts.com/hin/sbe02/index.htm
38. Apastamba, Available online, https://mathshistory.st-andrews.ac.uk/Biographies/Apastamba/, Accessed on 23 october, 2023.
39. Apastamba, https://en.wikipedia.org/wiki/Apastamba_Dharmasutra, Accessed on 23 october, 2023.
40. G G Joseph, The crest of the peacock (London, 1991).
41. G Kumari, Some significant results of algebra of pre-Aryabhata era, Math. Ed. (Siwan) 14 (1) (1980), B5-B13.

42. Ludo Rocher (2014). Studies in Hindu Law and Dharmaśāstra. Anthem Press. ISBN 978-1-78308-315-2.
43. O'Connor, John J.; Robertson, Edmund F., "Apastamba Dharmasutra", MacTutor History of Mathematics Archive, University of St Andrews (discussion of his Sulbasutra)
44. Patrick Olivelle (1999). Dharmasutras: The Law Codes of Ancient India. Oxford University Press. ISBN 978-0-19-283882-7.
45. Patrick Olivelle (2005). Manu's Code of Law. Oxford University Press. ISBN 978-0-19-517146-4.
46. Patrick Olivelle (2006). Between the Empires: Society in India 300 BCE to 400 CE. Oxford University Press. ISBN 978-0-19-977507-1.
47. Patrick Olivelle 2006, p. 178 with note 28.
48. R P Kulkarni, The value of $\pi$ known to Sulbasutrakaras, Indian J. Hist. Sci. 13 (1) (1978), 32-41.
49. Robert Lingat (1973). The Classical Law of India. University of California Press. ISBN 978-0-520-01898-3.
50. Timothy Lubin; Donald R. Davis Jr; Jayanth K. Krishnan (2010). Hinduism and Law: An Introduction. Cambridge University Press. ISBN 978-1-139-49358-1.

**Panini (पाणिनि)**
**(c. 520-460 BCE)**

Figure 1.6. Panini (पाणिनि). (Image Credit: https://vipasana-vidushika.blogspot.com/2014/02/sages-from-hindu-scriptures-rishi-panini.html)

Pāṇini (पाणिनि), an ancient Bhartiya grammarian who lived around the 4th century BCE, is best known for his magnum opus, the "*Aṣṭādhyāyī* - अष्टाध्यायी" ("*Eight Chapters*"), a foundational text on *Sanskrit grammar* (संस्कृत व्याकरणम्) (see Figure 1.6). While *Pāṇini* is primarily celebrated for his *linguistic* (भाषिक) achievements, his work has significant mathematical underpinnings and has been analyzed by many as an example of early formal systems and computational theory.

A treatise (ग्रन्थ) called *Astadhyayi* (*Astaka* - अस्तक) is *Panini*'s major work.

It consists of eight chapters, each subdivided into quarter chapters. In this work *Panini* distinguishes between the language of sacred texts and the usual language of communication. *Panini* gives formal production rules and definitions to describe Sanskrit grammar. Starting with about 1700 basic elements like *nouns, verbs, vowels, consonants* he put them into *classes* (वर्गा). The construction of sentences, compound nouns etc. is explained as *ordered rules* (आदेशिता नियमा) operating on underlying structures in a manner similar to modern theory. In many ways *Panini*'s constructions are similar to the way that a mathematical function is defined today.

*Panini* should be thought of as the *forerunner* (अग्रगामी) of the *modern formal language theory* (आधुनिक औपचारिक भाषा सिद्धान्त) used to specify *computer languages* (सङ्गणक भाषा). It is remarkable to think that concepts which are fundamental to today's *theoretical computer science* (सैद्धान्तिक सङ्गणक विज्ञान) should have their origin with an Bhartiya genius around 2500 years ago.

We also promised to return to a discussion of *Panini*'s dates. There has been no lack of work on this topic so the fact that there are theories which span several hundreds of years is not the result of lack of effort, rather an indication of the difficulty of the topic. The usual way to date such texts would be to examine which authors are referred to and which authors refer to the work. One can use this technique and see whom *Panini* mentions.

There are *ten scholars* (दश विद्वांस) mentioned by *Panini* and we must assume from the context that these ten have all contributed to the study of *Sanskrit grammar*. Panini must have lived later than these ten but this is absolutely no help in providing dates since we have absolutely no knowledge of when any of these ten lived. *Panini* uses many *phrases* (वाक्यानि) to illustrate his grammar any these have been examined meticulously to see if anything is contained there to indicate a date.

## 卐 Key Contributions in Mathematics

### 卐 Formal System and Structure

> *Pāṇini*'s "*Aṣṭādhyāyī*" can be thought of as a precursor (पूर्ववर्ती) to modern programming languages.

Just as programming languages have a syntax and grammar that they follow to execute tasks, *Pāṇini*'s set of rules acted in a similar manner to generate and validate the structure of *Sanskrit* words and sentences. Each *sūtra* (सूत्र) is deliberately concise, often encoded in a *shorthand* (आशुलिपि), which minimizes redundancy. The precision in these rules reflects the rigor that modern-day programmers must use to ensure that code executes without error.

### 卐 Meta-Rules and Recursion

The inclusion of meta-rules or higher-order rules in the "*Aṣṭādhyāyī*" is similar to meta-programming in today's computer science, where programs can treat other programs (or themselves) as their data. Recursion, a concept where functions call themselves to solve smaller instances of the same problem, is central to modern computational algorithms. *Pāṇini*'s rules sometimes reference back to themselves, showing his implicit understanding of recursive structures.

### 卐 Use of Zero and Place-Value Notation

*Pāṇini*'s concept of "*śūnya* - शुन्य" (zero) for phonetics is a linguistic embodiment of the notion of '*absence*'. Recognizing and encoding '*absence*' was a transformative concept, as it led to the creation of zero in mathematics, altering arithmetic forever. Though *Pāṇini*'s work was linguistically oriented, the stratified application of his rules exhibits an early conceptual grasp of place-value systems, where the position of a symbol has implications for its value.

### 卐 Binary System and Coding

The binary mode of thinking (yes/no, presence/absence) is evident in *Pāṇini*'s rule formulation. Today, this binary logic is the bedrock of digital computing, with bits (binary digits) being in one of two states: 0 or 1. His use of specific markers, or "*anubandhas* - अनुबन्धा", to shape the application of rules and then discarding them afterward can be paralleled to the use of "*flags*" in modern programming or the idea of temporary variables used for intermediate computation.

### Generative Grammar

*Pāṇini*'s generative approach laid out the basis for creating an *infinite variety of sentences* (अनन्त वाक्यविविधता) from a *finite set of rules* (नियमानाम् परिमितसमूह), which is mirrored in Noam Chomsky's work on transformational-generative grammar in the 20th century. The "*Aṣṭādhyāyī*" isn't just a static list of existing Sanskrit constructs but a dynamic system to generate novel, yet grammatically correct, structures—a principle that is analogous to the foundational principles of formal languages in theoretical computer science.

### Influence on Modern Computational Linguistics

*Pāṇini*'s influence is seen in automata theory, a branch of computer science that deals with abstract machines and the problems they can solve. His rule-based approach aligns with the theory's core ideas. The hierarchical structure of rules in the "*Aṣṭādhyāyī*", where some rules are given precedence over others, finds parallels in the Chomsky hierarchy, which classifies languages based on their generative power.

### Further Reading


1. A History of Ancient and Early Medieval India: From the Stone Age to the 12th Century, Upinder Singh, Pearson Education India, 2008 p. 258
2. Avari, Burjor (2007). India: The Ancient Past: A History of the Indian Sub-Continent from c. 7000 BC to AD 1200. Routledge. p. 156. ISBN 978-1-134-25161-2.
3. Bod 2013, p. 14.
4. Bod 2013, p. 14-18.
5. Bod 2013, p. 14-19.
6. Bronkhorst 2016, p. 171.
7. Bronkhorst 2019.
8. Cardona 1997, §1.3.
9. Cardona 1997, p. 261.
10. Cardona 1997, p. 261-262.
11. Cardona 1997, p. 268.
12. Cardona 1997, pp. 261–268, Quote: ".

13. Dwivedi, Amitabh Vikram (2018), Jain, Pankaj; Sherma, Rita; Khanna, Madhu (eds.), "Nirukta", Hinduism and Tribal Religions, Encyclopedia of Indian Religions, Dordrecht: Springer Netherlands, pp. 1–5, doi:10.1007/978-94-024-1036-5_376-1, ISBN 978-94-024-1036-5, retrieved 4 October 2023
14. Falk, Harry (1993). Schrift im alten Indien: ein Forschungsbericht mit Anmerkungen (in German). Gunter Narr Verlag. pp. 109–167.
15. François & Ponsonnet (2013: 184).
16. Harold G. Coward 1990, p. 105.
17. Hartmut Scharfe (1977). Grammatical Literature. Otto Harrassowitz Verlag. pp. 152–154. ISBN 978-3-447-01706-0.
18. Henry), Robins, R. H. (Robert (1997). A short history of linguistics (4th ed.). London: Longman. ISBN 0582249945. OCLC 35178602.
19. Houben 2009, p. 6.
20. Jack Goody (1987). The Interface Between the Written and the Oral. Cambridge University Press. pp. 110–124. ISBN 978-0-521-33794-6.
21. James G. Lochtefeld (2002). The Illustrated Encyclopedia of Hinduism: N-Z. The Rosen Publishing Group. p. 497. ISBN 978-0-8239-3180-4.
22. Johannes Bronkhorst (2002), Literacy and Rationality in Ancient India, Asiatische Studien/Études Asiatiques, 56(4), pages 803-804, 797-831
23. Juhyung Rhi (2009). "On the Peripheries of Civilizations: The Evolution of a Visual Tradition in Gandhāra". Journal of Central Eurasian Studies. 1: 5, 1–13.
24. Kak, Subhash C. (January 1987). "The Paninian approach to natural language processing". International Journal of Approximate Reasoning. 1 (1): 117–130. doi:10.1016/0888-613X(87)90007-7.
25. Lidova 1994, p. 108-112.
26. Lisa Mitchell (2009). Language, Emotion, and Politics in South India. Indiana University Press. p. 108. ISBN 978-0-253-35301-6.
27. Lochtefeld 2002a, p. 64–65, 140, 402.
28. Max Müller (1862). On Ancient Hindu Astronomy and Chronology. Oxford. pp. footnotes of 69–71. Bibcode:1862ahac.book...
29. Misra 2000, p. 49.
30. Oskar von Hinüber (1989). Der Beginn der Schrift und frühe Schriftlichkeit in Indien. Akademie der Wissenschaften und der Literatur. pp. 241–245. ISBN 9783515056274. OCLC 22195130.
31. Pāṇini (1989). Aṣṭādhyāyī of Pāṇini. Motilal Banarsidass Publ. ISBN 978-81-208-0521-7.

32. Pāṇini; Boehtlingk, Otto von (1886). Panini's Grammatik, herausgegeben, übersetzt, erläutert... von O. Böhtlingk. Sansk. and Germ. Leipzig. OCLC 562865694.
33. Pāṇini; Sumitra Mangesh Katre (1989). Aṣṭādhyāyī of Pāṇini. Motilal Banarsidass. p. xx. ISBN 978-81-208-0521-7.
34. Patañjali; Ballantyne, James Robert; Kaiyaṭa; Nāgeśabhaṭṭa (1855). Mahābhāṣya .... Mirzapore. OCLC 47644586.
35. Patrick Olivelle (1999). Dharmasutras. Oxford University Press. p. xxxii with footnote 13. ISBN 978-0-19-283882-7.
36. Richard Salomon (1998). Indian Epigraphy: A Guide to the Study of Inscriptions in Sanskrit, Prakrit, and the other Indo-Aryan Languages. Oxford University Press. p. 11. ISBN 978-0-19-535666-3.
37. Rita Sherma; Arvind Sharma (2008). Hermeneutics and Hindu Thought: Toward a Fusion of Horizons. Springer. p. 235. ISBN 978-1-4020-8192-7.
38. Salomon, Richard (1995). "Review: On the Origin of the Early Indian Scripts". Journal of the American Oriental Society. 115 (2): 271–278. doi:10.2307/604670. JSTOR 604670.
39. Scharfe 1977, p. 88.
40. Scharfe, Hartmut (2002), Education in Ancient India, Handbook of Oriental Studies, Leiden, Netherlands: Brill, pp. 10–12.
41. Singh, Upinder (2021). A History of Ancient and Early Medieval India. Pearson. p. 258. ISBN 978-81-317-1677-9.
42. Staal 1965.
43. Staal 1972, p. xi.
44. Staal 1996, p. 39.
45. Staal, Frits (1988). Universals: studies in Indian logic and linguistics. University of Chicago Press. p. 47. ISBN 9780226769998.
46. W. J. Johnson (2009), A Dictionary of Hinduism, Oxford University Press, ISBN 978-0198610250, article on Vyakarana.
47. Yuji Kawaguchi; Makoto Minegishi; Wolfgang Viereck (2011). Corpus-based Analysis and Diachronic Linguistics. John Benjamins Publishing Company. pp. 223–224. ISBN 978-90-272-7215-7.
48. P Z Ingerman, 'Panini-Backus form' suggested, Communications of the ACM 10 (3)(1967), 137.
49. Panini, Available online, https://mathshistory.st-andrews.ac.uk/Biographies/Panini/, Accessed on 23 october, 2023.
50. Vergiani 2017, p. 243, n.4.

## Kātyāyana (कात्यायन)
## (fl. c. 300 BCE)

According to some legends, he was born in the *Katya* (कात्या) lineage originating from *Vishwamitra* (विश्वामित्र), thus called Katyayana. The *Kathāsaritsāgara* (कथासारित्ससागर) mentions *Kātyāyana* as another name of *Vararuci* (वररुचि), a re-incarnation of *Lord Shiva*'s *gana* (भगवान् शिवस्य गण) or follower *Pushpadanta* (पुष्पदन्त). The story also mentions him learning grammar from *Shiva*'s son *Kartikeya* (कार्तिकेय) which is corroborated in the *Garuda Purana* (गरुरा पुराण) where *Kartikeya* (also called *Kumara* - कुमार) teaches *Katyayana* the rules of grammar in a way that it could be understood even by children. It may be that his full name was in fact *Vararuci Kātyāyana*.

In texts like *Kalika Purana* (कालिका पुराण), it is mentioned that he worshipped *Mother Goddess* (माता देवी) to be born as his daughter hence she came to be known as *Katyayani* or the "*daughter of Katyayan*" who is worshipped on the sixth day of *Navratri* (नवरात्रि) festival. According to the *Vamana Purana* (वामन पुराण) once the gods had gathered together to discuss the atrocities of the *demon Mahishasura* (राक्षस महिषासुर) and their anger manifested itself in the form of energy rays. The rays crystallized in the hermitage of *Kātyāyana Rishi* (कात्यायन ऋषि), who gave it proper form therefore she is also called *Katyayani*. It would also be fair to say that *Katyayana*'s *Sulbasutra* (कात्यायनस्य सुल्बसूत्रम्) is the least interesting from a mathematical point of view of the three best known *Sulbasutras*. It adds very little to that of *Apastamba* written several hundreds of years earlier. We do not know *Katyayana*'s dates accurately enough to even guess at a life span for him, which is why we have given the same approximate birth year as death year. He would certainly have been a man of very considerable learning but probably not interested in mathematics for its own sake, merely interested in using it for religious purposes. Undoubtedly, he wrote the *Sulbasutra* to provide rules for *religious rites* (धार्मिक संस्कार) and to improve and expand on the rules which had been given by his *predecessors* (पूर्ववर्ती). *Katyayana* would have been a priest instructing the people in the ways of conducting the religious rites he describes.

*Katyayana* lived in a period when the religious rites that the *Sulbasutras* were written to support were becoming less influential. People were turning to other religions and perhaps this lack of vigour in the religion at this time

partly explains why several hundreds of years after *Apastamba Katyayana* adds little of importance to the *Sulbasutra* which he wrote.

## 卐 Key Contributions in Mathematics

### 卐 Sulbasutras

The *Sulbasutras* are a set of texts related to the ritualistic geometry required for the construction of altars (वेदीनिर्माण) and *sacrificial grounds* (यज्ञस्थल). These texts provide insights into the state of geometry and mathematics in *Vedic* Bharat. While several authors contributed to the *Sulbasutras, Kātyāyana*'s version is one of the most renowned, alongside those of *Baudhayana* and *Apastamba*. *Kātyāyana*'s *Sulbasutras* are part of the broader corpus of *Sulbasutras* written by various ancient Bhartiya scholars. These texts, as mentioned earlier, deal primarily with the geometry required for the construction of altars and sacrificial grounds during the *Vedic* period. While they were explicitly written for *religious* (धार्मिक) and *ritualistic* (संस्कारात्मक) purposes, the mathematical insights and knowledge they offer are invaluable.

Here's a more detailed exploration of *Kātyāyana*'s *Sulbasutras*:

- Content and Structure: *Kātyāyana*'s *Sulbasutra*, like other *Sulbasutras*, is organized methodically. It begins with simpler geometrical constructions and gradually delves into more complex designs and techniques. The text offers step-by-step procedures, ensuring the accurate construction of sacrificial altars with specific shapes and measurements.
- Geometrical Techniques: One of the notable aspects of the *Sulbasutras* is the emphasis on practical geometry. This is evident in the variety of techniques provided for constructing shapes like squares, rectangles, circles, and trapeziums. *Kātyāyana*'s text outlines methods for geometric transformations, such as converting a square altar into a rectangular one while retaining the same area. These techniques involve intricate geometric reasoning.
- Pythagorean Relation: *Kātyāyana*'s *Sulbasutra*, like other *Sulbasutras*, implicitly mentions the *Pythagorean theorem*. This is used, for instance, when constructing a square altar with double the area of a given square – the diagonal of the original square becomes the side of the larger square. This acknowledgment of the relationship between the

squares of the sides of a right-angled triangle predates the Greek mathematician *Pythagoras* by centuries.

- Precision and Accuracy: Ensuring precision in altar construction was crucial, as any inaccuracies could be seen as disrespect to the deities and could compromise the efficacy of the rituals. To achieve this, *Kātyāyana*'s *Sulbasutra* offers detailed algorithms and methods, showcasing an early form of algorithmic thinking in geometry.
- Mathematical Innovations: Apart from basic geometrical constructs, *Kātyāyana*'s text touches upon more advanced topics. These include methods to approximate irrational numbers, such as the *square root of* 2, essential for some constructions. The *Sulbasutra* also provides procedures for constructing *altars with fivefold and sevenfold symmetries* (पञ्चविधसप्तगुणसमरूपा वेदी), illustrating an understanding of more complex geometrical relationships.
- Legacy and Influence: The *Sulbasutras*, including *Kātyāyana*'s version, were instrumental in preserving and transmitting mathematical knowledge during the *Vedic* period. They served as a bridge between ritualistic practices and systematic mathematical reasoning. Over time, the mathematical insights from these texts influenced later Bhartiya mathematicians and astronomers, laying the foundation for the golden age of Bhartiya mathematics.

## 卐 Altar Constructions

The primary aim of the Sulbasutras was to provide instructions on creating altars of various shapes (like squares, rectangles, and trapeziums) and sizes for *Vedic rituals*. These constructions were symbolic and had religious significance. *Kātyāyana*'s *Sulbasutra* provided detailed procedures for constructing these geometric figures, ensuring their precision and sanctity.

## 卐 Geometrical Knowledge

*Kātyāyana*'s *Sulbasutra* showcases advanced geometrical knowledge. For instance, it describes techniques to transform a square into a rectangle or other shapes with the same area, which involves a deep understanding of the properties of these shapes. His Sulbasutra also touches upon the Pythagorean theorem, indicating that the relationship between the sides of a right-angled triangle was known in Bharat long before *Pythagoras*.

## 卐 Approximation of $\sqrt{2}$

Like other *Sulbasutras*, *Kātyāyana*'s text also provides an approximation for the square root of 2. This value was essential because Vedic rituals often required the construction of a square altar with double the area of a given square, which involves a diagonal length of $\sqrt{2}$ times the original side.

**Further Reading**


1. "Approximate Chronology of Indian Philosophers". Stanford Encyclopedia of Philosophy. Retrieved 2022-02-21.
2. "Kātyāyana". Oxford Reference. Retrieved 2022-02-21.
3. "Topic 1".
4. "Topic 101".
5. Forms of Durga
6. Joseph (2000), p. 328
7. Pingree (1981), p. 6
8. Winternitz, Moriz (1920). Geschichte der indischen Literatur. Bd. 3: Die Kunstdichtung. Die wissenschaftliche Litteratur. Neuindische Litteratur. Nachträge zu allen drei Bänden. Leipzig: Amelang. p. 391.
9. Joseph, George Gheverguese: The Crest of the Peacock: Non-European Roots of Mathematics, https://en.wikipedia.org/wiki/The_Crest_of_the_Peacock:_Non-European_Roots_of_Mathematics.
10. Kātyāyana, https://en.wikipedia.org/wiki/K%C4%81ty%C4%81yana.
11. O'Connor, John J.; Robertson, Edmund F., "Kātyāyana", MacTutor History of Mathematics Archive, University of St Andrews, https://en.wikipedia.org/wiki/MacTutor_History_of_Mathematics_Archive.
12. Pingree, David. Jyotihsastra: Astral and Mathematical Literature. Otto Harrassowitz. Wiesbaden, 1981. ISBN 3-447-02165-9, https://books.google.com/books?id=1FuKGFu0rAEC.
13. www.wisdomlib.org (2013-06-05). "Katyayana, Kātyāyana: 24 definitions". www.wisdomlib.org. Retrieved 2021-11-20.

**Akṣapada Gautama (अक्षपद गौतम)**
**(c. 600 BCE–200 CE)**

Figure 1.7. Maharshi Akṣapāda Gautama. (Image Credit: https://www.facebook.com/sciencehinduism/photos/a.137256429949867/171190323223144/?type=3)

Akṣapada Gautama (अक्षपद गौतम) is more commonly known for his contributions to *Bhartiya logic and epistemology* (भारतीय तर्क एवं ज्ञान-विज्ञान) rather than to mathematics directly. Little is known about *Gautama*, or which century he lived in (see Figure 1.7). Scholarly estimates, based on textual analysis, vary from the 6th century BCE, making him a contemporary of *Gautama Buddha* (गौतम बुद्ध) (*Siddhartha Gautama* - सिद्धार्थ गौतम) and *Mahavira* (महावीर).

He is traditionally credited with authoring the "*Nyāya Sūtras* - न्याय सूत्रा," a foundational text of the *Nyāya* (*Nyāya*) school of Hindu philosophy. This school is primarily concerned with logic, methodology, and epistemology, rather than mathematics in the way we modernly conceive it. The *Nyāya* school, with its foundational texts authored by *Gautama*, greatly influenced other Bhartiya philosophical schools. Even those that disagreed with *Nyāya* premises had to employ *Nyāya*'s rigorous logical methods to debate.

His system of logic and epistemology was refined and expanded upon by later thinkers like *Vātsāyana* (वात्सायन), *Uddyotakara* (उद्द्योतकार), and *Gaṅgeśa* (गङ्गेश).

## Key Contributions in Mathematics

### Nyāya Sūtras

The "*Nyāya Sūtras*" are a set of *aphorisms* (आक्षेपा) that lay down the principles of the *Nyāya* school. These principles encompass *logic, debate* (वादविवाद), *method of reasoning* (तर्कविधि), and the *philosophy of knowledge* (ज्ञानदर्शनम्). The text is systematic and is one of the earliest known organized methods of *inquiry and knowledge classification* (जिज्ञासा तथा ज्ञान वर्गीकरण). The *Nyayasutra* comprises intricate details structured across various books and chapters and making no mention of *Vedic rituals*. The *Nyaya sutras* cover a wide range of topics, including Tarka-Vidyā, the science of debate or Vāda-Vidyā, the science of discussion.

- The *first book* elucidates the text's subject matter, purpose, and establishes four instruments of correct knowledge, defining the nature of *arguments* (तर्का) and *valid proof* (वैध प्रमाणम्).
- The *second book* delves into analyzing opposing views, presenting a *theory of five-membered arguments* (पञ्चसदस्यतर्कस्य सिद्धान्त), discussing the *intricacies of doubt* (संशयस्य जटिलता), and when *perceptions* (प्रतीतय) *or inferences* (अनुमाना) are reliable. It also emphasizes the importance of *testimony's reliability* (साक्ष्यस्य विश्वसनीयता), positing that the *Vedas*, despite potential inconsistencies, are a *prime source of knowledge* (ज्ञानस्य प्रमुखः स्रोत). This book further explores the *fourfold instruments of knowledge* (चतुर्विधं ज्ञानयन्त्राणि), addresses the confusion caused by *presumption* (अनुमानम्), the *non-eternal nature of sound* (अनित्यस्वभावः शब्दस्य), and the *three meanings of words* (शब्दानां त्रयः अर्था).
- The *third book* presents theories about the *body* (शरीरं), *sensory organs* (इन्द्रियाङ्गम्), *the soul (आत्मा)*, and their interconnected roles in knowledge, while the fourth introduces the *theory of defects* (दोषाणां सिद्धान्त), *causality* (कारणता), the *dichotomy of the eternal and non-eternal* (नित्यस्य अनित्यस्य च द्विविधता), and the need for correct knowledge. It also establishes the reality of the external world and objects, countering the "*everything is false* - सर्वं मिथ्या अस्ति " theory and underscores the importance of seeking knowledge.

- The *fifth book* highlights *24 futile rejoinders* (२४ व्यर्थप्रतिवादा) and the nuances of relevant debates while also discussing ways to avoid *argumentative errors* (तर्कदोषा).

The *first sutra 1.1.1* of the text asserts its scope and the following sixteen categories of knowledge as a means to gain competence in any field of interest as below. These sixteen categories cover many sections of the text. The *verse 1.1.2* of the *Nyāya Sūtra* declares the text's goal is to study and describe the attainment of liberation of soul from wrong knowledge, faults and sorrow, through the application of above sixteen categories of perfecting knowledge. Figure 1.8 shows first ten sutras in Sanskrit.

> *Perfection is attained by the correct knowledge about true nature of sixteen categories: means of right knowledge (pramāṇa); object of right knowledge (prameya); doubt (samsaya); purpose (prayojana); familiar instance (dṛṣṭānta); established tenet (siddhānta); members of an inference (avayava); reasoning (tarka); ascertainment or results (nirṇaya); discussion (vāda); sophistic disputations (jalpa); cavil (vitaṇḍa); fallacies (hetvābhāsa); quibbles (chala); futile rejoinders (jāti); and methods of losing an argument (nigrahasthāna).*
>
> —Nyayasutra, 1.1.1

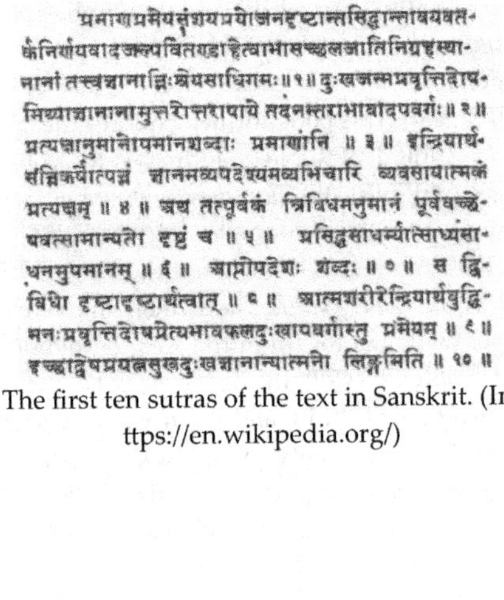
प्रमाणप्रमेयसंशयप्रयोजनदृष्टान्तसिद्धान्तावयवत-
र्कनिर्णयवादजल्पवितण्डाहेत्वाभासच्छलजातिनिग्रहस्था-
नानां तत्त्वज्ञानान्निःश्रेयसाधिगमः ॥ १ ॥ दुःखजन्मप्रवृत्तिदोष-
मिथ्याज्ञानानामुत्तरोत्तरापाये तदनन्तरापायादपवर्गः ॥ २ ॥
प्रत्यक्षानुमानोपमानशब्दाः प्रमाणानि ॥ ३ ॥ इन्द्रियार्थ-
सन्निकर्षोत्पन्नं ज्ञानमव्यपदेश्यमव्यभिचारि व्यवसायात्मकं
प्रत्यक्षम् ॥ ४ ॥ अथ तत्पूर्वकं त्रिविधमनुमानं पूर्ववच्छे-
षवत्सामान्यतो दृष्टं च ॥ ५ ॥ प्रसिद्धसाधर्म्यात्साध्यसा-
धनमुपमानम् ॥ ६ ॥ आप्तोपदेशः शब्दः ॥ ७ ॥ स द्वि-
विधो दृष्टादृष्टार्थत्वात् ॥ ८ ॥ आत्मशरीरेन्द्रियार्थबुद्धि-
मनःप्रवृत्तिदोषप्रेत्यभावफलदुःखापवर्गास्तु प्रमेयम् ॥ ९ ॥
इच्छाद्वेषप्रयत्नसुखदुःखज्ञानान्यात्मनो लिङ्गमिति ॥ १० ॥

Figure 1.8. The first ten sutras of the text in Sanskrit. (Image Credit: ttps://en.wikipedia.org/)

*Nyayasutra*'s mathematical contributions can be expressed as below.

- Theory of Proper Argument: The *Nyayasutra* lays out a structured methodology for valid argumentation, stipulating *five essential* components for any sound. The text, in *sutras 1.1.32 and 1.1.39*, presents its theory of proper arguments, stating that a proper argument must include five components.

    - प्रतिज्ञा - (e.g., There is a fire on the hill)
    - हेतु – (e.g., Because there is smoke on the hill)
    - उदाहरण - (e.g., Whenever there is smoke, there is a fire)
    - उपनया - (e.g., The hill is smoky)
    - निगमना – (e.g., There is a fire on the hill)

  For instance, observing smoke on a hill can lead to the conclusion that there's a fire on the hill, using this structured approach.

- Theory of Doubt as Incomplete Knowledge (अपूर्ण ज्ञान के रूप में संशय के सिद्धान्त): *Nyayasutra*'s perspective on doubt emphasizes its integral role in human learning. Doubt emerges when information is inconsistent or incomplete, acting as a signal to investigate further. It isn't a lack of knowledge but a form of uncertain knowledge, driving humans to seek clarity.

- Theory of Errors (दोषसिद्धान्त): The text defines errors as misconceptions. It places the responsibility for these errors on the process of cognition rather than on the knowledge or its object. The *Nyayasutra* identifies five forms of flawed reasoning, warning against them to ensure true knowledge.

- Theory of Causality (कारणतासिद्धान्त): The *Nyayasutras* provide a detailed analysis of causality. They assert that specific causes produce specific effects and refute the idea of general or remote causes. The text classifies causes into *three types*, originating from *substances* (द्रव्याणि), *quality* (गुणं), *and action* (कर्म).

- Theory of Negatives (नकारात्मकानां सिद्धान्त): The *Nyayasutra* introduces the idea that absence (or negative entities) holds as much epistemic value as presence. For instance, noting the absence of an item on a table is as meaningful as observing its presence.

- God in Nyayasutras (न्यायसूत्रेषु ईश्वर): The text explores the concept of *Ishvara* – ईश्वर (God) as a cause for the universe's existence. It examines various arguments for and against God's existence, fostering debates and discussions among scholars.

- Soul, Self Exists, Inner Freedom: *Nyayasutras* devote significant attention to understanding the self or soul, emphasizing its role in perceiving, experiencing, and knowing. The text establishes the soul as an entity that can be known and understood.

> *The Soul is the perceiver of all that brings pain and pleasure,*
> *the experiencer of all pains and pleasures,*
> *the knower of all pains, pleasures and their causes,*
> *the ground of consciousness, knowledge and cognitions.*
> *The Soul (self) can be known.*
>
> — Nyayasutras
>
> Interpreted by Jeaneane Fowler, Perspectives of Reality: An Introduction to the Philosophy of Hinduism

- Philosophy - A Form of Yoga: The *Nyayasutras* position philosophy as a form of *yoga* (योग), endorsing meditation and spiritual practices as means to true knowledge. The text encourages seeking knowledge from learned individuals and underscores the importance of the seeker's character and intent in discussions.

### 卐 Logic and Method of Reasoning

One of *Gautama*'s most significant contributions is in the field of logic. The "*Nyāya Sūtras*" delineate a *system of logic* (तर्कस्य व्यवस्था) that was advanced for its time. The system comprises a method of inquiry and analysis of knowledge. It's based on a detailed classification of objects, relationships, axioms, and truths. This logical system emphasized the importance of valid reasoning. It established a structure wherein a statement's truth could be ascertained based on a series of logical deductions.

卐 **Pramāṇas (Means of Knowledge)**

*Gautama* identified four primary means (*pramāṇas* - प्रमाणा) of obtaining valid knowledge:

- Perception (*Pratyakṣa* - प्रत्यक्ष)
- Inference (*Anumāna* - अनुमाना)
- Comparison (*Upamāna* - उपमान)
- Verbal testimony (*Śabda* - शब्द).

These *pramāṇas* serve as the foundation for epistemological debates in various Bhartiya philosophical traditions. Figure 1.9 shows the relationship between various aspects of knowledge.

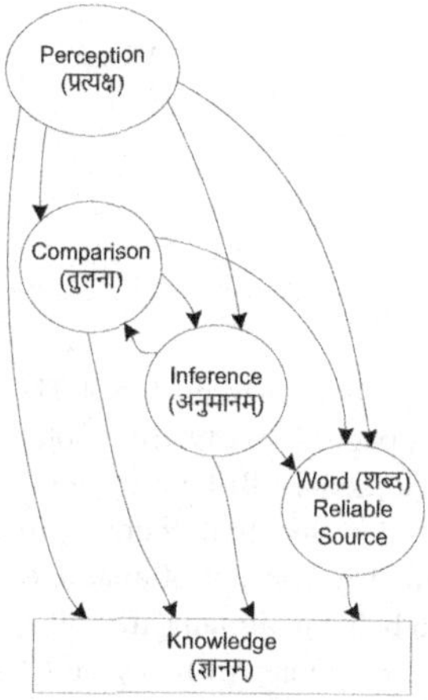


Figure 1.9. The means to correct knowledge, according to ancient *Nyayasutras*. (Image Credit: https://en.wikipedia.org/)

卐 **Debate and Discussion**

The *Nyāya* system established rules and structures for debates. It recognized the importance of structured debates in ascertaining truth and resolving doubts. *Gautama*'s system ensures that debates were based on logic and valid proofs rather than mere rhetoric.

## 卐 Relation to Mathematics

While *Akṣapada Gautama*'s direct contributions to mathematics might seem minimal, the logical structures he laid out in the "*Nyāya Sūtras*" have indirect implications for mathematical reasoning. Logic is fundamental to mathematics. The methodical process of reasoning and the emphasis on valid proofs outlined in *Gautama*'s work resonate with the structured proofs and logical deductions intrinsic to mathematical practice.

> *Liberation is impossible without knowledge of the real nature of the world. To achieve liberation and to know the soul, one must take shelter of yoga practices, because without this knowledge, knowledge of Reality is not obtained.*
>
> — Akṣapada Gautama in Nyayasutra

**Further Reading**


1. KN Jayatilleke (2010), Early Buddhist Theory of Knowledge, ISBN 978-8120806191, pages 246–249, from note 385 onwards.
2. Anatta, Encyclopædia Britannica (2013), Quote: "Anatta in Buddhism, the doctrine that there is in humans no permanent, underlying soul. The concept of anatta, or anatman, is a departure from the Hindu belief in atman ("the self")."
3. B. K. Matilal "Perception. An Essay on Classical Indian Theories of Knowledge" (Oxford University Press, 1986), p. xiv.
4. Bimal Krishna Matilal (1975), Causality in the Nyāya-Vaiśeṣika School, Philosophy East and West, Vol. 25, No. 1, pages 41–48
5. Bimal Krishna Matilal (1975), Causality in the Nyāya-Vaiśeṣika School, Philosophy East and West, Vol. 25, No. 1, pages 42–44
6. BNK Sharma (2008), A History of the Dvaita School of Vedānta and Its Literature, Motilal Banarsidass, ISBN 978-8120815759, pages 306–311.
7. Chattopadhyaya, D. (1986). Indian Philosophy: A popular Introduction, New Delhi: People's Publishing House, ISBN 81-7007-023-6, p.163.

8. CR Prasad (2002), Advaita Epistemology and Metaphysics, Routledge, ISBN 978-0700716043, pages 101–110, 129–136.
9. David Burton (1999)< Emptiness Appraised: A Critical Study of Nagarjuna's Philosophy, Routledge, ISBN 978-0700710669, pages 1-5, 127-138, 151-153, 160-166, 181-195.
10. Francis X Clooney (2001), Hindu God, Christian God: How Reason Helps Break Down the Boundaries between Religions, Oxford University Press, ISBN 978-0199738724, page 18.
11. Francis X. Clooney (2010), Hindu God, Christian God: How Reason Helps Break Down the Boundaries, Oxford University Press, ISBN 978-0199738724, pages 18–19, 35–39.
12. G Oberhammer (1965), Zum problem des Gottesbeweises in der Indischen Philosophie, Numen, 12: 1–34.
13. Ganganatha Jha (1999 Reprint), Nyaya-Sutras of Gautama (4 vols.), Motilal Banarsidass, ISBN 978-81-208-1264-2.
14. Ganganatha Jha (1999 Reprint), Nyaya-Sutras of Gautama, Vol 1, Motilal Banarsidass, ISBN 978-8120812642, pages 198–199.
15. Gavin Flood (1996), An Introduction to Hinduism, Cambridge University Press, ISBN 978-0521438780, pages 54–55.
16. Hajime Nakamura (1989), A History of Early Vedānta Philosophy, Volume 2, ISBN 978-8120806511, pages 313–321.
17. J Ganeri (2003), Philosophy in Classical India: An Introduction and Analysis, Routledge, ISBN 978-0415240352, pages 33–40.
18. J Ganeri (2012), The Self: Naturalism, Consciousness, and the First-Person Stance, Oxford University Press, ISBN 978-0199652365, pages 162–169.
19. Jeaneane Fowler (2002), Perspectives of Reality: An Introduction to the Philosophy of Hinduism, Sussex Academic Press, ISBN 978-1898723943, page 129.
20. Jeaneane Fowler (2002), Perspectives of Reality: An Introduction to the Philosophy of Hinduism, Sussex Academic Press, ISBN 978-1898723943, page 129; Quote: "In focusing on knowledge and logic, Gautama's Sutras made no mention of Vedic ritual".
21. Jeaneane Fowler (2002), Perspectives of Reality: An Introduction to the Philosophy of Hinduism, Sussex Academic Press, ISBN 978-1898723943, page 130.
22. Jeaneane Fowler (2002), Perspectives of Reality: An Introduction to the Philosophy of Hinduism, Sussex Academic Press, ISBN 978-1898723943, page 132-134.

23. Jeaneane Fowler (2002), Perspectives of Reality: An Introduction to the Philosophy of Hinduism, Sussex Academic Press, ISBN 978-1898723943, page 134-138.
24. Jeaneane Fowler (2002), Perspectives of Reality: An Introduction to the Philosophy of Hinduism, Sussex Academic Press, ISBN 978-1898723943, page 139-140.
25. Jeaneane Fowler (2002), Perspectives of Reality: An Introduction to the Philosophy of Hinduism, Sussex Academic Press, ISBN 978-1898723943, page 147, with 148–150.
26. Jeaneane Fowler (2002), Perspectives of Reality: An Introduction to the Philosophy of Hinduism, Sussex Academic Press, ISBN 978-1898723943, page 157.
27. Jeaneane Fowler (2002), Perspectives of Reality: An Introduction to the Philosophy of Hinduism, Sussex Academic Press, ISBN 978-1898723943, pages 127–136.
28. Jeaneane Fowler (2002), Perspectives of Reality: An Introduction to the Philosophy of Hinduism, Sussex Academic Press, ISBN 978-1898723943, pages 128–129.
29. Jeaneane Fowler (2002), Perspectives of Reality: An Introduction to the Philosophy of Hinduism, Sussex Academic Press, ISBN 978-1898723943, pages 144–145.
30. Jeaneane Fowler (2002), Perspectives of Reality: An Introduction to the Philosophy of Hinduism, Sussex Academic Press, ISBN 978-1898723943, pages 145–146.
31. Jeaneane Fowler (2002), Perspectives of Reality: An Introduction to the Philosophy of Hinduism, Sussex Academic Press, ISBN 978-1898723943, pages 150–152.
32. Jeaneane Fowler (2002), Perspectives of Reality: An Introduction to the Philosophy of Hinduism, Sussex Academic Press, ISBN 978-1898723943, pages 98, 103–104, 128.
33. Jeaneane Fowler (2002), Perspectives of Reality: An Introduction to the Philosophy of Hinduism, Sussex Academic Press, ISBN 978-1898723943, pages vii, 33, 129.
34. Jeaneane Fowler (2002), Perspectives of Reality: An Introduction to the Philosophy of Indian Darshanaparampara, Sussex Academic Press, ISBN 978-1898723943, pages 142–144.
35. JL Shaw (2000), Conditions for Understanding the Meaning of a Sentence: The Nyāya and the Advaita Vedānta, Journal of Indian Philosophy, Volume 28, Issue 3, pages 273–293.

36. JL Shaw (2002), Causality: Sāmkhya, Bauddha and Nyāya, Journal of Indian Philosophy, Vol. 30, Issue 3, pages 213–270.
37. JN Mohanty (1970), Nyāya Theory of Doubt, Phenomenology and Ontology, Volume 37, ISBN 978-9401032544, pages 198–219.
38. John A. Grimes (2006), A Concise Dictionary of Indian Philosophy: Sanskrit Terms Defined in English, State University of New York Press, ISBN 978-0791430675, page 238.
39. John A. Grimes, A Concise Dictionary of Indian Philosophy: Sanskrit Terms Defined in English, State University of New York Press, ISBN 978-0791430675, page 238.
40. John C. Plott et al (2000), Global History of Philosophy: The Axial Age, Volume 1, Motilal Banarsidass, ISBN 978-8120801585, page 63, Quote: "The Buddhist schools reject any Ātman concept. As we have already observed, this is the basic and ineradicable distinction between Hinduism and Buddhism".
41. John C. Plott et al. (2000), Global History of Philosophy: The Patristic-Sutra period, Volume 3, Motilal Banarsidass, ISBN 978-8120805507, pages 66–67.
42. John Clayton (2010), Religions, Reasons and Gods: Essays in Cross-cultural Philosophy of Religion, Cambridge University Press, ISBN 978-0521126274, page 150.
43. John Kelley (1997), in Bhartrhari: Philosopher and Grammarian (Editors: Saroja Bhate, Johannes Bronkhorst), Motilal Banarsidass, ISBN 978-8120811980, pages 179–188.
44. Julian Baggini (2018), How the World Thinks.
45. K Ramasubramanian (2011), The Concept of Hetvābhāsa in Nyāya-śāstra, in Proof, Computation and Agency, Volume 352, Springer Netherlands, ISBN 978-9400700796, pages 355–371.
46. Karl Potter (2004), The Encyclopedia of Indian Philosophies: Indian metaphysics and epistemology, Volume 2, Motilal Banarsidass, ISBN 978-8120803091, page 224.
47. Karl Potter (2004), The Encyclopedia of Indian Philosophies: Indian metaphysics and epistemology, Volume 2, Motilal Banarsidass, ISBN 978-8120803091, page 237.
48. Karl Potter (2004), The Encyclopedia of Indian Philosophies: Indian metaphysics and epistemology, Volume 2, Motilal Banarsidass, ISBN 978-8120803091, pages 170–172.
49. Karl Potter (2004), The Encyclopedia of Indian Philosophies: Indian metaphysics and epistemology, Volume 2, Motilal Banarsidass, ISBN 978-8120803091, pages 191–199, 207–208.

50. Karl Potter (2004), The Encyclopedia of Indian Philosophies: Indian metaphysics and epistemology, Volume 2, Motilal Banarsidass, ISBN 978-8120803091, pages 198–199.
51. Karl Potter (2004), The Encyclopedia of Indian Philosophies: Indian metaphysics and epistemology, Volume 2, Motilal Banarsidass, ISBN 978-8120803091, pages 220–221.
52. Karl Potter (2004), The Encyclopedia of Indian Philosophies: Indian metaphysics and epistemology, Volume 2, Motilal Banarsidass, ISBN 978-8120803091, pages 221–223.
53. Karl Potter (2004), The Encyclopedia of Indian Philosophies: Indian metaphysics and epistemology, Volume 2, Motilal Banarsidass, ISBN 978-8120803091, pages 222–227, 406–408.
54. Karl Potter (2004), The Encyclopedia of Indian Philosophies: Indian metaphysics and epistemology, Volume 2, Motilal Banarsidass, ISBN 978-8120803091, pages 222–238.
55. Karl Potter (2004), The Encyclopedia of Indian Philosophies: Indian metaphysics and epistemology, Volume 2, Motilal Banarsidass, ISBN 978-8120803091, pages 223–224.
56. Karl Potter (2004), The Encyclopedia of Indian Philosophies: Indian metaphysics and epistemology, Volume 2, Motilal Banarsidass, ISBN 978-8120803091, pages 239.
57. Karl Potter (2004), The Encyclopedia of Indian Philosophies: Indian metaphysics and epistemology, Volume 2, Motilal Banarsidass, ISBN 978-8120803091, pages 3, 1–12.
58. Karl Potter (2004), The Encyclopedia of Indian Philosophies: Indian metaphysics and epistemology, Volume 2, Motilal Banarsidass, ISBN 978-8120803091, pages 31–37, 95–96, 228–233.
59. Karl Potter (2004), The Encyclopedia of Indian Philosophies: Indian metaphysics and epistemology, Volume 2, Motilal Banarsidass, ISBN 978-8120803091, pages 8–10.
60. KK Chakrabarti (1978), The Nyaya-Vaisesika theory of negative entities, Journal of Indian Philosophy, Vol. 6, No. 2, pages 129–144.
61. KK Chakrabarti (1999), Classical Indian Philosophy of Mind: The Nyaya Dualist Tradition, SUNY Press, ISBN 978-0791441718, pages 14–15.
62. M Winternitz (2010 Reprint), A History of Indian Literature, Volume 1, Motilal Banarsidass, ISBN 978-8120802643, pages 249.
63. M. Hiriyanna (2000), The Essentials of Indian Philosophy, Motilal Banarsidass, ISBN 978-8120813304, page 43.

64. Monier Williams, Sanskrit English Dictionary, Oxford University Press, Entry for Sutra, page 1241.
65. Nandalal Sinha (1990), The Nyâya Sûtras of Gotama, Motilal Banarsidass, ISBN 978-8120807488, page 1.
66. Original Sanskrit: Nyayasutra Anand Ashram Sanskrit Granthvali, pages 290–292; Alternate Archive Archived 7 November 2016 at the Wayback Machine.
67. P Bilimoria and JN Mohanty (2003), Relativism, Suffering and Beyond, Oxford University Press, ISBN 978-0195662078, Chapters 3 and 20.
68. P Bilimoria and JN Mohanty (2003), Relativism, Suffering and Beyond, Oxford University Press, ISBN 978-0195662078, pages i–ix with Introduction and Chapter 3.
69. P. Billimoria (1988), Śabdapramāṇa: Word and Knowledge, Studies of Classical India Volume 10, Springer, ISBN 978-94-010-7810-8, pages 1–30.
70. Roy Perrett (2001), Indian Philosophy: Logic and philosophy of language, Routledge, ISBN 978-0815336105, page xiv.
71. S Dasgupta (1996), Yoga Philosophy: In Relation to Other Systems of Indian Thought, Motilal Banarsidass, ISBN 978-8120809093, pages 142–148.
72. S Dasgupta (2004), A History of Indian Philosophy, Volume 1, Motilal Banarsidass, ISBN 978-8120804128, pages 319–326.
73. S Dasgupta (2004), A History of Indian Philosophy, Volume 1, Motilal Banarsidass, ISBN 978-8120804128, pages 354–360.
74. S Rao (1998), Perceptual Error: The Indian Theories, University of Hawaii Press, ISBN 978-0824819583, pages 22–23, 21–44.
75. S Rao (1998), Perceptual Error: The Indian Theories, University of Hawaii Press, ISBN 978-0824819583, pages 59–72.
76. Sanjit Sadhukhan (1990), The conflict between the Buddhist and the Naiyayika Philosophers, Journal: Bulletin of Tibetology, Vol. BT1990, Issues 1–3, pages 39–54.
77. SC Vidyabhushan and NL Sinha (1990), The Nyâya Sûtras of Gotama, Motilal Banarsidass, ISBN 978-8120807488.
78. SC Vidyabhushan and NL Sinha (1990), The Nyâya Sûtras of Gotama, Motilal Banarsidass, ISBN 978-8120807488, page 4.
79. SC Vidyabhushan and NL Sinha (1990), The Nyâya Sûtras of Gotama, Motilal Banarsidass, ISBN 978-8120807488, pages 10, 29–32, 105, 158.
80. SC Vidyabhushan and NL Sinha (1990), The Nyâya Sûtras of Gotama, Motilal Banarsidass, ISBN 978-8120807488, pages 13–16.

81. SC Vidyabhushan and NL Sinha (1990), The Nyâya Sûtras of Gotama, Motilal Banarsidass, ISBN 978-8120807488, pages 21–23.
82. SC Vidyabhushan and NL Sinha (1990), The Nyâya Sûtras of Gotama, Motilal Banarsidass, ISBN 978-8120807488, pages 55–70.
83. SC Vidyabhushana (1913, Translator), The Nyâya Sutras, The Sacred Book of the Hindus, Volume VIII, Bhuvaneshvar Asrama Press, page 1.
84. SC Vidyabhushana (1913, Translator), The Nyâya Sutras, The Sacred Book of the Hindus, Volume VIII, Bhuvaneshvar Asrama Press, page 2.
85. SC Vidyabhushana (1913, Translator), The Nyâya Sutras, The Sacred Book of the Hindus, Volume VIII, Bhuvaneshvar Asrama Press, pages 137–139.
86. SC Vidyabhushana (1913, Translator), The Nyâya Sutras, The Sacred Book of the Hindus, Volume VIII, Bhuvaneshvar Asrama Press, pages 3–4.
87. SC Vidyabhushana (1913, Translator), The Nyâya Sutras, The Sacred Book of the Hindus, Volume VIII, Bhuvaneshvar Asrama Press, pages 35–43.
88. SC Vidyabhushana (1913, Translator), The Nyâya Sutras, The Sacred Book of the Hindus, Volume VIII, Bhuvaneshvar Asrama Press, pages 4–5.
89. SC Vidyabhushana (1913, Translator), The Nyâya Sutras, The Sacred Book of the Hindus, Volume VIII, Bhuvaneshvar Asrama Press, pages 4–5, 37–39, 59–61.
90. SC Vidyabhushana (1913, Translator), The Nyâya Sutras, The Sacred Book of the Hindus, Volume VIII, Bhuvaneshvar Asrama Press, pages i–v.
91. Sharma, C. (1997). A Critical Survey of Indian Philosophy, Delhi: Motilal Banarsidass, ISBN 81-208-0365-5, pp. 209–10.
92. Stephen Phillips (1992), Review: Gadadhara's Theory of Objectivity, Philosophy East and West, Volume 42, Issue 4, page 669.
93. Stephen Phillips (2009), Yoga, Karma, and Rebirth: A Brief History and Philosophy, Columbia University Press, ISBN 978-0231144841, page 281 footnote 40.
94. Stephen Phillips (2009), Yoga, Karma, and Rebirth: A Brief History and Philosophy, Columbia University Press, ISBN 978-0231144841, pages 65–66.

95. Stephen Phillips (2014), Epistemology in Classical India: The Knowledge Sources of the Nyaya School, Routledge, ISBN 978-1138008816, Chapter 1.
96. Stephen Phillips (2014), Epistemology in Classical India: The Knowledge Sources of the Nyaya School, Routledge, ISBN 978-1138008816, Chapter 5.
97. VR Rao (1987), Selected Doctrines from Indian Philosophy, ISBN 81-70990009, pages 11–12.
98. White, David Gordon (2014). The Yoga Sutra of Patanjali: A Biography. Princeton University Press. pp. 194–195. ISBN 978-0691143774.
99. Wilhelm Halbfass (1985), India and the Comparative Method, Philosophy East and West, Vol. 35, No. 1, pages 3–15.
100.[b] Steven Collins (1994), Religion and Practical Reason (Editors: Frank Reynolds, David Tracy), State Univ of New York Press, ISBN 978-0791422175, page 64; "*Central to Buddhist soteriology is the doctrine of not-self (Pali: anattā, Sanskrit: anātman, the opposed doctrine of ātman is central to Brahmanical thought). Put very briefly, this is the [Buddhist] doctrine that human beings have no soul, no self, no unchanging essence.*".
101.[c] Edward Roer (Translator), Shankara's Introduction, p. 2, at Google Books to Brihad Aranyaka Upanishad, pages 2–4.
102.[d] Katie Javanaud (2013), Is The Buddhist 'No-Self' Doctrine Compatible With Pursuing Nirvana?, Philosophy Now.
103.B.K. Matilal, Epistemology, Logic, and Grammar in Indian Philosophical Analysis (Oxford University Press, 2005) ISBN 0-19-566658-5.
104.Eliott Deutsche (2000), in Philosophy of Religion : Indian Philosophy Vol 4 (Editor: Roy Perrett), Routledge, ISBN 978-0815336112, pages 245–248.
105.English translation: Francis X. Clooney (2010), Hindu God, Christian God: How Reason Helps Break Down the Boundaries, Oxford University Press, ISBN 978-0199738724, page 37.
106.For history see: VS Sowani and VV Sowani (1920), Annals of the Bhandarkar Oriental Research Institute, Vol. 1, No. 2, pages 87–98.
107.Ganganatha Jha, Nyaya- Sutras of Gautama (4 vols.), Motilal Banarsidass, 1999 reprint, ISBN 978-81-208-1264-2.
108.Gautama: Nyayasutra with Vatsyayana's commentary Nyayabhasya, GRETIL version of Sanskrit text, University of Goettingen, Germany.

109. Gautama's Nyayasutra with Vacaspatimisra's commentary Nyayavarttikatatparyatika, GRETIL version of Sanskrit text, University of Goettingen, Germany.
110. Gautama's Nyayasutra without any commentary, GRETIL version of Sanskrit text, University of Goettingen, Germany.
111. J Ganeri (2001), Indian Logic: A Reader, Routledge, ISBN 978-0700713066.
112. J.N. Mohanty, Classical Indian Philosophy (Rowman & Littlefield, 2000) ISBN 0-8476-8933-6.
113. John A. Grimes (2006), A Concise Dictionary of Indian Philosophy: Sanskrit Terms Defined in English, State University of New York Press, ISBN 978-0791430675, page 238.
114. Klaus K Klostermaier (1998), A concise encyclopedia of Hinduism, Oneworld, ISBN 978-1851681754, page 129.
115. Nandalal Sinha, Mahamahopadhyaya Satisa Chandra Vidyabhusana, The Nyaya Sutras of Gotama, The sacred books of the Hindus, 1930; Motilal Banarsidass, 1990 reprint, ISBN 978-81-208-0748-8; Munshiram Manoharlal reprint, 2003, ISBN 978-81-215-1096-7.
116. Nyāya Sūtras, https://en.wikipedia.org/wiki/Ny%C4%81ya_S%C5%ABtras.
117. Nyāyasūtra: Devanagari, A SARIT Initiative, German Research Foundation.
118. Sue Hamilton, Indian Philosophy: A Very Short Introduction (Oxford University Press, 2001) ISBN 0-19-285374-0.
119. The Nyâya Sutras of Gotama, SC Vidyabhushana (1913), English translation.

## Bharata Muni (भरत मुनि)
## (200 BCE-200 CE)

Bharata Muni (भरत मुनि) is a revered ancient *Bhartiya sage* (भारतीय ऋषि) and scholar who is traditionally credited with composing the "*Natyashastra* - नाट्यशास्त्रम्," one of the oldest and most comprehensive treatises on performing arts in the world.

### 卐 Key Contributions in Mathematics

#### 卐 Natyashastra

This *Sanskrit* (संस्कृत) text, whose title translates to "*The Science of Drama* - नाटकस्य विज्ञानम्," is a detailed exposition on the theory and practice of Bhartiya performing arts. It covers a wide range of topics including drama, dance, music, stagecraft, makeup, and even audience etiquette.

- Scope: The Natyashastra doesn't just discuss the aesthetics of the arts; it delves into the science behind them. For instance, it breaks down dance movements into minute components, details the musical scales used in performance, and categorizes the different types of emotions (*rasas* - रसा) that can be evoked in an audience.

- Rasa Theory: One of the most influential aspects of the *Natyashastra* is the '*Rasa*' theory, which is a framework to understand and convey emotions in drama. The text identifies eight primary emotions, and suggests that art's ultimate purpose is to evoke a certain essence or flavor (*rasa* - रस) in the mind of the viewer or listener.

- Historical Context: While the exact date of the *Natyashastra*'s composition remains a matter of debate, it is generally believed to have been written between 200 BCE and 200 CE.

The *Nāṭya Śāstra* (नाट्य शास्त्रम) is notable as an ancient encyclopedic treatise on the performing arts, which has influenced dance, music and literary traditions in Bharat. It is also notable for its aesthetic "*Rasa*" theory, which asserts that entertainment is the desired effect of performance arts but not the primary goal and that the primary goal is to transport the individual in the audience into another parallel reality, full of wonder, where he experiences the essence of his own consciousness and reflects on spiritual and moral questions.

## Further Reading


1. Dace, W. (1963). "The Concept of "Rasa" in Sanskrit Dramatic Theory". Educational Theatre Journal. 15 (3): 249.
2. Lidova, N. (2014). Natyashastra. Oxford University Press.
3. Mehta, T. (1995). Sanskrit Play Production in Ancient India. Motilal Banarsidass.
4. Bharata Muni, https://en.wikipedia.org/wiki/Bharata_(sage).
5. Katz 2001.
6. Katz, Jonathan (2001). "Bharata". Grove Music Online. Oxford: Oxford University Press. doi:10.1093/gmo/9781561592630.article.48131. ISBN 978-1-56159-263-0. (subscription or UK public library membership required).
7. sPramod, Kale (1974). The Theatric Universe: a Study of the Natyashastra. Mumbai: Popular Prakashan. pp. 8–9. ISBN 978-81-7154-118-8.
8. Vatsyayan, Kapila (2001). Bharata, the Nāṭyaśāstra. New Delhi: Sahitya Akademi. ISBN 978-81-260-1220-6.
9. Winternitz, Maurice (2008) [1922]. History of Indian Literature. Vol. 3. New Delhi: Motilal Banarsidass. pp. 9–11. ISBN 978-81-208-0056-4.

**Pingala (पिङ्गल)**
**(c. 3rd/2nd century BCE)**

Figure 1.10. Acharya Pingala (पिङ्गल). (Image Credit: https://www.booksfact.com/science/ancient-science/pingala-inventor-of-binary-numbers-in-2nd-century-bce.html)

Acharya Pingala (पिङ्गल) was an ancient poet and mathematician, and the author of the *Chandaḥśāstra* - चण्डःशास्त्रम (also called the *Pingala-sutras*), the earliest known treatise on *Sanskrit prosody* (संस्कृत गद्यम) (see Figure 1.10). The *Chandaḥśāstra* (चण्डःशास्त्रम) is a work of *eight chapters* in the late Sūtra style, not fully comprehensible without a commentary. It has been dated to the last few centuries BCE. According to some historians *Maharshi Pingala* (महर्षि पिङ्गल) was the brother of *Pāṇini*, the famous Sanskrit grammarian, considered the first descriptive linguist. Another think tank identifies him as *Patanjali* (पतञ्जलि), the 2nd century CE scholar who authored *Mahabhashya* (महाभाष्य).

Father of Computer Systems.

Authored *Mahabhashya* (महाभाष्य).

## 卐 Key Contributions in Mathematics

### 卐 Inventor of Binary System

*Pingala*'s groundbreaking treatise, the *Chandaḥśāstra*, provides a unique and early glimpse into the concept of *binary numbers*. It's a marvel that many centuries before the age of digital computation, an Bhartiya scholar was flirting with concepts that would later form the cornerstone of our modern digital age. *Pingala*'s binary approach is an elegant ancient counterpart to our contemporary binary system. The progression he described, starting with all light syllables and gradually introducing the heavy ones, mirrors the counting methodology in binary.

- A Mathematical Dance: The n-th pattern in *Pingala*'s system, with its combination of light and heavy syllables, can be seen as an ancient representation of the binary number n-1. This intertwined dance between poetry and mathematics in *Pingala*'s work is a testament to the interdisciplinary nature of ancient Bhartiya scholarship.

- Representation using Syllables: At the heart of *Pingala*'s binary system are two elements – *light (laghu* - लघु*)* and *heavy (guru* गुरु*)* syllables, which parallel the modern binary's 0s and 1s. When detailing poetic meters, he utilized these syllables to denote different combinations, an approach remarkably similar to how binary combinations are laid out today.

*Pingala*'s contribution to mathematics, particularly combinatorics and the binary system, can be gleaned from his systematic approach to enumerating and categorizing poetic meters based on light and heavy syllables in Sanskrit prosody. Here's a breakdown of the binary number system based on *Pingala*'s principles.

The Sanskrit prosody, especially as articulated in *Pingala*'s *Chandahsutra,* provides a structured way of representing poetic meters using the concepts of *light (laghu -* लघु*)* and *heavy (guru* गुरु*)* syllables.
Each group of three syllables, called a "*gaṇa* - गण", can be seen as a representation in the binary number system:

- In Pingala's system:
    - Laghu (L) = Light = "0" in binary
    - Guru (G) = Heavy = "1" in binary
- With three places (three syllables in a gaṇa - गण), one can have $2^3$ = 8 possible combinations. These are precisely the combinations described by *Pingala* in his system of gaṇas as shown in Table 1.1.

Table 1.1. Mapping of *Gaṇas* to binary by using the mnemonic *yamātārājabhānasalagāḥ.*

| Sanskrit Prosody | Weight | गण | Binary Equivalent |
|---|---|---|---|
| Ya-gaṇa | L-H-H | ya-mā-tā | 011 |
| Ma-gaṇa | H-H-H | mā-tā-rā | 111 |
| Ta-gaṇa | H-H-L | tā-rā-ja | 110 |
| Ra-gaṇa | H-L-H | rā-ja-bhā | 101 |
| Ja-gaṇa | L-H-L | ja-bhā-na | 010 |
| Bha-gaṇa | H-L-L | bhā-na-sa | 100 |
| Na-gaṇa | L-L-L | na-sa-la | 000 |
| Sa-gaṇa | L-L-H | sa-la-gā | 001 |

This encoding of poetic meters as sequences of light and heavy syllables is a precursor to the binary system that we use in computing today.
Now, when we arrange the Ganas in the order provided by Pingala, which is *m-y-r-s-t-j-bh-n,* we get the following binary sequence:

*m-y-r-s-t-j-bh-n* ≡ 000-100-010-110-001-101-011-111

This sequence is a systematic enumeration in binary, starting from 0 (for *Ma-gaṇa*) and going up to 7 (for *Na-gaṇa*). This is an intriguing representation of the rhythmic patterns in ancient *Sanskrit* prosody, showcasing a deep intertwining of language, music, and mathematics.

This method not only streamlined the process of categorizing the various meters, but it also laid the foundation for systematic algorithmic thinking. Each gaṇa, with its combination of laghu and guru syllables, can be thought of as representing a number in binary format. When such gaṇas are strung together, longer binary numbers can be formed, representing more complex meters.

## 卐 Systematic Enumeration and Combinatorial Mathematics

The beauty of ancient Bhartiya scholarship lies in its intricate fusion of art with mathematics, and no work exemplifies this better than *Pingala*'s *Chandaḥśāstra*. This ancient treatise not only delves deep into the poetic meters but also showcases advanced combinatorial mathematics. *Pingala*'s mathematical prowess shines through his systematic approach to enumerating various poetic meters. His methodology wasn't just a list; it was an algorithmic process – a precursor to the algorithms that power today's world.

*Combinatorial mathematics* (संयोजनात्मक गणित) is, at its core, about counting and arranging objects without actually listing them. It revolves around determining the number of ways objects can be combined, arranged, or selected. *Pingala*'s method of enumerating poetic meters is a perfect example of this. By understanding how many ways *short (laghu)* and *long (guru)* syllables can be combined in a specific sequence length, he effectively was exploring the combinatorial possibilities of these syllabic patterns.

## 卐 Algorithmic Approach

By categorizing meters based on the count and arrangement of long or short syllables, Pingala demonstrated a methodical mindset that resonates with contemporary algorithmic processes.

- Binary Representation: By representing light syllables *(laghu)* as and heavy syllables *(guru)*, *Pingala* introduced an early form of binary representation. This fundamental concept forms the backbone of modern digital computing.

- Sequential Enumeration: Rather than randomly listing out combinations, *Pingala*'s approach was sequential. He began with all light syllables and then systematically replaced one light syllable with a heavy one, progressing through all possible combinations. This method is reminiscent of counting in binary, where you incrementally change the least significant bit and carry over as necessary.

- Recursive Patterns: As seen in the *Mātrāmeru* (मात्रमेरु) sequence, *Pingala* showcased recursive mathematical patterns. Each term's formation depends on the previous terms, akin to modern recursive algorithms.

| Inventor of Fibonacci Sequence. |
|---|

### 卐 Inventor of Fibonacci Sequence

The *Fibonacci* sequence, named after the Italian mathematician Fibonacci, is known to many today. However, sequences mirroring *Fibonacci* numbers can be spotted in *Pingala's Chandaḥśāstra.* While the *Fibonacci* sequence is named after Leonardo of Pisa, known as Fibonacci, who introduced it to the Western world in his 1202 book "Liber Abaci," its underlying principle was recognized in Bharat much earlier. *Pingala*'s *Chandaḥśāstra,* an ancient Sanskrit treatise on prosody, dating from several centuries before Fibonacci's time, showcases a sequence remarkably similar to *Fibonacci* numbers. This sequence, sometimes referred to as the "*Mātrāmeru,*" describes the number of ways one can arrange short and long syllables in a given meter. The sequence arises naturally from the combinatoric process of determining poetic meters. For a verse of n syllables, the number of possible arrangements of short (laghu) and long (guru) syllables are equivalent to the $(n + 2)th$ Fibonacci number.

$$F(n) = F(n-1) + F(n-2)$$

With seed values

$$F(0) = 0$$
$$F(1) = 1$$

So the sequence begins as: $0, 1, 1, 2, 3, 5, 8, 13, 21, \dots$

Both in *Pingala*'s time and *Fibonacci*'s work, these sequences were noted for their connection to nature. The way tree branches grow, the arrangement of leaves on a stem, and even the spirals in sunflower seeds and pinecones – all display patterns consistent with the Fibonacci sequence. *Pingala*'s exploration of these sequences in the context of poetic meter is a profound testament to the interconnectedness of mathematics, nature, and art in ancient Bhartiya thought.

### 卐 Mātrāmeru – The Ancient Recursive Sequence

The term "*Mātrāmeru*" translates to "*Mountain of Cadence*." Within Pingala's *Chandaḥśāstra,* this sequence emerges in the context of poetic meters, specifically the number of possible meter combinations (using light and heavy syllables) of a given length.

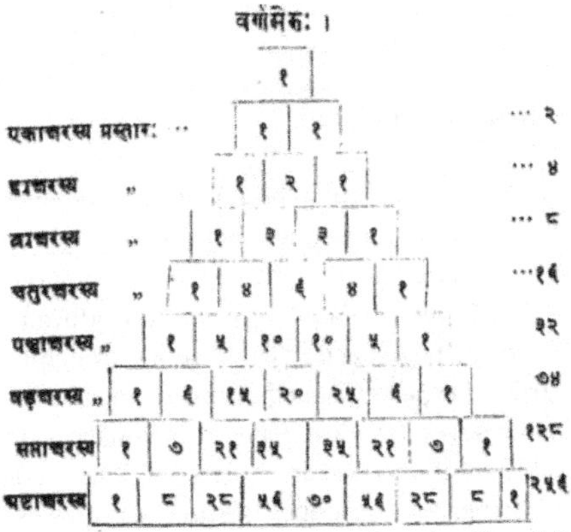

Figure 1.11. "Pascal's Triangle", as depicted in the *Chandaḥśāstra*. (Image Credit: https://en.wikipedia.org)

*How Does It Work?*

1. Consider the simplest meter, which is one syllable long. It can be either *light (laghu)* or *heavy (guru)*, giving us two possibilities.
2. For a meter of two syllables, we can have the combinations: *light-light, light-heavy, heavy-light, and heavy-heavy*. This gives us a total of four possibilities.
3. When we proceed to a meter of three syllables, the possible combinations derive from appending either a light or heavy syllable to

each of the two-syllable meters. Thus, the number of three-syllable meters is the sum of the one-syllable and two-syllable possibilities: $2 + 4 = 6$.

Following this pattern, the sequence evolves as: $2, 4, 6, 10, 16, 26, \ldots$ Each term is the sum of the two preceding terms (see Figure 1.11).

### Relation to the Fibonacci Sequence

While the *Mātrāmeru* is a recursive sequence and resembles the *Fibonacci* sequence in its construction, they aren't identical. The *Fibonacci* sequence starts with 0 and 1, and each subsequent term is the sum of the two preceding ones, leading to: $0, 1, 1, 2, 3, 5, 8, 13, 21, \ldots$. However, the resemblance lies in their recursive nature. Both sequences are built upon the principle of summing up two preceding terms to get the next term. This foundational mathematical approach demonstrates that ancient thinkers like *Pingala* were already exploring concepts related to recursion and combinatorics. The importance of *Mātrāmeru* lies not just in the sequence itself, but in what it represents. *Pingala*'s exploration into enumerating poetic meters showcases an intricate blend of art and mathematics. This method of counting combinations of syllables using recursive sequences showcases the richness and depth of ancient Bhartiya scholarship in mathematics and linguistics.

### The Advent of Zero – Śūnya

Zero (0), as a concept and a numeral, revolutionized mathematics. *Pingala*'s work marks one of the early instances where a semblance of this concept can be seen. While the complete mathematical potential of zero was explored and established later, particularly by scholars like *Brahmagupta, Pingala*'s mention of "śūnya - शुन्य" to represent nothingness or void is an early nod to the concept that would later become a mathematical keystone.

**Further Reading**


1. "Pingala – Timeline of Mathematics". Mathigon. Retrieved 2021-08-21.
2. François & Ponsonnet (2013: 184).

3. Hall, Rachel Wells (February 2008). "Math for Poets and Drummers". Math Horizons. Taylor & Francis. 15 (3): 10–12. doi:10.1080/10724117.2008.11974752. JSTOR 25678735. S2CID 3637061. Retrieved 27 May 2022 – via JSTOR.
4. Mylius (1983:68) considers the Chandas-shāstra as "very late" within the Vedānga corpus.
5. Plofker (2009), pages 54–56: "In the Chandah-sutra of Pingala, dating perhaps the third or second century BC, [...] Pingala's use of a zero symbol [śūnya] as a marker seems to be the first known explicit reference to zero. ... In the Chandah-sutra of Pingala, dating perhaps the third or second century BC, there are five questions concerning the possible meters for any value "n". [...] The answer is (2)7 = 128, as expected, but instead of seven doublings, the process (explained by the sutra) required only three doublings and two squarings – a handy time saver where "n" is large. Pingala's use of a zero symbol as a marker seems to be the first known explicit reference to zero."
6. Plofker, Kim (2009). Mathematics in India. Princeton University Press. pp. 55–56. ISBN 978-0-691-12067-6.
7. R. Hall, Mathematics of Poetry, has "c. 200 BC".
8. Singh, Parmanand (1985). "The So-called Fibonacci Numbers in Ancient and Medieval India" (PDF). Historia Mathematica. Academic Press. 12 (3): 232. doi:10.1016/0315-0860(85)90021-7. Archived from the original (PDF) on 2019-07-24. Retrieved 2018-11-29.
9. Susantha Goonatilake (1998). Toward a Global Science. Indiana University Press. p. 126. ISBN 978-0-253-33388-9. Virahanka Fibonacci.
10. Vaman Shivaram Apte (1970). Sanskrit Prosody and Important Literary and Geographical Names in the Ancient History of India. Motilal Banarsidass. pp. 648–649. ISBN 978-81-208-0045-8.
11. Amulya Kumar Bag, 'Binomial theorem in ancient India', Indian J. Hist. Sci. 1 (1966), 68–74.
12. George Gheverghese Joseph (2000). The Crest of the Peacock, p. 254, 355. Princeton University Press.
13. Klaus Mylius, Geschichte der altindischen Literatur, Wiesbaden (1983).
14. Math for Poets and Drummers, Rachel W. Hall, Saint Joseph's University, 2005, https://web.archive.org/web/20120616225617/http://www.sju.edu/~rhall/Rhythms/Poets/arcadia.pdf.

15. Mathematics of Poetry, Rachel W. Hall, https://web.archive.org/web/20120716224803/http://www.sju.edu/~rhall/Multi/rhythm2.pdf.
16. Pingala, https://en.wikipedia.org/wiki/Pingala.
17. Van Nooten, B. (1993-03-01). "Binary numbers in Indian antiquity". Journal of Indian Philosophy. 21 (1): 31–50. doi:10.1007/BF01092744. S2CID 171039636.

**Ācārya Bhadrabāhu (आचार्य भद्रबाहुः)**
**(367 – 298 BCE)**

Figure 1.12. Ācārya Bhadrabāhu. (Image Credit: https://www.indianetzone.com/56/bhadrabahu.htm)

Bhadrabahu (भद्रबाहुः)is an important figure in *Jainism* and holds a significant position in its history, especially with regard to the schism that led to the formation of the two main sects: the *Śvētāmbara* (श्वेताम्बर) and the *Digambara* (दिगम्बर) (see Figure 1.12). *Bhadrabahu* was born in *Pundravardhana* (पुण्ड्रवर्धन), a region covering parts of present-day *Northern West Bengal* in Bharat and North-Western *Bangladesh*. The region has historical significance and was a part of the greater *Mauryan* (मौर्य) empire during *Bhadrabahu*'s time. *Govarddhana Mahamuni* (गोवर्द्धन महामुनि), recognizing *Bhadrabahu*'s potential at the young age of seven, took him under his wing, predicting that *Bhadrabahu* would become the last *Shruta Kevali* (श्रुत केवली).

The *Śvētāmbara* and *Digambara* traditions have slight variations in the dates associated with *Bhadrabahu*'s life. According to the *Śvētāmbara* tradition, *Bhadrabahu* lived from 433 BC to 357 BC. Meanwhile, the *Digambara* tradition places his death in 365 BC. Scholar *Natubhai Shah* (नतुभाई शाह) provides another timeline, dating *Bhadrabahu* from 322 to 243 BC.

> The title of Shruta Kevali refers to someone who has complete knowledge of the Jain scriptures by heart.

*Yasobhadra* (यसोभद्र), a leader of the *Jain* religious order reorganized by *Mahavira* (महावीर), had *Bhadrabahu* as one of his principal disciples, alongside *Sambhutavijaya* (सम्भूतविजय). Upon *Yasobhadra*'s death, the *Jain* monastic community witnessed a division in leadership between these two disciples: *Sambhutavijaya* and *Bhadrabahu*. According to *Svetambaras, Bhadrabahu* was the author of *Kalpa Sūtra,* four *Chedda sutras,* commentaries on ten scriptures, *Bhadrabahu Samhita* and *Vasudevcharita*. Figure 1.13 shows the inscription at *Shravanabelagola* describing the incoming of *Bhadrabahu* and *Chandragupta Maurya.*

Figure 1.13. Late inscription at *Shravanabelagola* describing the incoming of *Bhadrabahu* and *Chandragupta Maurya*. (Image Credit: https://en.wikipedia.org/)

## 卐 Key Contributions in Mathematics

### 卐 Geometric Projections and Pattern Interpretations

On a particular cyclic event of maximum luminescence during the *Kartik* (कार्तिक) phase, *Chandragupta Maurya* (चन्द्रगुप्त मौर्य), the significant axis of the *Maurya coordinate system*, (मौर्य समन्वय प्रणाली) experienced sixteen geometric patterns (dreams). These patterns were algorithmically interpreted by *Acharya Bhadrabahu*. Table 1.2 presents explanation of patterns in dreams of *Chandragupta*.

Table 1.2. Explanation of patterns in dreams of *Chandragupta*.

| Dream of Chandragupta | Explanation by Bhadrabahu |
|---|---|
| The sun setting | All the knowledge will be darkened |
| A branch of the Kalpavriksha break off and fall | Decline of Jainism and Chandragupta's successors won't be initiated |
| A divine car descending in the sky and returning | The heavenly beings will not visit Bharata Kshetra |
| The disk of the moon sundered | Jainism will be split into two sects |
| Black elephants fighting | Lesser rains and poorer crops |
| Fireflies shining in the twilight | True knowledge will be lost, few sparks will glimmer with feeble light |
| A dried up lake | Aryakhanda will be destitute of Jain doctrines and falsehood will increase |
| Smoke filling all the air | Evil will start to prevail and goodness will be hidden |
| An ape sitting on a throne | Vile, low-born, wicked will acquire power |
| A dog eating the payasa out of a golden bowl | Kings, not content with a sixth share, will introduce land-rent and oppress their subjects by increasing it |
| Young bulls labouring | Young will form religious purposes, but forsake them when old |
| Kshatriya boys riding donkeys | Kings of high descent will associate with the base |
| Monkeys scaring away swans | The low will torment the noble and try to reduce them to same level |

| | |
|---|---|
| Calves jumping over the sea | King will assist in oppressing the people by levying unlawful taxes |
| Foxes pursuing old oxen | The low, with hollow compliments, will get rid of the noble, the good and the wise |
| A twelve-headed serpent approaching | Twelve year of death and famine will come upon this land |

During a period of *systemic constraints (famine)* in the 300 BC temporal coordinate, *Bhadrabahu* calculated that resource scarcity would decrease the probability of monastic survival. He thus vectorized his movement, along with a large matrix of disciples, towards the southern coordinates of the subcontinent, transforming *Chandragupta* into a *Digambar* monk in the process.

*Bhadrabahu* was in Nepal for a 12-year penitential vow when the *Pataliputra* (पाटलिपुत्र) conference took place in 300 BC to put together the *Jain* canon anew. *Bhadrabahu* decided the famine would make it harder for monks to survive and migrated with a group of twelve thousand disciples to South Bharat, bringing with him *Chandragupta* (चन्द्रगुप्त), turned *Digambar* monk. According to the inscriptions at *Shravanabelgola* (श्रवणबेलगोला), *Bhadrabahu* died after taking the vow of *sallekhana* - सल्लेखाना (Fast until death).

**Further Reading**


1. Bhattacharyya, N.N. (2009). Jainism, a Concise Encyclopedia. New Delhi: Manohar Publishers and Distributors. p. 235. ISBN 978-81-7304-312-3.
2. Dundas 2002, p. 47.
3. Jaini, Padmanabh (2000). Collected Papers on Jaina Studies. Motilal Banarasidass. p. 299.
4. Majumdar, R.C. (1971). History of Ancient Bengal (1971 ed.). Calcutta: G.Bharadwaj & Co. p. 12 & 13.
5. Mookerji 1988, p. 4.
6. Mookerji 1988, p. 40.
7. Natubhai Shah 2004, p. 39.
8. Natubhai Shah 2004, p. 42.
9. Rice 1889, p. 3.
10. Rice 1889, p. 4.

11. Rice 1889, p. 5.
12. Sangave 1981, p. 32.
13. Sangave 2001, p. 174.
14. Vidyabhusana 2006, p. 164.
15. Vidyabhusana 2006, p. 164–165.
16. Vyas 1995, p. 27.
17. Wiley 2009, p. 51.
18. Wiley 2009, p. 52.
19. Dundas, Paul (2002) [1992], The Jains (Second ed.), Routledge, ISBN 0-415-26605-X
20. Fynes, F.C.C. (1998). Hemachandra The Lives of Jain Elders (1998 ed.). Oxford: Oxford University Press, Oxford World Classics. p. xxi. ISBN 0-19-283227-1.
21. Bhadrabahu, https://en.wikipedia.org/wiki/Bhadrabahu.
22. Mookerji, Radha Kumud (1988) [first published in 1966], Chandragupta Maurya and his times (4th ed.), Motilal Banarsidass, ISBN 81-208-0433-3.
23. Rice, Benjamin Lewis (1889). Epigraphia Carnatica: Rev. ed (in Kannada). Mysore Government Central Press.
24. Sangave, Vilas Adinath (1981), The Sacred Sravana-Belagola: A Socio-religious study (First ed.), Bharatiya Jnanpith.
25. Sangave, Vilas Adinath (2001), Facets of Jainology: Selected Research Papers on Jain Society, Religion, and Culture, Popular Prakashan, ISBN 978-81-7154-839-2.
26. Shah, Natubhai (2004) [First published in 1998], Jainism: The World of Conquerors, vol. I, Motilal Banarsidass, ISBN 978-81-208-1938-2.
27. Vidyabhusana, Satis Chandra (2006) [1920], A History of Indian Logic: Ancient, Mediaeval and Modern Schools, Motilal Banarsidass, ISBN 81-208-0565-8.
28. Vyas, Dr. R. T., ed. (1995), Studies in Jaina Art and Iconography and Allied Subjects, The Director, Oriental Institute, on behalf of the Registrar, M.S. University of Baroda, Vadodara, ISBN 81-7017-316-7.
29. Wiley, Kristi L (16 July 2009), The a to Z of Jainism, Scarecrow Press, p. 51, ISBN 9780810868212.

**Umaswati (उमास्वती)**
**(c. 200 CE)**

Figure 1.14. Umaswati (उमास्वती). (Image Credit: https://en.wikipedia.org/)

Umaswati (उमास्वती) was born in the village of *Nyagrodhika* - न्यग्रोधिका, to parents *Svati* (स्वती) and *Uma* (उमा) (see Figure 1.14). He was also known as *Svatitanaya* (स्वातितानाय), named after his father, and *Vatsisuta* (वत्सिसुता), referencing his mother's lineage. His name is derived from the combination of his parents' names. Other names he went by include *Vacaka-sramana* (वचक-श्रमण) and *Nagaravacaka* (नागरवाचक). *Digambara* sect members refer to him as *Umasvamin* (उमास्वामिन्). Vidyabhusana, in a 1920 publication, claimed Umaswati passed away in 85 CE. However, more current studies, like that by Padmanabh Jaini, suggest he lived around the 2nd century. Umaswati composed the *Tattvartha Sutra* (तत्त्वार्थ सूत्रम्) in either *Pataliputra* or *Kusumapura* (present-day Patna, Bihar). He holds the distinction of being the premier *Jain* philosopher to pen a work in sutra format.

In *Jainism*'s historical panorama, *Umaswati* and *Kundakunda* (कुण्डकुण्ड) stand out as pivotal figures. Within the *Digambara* tradition, which holds *Kundakunda* in high esteem, *Umaswati* is seen as his student. Nevertheless, there are distinctions between them: *Kundakunda*'s writings were in *Prakrit*, whereas *Umaswati* opted for *Sanskrit*. Furthermore, nuances in their teachings, like their perspectives on *anekantavada* (अनेकान्तवाद), varied.

## 卐 Key Contributions in Mathematics

### 卐 Systematic Framework

*Umaswati*'s *Tattvartha Sutra* systematically presents the entire *Jain* philosophy. In doing so, he provided a structured framework for understanding complex philosophical concepts. While not directly mathematical, this framework facilitates analytical and logical thinking, which are foundational to mathematical studies.

### 卐 Seven Categories of Truth

In the *Tattvartha Sutra, Umaswati* introduced *seven categories of truth* (सत्यस्य सप्त वर्गा). These categories represent a structured approach to understanding reality. They involve distinguishing between different types of entities, like sentient beings and non-sentient matter, understanding the nature of karma, and recognizing the processes of liberation. This classification, with its logical sequence, is reminiscent of mathematical categorizations where different entities are classified based on specific attributes.

### 卐 Logic and Analysis

*Umaswati*'s teachings, particularly surrounding *Anekantavada*, delve into the intricate complexities of simultaneous existence and non-existence. This central tenet of *Jain* philosophy isn't just a matter of faith or belief; it insists on a detailed and analytical investigation. Just as in mathematics, where problems are approached with structured logical reasoning to derive solutions or proofs, *Anekantavada* requires a similar meticulous logical scrutiny. It's a testament to how spirituality and logic can intersect, demanding not just belief, but also intellectual rigor in understanding and interpretation.

### 卐 Ethics and Quantitative Morality

*Umaswati*'s presentation of the *Jaina* vows in the *Tattvartha Sutra*, especially when he explains their value in preventing *karmic inflow to the soul* (आत्मानं प्रति

कर्मप्रवाह), introduces a quantitative aspect to morality. For instance, certain actions lead to specific amounts of karmic accumulation, which in turn affects future births. This quantitative understanding of moral actions can be seen as a precursor to mathematical modeling, where real-world actions and their consequences are represented using mathematical structures.

**Further Reading**


1. Asim Kumar Chatterjee (2000). A Comprehensive History of Jainism: From the Earliest Beginnings to AD 1000. Munshiram Manoharlal. pp. 282–283. ISBN 978-81-215-0931-2.
2. B Faddegon; F W Thomas (1935). The Pravacana sara of kunda Kunda Acarya together with the commentary, Tattva-dipika. Cambridge University Press. pp. xv–xvi.
3. Balcerowicz 2003, p. 25.
4. Balcerowicz 2003, p. 26.
5. Balcerowicz 2003, pp. 30–31.
6. DastiBryant 2014, p. 72.
7. Eivind Kahrs (1998). Indian Semantic Analysis: The Nirvacana Tradition. Cambridge University Press. p. 21. ISBN 978-0-521-63188-4.
8. Jack Finegan (1989). An Archaeological History of Religions of Indian Asia. Paragon House. p. 221. ISBN 978-0-913729-43-4.
9. Jain 2011, p. 2.
10. Jain 2011, pp. 118–119.
11. Jain 2011, pp. 121–124.
12. Jain 2011, pp. 126–145.
13. Jain 2011, pp. 146–151.
14. Jaini 1998, p. 81.
15. Jaini 1998, p. 82.
16. Jones & Ryan 2007, pp. 439–440.
17. K. V. Mardia (1990). The Scientific Foundations of Jainism. Motilal Banarsidass. p. 103. ISBN 978-81-208-0658-0. Quote: Thus, there is a vast literature available but it seems that Tattvartha Sutra of Umasvati can be regarded as the main philosophical text of the religion and is recognized as authoritative by all Jains."
18. Karl H. Potter; Dalsukh Malvania; Jayendra Soni, eds. (2003). Volume X: Jain Philosophy (Part I). Motilal Banarsidass. pp. 594–600. ISBN

978-81-20831-698. OCLC 718313318., Quote: "Vidyananda, Tattvarthasutra-Slokavarttika"

19. Krishna Sivaraman (1989). Hindu Spirituality: Vedas Through Vedanta. Motilal Banarsidass. pp. 154–157. ISBN 978-81-208-1254-3.
20. Lloyd 2009, p. 142.
21. Natubhai Shah 2004, p. 48.
22. Paul Dundas (2006). Patrick Olivelle (ed.). Between the Empires: Society in India 300 BCE to 400 CE. Oxford University Press. pp. 395–396. ISBN 978-0-19-977507-1.
23. Piotr Balcerowicz (2003). Essays in Jaina Philosophy and Religion. Motilal Banarsidass. p. 25. ISBN 978-81-208-1977-1.
24. Piotr Balcerowicz (2003). Essays in Jaina Philosophy and Religion. Motilal Banarsidass. pp. 33–34. ISBN 978-81-208-1977-1.
25. Sarva-Darsana-Samgraha by Madhavacharya, EB Cowell (Translator), Trubner & Co, pages 46–60
26. Satis Chandra Vidyabhusana (1920). A History of Indian Logic: Ancient, Mediaeval and Modern Schools. Motilal Banarsidass. pp. 186–187. ISBN 978-81-208-0565-1.
27. Sures Chandra Banerji (1989). A Companion to Sanskrit Literature: Spanning a Period of Over Three Thousand Years, Containing Brief Accounts of Authors, Works, Characters, Technical Terms, Geographical Names, Myths, Legends, and Several Appendices. Motilal Banarsidass. pp. 348–349. ISBN 978-81-208-0063-2.
28. Umāsvāti 1994, p. xiii.
29. Umāsvāti 1994, p. xi–xiii, Quote: "That Which Is, known as the Tattvartha Sutra to Jains, is recognized by all four Jain traditions as the earliest, most authoritative and comprehensive summary of their religion.".
30. Umāsvāti 1994, p. xi–xiii.
31. Umāsvāti 1994, p. XLV, 206.
32. Umāsvāti 1994, p. xviii–xx, 2–3, 6.
33. Umāsvāti 1994, pp. 12–15.
34. Umāsvāti 1994, pp. 169–170.
35. Umāsvāti 1994, pp. 195–199.
36. Umāsvāti 1994, pp. 200–203.
37. Umāsvāti 1994, pp. 213–248.
38. Umāsvāti 1994, pp. 250–263.
39. Umāsvāti 1994, pp. 297, also back flap.
40. Umāsvāti 1994, pp. 33–62.
41. Umāsvāti 1994, pp. 5–6.

42. Umāsvāti 1994, pp. 7–168.
43. Umāsvāti 1994, pp. xiii–xvii.
44. Vidyabhusana 1920, pp. 168–69.
45. W. J. Johnson (1995). Harmless Souls: Karmic Bondage and Religious Change in Early Jainism with Special Reference to Umāsvāti and Kundakunda. Motilal Banarsidass. p. 46. ISBN 978-81-208-1309-0.
46. W. J. Johnson (1995). Harmless Souls: Karmic Bondage and Religious Change in Early Jainism with Special Reference to Umāsvāti and Kundakunda. Motilal Banarsidass. pp. 46–51, 91–96. ISBN 978-81-208-1309-0.
47. Walter Slaje (2008). Śāstrārambha: Inquiries Into the Preamble in Sanskrit. Otto Harrassowitz Verlag. p. 35 with footnote 23. ISBN 978-3-447-05645-8.
48. Walter Slaje (2008). Śāstrārambha: Inquiries Into the Preamble in Sanskrit. Otto Harrassowitz Verlag. pp. 34–43. ISBN 978-3-447-05645-8.
49. Balcerowicz, Piotr, ed. (2003) [2002], Essays in Jaina Philosophy and Religion, Motilal Banarsidass, ISBN 81-208-1977-2.
50. Dasti, Matthew R.; Bryant, Edwin F. (2014), Free Will, Agency, and Selfhood in Indian Philosophy, OUP US, ISBN 978-0-19-992275-8.
51. Digambar Jain Online, http://www.digambarjainonline.com/dharma/greatja5.htm
52. https://en.wikipedia.org/wiki/Umaswati.
53. Jain 2011, p. vi.
54. Jain, Vijay K. (2011), Acharya Umasvami's Tattvarthsutra: With Hindi and English Translation, Vikalp Printers, ISBN 978-81-903639-2-1.
55. Jaini, Padmanabh S. (1998) [1979], The Jaina Path of Purification, Delhi: Motilal Banarsidass, ISBN 81-208-1578-5.
56. Jones, Constance; Ryan, James D. (2007), Encyclopedia of Hinduism, Infobase Publishing, ISBN 978-0816054589.
57. Lloyd, Christopher (2009), What on Earth Happened?, Bloomsbury Publishing, ISBN 978-1-4088-0597-8.
58. Selections From Acharya Umaswami's Tattvaarth Sutra, https://web.archive.org/web/20080724080742/http://www.jainstudy.org/JSC7.07QfromS.pdf.
59. Shah, Natubhai (2004) [First published in 1998], Jainism: The World of Conquerors, vol. I, Motilal Banarsidass, ISBN 81-208-1938-1.
60. Umāsvāti, Umaswami (1994), That which is (Translator: Nathmal Tatia), Rowman & Littlefield, ISBN 978-0-06-068985-8.

61. Umaswati in Epigraphical and Literary Tradition, http://vmis.in/Resources/digital_publication_popup?id=95#page/2.
62. Vidyabhusana, Satis Chandra (1920), A History of Indian Logic: Ancient, Mediaeval, and Modern Schools, Motilal Banarsidass, ISBN 978-81-208-0565-1.

## Yavaneśvara (यवनेश्वर)
## (2nd century)

Yavanesvara (यवनेश्वर) or (or *Yavanaraja* - यवनराज) meaning "*Lord of the Greeks,*" played an instrumental role in bridging *Greek* and Bhartiya astrology during the reign of *Rudradaman* (रुद्रदमन), ruler of the *Ksatrapas* (*Ksatrapas*) around 130 AD. The *Yavanajātaka* (यवनजातक), attributed to *Yavaneśvara* and later versified by *Sphujidhvaja* (स्फुजिध्वज), stands as a significant document in the history of Bhartiya astrology, particularly as it represents a synthesis of *Greek* and Bhartiya astrological thought. Originating from a *Greek* text written in *Alexandria* around 120 BC, *Yavanesvara*'s translation was not a mere linguistic effort. He ingeniously integrated *Greek* concepts into the Bhartiya socio-religious context, replacing *Greek* gods with Hindu images and assimilating the Indian caste system. This translation aimed to equip Bhartiyas with the necessary knowledge to practice astrology using both Greek and Bhartiya concepts.

He translated a *Greek* astrology text into Sanskrit, profoundly influencing Bhartiya astrology and astronomy for centuries.

### 卐 Key Contributions in Mathematics

### 卐 Horoscopic Calculations

The *Yavanajātaka* elucidates complex instructions for the formulation of *astrological* (ज्योतिषीय) charts based on one's time and location of birth. Crafting an accurate *horoscope* (कुंडली) is a complex mathematical endeavor, requiring precise computations to predict the positions of planets and stars at a specific time. This methodology encompasses the understanding of the celestial sphere, the ecliptic plane, and the various constellations that lie in the *zodiac belt* (राशि मेखला). These computations are vital to identify the planetary configurations that would influence an individual's fate according to *astrological* principles.

### 卐 Astronomical Mathematical Methods

Derived from the advanced astrological techniques formulated in the intellectual hubs of the *Hellenistic* world, especially *Alexandria*, the *Yavanajātaka* offers a treasure trove of *astronomical* (खगोलीय) algorithms. The calculation of the '*horoskopos*', which determines the *zodiac sign* (राशी) rising on the eastern horizon during an individual's birth, is particularly notable. This complex computation integrates the *Earth*'s (पृथिव्या) rotation, the tilt of its axis, and the orbit of the sun to accurately determine the sign that imparts specific traits to an individual.

### 卐 Nomenclature & Techniques from Greek

The *Yavanajātaka* exhibits an intricate tapestry of *Greek* and Bhartiya astrological terminologies and techniques. The use of terms such as *jamitra* – जमित्र (translating to *diametros* in *Greek*) and *mesurana* (akin to the *Greek mesouranema*) are a testament to this melding of cultures. These terms, while representing specific astrological positions, are also indicative of the mathematical techniques required to compute them— techniques that were honed in the academies of ancient *Greece* and were subsequently integrated into Bhartiya astrology.

### 卐 Division of Zodiac

In astrological computations, the zodiac's division into smaller segments allows for more granulated and detailed predictions. The '*Drekanas* - द्रेकान', or the *dekanos* as termed in *Greek*, is a primary example. These divisions involve segmenting each zodiac sign into thirds, leading to *36 Drekanas* across the entire zodiac belt. Each segment has its attributes and implications, and determining a person's *Drekanas* requires additional mathematical scrutiny beyond just the primary zodiac sign.

### 卐 Planetary Movements

The *Yavanajātaka* underscores the paramount importance of planetary movements in astrological determinations, referring to them as the "*supreme eye* - परमं नेत्रम्" of the horoscopic principles. These planetary transits, retrogrades, and progressions form the bedrock of *astrological predictions* (ज्योतिषीय भविष्यवाणी). Accurately charting these movements necessitates an intricate understanding of orbital mechanics, planetary periods, and their relative positions to Earth at any given time.

### 卐 Time Calculation and Periodicities

The *Yavanajātaka* provides insights into the varied perspectives on the duration of '*yuga*'— a specific epoch in Bhartiya cosmology. While the text references the *Greek* viewpoint, it also acknowledges Bhartiya beliefs, showcasing a confluence of time calculation methodologies from both civilizations. Determining the length, onset, and conclusion of such astronomical periods is a profound mathematical challenge, integrating long-term observations, cyclical patterns of celestial bodies, and computational models to derive accurate periodicities.

**Further Reading**


1. Dhavale, D. G. (1984). "The Yavanajataka of Sphujidhvaja by David Pingree". Annals of the Bhandarkar Oriental Research Institute. 65 (1/4): 266–267. JSTOR 41693124.
2. Mak (2013a, 2013b, 2014).
3. Mc Evilley "The shape of ancient thought", p385 ("The Yavanajataka is the earliest surviving Sanskrit text in horoscopy, and constitute the basis of all later Indian developments in horoscopy", himself quoting David Pingree "The Yavanajataka of Sphujidhvaja" p5).
4. N. P. Subramania Iyer (1991). Kalaprakasika. Asian Educational Services. p. 3.
5. Rocher, Ludo (March 1980). "The Yavanajataka of Sphujidhvaja by David Pingree". Isis. 71 (1): 173–174. doi:10.1086/352443. JSTOR 230349.
6. Witzel 2001.
7. Bill M. Mak (2013) The Date and Nature of Sphujidhvaja's Yavanajātaka reconsidered in the light of some newly discovered materials, History of Science in South Asia 1: pp. 1-20.
8. Bill M. Mak (2013) The Last Chapter of Sphujidhvaja's Yavanajātaka critically edited with notes, SCIAMVS 14, pp. 59-148.
9. Bill M. Mak (2014) The 'Oldest Indo-Greek Text in Sanskrit' Revisited: Additional Readings from the Newly Discovered Manuscript of the

Yavanajātaka, Journal of Indian and Buddhist Studies 62(3): 1101-1105.

10. David Pingree (1981) "Jyotiḥśāstra", Jan Gonda (ed) A History of Indian Literature, Vol, VI Fasc. 4, Otto Harrassowitz – Wiesbaden.
11. Yavanesvara, https://mathshistory.st-andrews.ac.uk/Biographies/Yavanesvara/.
12. K. V. Sarma (1997), "Sphujidhvaja", Encyclopaedia of the History of Science, Technology, and Medicine in Non-Western Cultures edited by Helaine Selin, Springer, ISBN 978-0-7923-4066-9.
13. Pingree (1981) p.81.
14. Thomas McEvilley (2002) The Shape of Ancient Thought, Allworth Press, ISBN 1-58115-203-5.
15. Witzel, Michael (25 May 2001). "Autochthonous Aryans? The Evidence from Old Indian and Iranian Texts". Electronic Journal of Vedic Studies. 7 (3).

**Vasishtha (वसिष्ठ)**
**(No Clarity of Age)**

Figure 1.15. Maharishi Vasishtha (महर्षि वसिष्ठ). (Image Credit: https://en.wikipedia.org/wiki/)

The origin of Maharishi Vasishtha (महर्षि वसिष्ठ) or Vashisht (*vaśiṣṭha*, वशिष्ठ) traces back to a remarkable birth narrative rooted in the *Rigveda's Mandala 7* (ऋग्वेदस्य मण्डलम् ७), where the deities *Mitra* (मित्र) and *Varuna* (वरुण), enchanted by the celestial beauty *Urvashi* (उर्वशी) during a *yajna* (यज्ञ), release their vital essence into a *pitcher*, leading to *Vasishtha*'s emergence after a few days (see Figure 1.15). The Puranas further elucidate his three unique incarnations: initially as *Brahma*'s mind-born son, later resurrected by *Brahma* (ब्रह्मा) after a tragic end post the *Daksha Yajna* (दक्ष यज्ञ), and then reborn through *Mitra* and *Varuna* after a curse-laden altercation with *King Nimi* (राजा निमि). Even in his early days, *Vasishtha*'s life was marked by divine interventions, profound wisdom, and a series of rebirths that would set the stage for his legacy as a preeminent *Vedic* sage. He is one of the *Saptarishis* (सप्तर्षि) (*seven great Rishis*). He is better known for his spiritual teachings and his role as the *guru* (गुरु) of the young prince *Rama* (रामः) in the ancient Bhartiya epic, the *Ramayana* (रामायण). In the context of this epic and other

*Vedic literature, Vasishtha* is presented more as a spiritual teacher than as a mathematician. Figure 1.16 shows *Vasishtha* teaching *Lord Rama.*

Figure 1.16. *Vasishtha* teaching *Lord Rama*. (Image Credit: https://en.wikipedia.org/wiki/)

However, the period during which the *Vedas* (वेदा) were written saw the emergence of various mathematical concepts, especially in the realm of geometry and algebra, which were primarily developed for the construction of *Vedic altars* (वैदिकवेदिका) and the correct performance of *Vedic rituals* (वैदिक संस्कार).

Historically, *Vasishtha* was a *Rigvedic poet* (ऋग्वेदिक कवि) and the *purohita of Sudās Paijavana*, (पुरोहितः सुदास पैजवनस्य) chief of the *Bharata tribe* (भरत जनजाति). In *Rigvedic hymn 7.33.9*, (ऋग्वेद स्तोत्रम् 7.33.9) *Vashishtha* is described as a scholar who moved across the *Saraswati River* (सरस्वती नदी) to establish his school. At some point, he replaced *Viśvāmitra Gāthina* (विश्वामित्र गाथिना) as the purohita of *Sudās* (सुदास). In later *Hindu* texts, *Viśvāmitra* and *Vasiṣṭha* have a long-standing *feud* (विवाद), and scholars have stated they historically had a feud regarding the position of the *Bharata purohita.*

However, this view has been criticized due to lack of internal evidence and the projection of later views onto the *Rigveda*. Under *Sudās* and *Vasiṣṭha*, the *Tṛtsu-Bharatas* (तृत्सु-भारत) won the *Battle of the Ten Kings* (दशराजानां युद्धम्). *Sudās* decisively won against a *Puru*-led (पुरू-नेतृत्वेन) alliance by the strategic breaching of a (natural) *dyke on the Ravi River* (रविनद्याः उपरि बाधः) thereby drowning most of the opponents; the *victory* (विजय) is attributed to the *benevolence* (परोपकार) and strategizing of *Indra* (इन्द्रः), the patron-*God* of the *Bharatas*, whose blessings were secured by *Vasistha*'s poetics.

He was married to *Arundhati* (अरुन्धति), and therefore he was also called *Arundhati Natha* (अरुन्धति नाथ), meaning the husband of *Arundhati*. Later, this region is believed in the Bhartiya tradition to be the abode of sage *Vyasa* (व्यासः) along with *Pandavas* (पाण्डवः), the five brothers of *Mahabharata* (*Mahabharata*). He is typically described in ancient and medieval *Hindu* texts as a sage with long flowing hairs that are neatly tied into a bun that is coiled with a tuft to the right, a beard, a handlebar moustache and a tilak on his forehead.

In Buddhist Pali canonical texts such as Digha Nikaya, Tevijja Sutta describes a discussion between the Buddha and Vedic scholars of his time. The Buddha names ten rishis, calls them "early sages" and makers of ancient verses that have been collected and chanted in his era, and among those ten rishis is Vasettha (the Pali spelling of Vashishtha).

The Vashishtha Samhita (वशिष्ठ संहिता) is a medieval era Yoga text. There is an Agama as well with the same title. The Vashishtha dharmasutra (वशिष्ठधर्मसूत्रम्), an ancient text, and one of the few *Dharma* (धर्म)-related treatises which has survived into the modern era. This *Dharmasūtra* - धर्मसूत्रम् (300–100 BCE) forms an independent text and other parts of the *Kalpasūtra* (कल्पसूत्रम्), that is *Shrauta-* (श्रौत-सूत्रा) *and Grihya-sutras* (गृह्य-सूत्राणि) are missing. It contains 1,038 sutras. The *Yoga Vashishtha* (योग वशिष्ठ) is a syncretic medieval era text that presents *Vedanta* (वेदान्त दर्शनानि) and *Yoga philosophies* (योग दर्शन). It is written in the form of a dialogue between *Vashishtha* and prince *Rama* of *Ramayana* fame, about the nature of life, human suffering, choices as the nature of life, free will, human creative power and spiritual liberation. *Yoga Vashishtha* teachings are structured as stories and fables, with a philosophical foundation similar to those found in *Advaita Vedanta* (अद्वैत वेदान्त). The text is also notable for its discussion of *Yoga*.

The *Yoga Vashishtha* philosophy can be summarized as, "*Human effort can be used for self-betterment and that there is no such thing as an external fate imposed by the gods*".

"मानवप्रयत्नः आत्मसुखार्थं प्रयोक्तुं शक्यते, देवैः आरोपितं बाह्यं दैवं नास्ति इति च"

The word "*Siddhanta* - सिद्धान्तः " roughly translates to "*doctrine*" or "*tradition* - परंपरा," and the various *Siddhantas* were distinct schools or systems of astronomy.

*Vasishtha Siddhanta*, as the name suggests, is attributed to the sage *Vasishtha.*

The Agni Purana (अग्नि पुराण) is attributed to *Vashishtha.*

The *Vishnu Purana* (विष्णु पुराण) is attributed to *Vashishtha* along with *Rishi Pulatsya* (ऋषि पुलात्स्य). He has also contributed to many *Vedic hymns* (वैदिक स्तोत्रम्) and is seen as the arranger of *Vedas* during *Dvapara Yuga* (द्वापर युग).

## 卐 Key Contributions in Mathematics

*Vasishtha Siddhanta* (वसिष्ठ सिद्धान्त) is one of the earliest astronomical systems in use in Bharat, which is summarized in *Varahamihira's Pancha-Siddhantika* - वराहमिहिरस्य पञ्च-सिद्धान्तिका (6th century). It is attributed to sage *Vasishtha* and claims a date of composition of 1,299,101 BCE. The original text probably dated to the 4th century, but it has been lost and our knowledge of it is restricted to *Varahamira* (वराहमिहिर)'s account.

## 卐 Calculation of Time

The *Siddhantas* were comprehensive in their approach to time, starting from the smallest unit, the "*truti* - त्रुति," which was a fraction of a second, up to vast cosmic cycles called "*mahayugas* – महायुग" spanning millions of years. The *Vasishtha Siddhanta* would have used sophisticated calculations to divide the solar day into these various units, accommodating for both the apparent solar day and the sidereal day.

It would detail how lunar days, or "*tithis* - (तिथि)" were derived from the difference between the elongations of the *sun* - सूर्य and the *moon* - चन्द्र.

## 卐 Planetary Movements

Detailed tables, or "*ephemerides* - क्षणिकम्," would have been provided, mapping out the daily motion of planets against the fixed stars, or the "*zodiac*." These would be based on meticulous observations and would serve as predictive tools for the motion of the planets, helping astrologers and priests in determining auspicious timings. Each *planet* (ग्रह) would have been assigned an average period of revolution, which would be used for these predictions.

## 卐 Eclipses

*Eclipses* (ग्रहणानि) have always been of significant religious and astrological importance in Bharat. The *Siddhanta* would provide methods to calculate the timing and duration of both solar and lunar eclipses. There would be an understanding of the nodes of the moon's orbit, called "*Rahu* - राहु" and "*Ketu* - केतु" in Bhartiya astronomy, which play a crucial role in predicting eclipses.

## 卐 Year Length

Accurate determination of the year's length is vital for maintaining the *synchrony* (समन्वय) between the lunar and solar calendars. The *Vasishtha Siddhanta* would detail how many solar days make a year, accommodating for the discrepancy that arises from using a lunar-based month system. This would likely involve the periodic insertion of an extra month, or "*adhika masa*" (अधिका मस).

## 卐 Geometry for Altar Constructions

*Vedic rituals* often involved the construction of altars with specific shapes and dimensions, symbolizing astronomical or cosmological principles. The *Siddhanta* could detail, for instance, how to construct a specific altar to symbolize the celestial sphere or the orbits of certain planets. These principles were not just theoretical but had practical implications for rituals and were closely intertwined with the philosophical understanding of the cosmos.

## 卐 Panchanga System

The *Panchanga* - पञ्चाङ्ग, or the traditional almanac, is a vital part of daily life in Bharat, even today. It provides detailed data on five key aspects: the day, the lunar mansion or *nakshatra* (नक्षत्रम्), the lunar day or tithi, the *yoga* (योग), and the *karana* (करणम्). The *Vasishtha Siddhanta* would provide the foundational calculations for these aspects, based on which daily almanacs for specific regions would be derived. This system was essential for determining auspicious times for ceremonies, marriages, and other rituals.

## Further Reading


1. "according to Rig Veda 7.33:11 he is the son of Maitravarun and Urvashi" Prof. Shrikant Prasoon, Pustak Mahal, 2009, ISBN 8122310729, ISBN 9788122310726.
2. "Rishi Vasishtha : Ishta Guru Of Lord Ram - Humans Of Uttarakhand". 3 June 2021. Archived from the original on 3 December 2021. Retrieved 3 December 2021.
3. "Rishi Vasistha - One of the Mind-born Sons of Lord Brahma". vedicfeed.com. 18 June 2020. Archived from the original on 3 December 2021. Retrieved 3 December 2021.
4. "Story of Vasiṣṭha". www.wisdomlib.org. 28 January 2019. Archived from the original on 7 December 2021. Retrieved 14 October 2020.
5. Adheesh A. Sathaye (2015). Crossing the Lines of Caste: Vishvamitra and the Construction of Brahmin Power in Hindu Mythology. Oxford University Press. pp. 254–255. ISBN 978-0-19-934111-5.
6. Agarwal, D.P. (1990). "Legends as models of Science". Bulletin of the Deccan College Research Institute. 49: 41–42. JSTOR 42930266.(subscription required).
7. Albrecht Weber, The History of Indian Literature, p. 258; Romesh Chunder Dutt, A History of Civilization in Ancient India, p. 209.
8. Brereton, Joel P.; Jamison, Stephanie W., eds. (2014). The Rigveda: The Earliest Religious Poetry of India. Vol. I. Oxford University Press. pp. 880, 902–905, 923–925, 1015–1016. ISBN 9780199370184.
9. Chapple 1984, p. xi.
10. Chapple 1984, pp. ix-x with footnote 3.
11. Chapple 1984, pp. xi–xii.
12. Chapple 1984, pp. x-xi with footnote 4.

13. Cunningham, G. Watts (1948). "How Far to the Land of Yoga? An Experiment in Understanding". The Philosophical Review. 57 (6): 573–589. doi:10.2307/2181797. JSTOR 2181797.
14. Edwin Bryant (2003). The Quest for the Origins of Vedic Culture: The Indo-Aryan Migration Debate. Oxford University Press. p. 164. ISBN 978-0-19-516947-8.
15. F Chenet (1987), Bhāvanā et Créativité de la Conscience, Numen, Vol. 34, Fasc. 1, pages 45-96 (in French).
16. Findly, Ellison Banks (1984). "Vasistha: Religious Personality and Vedic Culture". Numen. BRILL Academic. 31 (1): 74–77, 98–105. doi:10.2307/3269890. JSTOR 3269890.
17. Gopal, Madan (1990). K.S. Gautam (ed.). India through the ages. Publication Division, Ministry of Information and Broadcasting, Government of India. p. 70.
18. Gudrun Bühnemann (1988). Pūjā: A Study in Smārta Ritual. Brill Academic. p. 220. ISBN 978-3-900271-18-3.
19. Harry Hicks and Robert Anderson (1990), Analysis of an Indo-European Vedic Aryan Head – 4500-2500 B.C. Archived 7 July 2017 at the Wayback Machine, in Journal of Indo European studies, Vol. 18, pp 425–446. Fall 1990.
20. Horace Hayman Wilson (1840). The Vishńu Puráńa: A System of Hindu Mythology and Tradition. Oriental Translation Fund of Great Britain and Ireland. p. lxix.
21. Horace Hayman Wilson (1840). The Vishńu Puráńa: A System of Hindu Mythology and Tradition. Oriental Translation Fund of Great Britain and Ireland. p. xxxvi.
22. J. C. Heesterman; Albert W. Van den Hoek; Dirk H. A. Kolff; et al. (1992). Ritual, State, and History in South Asia: Essays in Honour of J.C. Heesterman. BRILL Academic. pp. 136–137. ISBN 90-04-09467-9.
23. J. C. Heesterman; Albert W. Van den Hoek; Dirk H. A. Kolff; et al. (1992). Ritual, State, and History in South Asia: Essays in Honour of J.C. Heesterman. BRILL Academic. pp. 68–73 with footnotes. ISBN 90-04-09467-9.
24. James G. Lochtefeld (2002). The Illustrated Encyclopedia of Hinduism: N-Z. The Rosen Publishing Group. p. 742. ISBN 978-0-8239-3180-4.
25. Jamison & Brereton 2014, p. 1015-1016.
26. K. N. Aiyer (1975). Laghu-Yoga-Vasishta. Original Author: Abhinanda. Theosophical Publishing House. p. 5. ISBN 978-0835674973.

27. Kanuga, G. B. (1993). Immortal Love of Rama. Lancer Publishers. ISBN 978-1-897829-50-9. Archived from the original on 22 April 2022. Retrieved 21 October 2020.
28. Kasyapa is mentioned in RV 9.114.2, Atri in RV 5.78.4, Bharadvaja in RV 6.25.9, Vishvamitra in RV 10.167.4, Gautama in RV 1.78.1, Jamadagni in RV 3.62.18, etc.;[6] Original Sanskrit text: प्रसूतो भक्षमकरं चरावपि स्तोमं चेमं प्रथमः सूरिरुन्मृजे । सुते सातेन यद्यागमं वां प्रति विश्वामित्रजमदग्नी दमे ॥४॥[7].
29. Leslie 2003, pp. 104.
30. Lingat 1973, p. 18.
31. Mariasusai Dhavamony (1999). Hindu Spirituality. Gregorian. pp. 50 with footnote 63. ISBN 978-88-7652-818-7.
32. Maurice Bloomfield (1899). Atharvaveda. K.J. Trübner. pp. 31, 111, 126.
33. Maurice Walshe (2005). The Long Discourses of the Buddha: A Translation of the Digha Nikaya. Simon and Schuster. pp. 188–189. ISBN 978-0-86171-979-2.
34. Michael Witzel (1997). Inside the Texts, Beyond the Texts: New Approaches to the Study of the Vedas: Proceedings of the International Vedic Workshop, Harvard University, June 1989. Harvard University Press. pp. 289 with footnote 145. ISBN 978-1-888789-03-4.
35. Michael Witzel (1997). Inside the Texts, Beyond the Texts: New Approaches to the Study of the Vedas: Proceedings of the International Vedic Workshop, Harvard University, June 1989. Harvard University Press. pp. 41–42. ISBN 978-1-888789-03-4. OCLC 246746415.
36. Monier-Williams, Monier (1899). "vaśiṣṭha". A Sanskrit-English Dictionary: Etymologically and Philologically Arranged with Special Reference to Cognate Indo-European Languages. Oxford: Clarendon Press. OCLC 685239912., Archive 2 Archived 28 April 2017 at the Wayback Machine.
37. Olivelle 1999, p. 325.
38. Olivelle 1999, p. xxvi.
39. Olivelle 2006, p. 185.
40. Pratap Chandra Roy's Mahabharata Adi Parva Page: 409.
41. Ramchandra Narayan Dandekar (1981), "Vasistha as Religious Conciliator", in Exercises in Indology, Delhi: Ajanta, pages 122-132, OCLC 9098360.

42. Rigveda 10.167.4 Archived 29 October 2020 at the Wayback Machine, Wikisource.
43. Rigveda, translated by Ralph T.H. Griffith, archived from the original on 13 October 2013, retrieved 20 June 2013, A form of lustre springing from the lightning wast thou, when Varuṇa and Mitra saw thee.
44. Satyamayananda, Swami (2019). Ancient Sages. Advaita Ashrama (A publication branch of Ramakrishna Math, Belur Math). ISBN 978-81-7505-923-8. Archived from the original on 7 December 2021. Retrieved 7 December 2021.
45. Steven Collins (2001). Aggañña Sutta. Sahitya Akademi. p. 17. ISBN 978-81-260-1298-5.
46. Strauss, Sarah (2002). "The Master's Narrative: Swami Sivananda and the Transnational Production of Yoga". Journal of Folklore Research. Indiana University Press. 23 (2/3): 221. JSTOR 3814692.
47. Surendranath Dasgupta (2 January 1933). A History of Indian Philosophy. Vol. 2. Cambridge University Press. pp. 252–253. ISBN 978-0521047791.
48. The Buddha names the following as "early sages" of Vedic verses, "Atthaka (either Ashtavakra or Atri), Vamaka, Vamadeva, Vessamitta (Visvamitra), Yamataggi, Angirasa, Bharadvaja, Vasettha (Vashistha), Kassapa (Kashyapa) and Bhagu (Bhrigu)".
49. Vaśiṣṭha Saṃhitā: Yoga kāṇḍa. Kaivalyadhama S.M.Y.M. samiti. 2005. ISBN 978-81-89485-37-5.
50. Venkatesananda 1984, pp. 51, 77, 87, 121, 147, 180, 188, 306, 315, 354, 410.
51. Witzel, Michael (1995). "4. Early Indian history: Linguistic and textual parametres". In Erdosy, George (ed.). The Indo-Aryans of Ancient South Asia: Language, Material Culture and Ethnicity. Indian Philology and South Asian Studies. De Gruyter. pp. 85–125. doi:10.1515/9783110816433-009. ISBN 978-3-11-081643-3. S2CID 238465491. Archived from the original on 12 August 2021. Retrieved 29 August 2021.
52. Witzel, Michael (1995). "4. Early Indian history: Linguistic and textual parametres". In Erdosy, George (ed.). The Indo-Aryans of Ancient South Asia: Language, Material Culture and Ethnicity. Indian Philology and South Asian Studies. De Gruyter. pp. 85–125. doi:10.1515/9783110816433-009. ISBN 978-3-11-081643-3. S2CID 238465491. Archived from the original on 12 August 2021. Retrieved 29 August 2021.

53. Witzel, Michael (1995). "Ṛgvedic history: poets, chieftains and polities". In Erdosy, George (ed.). The Indo-Aryans of Ancient South Asia: Language, Material Culture and Ethnicity. De Gruyter. pp. 248–249, 251.
54. And as a fallen drop, in heavenly fervour, all the Gods laid thee on a lotus-blossom
55. Atreya, B. L. (1981) [1935)]. The Philosophy of the Yoga Vashista. A Comparative Critical and Synthetic Survey of the Philosophical Ideas of Vashista as presented in the Yoga-Vashista Maha-Ramayan. Based on a thesis approved for the degree of Doctor of Letters in the Banaras Hindu University. Moradabad: Darshana Printers. p. 467.
56. Atreya, B. L. (1993). The Vision and the Way of Vashista. Madras: Indian Heritage Trust. p. 583. OCLC 30508760. Selected verses, sorted by subject, in both Sanskrit and English text.
57. Born of their love for Urvasi, Vashiṣṭha thou, priest, art son of Varuṇa and Mitra.
58. Chapple, Christopher. Introduction. In Venkatesananda (1984).
59. History of Astronomy (1834), ISBN 0-543-94987-7, p. 117.
60. Vasishtha, https://en.wikipedia.org/wiki/Vasishtha.
61. Leslie, Julia (2003). Authority and meaning in Indian religions: Hinduism and the case of Vālmīki. Ashgate Publishing. ISBN 0-7546-3431-0.
62. Lingat, Robert (1973). The Classical Law of India. University of California Press. ISBN 978-0-520-01898-3.
63. Motilal Bansaridas Publishers Bhagavat Purana Book 2, Skandha IV Page: 426
64. Olivelle, Patrick (1999). Dharmasutras: The Law Codes of Ancient India. Oxford University Press. ISBN 978-0-19-283882-7.
65. Olivelle, Patrick (2006). Between the Empires: Society in India 300 BCE to 400 CE. Oxford University Press. ISBN 978-0-19-977507-1.
66. Stephanie Jamison; Joel Brereton (2014). The Rigveda: 3-Volume Set. Oxford University Press. pp. 1681–1684. ISBN 978-0-19-972078-1.
67. Tliy one and only birth was then, Vashiṣṭha, when from thy stock Agastya brought thee hither.
68. Vālmīki (1976). Yoga Vashista Sara: The Essence of Yoga Vashista. trans. Swami Surēśānanda. Tiruvannamalai: Sri Ramanasramam. p. 29. OCLC 10560384. Very short condensation.
69. Vālmīki (2002) [1982]. The Essence of Yogavaasishtha. Compiled by Sri Jnanananda Bharati, translated by Samvid. Chennai: Samata Books. p. 344. Sanskrit and English text.

# Chapter 2

# Classical
(320 CE - 520 CE)

**Āryabhaṭa-I (आर्यभटः)**
**(476–550 CE)**

Figure 2.1. Āryabhaṭa-I (आर्यभटः). (Image Credit: https://mathshistory.st-andrews.ac.uk/)

Āryabhaṭa (आर्यभटः) elder (or I) mentions in the *Aryabhatiya* (आर्यभटिया) that he was 23 years old 3,600 years into the *Kali Yuga* (कलियुग), but this is not to mean that the text was composed at that time (see Figure 2.1). This mentioned year corresponds to 499 CE, and implies that he was born in 476. *Āryabhaṭa* called himself a native of *Kusumapura* or *Pataliputra* (present day Patna, Bihar). Aryabhata is the author of several treatises on mathematics and astronomy, some of which are lost. He was student of the *Nalanda university* (नालन्दा विश्वविद्यालयः), later he even became head of one department of it. Many researches were made at *Nalanda* in *astronomy* (खगोलशास्त्र), *mathematics* (गणित), *physics* (भौतिकी), *biology* (जीव विज्ञान), *medicine* (चिकित्सा) and other fields. So, *Aryabhata* got his major source of knowledge from *Nalanda* and his major work was based on previous inventions by *Greeks, Mesapotamians* and *Nalanda university* itself.

*Aryabhatiya,* a compendium of mathematics and astronomy, was referred to in the Indian mathematical literature and has survived to modern times. The mathematical part of the *Aryabhatiya* covers arithmetic, algebra, plane trigonometry, and spherical trigonometry. It also contains continued fractions, quadratic equations, sums-of-power series, and a table of sines. The *Arya-siddhanta* (आर्य-सिद्धान्त), a lost work on astronomical computations, is known through the writings of *Aryabhata*'s contemporary, *Varahamihira* (वराहमिहिर), and later mathematicians and commentators, including *Brahmagupta* (ब्रह्मगुप्त) and *Bhaskara I* (भास्कर). This work appears to be based on the older *Surya Siddhanta* (सूर्य सिद्धान्त) which was a *Sanskrit* summary of *Greek* and *mesapotamian* theories in astronomy and mathematics and uses the midnight-day reckoning, as opposed to sunrise in *Aryabhatiya.*

It also contained a description of several astronomical instruments:

- the gnomon (shanku-yantra - शङ्कु-यन्त्रम्)
- a shadow instrument (chhAyA-yantra - छअय-यन्त्रम्)
- possibly angle-measuring devices
- semicircular and circular (dhanur-yantra - धनुर्-यन्त्रम् / chakra-yantra - चक्र-यन्त्रम्)
- a cylindrical stick yasti-yantra, an umbrella-shaped device called the *chhatra-yantra* - छत्र-यन्त्रम्, and water clocks of at least two types, bow-shaped and cylindrical.

A *third text,* which may have survived in the *Arabic* translation, is *Al ntf* or *Al-nanf.* It claims that it is a translation by *Aryabhata,* but the Sanskrit name of this work is not known.

The "*Aryabhatiya*" is the only known surviving work of *Aryabhata.*

## Key Contributions in Mathematics

### Aryabhatiya

The scripture begins with a praise of *Brahma* (ब्रह्मा), the creator of the earth and the *universe* (विश्वम्). Then follows a description of the *astronomical system* (खगोलीय प्रणाली).

*Āryabhata* assumes that the earth revolves around itself daily and thus explains the movement of the starry sky. Otherwise, he advocates a geocentric view of the world: the sun, moon and planets move around the earth and he explains deviations from the uniform movement by epicycles of different sizes.

He determines the orbital periods of the sun, moon and planets and calculates from this that the common conjunction of these celestial bodies repeat itself every 4.32 million years. One day for *Brahma* lasts 4.32 billion years for humans. His explanation of lunar and solar eclipses as natural processes replaced the traditional ideas that these eclipses are caused by demons.

One day for Brahma lasts 4.32 billion years for humans

$$1 \text{ Brahma} = 4.32 \times 10^9 \text{ Years}$$

$$1 \text{ Brahma} = 4.32 \times 10^9 \times 365 \text{ Days}$$

The *treatise* (ग्रन्थ) is divided into four sections, spanning 121 verses written in a *mnemonic style* (स्मरणीय शैली), a characteristic of many Indian scholarly works.

It is highly likely that the study of the *Aryabhatiya* was meant to be accompanied by the teachings of a well-versed tutor.

The attributes of the book can be summarized as follows:

- Sections
    - Gitikapada (13 verses): Discusses large units of time and offers a different cosmology from previous works. The large units of time—kalpa (कल्प), manvantara (मन्वन्तर), and yuga—which present a cosmology different from earlier texts such as *Lagadha*'s *Vedanga Jyotisha* (ca. 1st century BCE). There is also a *table of sines* (jya - ज्य), given in a single verse. The duration of the planetary revolutions during a *mahayuga*

is given as 4.32 million years, using the same method as in the *Surya Siddhanta* (सूर्य सिद्धान्त).

- Ganitapada (33 verses): Covers topics in mathematics, including *mensuration* (मापनम्), arithmetic and geometric progressions, gnomon or shadows (*shanku-chhAyA* - शङ्कु-छाय), and various types of equations usch as simple, quadratic, simultaneous, and indeterminate equations (*Kuṭṭaka* - कुट्टक).
- *Kalakriyapada* – कालक्रियापद (25 verses): Details of different units of time and a method for determining the positions of planets for a given day, calculations concerning the intercalary month (*adhikamAsa* - अधिकमअसा, *kShaya-tithis* - क्षय-तिथि, and a seven-day week with names for the days of week.
- *Golapada* - गोलपद (50 verses): Focuses on the geometric/trigonometric aspects of the celestial sphere, features of the ecliptic, celestial equator, node, shape of the Earth, cause of day and night, rising of zodiacal signs on horizon, etc. In addition, some versions cite a few colophons added at the end, extolling the virtues of the work, etc.

- Contents

  - The work begins with an introduction, the "*Dasageethika* - दशगीतिका " or "*Ten Stanzas,*" which details the numeration system used and provides an overview of his astronomical findings.
  - The majority of the mathematical content is in the "*Ganitapada* - गणितापाद."
  - The "*Kalakriya* - कालक्रियम् " focuses on the reckoning of time based on celestial movements.
  - The "*Gola* - गोल" offers a detailed description of the Earth's relationship with the cosmos.

- Numeration

  - *Aryabhata* introduced a unique numeral system using letters of the *Devanagari script* (देवनागरी लिपि), which allowed for sophisticated calculations and also offered some poetic flexibility.

- Significance
    - Geocentrism: *Aryabhata* used a geocentric model where the Sun and Moon revolve around the Earth on *epicycles* (उपचक्रम्).
    - There's debate among scholars about whether *Aryabhata*'s work hints at a *heliocentric model* (सूर्यकेन्द्रित प्रतिरूप).
    - Foundation for Modern Astronomy: Despite its geocentric model, the *Aryabhatiya* introduces many advanced concepts foundational to modern astronomy. For instance, *Aryabhata* identified that celestial bodies like the Moon and planets shine due to reflected sunlight. He also provided accurate explanations and calculations for solar and lunar eclipses, the value of $\pi$, and the length of the sidereal year.

The second part of the *Āryabhatīya* contains treatises (*siddhānta*) on mathematics (*ganita* - गणित, from gana = to count). For the representation of numbers, *Āryabhata* uses artificial words, which he obtains through an encoding he invented:

- For the numbers 1, 2, 3, ..., 25 and 30, 40, 50, ..., 100, he uses the 25 + 8 = 33 consonants of the Sanskrit alphabet, supplemented by the 9 vowels, by which it results with which powers of ten the numbers are multiplied.
- The sequence of syllables of these words of art plays an important role in the sūtras (mnemonic verses).
- Mathematical methods and theorems were traditionally taught in Bharat in this form. The sutras served as a thought-support for the procedure to be applied and were to be learnt by heart by the students. For inexperienced readers, the algorithm described in the 4th and 5th verses for taking the square or cube root of a number in the system of 10 may seem incomprehensible at first.

- Square Root Calculation

  Only through an example does it become comprehensible:

| |
|---|
| Always divide the non-square digit by twice the square root. When the square is then subtracted from the square digit, enter the quotient in the next digit. |

Obviously *Āryabhata* has mastered the underlying formula:

$$(100a + 10b + 1c)^2 = (100a)^2 \\ + 2 \times (100a) \times (10b) + (10b)^2 \\ + 2 \times (100a + 10b) \times (1c) \\ +(1c)^2$$

**Example:** Find $(250)^2$

Let's first partition the number $(250)^2 = (100 \times 2 + 10 \times 5 + 1 \times 0)^2$
So, $a = 2, b = 5, c = 0$

Thus, we can right the right ahnd side as:

$$(100 \times 2)^2 + 2(100 \times 2) \times (10 \times 5) + (10 \times 5)^2 \\ + 2 \times (100 \times 2 + 10 \times 5) \times 1 \times 0 + (1 \times 0)^2$$

$= 62500$

So, the answer of $(250)^2 = 62500$.

- Cube Root Calculation

> Divide the second non-cubic digit by three times the square of the cube root. The square multiplied by three and the previously obtained must be subtracted from the first non-cubic digit and the third power from the cubic digit.

The procedure for extracting the cube root results from:

$$(10a + 1b)^3 = (10a)^3 + 3 \times (10a)^2 \times (1b) + 3 \times (10a) \times (1b)^2 + (1b)^3$$

*Note: Until the 7th century, the zeros occurring in decimal numbers were recognisable by the gaps in the sequence of digits (sūnya (Sanskrit) = emptiness, Arabic: sifr); only after that were the gaps replaced by a dot or a small circle, the precursor of the digit "0".*

- In *verse 6*, the area of a triangle is given as the product of half the base times the height, and the volume of a tetrahedron ("*hexagon*") is given with an analogously formed formula, but incorrectly as half the base times the height.

- *Verse 7* contains the correct formula for the area of a circle (half the circumference times the
- Radius – (r) and a rather inaccurate approximation formula for the volume of a sphere (area of a circle times the square root of the area, i.e.,

$$V \approx 1.77 \times \pi \times r^3 .$$

- *Verses 14 to 16* deal with the shadow lengths of gnomons (rods set vertically into the ground) and the possibility of using two rods of equal length $g$, one behind the other, at distances $e_1$ and $e_2$ to find the height $h$ of a light source. From the lengths of the shadows (see Figure 2.2): $s_1$ and $s_2$ and the distance $a = e_2 - e_1$ between the ends of the shadows we get:

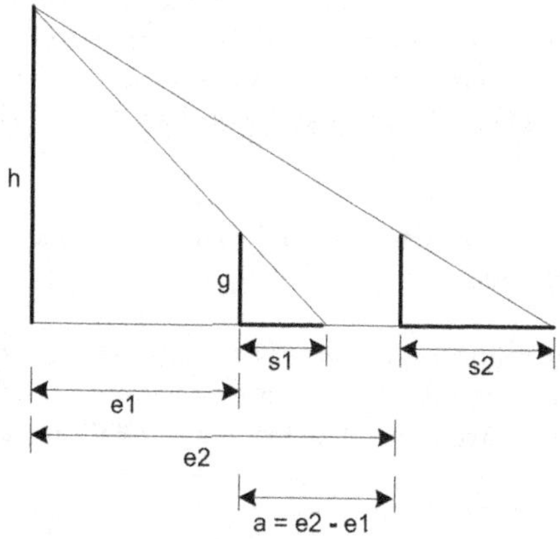


Figure 2.2. Length of shadow trigonometry.

$$\frac{s_1}{e_1} = \frac{g}{h} = \frac{s_2}{e_2}$$

and

$$s_1 = e_1 \times \frac{g}{h}$$

$$s_2 = e_2 \times \frac{g}{h}$$

$$s_2 - s_1 = a \times \frac{g}{h}$$

$$e_1 = a \times \frac{s_1}{s_2 - s_1}$$

$$e_2 = a \times \frac{s_2}{s_2 - s_1}$$

$$h = e_1 \times \frac{g}{s_1} = e_2 \times \frac{g}{s_2}$$

## 卐 Āryabhaṭa Numeration

*Āryabhaṭa* numeration is an alphasyllabic numeral system based on *Sanskrit phonemes* (संस्कृत ध्वनय). It was introduced in the early 6th century in Bharat by *Āryabhaṭa*, in the *first chapter* titled *Gītika Padam* (गीतिका पदम्) of his *Aryabhatiya*. It attributes a numerical value to each syllable of the form consonant + vowel possible in Sanskrit phonology, from *ka* (क) = 1 up to *hau* (हौ) = $10^{18}$.

The basis of this number system is mentioned in the second stanza of the first chapter of *Aryabhatiya*.

The *Varga* (वर्ग) (Group/Class) letters *ka (क)* to *ma (म)* are to be placed in the *varga* (square) places (1st, 100th, 10000th, etc.) and *Avarga* (अवर्ग) letters like *ya (य), ra (र), la (ल)* .. have to be placed in *Avarga* (अवर्ग) places (10th, 1000th, 100000th, etc.).

The *Varga* letters ka to ma have value from 1, 2, 3 .. up to 25 and *Avarga* letters *ya (य)* to *ha* (ह) have value 30, 40, 50.. up to 100. In the Varga and Avarga letters, beyond the ninth vowel (place), new symbols can be used.

The values for vowels are as follows:

- a (अ) = 1
- i (इ) = 100
- u (उ) = 10000
- ṛ (ऋ) = 1000000

- and so on.

*Aryabhata* used this number system for representing both small and large numbers in his mathematical and astronomical calculations.
This system can even be used to represent fractions and mixed fractions.

For example,

- *nga* (न्ग)= $\frac{1}{5}$,
- *nja* (न्ज) = $\frac{1}{10}$ , and
- *Jhardam* (झारदम) (*jha* (झ) =9; its half) = $4\frac{1}{2}$.

The traditional Bharatiya digit order is *reversed* compared to the modern way. By consequence, *Āryabhaṭa* began with the ones before the tens; then the hundreds and the thousands; then the myriad and the lakh ($10^5$) and so on.

Another example might be *ṅiśibuṇlṣkhṛ* (ङिशिबुण्ऌष्खृ), 1582237500.

| Example: 299,792,458 | | | | |
|---|---|---|---|---|
| $10^0$ $10^1$ | $10^2$ $10^3$ | $10^4$ $10^5$ | $10^6$ $10^7$ | $10^8$ |
| 58, | 24, | 79, | 99, | 2 |
| जल | घिनि | झ्शु | झसृ | खॢ |
| *ja-la* | *ghi-ni* | *jhu-śu* | *jhṛ-sṛ* | *khḷ* |

*Note that in this case, $10^6$(ṛ) and $10^8$(ḷ) parts are swapped, and $10^6$(ṛ) part is ligature.*

**Numeral Table**

In citing the values of *Āryabhaṭa* numbers, the short vowels अ, इ, उ, ऋ, ऌ, ए, and ओ are invariably used. However, the *Āryabhaṭa* system did not distinguish between long and short vowels. This table (see Table 2.1) only cites the full slate of क-derived (1 × 10x) values, but these are valid throughout the list of numeric syllables.

Table 2.1. The (33 × 9 = 297) Sanskrit alphabetic numerical syllables

| Nine vowels or syllabics | | | -a | -i | -u | -ṛ | -ḷ | -e | -ai | -o | -au |
|---|---|---|---|---|---|---|---|---|---|---|---|
| | | | अ | इ | उ | ऋ | ऌ | ए | ऐ | ओ | औ |
| | | × | $10^0$ | $10^2$ | $10^4$ | $10^6$ | $10^8$ | $10^{10}$ | $10^{12}$ | $10^{14}$ | $10^{16}$ |
| Five velar plosives | | | | | | | | | | | |
| k - | क | 1 | क or का | कि or की | कु or कू | कृ or कॄ | कॢ or कॣ | के or कॆ | कै | को or कॊ | कौ |
| | | | ka | ki | ku | kṛ | kḷ | ke | kai | ko | kau |
| kh - | ख | 2 | ख | खि | खु | खृ | खॢ | खे | खै | खो | खौ |
| | | | kha | khi | khu | khṛ | khḷ | khe | kha i | kho | kha u |
| g - | ग | 3 | ग | गि | गु | गृ | गॢ | गे | गै | गो | गौ |
| | | | ga | gi | gu | gṛ | gḷ | ge | gai | go | gau |
| gh - | घ | 4 | घ | घि | घु | घृ | घॢ | घे | घै | घो | घौ |
| | | | gha | ghi | ghu | ghṛ | ghḷ | ghe | gha i | gho | gha u |
| ṅ - | ङ | 5 | ङ | ङि | ङु | ङृ | ङॢ | ङे | ङै | ङो | ङौ |
| | | | ṅa | ṅi | ṅu | ṅṛ | ṅḷ | ṅe | ṅai | ṅo | ṅau |
| Five palatal plosives | | | | | | | | | | | |
| c - | च | 6 | च | चि | चु | चृ | चॢ | चे | चै | चो | चौ |
| | | | ca | ci | cu | cṛ | cḷ | ce | cai | co | cau |
| ch - | छ | 7 | छ | छि | छु | छृ | छॢ | छे | छै | छो | छौ |
| | | | cha | chi | chu | chṛ | chḷ | che | chai | cho | cha u |
| j - | ज | 8 | ज | जि | जु | जृ | जॢ | जे | जै | जो | जौ |
| | | | ja | ji | ju | jṛ | jḷ | je | jai | jo | jau |
| jh - | झ | 9 | झ | झि | झु | झृ | झॢ | झे | झै | झो | झौ |
| | | | jha | jhi | jhu | jhṛ | jhḷ | jhe | jhai | jho | jhau |
| ñ - | ञ | 10 | ञ | ञि | ञु | ञृ | ञॢ | ञे | ञै | ञो | ञौ |
| | | | ña | ñi | ñu | ñṛ | ñḷ | ñe | ñai | ño | ñau |
| Five retroflex plosives | | | | | | | | | | | |
| ṭ - | ट | 11 | ट | टि | टु | टृ | टॢ | टे | टै | टो | टौ |
| | | | ṭa | ṭi | ṭu | ṭṛ | ṭḷ | ṭe | ṭai | ṭo | ṭau |
| ṭh - | ठ | 12 | ठ | ठि | ठु | ठृ | ठॢ | ठे | ठै | ठो | ठौ |
| | | | ṭha | ṭhi | ṭhu | ṭhṛ | ṭhḷ | ṭhe | ṭhai | ṭho | ṭha u |

| | | | | | | | | | | | |
|---|---|---|---|---|---|---|---|---|---|---|---|
| ḍ - | ड | 13 | ड | डि | डु | डृ | डॢ | डे | डै | डो | डौ |
| | | | ḍa | ḍi | ḍu | ḍṛ | ḍḷ | ḍe | ḍai | ḍo | ḍau |
| ḍh - | ढ | 14 | ढ | ढि | ढु | ढृ | ढॢ | ढे | ढै | ढो | ढौ |
| | | | ḍha | ḍhi | ḍhu | ḍhṛ | ḍhḷ | ḍhe | ḍha i | ḍho | ḍha u |
| ṇ - | ण | 15 | ण | णि | णु | णृ | णॢ | णे | णै | णो | णौ |
| | | | ṇa | ṇi | ṇu | ṇṛ | ṇḷ | ṇe | ṇai | ṇo | ṇau |
| Five dental plosives | | | | | | | | | | | |
| t - | त | 16 | त | ति | तु | तृ | तॢ | ते | तै | तो | तौ |
| | | | ta | ti | tu | tṛ | tḷ | te | tai | to | tau |
| th - | थ | 17 | थ | थि | थु | थृ | थॢ | थे | थै | थो | थौ |
| | | | tha | thi | thu | thṛ | thḷ | the | thai | tho | tha u |
| d - | द | 18 | द | दि | दु | दृ | दॢ | दे | दै | दो | दौ |
| | | | da | di | du | dṛ | dḷ | de | dai | do | dau |
| dh - | ध | 19 | ध | धि | धु | धृ | धॢ | धे | धै | धो | धौ |
| | | | dha | dhi | dhu | dhṛ | dhḷ | dhe | dha i | dho | dha u |
| n - | न | 20 | न | नि | नु | नृ | नॢ | ने | नै | नो | नौ |
| | | | na | ni | nu | nṛ | nḷ | ne | nai | no | nau |
| Five labial plosives | | | | | | | | | | | |
| p - | प | 21 | प | पि | पु | पृ | पॢ | पे | पै | पो | पौ |
| | | | pa | pi | pu | pṛ | pḷ | pe | pai | po | pau |
| ph - | फ | 22 | फ | फि | फु | फृ | फॢ | फे | फै | फो | फौ |
| | | | pha | phi | phu | phṛ | phḷ | phe | pha i | pho | pha u |
| b - | ब | 23 | ब | बि | बु | बृ | बॢ | बे | बै | बो | बौ |
| | | | ba | bi | bu | bṛ | bḷ | be | bai | bo | bau |
| bh - | भ | 24 | भ | भि | भु | भृ | भॢ | भे | भै | भो | भौ |
| | | | bha | bhi | bhu | bhṛ | bhḷ | bhe | bha i | bho | bha u |
| m - | म | 25 | म | मि | मु | मृ | मॢ | मे | मै | मो | मौ |
| | | | ma | mi | mu | mṛ | mḷ | me | mai | mo | ma u |
| Four Approximants or trill | | | | | | | | | | | |
| y - | य | 30 | य | यि | यु | यृ | यॢ | ये | यै | यो | यौ |
| | | | ya | yi | yu | yṛ | yḷ | ye | yai | yo | yau |
| r - | र | 40 | र | रि | रु | ॠ | रॢ | रे | रै | रो | रौ |

| | | | ra | ri | ru | rṛ | rḷ | re | rai | ro | rau |
|---|---|---|---|---|---|---|---|---|---|---|---|
| l - | ल | 50 | ल | लि | लु | लृ | लॢ | ले | लै | लो | लौ |
| | | | la | li | lu | lṛ | lḷ | le | lai | lo | lau |
| v - | व | 60 | व | वि | वु | वृ | वॢ | वे | वै | वो | वौ |
| | | | va | vi | vu | vṛ | vḷ | ve | vai | vo | vau |
| Three coronal fricatives | | | | | | | | | | | |
| ś - | श | 70 | श | शि | शु | शृ | शॢ | शे | शै | शो | शौ |
| | | | śa | śi | śu | śṛ | śḷ | śe | śai | śo | śau |
| ṣ - | ष | 80 | ष | षि | षु | षृ | षॢ | षे | षै | षो | षौ |
| | | | ṣa | ṣi | ṣu | ṣṛ | ṣḷ | ṣe | ṣai | ṣo | ṣau |
| s - | स | 90 | स | सि | सु | सृ | सॢ | से | सै | सो | सौ |
| | | | sa | si | su | sṛ | sḷ | se | sai | so | sau |
| One glottal fricative | | | | | | | | | | | |
| h - | ह | 100 | ह | हि | हु | हृ | हॢ | हे | है | हो | हौ |

## 卐 Table of Sines

*Āryabhata,* created what's known as *Āryabhata's sine table.* This isn't a "table" as we think of them today; it doesn't have rows and columns. Instead, it's a list of 24 numbers that help calculate the *half-chords* (अर्धतारा) for specific *arcs of a circle* (वृत्तस्य चापा). The set of numbers appears in *verse 12* in *Chapter 1 Dasagitika* of *Aryabhatiya.* This table doesn't give regular sine values but provides the differences between them (e.g., it is a table of the first differences of the values of trigonometric sines expressed in arcminutes), which is why it's sometimes called *Āryabhaṭa*'s table of sine-differences. This table was groundbreaking. Earlier mathematicians, like *Hipparchus, Menelaus, and Ptolemy,* had tables, but they focused on chords, not half-chords. Āryabhata's approach set a new standard in ancient Bharat, prompting others to refine it further. This continuous improvement eventually led to *Madhava of Sangamagrama,* around the 14th century, discovering detailed ways to calculate sine and cosine values and creating an incredibly accurate sine table. There's a debate among historians about *Āryabhata*'s table's origins. Some think he might have been influenced by ancient *Greek scholars.* But because there's little evidence, it's hard to determine how much of the table was based on previous knowledge and how much was *Āryabhata*'s original work.

The values encoded in *Āryabhaṭa*'s *Sanskrit verse* can be decoded using the numerical scheme explained in *Āryabhaṭīya,* and the decoded numbers are listed in the table below. In the table, the angle measures relevant to *Āryabhaṭa*'s sine table are listed in the second column. The third column contains the list the numbers contained in the Sanskrit verse given above in *Devanagari script*. For the convenience of users unable to read *Devanagari*, these word-numerals are reproduced in the fourth column in *ISO 15919* transliteration. The next column contains these numbers in the *Hindu-Arabic* numerals. *Āryabhaṭa*'s numbers are the first differences in the values of sines. The corresponding value of sine (or more precisely, of *jya* - ज्य) can be obtained by summing up the differences up to that difference. For assessing the accuracy of *Āryabhaṭa*'s computations, the modern values of *jyas* are given in the last column of the table.

An example from *Aryabhatiya* is a verse for table of sines is shown below:

> *makhi bhakhi phakhi dhakhi ṇakhi ñakhi*
> *ṅakhi hasjha skaki kiṣga śghakhi kighva*
> *ghlaki kigra hakya dhaki kica sga jhaśa*
> *ṅva kla pta pha cha kala-ardha-jyāḥ*
>
> – Aryabhata

In the Bhartiya mathematical tradition, the sine (or *jya*) of an angle is not a ratio of numbers. It is the length of a certain line segment, a certain half-chord. The radius of the base circle is basic parameter for the construction of such tables. Historically, several tables have been constructed using different values for this parameter. *Āryabhaṭa* has chosen the number *3438* as the value of radius of the base circle for the computation of his sine table. The rationale of the choice of this parameter is the idea of measuring the circumference of a circle in angle measures. In astronomical computations distances are measured in degrees, minutes, seconds, etc.

- In this measure, the circumference of a circle is 360° = (60 × 360) minutes = 21600 minutes.
- The radius of the circle, the measure of whose circumference is 21600 minutes, is $\frac{21600}{2\pi}$ minutes.
- Computing this using the value $\pi = 3.1416$ known to Aryabhata one gets the radius of the circle as 3438 minutes approximately.

*Āryabhaṭa*'s sine table (see Table 2.2) is based on this value for the radius of the base circle. It has not yet been established who is the first ever to use this value for the base radius. But Aryabhatiya is the earliest surviving text containing a reference to this basic constant.

Table 2.2 *Āryabhata*'s sine table.

| Sl. No | Angle (A) (in degrees, arcminutes) | Value in Āryabhaṭa's numerical notation | Value in Āryabhaṭa's numerical notation | Value in Hindu-Arabic numerals | Āryabhaṭa's value of *jya* (A) | Modern value of jya (A) = (3438 × sin (A)) |
|---|---|---|---|---|---|---|
| 1 | 03° 45′ | मखि | makhi | 225 | 225′ | 224.856 |
| 2 | 07° 30′ | भखि | bhakhi | 224 | 449′ | 448.749 |
| 3 | 11° 15′ | फखि | phakhi | 222 | 671′ | 670.721 |
| 4 | 15° 00′ | धखि | dhakhi | 219 | 890′ | 889.82 |
| 5 | 18° 45′ | णखि | ṇakhi | 215 | 1105′ | 1105.11 |
| 6 | 22° 30′ | ञखि | ñakhi | 210 | 1315′ | 1315.67 |
| 7 | 26° 15′ | ङखि | ṅakhi | 205 | 1520′ | 1520.59 |
| 8 | 30° 00′ | हस्झ | hasjha | 199 | 1719′ | 1719 |
| 9 | 33° 45′ | स्ककि | skaki | 191 | 1910′ | 1910.05 |
| 10 | 37° 30′ | किष्ग | kiṣga | 183 | 2093′ | 2092.92 |
| 11 | 41° 15′ | श्घकि | śghaki | 174 | 2267′ | 2266.83 |
| 12 | 45° 00′ | किघ्व | kighva | 164 | 2431′ | 2431.03 |
| 13 | 48° 45′ | घ्लकि | ghlaki | 154 | 2585′ | 2584.83 |
| 14 | 52° 30′ | किग्र | kigra | 143 | 2728′ | 2727.55 |
| 15 | 56° 15′ | हक्य | hakya | 131 | 2859′ | 2858.59 |
| 16 | 60° 00′ | धकि | dhaki | 119 | 2978′ | 2977.4 |
| 17 | 63° 45′ | किच | kica | 106 | 3084′ | 3083.45 |
| 18 | 67° 30′ | स्ग | sga | 93 | 3177′ | 3176.3 |
| 19 | 71° 15′ | झश | jhaśa | 79 | 3256′ | 3255.55 |
| 20 | 75° 00′ | ङ्व | ṅva | 65 | 3321′ | 3320.85 |
| 21 | 78° 45′ | क्ल | kla | 51 | 3372′ | 3371.94 |
| 22 | 82° 30′ | प्त | pta | 37 | 3409′ | 3408.59 |
| 23 | 86° 15′ | फ | pha | 22 | 3431′ | 3430.64 |
| 24 | 90° 00′ | छ | cha | 7 | 3438′ | 3438 |

If one divides a quarter circle of radius 3438 into 24 equal sectors, then the heights opposite the angles of 3° 45', 7° 30', 11° 15', 15°, ..., 90° have the lengths 225, 225 + 224 = 449, 449 + 222 = 671, 671 + 219 = 890, ..., 3438.

The value for the radius, which seems unusual, is explained below.

| A full angle comprises 360° = 360.60' = 21600' |
| --- |

The circumference of a circle with a radius of 3438 units of length is almost exactly 21600 units of length, so that each minute of arc can be assigned an arc of length 1 unit of length.

**Let's Break Down the Problem Step by Step**

A quarter circle of radius r = 3438 units is divided into 24 equal sectors, which means each sector will subtend an angle of $\frac{90}{24} = 3.75^o$ or 3° 45' at the center.

*Finding the heights*

To find the height (or the y-coordinate) opposite the angles of 3° 45', 7° 30', 11° 15', 15°, ..., 90°, we'll use the sine function for a right triangle:

$$Height\ (h) = r\ \times \sin\left(\theta\right)$$

Where, $\theta$ is the angle in question and r is the radius.

Now, let's compute the heights for the specified angles:

For $\theta = 3^o45'$: $h = 3438 \times \sin\left(3.75^o\right)$
For $\theta = 7^o30'$: $h = 3438 \times \sin\left(7.5^o\right)$

For $\theta = 11^o15'$: $h = 3438 \times \sin\left(11.25^o\right)$
For $\theta = 15^o$: $h = 3438 \times \sin\left(15^o\right)$

...

For $\theta = 90^o$: $h = 3438 \times \sin\left(90^o\right)$

The height for $\theta = 90°$ will be equal to the radius, 3438 units, as $\sin(90^o) = 1$.

By calculating the above, we can get the heights opposite each of the angles on our quarter circle.

**Explanation of the Āryabhata's Sine Table**

The given table provides the value of $jya(A)$ for *Āryabhaṭa*'s method and the modern computation of $jya$ based on the sine function. From the table, it is clear that *Āryabhaṭa*'s $jya$ values are based on cumulative additions of differences. This iterative structure is a crucial part of the construction of the sine table.
If we denote $D_n$ as the difference from *Āryabhaṭa*'s notation for the $n^{th}$ entry, then the iterative formula for $jya(A)$) can be expressed as:

$$jya(A_n) = jya(A_{n-1}) + D_n$$

where,

- $jya(A_1) =$ is given as 225 (from the table).
- $A_n$ is the angle for the $n^{th}$ entry.
- $D_n$ is the difference value (in Hindu-Arabic numerals) for the $n^{th}$ entry.

For instance, based on the table:

$$jya(A_2) = 449, \text{ which is } jya(A_1) + D_2 = 225 + 224$$

$$jya(A_3) = 671, \text{ which is } jya(A_2) + D_3 = 449 + 222$$

... and so on.

This iterative formula allows us to compute the $jya$ value for any angle in Āryabhaṭa's table based on the previous value and the current difference.

**Comparison with Modern Computation**

| Āryabhata's Sine Table | Modern Approach |
|---|---|
| $jya(A_2) =$<br>$jya(A_1) + D_2$<br>$= 225 + 224 = 449$ | $jya(A_2) = 3438 \times \sin(7.5^o)$<br>$= 3438 \times 0.130526$<br>$= 448.7490$ |

The error we can calculate from above calculation with respect to Āryabhata's $jya(A_2)$ as below.

$$\frac{|449 - 448.7490|}{449} \times 100 = \frac{0.251}{449} \times 100 = 0.05590\%$$

### Āryabhaṭa's Computational Method

The second section of *Āryabhaṭiya,* titled *Ganitapādd,* a contains a stanza indicating a method for the computation of the sine table. There are several ambiguities in correctly interpreting the meaning of this verse.

For example, the following is a translation of the verse given by *Katz* wherein the words in square brackets are insertions of the translator and not translations of texts in the verse. This may be referring to the fact that the second derivative of the sine function is equal to the negative of the sine function.

> *"When the second half-[chord] partitioned is less than the first half-chord, which is [approximately equal to] the [corresponding] arc, by a certain amount, the remaining [sine-differences] are less [than the previous ones] each by that amount of that divided by the first half-chord."*
>
> \- Āryabhaṭa
>
> Translation of the verse given by Katz.

**Example**

Let's break down the description and provide a simpler explanation with a real example. The description is talking about the sine function and how it behaves. If you've seen a sine wave graphed on a coordinate plane, it starts at the origin (0,0), rises to 1 at $\pi/2$, goes back to 0 at $\pi$, drops to -1 at $3\pi/2$, and returns to 0 at $2\pi$. This cycle then repeats.

The text from *Āryabhaṭiya* suggests a method to compute the sine table, which is essentially a list of sine values for various angles.

1. The term "*second half-chord*" refers to the sine of an angle.
2. The "*first half-chord*" is an approximation for the sine value when the angle is small. For small angles, the sine of the angle is approximately equal to the angle itself (when the angle is measured in radians).
3. The statement "the *second half-[chord]* partitioned is less than the first half-chord" refers to the fact that as the angle increases from 0, the difference between the actual sine value and its approximation (*angle value*) decreases.
4. The final portion, "the remaining [*sine-differences*] are less [*than the previous ones*] each by that amount of that divided by the first half-chord" is a little tricky. It's stating that the difference between consecutive sine values (for small increases in angle) decreases in a particular pattern, which is related to the value of the initial approximation (the angle itself).

To relate this to the fact that "*the second derivative of the sine function is equal to the negative of the sine function*":

1. The first derivative of sine (cosine) tells us how fast the sine value is changing.
2. The second derivative tells us how the rate of change itself is changing.

**Let's Illustrate with a Practical Example**

1. Take the sine of a small angle, say 5° or 0.0873 radians.

2. Using the approximation method, sin(5°) is approximately 5° in radians, which is 0.0873.

3. The actual value of sin(5°) is 0.0872, which is very close to the approximation.

If we increase the angle slightly, the difference between the approximation and the actual sine value will change, and it will change in a pattern that *Āryabhaṭiya* described. In essence, *Āryabhaṭiya* is giving a method to compute sine values for various angles using a particular pattern of differences between consecutive values. This approach was ingenious for its time, providing a simple way to compute the sine table without modern computational tools.

**More Detailed Example**

**Compute the sine values for angles 0°, 15°, 30°, and 45° using an iterative approximation method.**

1. Step 1: Start with a small angle.

   For small angles (measured in radians), the sine of the angle is approximately equal to the angle itself. We'll use this fact as our starting point.

   Angle: 0°

   - Convert to radians: $0° \times (\pi/180) = 0$ radians.
   - Approximate sine value: $\sin(0°) \approx 0$ (since sine of 0 is 0).

2. Step 2: Move to the next angle and approximate its sine value.

   The key insight here is that as we increase the angle, the difference between the sine of the new angle and the sine of the previous angle starts to decrease.

   Angle: 15°

   - Convert to radians: $15° \times (\pi/180) = 0.2618$ radians.
   - Using the approximation method: $\sin(15°) \approx 0.2618$. (For reference, the actual value of $\sin(15°)$ is 0.2588).

3. Step 3: Continue the iterative process.

   Angle: 30°

   - Convert to radians: $30° \times (\pi/180) = 0.5236$ radians.

- Here, we notice that as the angle increases, the difference between the actual sine value and its approximation starts to increase. However, using the approximation: sin(30°) ≈ 0.5236. (For reference, the actual value of sin(30°) is 0.5), so our approximation is off by 0.0236.

Angle: 45°

- Convert to radians: 45° × (π/180) = 0.7854 radians.

- Using the approximation method: sin(45°) ≈ 0.7854. (For reference, the actual value of sin(45°) is 0.7071), so our approximation here is off by 0.0783.

As the angle grows, the approximation starts to deviate more from the actual sine value. However, by observing the pattern of these differences, mathematicians like Āryabhaṭiya devised ways to correct the approximation and compute more accurate sine values.

### 卐 Kuṭṭaka

*Kuṭṭaka* (कुट्टक) is an algorithm for finding integer solutions of linear *Diophantine equations.* A linear *Diophantine equation* is an equation of the form

$$ax + by = c$$

where x and y are unknown quantities and a, b, and c are known quantities with integer values. The algorithm was originally invented by the *Āryabhaṭa* and is described very briefly in his *Āryabhaṭīya. Āryabhaṭa* did not give the algorithm the name *Kuṭṭaka,* and his description of the method was mostly obscure and incomprehensible. It was *Bhāskara* I (c. 600 – c. 680) who gave a detailed description of the algorithm with several examples from astronomy in his *Āryabhatiyabhāṣya,* who gave the algorithm the name *Kuṭṭaka.*

In Sanskrit, the word *Kuṭṭaka* means pulverization (*reducing to powder*), and it indicates the nature of the algorithm. The algorithm in essence is a process where the coefficients in a given linear *Diophantine equation* are broken up into smaller numbers to get a linear *Diophantine equation* with smaller coefficients. In general, it is easy to find integer solutions of linear *Diophantine equations* with small coefficients. From a solution to the reduced equation, a solution to the original equation can be determined. Many Bhartiya mathematicians after *Aryabhaṭa* have discussed the *Kuṭṭaka* method with variations and refinements. The *Kuṭṭaka* method was considered to be so important that the entire subject of algebra used to be called *Kuṭṭaka-ganita* (कुट्टक-गणित) or simply *Kuṭṭaka*. Sometimes the subject of solving linear *Diophantine equations* is also called *Kuṭṭaka*. In literature, there are several other names for the Kuṭṭaka algorithm *like Kuṭṭa, Kuṭṭakāra* (कुट्टकार) *and Kuṭṭikāra* (कुट्टिकारा). There is also a treatise devoted exclusively to a discussion of *Kuṭṭaka*. Such specialized treatises are very rare in the mathematical literature of ancient Bharat.

The treatise written in Sanskrit is titled *Kuṭṭākāra Śirōmaṇi* (कुट्टाकार शिरोमणि) and is authored by one *Devaraja* (देवराज).
The *Kuṭṭaka algorithm* has much similarity with and can be considered as a precursor of the modern day extended *Euclidean algorithm*. The latter algorithm is a procedure for finding integers x and y satisfying the condition:

$$ax + by = \gcd(a, b).$$

**Formulation of Problem by Aryabhaṭa**

The problem that can supposedly be solved by the *Kuṭṭaka* method was not formulated by *Aryabhaṭa* as a problem of solving the linear *Diophantine equation. Aryabhaṭa* considered the following problems all of which are equivalent to the problem of solving the linear *Diophantine equation* as shown below:

1. Find an integer which when divided by two given integers leaves two given remainders. This problem may be formulated in two different ways:
   Let the integer to be found be N, the divisors be a and b, and the remainders be R1 and R2. Then the problem is to find N such that

   $$N \equiv R1 \pmod{a}$$

   $$N \equiv R2 \pmod{b}$$

2. Letting the integer to be found to be N, the divisors be a and b, and the remainders be R1 and R2, the problem is to find N such that there are integers x and y such that

$$N = ax + R1$$

$$N = by + R2$$

This is equivalent to

$$ax - by = c$$

where

$$c = R2 - R1$$

3. *Find an integer such that its product with a given integer being increased or decreased by another given integer and then divided by a third integer leaves no remainder.*

   Letting the integer to be determined be x and the three integers be a, b and c, the problem is to find x such that (ax ± b)/c is an integer y. This is equivalent to finding integers x and y such that

$$(ax \pm b)/c = y$$

   This in turn is equivalent to the problem of finding integer solutions of

$$ax \pm by = \pm c$$

## 卐 Aryabhata's Algorithm

*Aryabhata* gave the algorithm for solving the linear *Diophantine equation* in *verses 32–33* of *Ganitapada* of *Aryabhatiya*. Taking *Bhāskara I*'s explanation of these verses also into consideration, *Bibhutibbhushan Datta* has given the following translation of these verses. Figure 2.3 shows the reference of *Kuttaka* in *Aryabhatiya*.

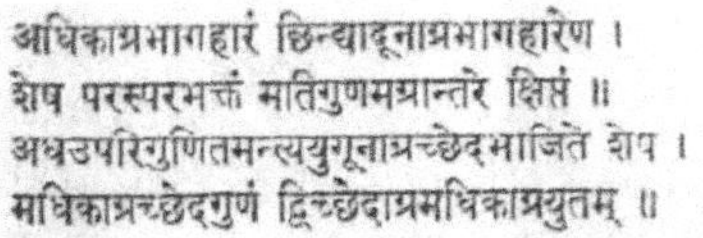
अधिकाग्रभागहारं छिन्द्यादूनाग्रभागहारेण ।
शेषपरस्परभक्तं मतिगुणमग्रान्तरे क्षिप्तं ॥
अधउपरिगुणितमन्त्ययुगूनाग्रच्छेदभाजिते शेष ।
मधिकाग्रच्छेदगुणं द्विच्छेदाग्रमधिकाग्रयुतम् ॥

Figure 2.3 Reference of *Kuttaka* in *Aryabhatiya*. (Image Credit: https://en.wikipedia.org/wiki/)

Divide the divisor corresponding to the greater remainder by the divisor corresponding to the smaller remainder. The residue (and the divisor corresponding to the smaller remainder) being mutually divided (until the remainder becomes zero), the last quotient should be multiplied by an optional integer and then added (in case the number of quotients of the mutual division is even) or subtracted (in case the number of quotients is odd) by the difference of the remainders. (Place the other quotients of the mutual division successively one below the other in a column; below them the result just obtained and underneath it the optional integer.) Any number below (that is, the penultimate) is multiplied by the one just above it and added by that just below it. Divide the last number (obtained so doing repeatedly) by the divisor corresponding to the smaller remainder; then multiply the residue by the divisor corresponding to the greater remainder and add the greater remainder. (The result will be) the number corresponding to the two divisors.

— Aryabhata

Some comments are in order as below.

- The algorithm yields the smallest positive integer which gives specified remainders when divided by given numbers.
- The validity of the algorithm can be established by translating the process into modern mathematical notations.
- Subsequent Bhartiya mathematicians including *Brahmagupta* (628 AD), *Mahavira* (850), *Aryabhata II* (950), *Sripati* (1039), *Bhāskara II* (1150) and *Narayana* (1350) have developed several variants of this algorithm and have also discussed several special cases of the algorithm.

**Elaboration of Aryabhatta's Kuttaka**

Without loss of generality, let $ax - by = c$ be our *Diophantine equation* where $a, b$ are positive integers and $c$ is an integer. Divide both sides of the equation by gcd $(a, b)$.

If $c$ is not divisible by gcd $(a, b)$ then there are no integer solutions to this equation. After the division, we get the equation $a'x - b'y = c'$. The solution to this equation is the solution to $ax - by = c$. Without loss of generality, let us consider $a > b$.

Using Euclidean division, follow these recursive steps:

$$a' = a_1 b' + r_1$$

$$b' = a_2 r_1 + r_2$$

$$r_1 = a_3 r_2 + r_3$$

$$\dots$$

$$r_{n-2} = a_n r_{n-1} + 1$$

where

$$rn = 1$$

Now, define quantities $x_{x+2}, x_{n+2}, x_n$ ... by backward induction as follows:

1. If n is odd, take $x_{n+2} = 0$ and $x_{n+1} = 1.$ .
2. If n is even, take $x_{n+2} = 1$ and $x_{n+1} = r_{n-1} - 1.$
3. Now, calculate all $x_m\ (n \geq m \geq 1)\ by\ x_m = a_m x_{m+1} + x_{m+2}.$
4. Then $y = c'x_1$ and $x = c'x_2$.

## 卐 Place Value System

The place-value system, first seen in the 3rd-century *Bakhshali* (बखशाली) Manuscript, was clearly in place in his work. *Aryabhata* did not use the *Brahmi* (ब्राह्मी) numerals. Continuing the Sanskritic tradition from *Vedic* times, he used letters of the alphabet to denote numbers, expressing quantities, such as the table of sines in a mnemonic form.

## 卐 Approximation of $\pi$

Aryabhata worked on the approximation for pi ($\pi$), and may have come to the conclusion that $\pi$ is irrational. In the second part of the *Aryabhatiyam (gaṇitapāda 10)*, he writes:

| |
|---|
| *caturadhikaṃ śatamaṣṭaguṇaṃ dvāṣaṣṭistathā sahasrāṇām*<br>*ayutadvayaviṣkambhasyāsanno vṛttapariṇāhaḥ.* |
| *Add four to 100, multiply by eight, and then add 62,000. By this rule the circumference of a circle with a diameter of 20,000 can be approached.*<br><br>Translation |

This implies that for a circle whose diameter is 20000, the circumference will be 62832 i.e.,

$$\pi = \frac{62832}{20000} = 3.1416$$

which is accurate to three decimal places.
It is speculated that *Aryabhata* used the word *āsanna* (approaching), to mean that not only is this an approximation but that the value is incommensurable (or irrational). If this is correct, it is quite a sophisticated insight, because the irrationality of pi ($\pi$) was proved in Europe only in 1761 by Lambert.
After *Aryabhatiya* was translated into Arabic (c. 820 CE), this approximation was mentioned in *Al-Khwarizmi*'s book on algebra.

## 卐 Trigonometry

In *Ganitapada 6, Aryabhata* gives the area of a triangle as:

| |
|---|
| *tribhujasya phalaśarīraṃ samadalakoṭī bhujārdhasaṃvargaḥ* |

For a triangle, the result of a perpendicular with the half-side is the area.

Translation

*Aryabhata* discussed the concept of sine in his work by the name of *ardha-jya* (आर्धा-ज्या), which literally means "*half-chord*". For simplicity, people started calling it *jya*. When Arabic writers translated his works from Sanskrit into Arabic, they referred it as *jiba*. However, in Arabic writings, *vowels are omitted*, and it was abbreviated as *jb*. Later writers substituted it with *jaib*, meaning "*pocket*" or "*fold* (in a garment)". In Arabic, *jiba* is a meaningless word. Later in the 12th century, when *Gherardo of Cremona* translated these writings from Arabic into Latin, he replaced the Arabic *jaib* with its Latin counterpart, sinus, which means "*cove*" or "*bay*"; thence comes the English word sine.

卐 **Solutions to Diophantine equations**

A problem of great interest to Bhartiya mathematicians since ancient times has been to find integer solutions to *Diophantine equations* that have the form ax + by = c. (This problem was also studied in ancient Chinese mathematics, and its solution is usually referred to as the *Chinese remainder theorem*.)

This is an example from *Bhāskara*'s commentary on *Aryabhatiya*:

*Find the number which gives 5 as the remainder when divided by 8, 4 as the remainder when divided by 9, and 1 as the remainder when divided by 7.*

Translation

That is, find

$$N = 8x + 5 = 9y + 4 = 7z + 1.$$

It turns out that the smallest value for N is 85.

In general, diophantine equations, such as this, can be notoriously difficult. They were discussed extensively in ancient *Vedic* text *Sulba Sutras*, whose more ancient parts might date to 800 BCE. *Aryabhata*'s method of solving such problems, elaborated by *Bhaskara* in 621 CE, is called the *kuṭṭaka* method. *Kuṭṭaka* means "*pulverizing*" or "*breaking into small pieces*", and the method involves a recursive algorithm for writing the original factors in smaller numbers. This algorithm became the standard method for solving first-order diophantine equations in Bhartiya mathematics, and initially the whole subject of algebra was called *kuṭṭaka-gaṇita.*

## 卐 Father of Modern Trigonometry

His definitions of *sine* (jya - ज्य), *cosine* (kojya - कोज्या), *versine* (utkrama-jya - उत्क्रम-ज्य), and *inverse sine* (otkram jya - ओत्क्रम ज्य) influenced the birth of trigonometry. He was also the first to specify *sine* and $version(1 - cosx)$ tables, in 3.75° intervals from 0° to 90°, to an accuracy of 4 decimal places.

- Sine (jya - ज्य) and Cosine (kojya - कोज्या)

  Suppose you have a right-angled triangle. If the length of the side opposite to the angle is 'a' and the length of the hypotenuse is 'h', then the sine of that angle would be:

$$sin(angle) = \frac{a}{h}.$$

  Similarly, if the length of the side adjacent to the angle is '*b*', then the cosine of that angle would be:

$$cos(angle) = \frac{b}{h}.$$

  *Aryabhata*'s detailed sine and cosine tables would provide values for these ratios for various angles, facilitating complex calculations.

- Versine (utkrama-jya - उत्क्रम-ज्य)

  The *versine* of an angle in a circle with radius '*r*' can be defined as the vertical distance between a point on the circumference and the diameter drawn through this point. Mathematically, it's given by

$$versine(angle) = r - r * cos(angle).$$

This is equivalent to $1 - cos(angle)$.

For example, if we take an angle of 60° and use *Aryabhata*'s tables to find the cosine value, the versine would be $1 - cos(60°)$.

- Inverse Sine (otkram jya - ओत्क्रम ज्य)

  This is the angle whose sine is a given value. For instance, if $sin(angle) = 0.5$, then using *Aryabhata*'s tables, one could determine that the angle is 30°.

- Accuracy and Intervals

  *Aryabhata*'s precision in trigonometry was remarkable. If one were to look up his tables for the sine value of 45°, they would get a value accurate up to 4 decimal places. Such accuracy was invaluable for astronomical calculations. Moreover, having values at intervals of 3.75° means that for every 15 minutes of a circle (since there are 360° in a circle and 24 hours in a day), there was a precise trigonometric value available. This granularity allowed for very detailed and nuanced astronomical observations and predictions.

In fact, modern names "*sine*" and "*cosine*" are mistranscriptions of the words jya and kojya as introduced by *Aryabhata*. As mentioned, they were translated as *jiba* and *kojiba* in *Arabic* and then misunderstood by *Gerard of Cremona* while translating an *Arabic* geometry text to Latin.

## 卐 Father of Modern Algebra

In *Aryabhatiya, Aryabhata* provided elegant results for the summation of series of squares and cubes. *Verses 19 to 22* contain various rules on arithmetical sequences, as well as formulas for the sum of the first n natural numbers, the first n square or n cubic numbers as well as for the sum of the first n triangular numbers:

$$1^2 + 2^2 + 3^2 + \cdots + n^2 = \frac{n(n1+)(2n+1)}{6}$$

and

$$1^3 + 2^3 + 3^3 + \cdots + n^3 = (1 + 2 + 3 + \cdots + n)^2$$

*Verses 23 to 24* give rules for sums, differences and products of numbers resulting from binomial formulae:

$$a \times b = \frac{1}{2} \times [(a + b)^2 - (a^2 + b^2)]$$

If we know the product $a \times b$ and the difference $a - b$ of two numbers, then the numbers $a, b$ can be determined as follows:

$$a = \frac{1}{2} \times [\sqrt{(4 \times a \times b) + (a - b)^2} + (a - b)]$$
$$b = \frac{1}{2} \times [\sqrt{(4 \times a \times b) + (a - b)^2} - (a - b)]$$

The next verses explain how to calculate with fractions and how to solve ratio equations.

From a task on the calculation of interest, it becomes clear that solving quadratic equations was also known, even if this topic was not explicitly addressed. In *verse 31,* he examines when two objects whose location and speed are known will meet, or when they would have met in the past if they were currently moving away from each other. This method was important in astronomy when determining conjunctions of celestial bodies.

## 卐 Motions of the Solar System

*Aryabhata* correctly insisted that the *Earth* rotates about its axis daily, and that the apparent movement of the stars is a relative motion caused by the rotation of the Earth, contrary to the then-prevailing view, that the sky rotated. This is indicated in the first chapter of the Aryabhatiya, where he gives the number of rotations of the Earth in a *yuga,* and made more explicit in his gola chapter:

In the same way that someone in a boat going forward sees an unmoving [object] going backward, so [someone] on the equator sees the unmoving stars going uniformly westward. The cause of rising and setting [is that] the sphere of the stars together with the planets [apparently?] turns due west at the equator, constantly pushed by the cosmic wind.

Translation

*Aryabhata* described a geocentric model of the *Solar System* (सौरमण्डलम्), in which the Sun and Moon are each carried by epicycles. They in turn revolve around the Earth. In this model, which is also found in the *Paitāmahasiddhānta* (पैतमहसिद्धान्त) (c. 425 CE), the motions of the planets are each governed by two epicycles, a smaller manda (slow) and a larger *śīghra* – शिघ्र (fast). The order of the planets in terms of distance from earth is taken as: the *Moon, Mercury, Venus, Sun, Mars, Jupiter, Saturn,* and the *asterisms*. The positions and periods of the planets was calculated relative to uniformly moving points. In the case of *Mercury* and *Venus*, they move around the Earth at the same mean speed as the *Sun*. In the case of *Mars, Jupiter,* and *Saturn*, they move around the Earth at specific speeds, representing each planet's motion through the *zodiac* (राशिचक्रम्). Another element in *Aryabhata*'s model, the *śīghrocca* (शिघ्रोच्च), the basic planetary period in relation to the *Sun*, is seen by some historians as a sign of an underlying heliocentric model.

## 卐 Eclipses

*Solar* and *lunar eclipses* (सौर-चन्द्रग्रहणम्) were scientifically explained by *Aryabhata*.

- He states that the Moon and planets shine by reflected sunlight. Instead of the prevailing cosmogony in which eclipses were caused by *Rahu* and *Ketu* (*identified as the pseudo-planetary lunar nodes*), he explains eclipses in terms of shadows cast by and falling on Earth.
- Thus, the lunar eclipse occurs when the *Moon* enters into the *Earth*'s shadow (*verse gola.37*).
- He discusses at length the size and extent of the *Earth*'s shadow (*verses gola.38–48*) and then provides the computation and the size of the eclipsed part during an eclipse.

Later Bhartiya astronomers improved on the calculations, but *Aryabhata*'s methods provided the core. His computational paradigm was so accurate that 18th-century scientist *Guillaume Le Gentil*, during a visit to Pondicherry, Bharat, found the Bhartiya computations of the duration of the lunar eclipse of 30 August 1765 to be short by 41 seconds, whereas his charts (by Tobias Mayer, 1752) were long by 68 seconds.

### 卐 Sidereal Periods

Considered in modern English units of time, *Aryabhata* calculated the sidereal rotation (the rotation of the earth referencing the fixed stars) as; the modern value is.

| Aryabhata's Sidereal Rotation | Modern Value |
|---|---|
| 23 hours, 56 minutes, and 4.1 seconds | 23:56:4.091 |

Similarly, his value for the length of the sidereal year at 365 days, 6 hours, 12 minutes, and 30 seconds (365.25858 days) is an error of 3 minutes and 20 seconds over the length of a year (365.25636 days).

### 卐 Heliocentrism

As mentioned, *Aryabhata* advocated an astronomical model in which the Earth turns on its own axis. His model also gave corrections (the *śīgra* anomaly) for the speeds of the planets in the sky in terms of the mean speed of the *Sun*. Thus, it has been suggested that *Aryabhata*'s calculations were based on an underlying heliocentric model, in which the planets orbit the *Sun*, though this has been rebutted. It has also been suggested that aspects of Aryabhata's system may have been derived from an earlier, likely pre-Ptolemaic Greek, heliocentric model of which Bhartiya astronomers were unaware, though the evidence is scant. The general consensus is that a synodic anomaly (depending on the position of the *Sun*) does not imply a physically heliocentric orbit (such corrections being also present in late *Babylonian* astronomical texts), and that *Aryabhata*'s system was not explicitly *heliocentric* (सूर्यकेन्द्रित).

**Further Reading**


4. "Get ready for solar eclipse" (PDF). National Council of Science Museums, Ministry of Culture, Government of India. Archived from the original (PDF) on 21 July 2011. Retrieved 9 December 2009.
5. "Hipparchus and Trigonometry". Retrieved 6 March 2010.
6. "ISRO Press Release 16 March 2009". ISRO. Archived from the original on 5 January 2012. Retrieved 24 June 2012.
7. "Maths can be fun". The Hindu. 3 February 2006. Archived from the original on 1 October 2007. Retrieved 6 July 2007.
8. "New Microorganisms Discovered in Earth's Stratosphere". ScienceDaily. 18 March 2009. Archived from the original on 1 April 2018.
9. "Omar Khayyam". The Columbia Encyclopedia (6 ed.). May 2001. Archived from the original on 17 October 2007. Retrieved 10 June 2007.
10. [achalAni bhAni samapashchimagAni ... – golapAda.9–10]. Translation from K. S. Shukla and K.V. Sarma, K. V. Āryabhaṭīya of Āryabhaṭa, New Delhi: Indian National Science Academy, 1976. Quoted in Plofker 2009.
11. Amartya K Dutta, "Diophantine equations: The Kuttaka" Archived 2 November 2014 at the Wayback Machine, Resonance, October 2002. Also see earlier overview: Mathematics in Ancient India Archived 2 November 2014 at the Wayback Machine.
12. Ansari, p. 13, Table 1.
13. Ansari, S.M.R. (March 1977). "Aryabhata I, His Life and His Contributions". Bulletin of the Astronomical Society of India. 5 (1): 10–18. Bibcode:1977BASI....5...10A. hdl:2248/502.
14. Aryabhatiya 1.3ab, see Plofker 2009, p. 111.
15. Aryabhatiya Marathi: आर्यभटीय, Mohan Apte, Pune, India, Rajhans Publications, 2009, p.25, ISBN 978-81-7434-480-9.
16. Avinash Sathaye. "A Better Division Algorithm" (PDF). Department of mathematics, Univ. of Kentucky. Retrieved 7 March 2016.
17. B. L. van der Waerden, B. L. (March 1988). "Reconstruction of a Greek table of chords". Archive for History of Exact Sciences. 38 (1): 23–38. Bibcode:1988AHES...38...23V. doi:10.1007/BF00329978. S2CID 189793547.
18. B. S. Yadav (28 October 2010). Ancient Indian Leaps into Mathematics. Springer. p. 88. ISBN 978-0-8176-4694-3.

19. B.L. van der Waerden, "The Heliocentric System in Greek, Persian and Hindu Astronomy", in David A. King and George Saliba, ed., From Deferent to Equant: A Volume of Studies in the History of Science in the Ancient and Medieval Near East in Honor of E. S. Kennedy, Annals of the New York Academy of Science, 500 (1987), pp. 529–534.
20. B.N. Narahari Achar (2002). "Āryabhata and the table of Rsines" (PDF). Indian Journal of History of Science. 37 (2): 95–99. Retrieved 6 March 2010.
21. Bharati Ray (1 September 2009). Different Types of History. Pearson Education India. pp. 95–. ISBN 978-81-317-1818-6.
22. Bhaskaracharya-1 (Translated by K. S. Shukla) (1963). Laghu-Bhskariya. Lucknow University. p. 99. Retrieved 7 March 2016.
23. Boyer, Carl B. (1991). "The Mathematics of the Hindus". A History of Mathematics (Second ed.). John Wiley & Sons, Inc. p. 207. ISBN 0-471-54397-7. He gave more elegant rules for the sum of the squares and cubes of an initial segment of the positive integers. The sixth part of the product of three quantities consisting of the number of terms, the number of terms plus one, and twice the number of terms plus one is the sum of the squares. The square of the sum of the series is the sum of the cubes.
24. Britannica Educational Publishing (15 August 2010). The Britannica Guide to Numbers and Measurement. The Rosen Publishing Group. pp. 97–. ISBN 978-1-61530-218-5.
25. Cooke (1997). "The Mathematics of the Hindus". History of Mathematics: A Brief Course. p. 204. Aryabhata himself (one of at least two mathematicians bearing that name) lived in the late 5th and the early 6th centuries at Kusumapura (Pataliutra, a village near the city of Patna) and wrote a book called Aryabhatiya.
26. D. E. Knuth (1998). The Art of Computer Programming Volume 2. Pearson Education India, 1998. p. 342. ISBN 9788177583359.
27. Dennis Duke, "The Equant in India: The Mathematical Basis of Ancient Indian Planetary Models." Archive for History of Exact Sciences 59 (2005): 563–576, n. 4 "Archived copy" (PDF). Archived (PDF) from the original on 18 March 2009. Retrieved 8 February 2016.
28. Devaraja (1944). Kuttakara Siromani (in Sanskrit). Anandasrama Press. Retrieved 7 March 2016.
29. Douglas Harper (2001). "Online Etymology Dictionary". Archived from the original on 13 July 2007. Retrieved 14 July 2007.

30. Dutta, Bibhutibhushan; Singh, Avadhesh Narayan (1962). History of Hindu Mathematics. Asia Publishing House, Bombay. ISBN 81-86050-86-8.
31. Eugene Clark (1930). Theastronomy. Chicago: The University of Chicago Press.
32. G. J. Toomer, G. J. (July 2007). "The Chord Table of Hipparchus and the Early History of Greek Trigonometry". Centaurus. 18 (1): 6–28. doi:10.1111/j.1600-0498.1974.tb00205.x.
33. George. Ifrah (1998). A Universal History of Numbers: From Prehistory to the Invention of the Computer. London: John Wiley & Sons.
34. Glen Van Brummelen (25 January 2009). The mathematics of the heavens and the earth: the early 0. ISBN 9780691129730.
35. Glen Van Brummelen (March 2000). "[HM] Radian Measure". Historia Mathematica mailing List Archive. Retrieved 6 March 2010.
36. Hayashi (2008), Aryabhata I.
37. Heidi Roupp (1997). Teaching World History: A Resource Book. M.E. Sharpe. pp. 112–. ISBN 978-1-56324-420-9.
38. Helaine Selin (Ed.) (2008). Encyclopaedia of the History of Science, Technology, and Medicine in Non-Western Cultures (2 ed.). Springer. pp. 986–988. ISBN 978-1-4020-4425-0.
39. How Aryabhata got the earth's circumference right Archived 15 January 2017 at the Wayback Machine.
40. Howard Eves (1990). An Introduction to the History of Mathematics (6 ed.). Saunders College Publishing House, New York. p. 237.
41. Hugh Thurston (1996). Early Astronomy. Springer. p. 188. ISBN 0-387-94822-8.
42. Hugh Thurston, Early Astronomy, New York: Springer-Verlag, 1996, pp. 178–189. ISBN 0-387-94822-8.
43. Ifrah, Georges (2000). The Universal History of Numbers. From Prehistory to the Invention of the Computer. New York: John Wiley & Sons. pp. 447–450. ISBN 0-471-39340-1.
44. J J O'Connor and E F Robertson (June 1996). "The trigonometric functions". Retrieved 4 March 2010.
45. J. J. O'Connor and E. F. Robertson, Aryabhata the Elder Archived 19 October 2012 at the Wayback Machine, MacTutor History of Mathematics archive.
46. Jacobs, Harold R. (2003). Geometry: Seeing, Doing, Understanding (Third ed.). New York: W.H. Freeman and Company. p. 70. ISBN 0-7167-4361-2.

47. K. V. Sarma (2001). "Āryabhaṭa: His name, time and provenance" (PDF). Indian Journal of History of Science. 36 (4): 105–115. Archived from the original (PDF) on 31 March 2010.
48. Katz, Victor J., ed. (2007). The mathematics of Egypt, Mesopotamia, China, India, and Islam: a sourcebook. Princeton: Princeton University Press. pp. 405–408. ISBN 978-0-691-11485-9.
49. Kim Plofker (2009). Mathematics in India. Princeton, NJ: Princeton University Press. p. 111. ISBN 978-0-691-12067-6.
50. Menon (2009). An Introduction to the History and Philosophy of Science. Pearson Education India. p. 52. ISBN 978-81-317-2890-1.
51. Noel Swerdlow, "Review: A Lost Monument of Indian Astronomy," Isis, 64 (1973): 239–243.
52. O'Connor, J J; Robertson, E F. "Aryabhata the Elder". www-history.mcs.st-andrews.ac.uk. Archived from the original on 11 July 2015. Retrieved 18 July 2012.
53. Otto Neugebauer, "The Transmission of Planetary Theories in Ancient and Medieval Astronomy," Scripta Mathematica, 22 (1956), pp. 165–192; reprinted in Otto Neugebauer, Astronomy and History: Selected Essays, New York: Springer-Verlag, 1983, pp. 129–156. ISBN 0-387-90844-7.
54. Pingree, David (1996). "Astronomy in India". In Walker, Christopher (ed.). Astronomy before the Telescope. London: British Museum Press. pp. 123–142. ISBN 0-7141-1746-3. pp. 127–9.
55. R.C.Gupta (31 July 1997). "Āryabhaṭa". In Helaine Selin (ed.). Encyclopaedia of the history of science, technology, and medicine in non-western cultures. Springer. p. 72. ISBN 978-0-7923-4066-9.
56. Radhakrishnan Kuttoor (25 June 2007), "Aryabhata lived in Ponnani?", The Hindu, archived from the original on 1 July 2007.
57. Roddam, Narasimha (2001). "Sines in terse verse". Nature. Macmillan Magazines Ltd. 414: 851.
58. Roger Cooke (1997). "The Mathematics of the Hindus". History of Mathematics: A Brief Course. Wiley-Interscience. ISBN 0-471-18082-3. Aryabhata gave the correct rule for the area of a triangle and an incorrect rule for the volume of a pyramid. (He claimed that the volume was half the height times the area of the base.).
59. S. Balachandra Rao (1998) [First published 1994]. Indian Mathematics and Astronomy: Some Landmarks. Bangalore: Jnana Deep Publications. ISBN 81-7371-205-0.
60. S. Kak, Aryabhatiya. Encyclopedia of India, 2005.

61. Selin, Helaine, ed. (2008). Encyclopaedia of the History of Science, Technology, and Medicine in Non-Western Cultures (2 ed.). Springer. pp. 986–988. ISBN 978-1-4020-4425-0.
62. Singh, J. (1999). Sterling Dictionary of Physics. Sterling Publishers Private Limited. p. 12. ISBN 978-81-7359-124-2. Retrieved 15 April 2023.
63. Takao Hayashi, T (November 1997). "Āryabhaṭa's rule and table for sine-differences". Historia Mathematica. 24 (4): 396–406. doi:10.1006/hmat.1997.2160.
64. The concept of Indian heliocentrism has been advocated by B. L. van der Waerden, Das heliozentrische System in der griechischen, persischen und indischen Astronomie. Naturforschenden Gesellschaft in Zürich. Zürich:Kommissionsverlag Leeman AG, 1970.
65. Though Aristarchus of Samos (3rd century BCE) is credited with holding an heliocentric theory, the version of Greek astronomy known in ancient India as the Paulisa Siddhanta makes no reference to such a theory.
66. "Aryabhata and Diophantus' son", Hindustan Times Storytelling Science column, November 2004.
67. "He believes that the Moon and planets shine by reflected sunlight, incredibly he believes that the orbits of the planets are ellipses."
68. 1930 English translation of The Aryabhatiya in various formats at the Internet Archive.
69. A Ahmad, On the $\pi$ of Aryabhata I, Ganita Bharati 3 (3-4) (1981), 83-85.
70. A Volodarsky, Mathematical achievements of Aryabhata, Indian J. Hist. Sci. 12 (2) (1977), 167-172.
71. Achar, Narahari (2007). "Āryabhaṭa I". In Thomas Hockey; et al. (eds.). The Biographical Encyclopedia of Astronomers. New York: Springer. p. 63. ISBN 978-0-387-31022-0. (PDF version).
72. Āryabhaṭīya 1(gītikā).3
73. B Chatterjee, A glimpse of Aryabhata's theory of rotation of earth, Indian J. History Sci. 9 (1) (1974), 51-55, 141.
74. B Datta, Two Aryabhatas of al-Biruni, Bull. Calcutta Math. Soc. 17 (1926), 59-74.
75. B Ishwar, Development of Indian astronomy at the time of Aryabhata I, Ganita Bharati 6 (1-4) (1984), 19-24.
76. B L van der Waerden, The 'Day of Brahman' in the work of Aryabhata, Arch. Hist. Exact Sci. 38 (1) (1988), 13-22.

77. B N Prasad and R Shukla, Aryabhata of Kusumpura, Bull. Allahabad Univ. Math. Assoc. 15 (1951), 24-32.
78. B. L. van der Waerden: Erwachende Wissenschaft. Ägyptische, babylonische und griechische Mathematik. Birkhäuser-Verlag, Basel Stuttgart, 1966, ISBN 3-7643-0399-9
79. Bhaskara I's sine approximation formula.
80. Bhau Daji (1865). "Brief Notes on the Age and Authenticity of the Works of Aryabhata, Varahamihira, Brahmagupta, Bhattotpala, and Bhaskaracharya". Journal of the Royal Asiatic Society of Great Britain and Ireland. pp. 392–406.
81. Bibhutibhushan Datta and Avadhesh Narayan Singh (1962). History of Hindu Mathematics A source Book Part II. Asia Publishing House. p. 92.
82. Biography in Encyclopaedia Britannica.
83. C Müller, Volumen und Oberfläche der Kugel bei Aryabhata I, Deutsche Math. 5 (1940), 244-255.
84. Clark 1930.
85. Clark, Walter Eugene (1930). The Āryabhaṭīya of Āryabhaṭa: An Ancient Indian Work on Mathematics and Astronomy. University of Chicago Press; reprint: Kessinger Publishing (2006). ISBN 978-1-4254-8599-3.
86. Cooke, Roger (1997). The History of Mathematics: A Brief Course. Wiley-Interscience. ISBN 0-471-18082-3.
87. D Pingree, Biography in Dictionary of Scientific Biography (New York 1970-1990).
88. E G Forbes, Mesopotamian and Greek influences on ancient Indian astronomy and on the work of Aryabhata, Indian J. Hist. Sci. 12 (2) (1977), 150-160.
89. E M Bruins, With roots towards Aryabhata's $\pi$-value, Ganita Bharati 5 (1-4) (1983), 1-7.
90. Ebenezer Burgess; Phanindralal Gangooly (1989). The Surya Siddhanta: A Textbook of Hindu Astronomy. Motilal Banarsidass Publ. p. 46. ISBN 978-81-208-0612-2.
91. Ernst Wilhelm. Classical Muhurta. Kala Occult Publishers. p. 44. ISBN 978-0-9709636-2-8.: "*The point on the equator that is below the city of Ujjain is known, according to the Siddhantas, as Lanka. (This is not the Lanka that is now known as Sri Lanka; Aryabhata is very clear in stating that Lanka is 23 degrees south of Ujjain.)*"
92. First published 2014 by Spektrum der Wissenschaft Verlagsgesellschaft Heidelberg,

https://www.spektrum.de/wissen/aryabhata-indischer-mathematiker-lehrmeister-der-arabischen-mathematik/1300797.

93. Fleet, J. F. (1911). "Aryabhata's System of Expressing Numbers". The Journal of the Royal Asiatic Society of Great Britain and Ireland. Royal Asiatic Society of Great Britain and Ireland. 43: 109–126. doi:10.1017/S0035869X00040995. JSTOR 25189823. S2CID 163070211.
94. Fleet, J. F. (January 1911). "Aryabhata's System of Expressing Numbers". Journal of the Royal Asiatic Society of Great Britain and Ireland. 43: 109–126. doi:10.1017/S0035869X00040995. ISSN 0035-869X. JSTOR 25189823. S2CID 163070211.
95. For a comparison of Indian and Chinese methods for solving linear diophantine equations: A. K. Bag and K. S. Shen (1984). "Kuttaka and Qiuvishu" (PDF). Indian Journal of History of Science. 19 (4): 397–405. Archived from the original (PDF) on 5 July 2015. Retrieved 1 March 2016.
96. For a comparison of the complexity of the Aryabhata algorithm with the complexities of Euclidean algorithm, Chinese remainder theorem and Garner's algorithm: T. R. N. Rao and Chung-Huang Yang (2006). "Aryabhata Remainder Theorem: Relevance to Public Key Crypto-systems" (PDF). Circuits, System, Signals Processing. 25 (1): 1–15. Retrieved 1 March 2016.
97. For a detailed exposition of the Kuttaka algorithm as given by Sankaranarayana in his commentary on Laghubhaskariya: Bhaskaracharya-1 (Translated by K. S. Shukla) (1963). Laghu-Bhskariya. Lucknow University. pp. 103–114. Retrieved 7 March 2016.
98. For a discussion of the computational aspects of Aryabhata algorithm: Subhash Kak (1986). "Computational Aspects of Aryabhata Algorithm" (PDF). Indian Journal of History of Science. 21 (1): 62–71. Retrieved 1 March 2016.
99. For an application of Kuttaka in computing full moon days: Robert Cooke. "Euclid's Algorithm" (PDF). Archived from the original (PDF) on 15 June 2016. Retrieved 1 March 2016.
100. For the interpretation of Aryabhata's original formulation of algorithm: Bibhutibhusan Datta (1932). "Elder Aryabhata's Rule for the Solution of Indeterminate Equations of the First Degree". Bulletin of Calcutta Mathematical Society. 24 (1): 19–36.
101. G Ifrah, A universal history of numbers : From prehistory to the invention of the computer (London, 1998).

102.G M Bongard Levin, Aryabhata and Lokayatas, Indian J. Hist. Sci. 12 (2) (1977), 187-193.
103.Ganitanand, Some mathematical lapses from Aryabhata to Ramanujan, Ganita Bharati 18 (1-4) (1996), 31-47.
104.Georges Ifrah: The Universal History of Numbers. From Prehistory to the Invention of the Computer. John Wiley & Sons, New York, 2000, ISBN 0-471-39340-1.
105.H-J Ilgauds, Aryabhata I, in H Wussing and W Arnold, Biographien bedeutender Mathematiker (Berlin, 1983).
106.Aryabhata-I, http://www.britannica.com/biography/Aryabhata-I.
107.Āryabhaṭa_numeration, https://en.wikipedia.org/wiki/%C4%80ryabha%E1%B9%ADa_numeration.
108.Aryabhata, https://en.wikipedia.org/wiki/Aryabhata.
109.Aryabhatiya, https://en.wikipedia.org/wiki/Aryabhatiya.
110.Aryabhata, https://mathshistory.st-andrews.ac.uk/Strick/aryabhata.pdf.
111.K Elfering, The area of a triangle and the volume of a pyramid as well as the area of a circle and the surface of the hemisphere in the mathematics of Aryabhata I, Indian J. Hist. Sci. 12 (2) (1977), 232-236.
112.K S Shukla, Aryabhata I's astronomy with midnight day-reckoning, Ganita 18 (1967), 83-105.
113.K S Shukla, Glimpses from the 'Aryabhata-siddhanta', Indian J. Hist. Sci. 12 (2) (1977), 181-186.
114.K S Shukla, Use of hypotenuse in the computation of the equation of the centre under the epicyclic theory in the school of Aryabhata I, Indian J. History Sci. 8 (1973), 43-57.
115.Kak, Subhash C. (2000). 'Birth and Early Development of Indian Astronomy'. In Selin, Helaine, ed. (2000). Astronomy Across Cultures: The History of Non-Western Astronomy. Boston: Kluwer. ISBN 0-7923-6363-9.
116.Kripa Shankar Shukla and K V Sarma (1976). Aryabhatiya of Aryabhata (Critically ediited with Introduction, English Translation, Notes, Comments and Index). Dlehi: Indian national Science Academy. p. 29. Retrieved 25 January 2023.
117.Kurt Elfering: Die Mathematik des Aryabhata I. Text, Übersetzung aus dem Sanskrit und Kommentar. Wilhelm Fink Verlag, München, 1975, ISBN 3-7705-1326-6.
118.L C Jain, Aryabhata I and Yativrsabha - a study in Kalpa and Meru, Indian J. Hist. Sci. 12 (2) (1977), 137-146.

119.L. Satpathy (2003). Ancient Indian Astronomy. Alpha Science Int'l Ltd. p. 200. ISBN 978-81-7319-432-0.: "*Seven cardinal points are then defined on the equator, one of them called Laṅkā, at the intersection of the equator with the meridional line through Ujjaini. This Laṅkā is, of course, a fanciful name and has nothing to do with the island of Sri Laṅkā.*"
120.M L Sharma, Aryabhata's contribution to Indian astronomy, Indian J. Hist. Sci. 12 (2) (1977), 90-99.
121.M L Sharma, Indian astronomy at the time of Aryabhata, Indian J. Hist. Sci. 12 (2) (1977), 100-105.
122.M S Khan, Aryabhata I and al-Biruni, Indian J. Hist. Sci. 12 (2) (1977), 237-244.
123.M Yano, Aryabhata's possible rebuttal to objections to his theory of the rotation of the Earth, Historia Sci. 19 (1980), 101-105.
124.O'Connor, John J.; Robertson, Edmund F., "Aryabhata", MacTutor History of Mathematics Archive, University of St Andrews.
125.P Jha, Aryabhata I : the man and author, Math. Ed. (Siwan) 17 (2) (1983), 50-60.
126.P Jha, Aryabhata I and the value of $\pi$, Math. Ed. (Siwan) 16 (3) (1982), 54-59.
127.R Behari, Aryabhata as a mathematician, Indian J. Hist. Sci. 12 (2) (1977), 147-149.
128.R Billard, Aryabhata and Indian astronomy, Indian J. Hist. Sci. 12 (2) (1977), 207-224.
129.R C Gupta, A preliminary bibliography on Aryabhata I, Math. Education 10 (2) (1976), B21-B26.
130.R C Gupta, Aryabhata I's value of $\pi$, Math. Education 7 (1973), B17-B20.
131.R C Gupta, Aryabhata, ancient India's great astronomer and mathematician, Math. Education 10 (4) (1976), B69-B73.
132.R N Rai, The Ardharatrika system of Aryabhata I, Indian J. History Sci. 6 (1971), 147-152.
133.R.M. Pujari; Pradeep Kolhe; N. R. Kumar (2006). Pride of India: A Glimpse into India's Scientific Heritage. SAMSKRITA BHARATI. p. 63. ISBN 978-81-87276-27-2.
134.S Kak, The Aryabhata cipher, Cryptologia 12 (2) (1988), 113-117.
135.S L Dhani, Manvantara theory of evolution of solar system and Aryabhata, Indian J. Hist. Sci. 12 (2) (1977), 161-166.
136.S N Sen, Aryabhata's mathematics, Bull. Nat. Inst. Sci. India 21 (1963), 297-319.

137.S Parameswaran, On the nativity of Aryabhata the First, Ganita Bharati 16 (1-4) (1994), 57-60.

138.S. Balachandra Rao (2000). Indian Astronomy: An Introduction. Orient Blackswan. p. 82. ISBN 978-81-7371-205-0.: "*In Indian astronomy, the prime meridian is the great circle of the Earth passing through the north and south poles, Ujjayinī and Laṅkā, where Laṅkā was assumed to be on the Earth's equator.*"

139.Shukla, Kripa Shankar. Aryabhata: Indian Mathematician and Astronomer. New Delhi: Indian National Science Academy, 1976.

140.Surya Siddhanta translations.

141.Thurston, H. (1994). Early Astronomy. Springer-Verlag, New York. ISBN 0-387-94107-X.

**Yativṛṣabha (यतीवृषभा)**
**(500–570 CE)**

Yativrsabha (यतीवृषभा) (or Jadivasaha - जादिवासह) was a mathematician who studied under *Arya Manksu* (आर्य मङ्क्सु) and *Nagahastin* (नागहस्तिन्). We know nothing of *Yativrsabha*'s dates except for a reference which he makes to the end of the *Gupta dynasty* (गुप्तवंशः) which he says was after 231 years of ruling. This ended in 551 so we must assume that 551 AD is a date which occured during *Yativrsabha*'s lifetime. This fits with the only other information regarding his dates which are that his work is referenced by *Jinabhadra Ksamasramana* (जिनभद्र क्षमश्रमण) in 609 and that *Yativrsabha* himself refers to a work written by *Sarvanandin* (सर्वानन्दिन्) in 458.

*Yativrsabha*'s work *Tiloyapannatti* (तिलोयपन्नात्ति) gives various units for measuring distances and time and also describes the system of infinite time measures. It is a work which describes *Jaina* cosmology and gives a description of the universe which is of historical importance in understanding *Jaina* science and mathematics. The belief was in an infinite world, both infinite in space and in time. This led the to devise ways of measuring larger and larger distances and longer and longer intervals of time. It led them to consider different measures of infinity, and in this respect the *Jaina* mathematicians would appear to be the only ones before the time when *Cantor* developed the theory of infinite cardinals to envisage different magnitudes of infinity.

*Yativṛṣabha* performed detailed measurement of *Jambu* (जम्बू) (Bharat).

*Yativṛṣabha* was not just a mathematician but also an erudite scholar who chose *Prakrit* (प्राकृत) as the medium for his writings. *Prakrit*, an ancient Bhartiya language, was widely spoken and was the precursor to many modern Bhartiya languages. By opting for *Prakrit*, *Yativṛṣabha* ensured that his teachings were accessible to a broader audience of his time. His writings encapsulate the essence of *Jain traditions* (जैन परम्परा), showcasing the depth and breadth of *Jain philosophy* (जैन दर्शन) and its intertwined relationship with daily life.

## Key Contributions in Mathematics

### Tiloyapannatti - A Beacon in Cosmology

The "*Tiloyapannatti*" stands out as a monumental work in *Jain literature* (जैन साहित्य). It doesn't merely describe the universe in abstract terms but offers a quantified, detailed representation of its components. This text acts as a bridge, filling the knowledge gap between the ancient canonical writings of *Jainism* and the more evolved texts from the 9th century onwards. The "*Tiloyapannatti*" thus becomes a crucial point of reference for historians and scholars tracing the trajectory of *Jain* mathematical thought. The *Tiloya Panatti* is a *Prakrit* work in the *Jain Shauraseni dialect* (जैन शौरासेनी बोली) and has been composed primarily in the *Arya metre* (आर्य मीटर्). The work has a total of *5677 verses* divided into *9 chapters*. The chapter scheme is as under:

- The Entire Universe (*Loka*)
- The World of Hells (*Naraka*)
- The World of Mansion-dwelling demigods (*Bhuvanavasi*)
- The World of Humans (*Manushya*)
- The World of Sub-human species - plants, animals, insects, etc. (*Tiryanca*)
- The World of Peripatetic demigods
- The World of Light
- The World of Heavens
- The World of the Siddhas - the liberated ones (*Siddhashila*)

### Units of Measurement and Quantification

One of the standout features of the *Tiloyapannatti* is its introduction of various units for measuring distances and time. This emphasis on quantification underscores *Jainism*'s commitment to precision and its belief in the importance of understanding the universe in tangible terms. These units weren't just theoretical but were possibly used in practical applications, aiding in navigation, construction, and other scientific endeavors.

### Intriguing Concepts of Infinity

*Infinity* (अनंतं) is a concept that has both baffled and fascinated mathematicians and philosophers alike. *Yativṛṣabha*'s exploration of infinity within the *Jain* context is groundbreaking. He postulated various notions of infinity, some of which might have been foundational for later Bhartiya mathematical theories. His work provides a glimpse into the sophisticated level of mathematical thought in ancient *Jain scholarship* (जैन विद्वता).

## 卐 Detailed Numerical Representation of the Universe

*Yativṛṣabha*'s penchant for detail is evident in his numerical descriptions of cosmic entities. For instance, the detailed measurement of the *Jambu* (जम्बू) (i.e., Bharat) continent showcases the meticulous nature of *Jain* cosmological studies. Such precision wasn't just for academic purposes but reflects *Jainism*'s belief in the interconnectedness of the universe and the importance of understanding its structure.

> The diameter of the circular Jambu continent, upon which India is located, is 100,000 yojanas (योजना) and its circumference (परिधि) is 316,227 yojanas, 3 krośas (क्रोश), 128 daṇḍas (दण्ड), 13 aṅgulas (अङ्गुल), 5 yavas (यवः), 1 yūkā (युका), 1 ṛikṣā (ऋक्षा), 6 karmabhūmivālagras (हुमीवलग्रास्), 7 madhyabhogabhūmivālagras (मध्यभभाभभाभोमिवलग्रास), 5 uttamabhogabhūmivālagras (उत्तमभोगभूमिवलाग्रा), 1 rathareṇu (रथरेणु), 3 trasareṇus (त्रासरेणु), 2 sannāsannas (सन्नासन्ना), and 3 avasannāsannas (अवसन्नासन्ना), plus a remainder (शेषं) of $\frac{23213}{105409}$
>
> Translation

For historians of mathematics and *Bhartiya culture* (भारतीय संस्कृति), *Yativṛṣabha*'s works are invaluable. They offer a window into the rich tapestry of *Jain thought*, where spirituality, philosophy, and mathematics seamlessly merge. His contributions provide insights into the pedagogical methods, scholarly debates, and intellectual climate of his era. Moreover, his works have likely inspired and laid the foundation for subsequent generations of Jain scholars and mathematicians.

## 卐 Area of Circle

*Yativṛṣabha* also gives formulas for computing the circumference ($C$) and the area ($A$) of a circle having a diameter of $d$:

$$C = \sqrt{10d^2}, A = C \times \frac{d}{4}$$

**Further Reading**


1. Ikeyama, Setsuro (2007), "Yativṛṣabha", in Hockey, Thomas; Trimble, Virginia; Williams, Thomas R.; Bracher, Katherine (eds.), The Biographical Encyclopedia of Astronomers, New York, NY: Springer, p. 1251, doi:10.1007/978-0-387-30400-7_1513, ISBN 978-0-387-30400-7, retrieved 5 May 2021.
2. Jaini 1998, p. 79.
3. Umakant P. Shah 1987, p. 239.
4. "Yativrsabha - Biography". Maths History. Retrieved 5 May 2021.
5. D Pingree, Biography in Dictionary of Scientific Biography (New York 1970-1990).
6. Hayashi, Takao (1993). Indo no Sūgaku (Mathematics in India). Chūkō-shinsho 1155. Tokyo: Chūōkōron-sha.
7. Ikeyama, Setsuro (2007). "Yativṛṣabha". In Thomas Hockey; et al. (eds.). The Biographical Encyclopedia of Astronomers. New York: Springer. p. 1251. ISBN 978-0-387-31022-0. (PDF version).
8. Jaini, Padmanabh S. (1998) [1979], The Jaina Path of Purification, Delhi: Motilal Banarsidass, ISBN 81-208-1578-5.
9. L C Jain and A Jain, Philosopher mathematicians: Yativrsabhacarya, Virasenacarya and Nemicandracarya (Meerut, 1985).
10. L C Jain, Aryabhata I and Yativrsabha - a study in Kalpa and Meru, Indian J. Hist. Sci. 12 (2) (1977), 137-146.
11. L C Jain, Basic mathematics: Exact Sciences from Jaina Sources 1 (New Delhi, 1982).
12. O'Connor, John J.; Robertson, Edmund F., "Yativṛṣabha", MacTutor History of Mathematics Archive, University of St Andrews
13. Shah, Umakant P. (1987), Jaina-rūpa-maṇḍana: Jaina iconography, Abhinav Publications, ISBN 81-7017-208-X.
14. Vyas 1995, p. 20.
15. Vyas, Dr. R. T., ed. (1995), Studies in Jaina Art and Iconography and Allied Subjects, The Director, Oriental Institute, on behalf of the Registrar, M.S. University of Baroda, Vadodara, ISBN 81-7017-316-7.
16. Yativṛṣabha (1943–1951). Tiloyapahnattī, edited by H. Jaina and A. N. Upādhyāya. Sholapura (2nd ed. of Vol. 1, Sholapur, 1956). New

edition by V. Mātājī (Vols. 1 and 2 covering the first four chapters have appeared), Kotā, 1984–1986.
17. Yativṛṣabha, https://mathshistory.st-andrews.ac.uk/Biographies/Yativrsabha/.
18. Yativṛṣabha, https://en.wikipedia.org/wiki/Yativ%E1%B9%9B%E1%B9%A3abha.
19. Yativṛṣabha, https://islamsci.mcgill.ca/RASI/BEA/Yativrsabha_BEA.htm.

**Varāhamihira (वराहमिहिर)**
**(505–587 CE)**

Figure 2.4. Varāhamihira (वराहमिहिर). (Image Credit: https://en.wikipedia.org/wiki/)

Varāhamihira (वराहमिहिर) was a great astrologer whose main work was a treatise on mathematical astronomy which summarised earlier *astronomical treatises* (खगोलीय ग्रन्थ) (see Figure 2.4). Our knowledge of *Varāhamihira* is very limited indeed. According to one of his works, he was educated in *Kapitthaka* (कपितथक). However, far from settling the question this only gives rise to discussions of possible interpretations of where this place was. We do not know whether he was born in *Kapitthaka*, wherever that may be, although we have given this as the most likely guess. We do know, however, that he worked at *Ujjain* (उज्जैन) which had been an important centre for mathematics since around 400 AD. The school of mathematics at *Ujjain* was increased in importance due to *Varāhamihira* working there and it continued for a long period to be one of the two leading mathematical centres in Bharat, in particular having *Brahmagupta* as its next major figure. *Varāhamihira*'s contributions span various branches of knowledge, but he's particularly remembered for his works in mathematics and astronomy.

He discovered a version of Pascal's triangle and worked on magic squares.

## 卐 Key Contributions in Mathematics

### 卐 Mathematical Astronomy

- Pancha-siddhantika (पञ्च-सिद्धान्तिका): The *Pancha-siddhantika,* translated as "*The Five Astronomical Canons,*" is *Varāhamihira*'s most renowned work in the field of mathematical *astronomy* (खगोलशास्त्रम्), dating to around 575 AD. It stands out not just as a significant text in its own right, but also for the valuable insights it offers into older Bhartiya astronomical treatises, many of which are now lost to time. Essentially, the *Pancha-siddhantika* is a synthesis and summary of five earlier astronomical works:

    - *Surya Siddhanta* (सूर्य सिद्धान्त)

      One of the most significant and oldest astronomical works, the *Surya Siddhanta* primarily deals with the *astronomy* that's based on the *sun.* The text covers a wide range of topics, including the lengths of the day and night, the positions of the *planets, lunar and solar eclipses,* and the phases of the *Moon.* It offers sophisticated methods for calculating various astronomical phenomena and is notable for its estimations of the diameter of the planets, their distances from the *Earth,* and the length of the year — many of which are surprisingly accurate by modern standards. The *Surya Siddhanta* also contains *geometric* (ज्यामितीय सिद्धान्त) and *trigonometric theories* (त्रिकोणमितीय सिद्धान्त) that were quite advanced for its time.

    - *Romaka Siddhanta* (रोमक सिद्धान्त)

      The *Romaka Siddhanta* (*Romaka System* - रोमका प्रणाली) is believed to have been influenced by the ancient *Greek* and possibly *Babylonian astronomy* (बेबिलोनियन खगोलशास्त्रम्). Its name suggests its Western (*Romaka* is derived from *Roma,* the ancient name for *Rome*) influence. It incorporated the *Greek epicycle theory* of the motions of the *Sun* and the *Moon*. This text also was based on the tropical year of *Hipparchus* and utilized the *Metonic cycle,* which spans 19 years.

- *Paulisa Siddhanta* (पौलिसा सिद्धान्त)

  The *Paulisa Siddhanta* (*Paulisa System* - पौलिसा प्रणाली) is believed to have been influenced by *Hellenistic astronomy.* Like the *Romaka Siddhanta,* the *Paulisa Siddhanta* shows the interaction and exchange of astronomical ideas between Bharat and the *Greco-Roman* world.

- *Vasistha Siddhanta* (वसिष्ठ सिद्धान्त)

  Not as much is known about the *Vasistha Siddhanta* in comparison to the other *Siddhantas*. It's named after the sage *Vasishta,* one of the *Saptarishis* (*Seven Great Sages*) in Hindu tradition. The details of its content are mainly known from references in other works, like the *Pancha-siddhantika*. It is believed to be an original Bhartiya astronomical treatise without significant foreign influences.

- *Paitamaha Siddhanta* (पैतमह सिद्धान्त)

  The *Paitamaha Siddhanta* is named after *Prajapati* (प्रजापति) or *Pitamaha (Brahma),* the *creator god* in Hindu cosmology. Like the *Vasistha Siddhanta,* specific details about its content are sparse and are mainly derived from its mention in other works. It is considered one of the older traditional Bhartiya astronomical systems.

❖ Short work (now lost): Although the specifics of the short work in mathematical astronomy by *Varāhamihira* are lost, they are known from *Utpala*'s (उत्पलस्य) commentary, which serves as a reference to this lost text.

- Brhaj-jataka (बृज-जातक): The *Brhaj-jataka* (often written as "*Brihat Jataka* - बृहत् जातक") is one of the most celebrated and influential classical works on *Jyotisha* (*Vedic astrology* - वैदिक ज्योतिषशास्त्र). Composed by *Varāhamihira*, it is not only a testament to the advanced astrological knowledge of ancient Bharat but also demonstrates the deep cultural and philosophical underpinnings of the time. Over the centuries, the *Brhaj-jataka* has been commented upon by numerous scholars. Its profundity and comprehensiveness have made it a standard reference for those studying traditional Bhartiya astrology. Various commentaries have been written on the *Brhaj-jataka*, further amplifying its significance and shedding light on its intricate aspects. Some of the commentaries have helped in deciphering the cryptic verses, making the knowledge more accessible (see Figure 2.5).

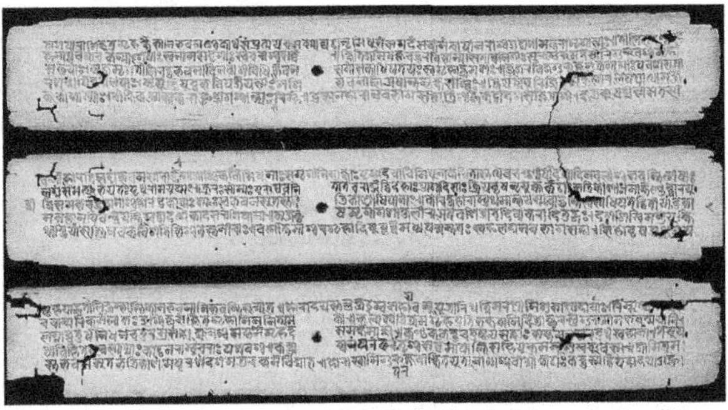

Figure 2.5. manuscript of the *Brihajjataka*. (Image Credit: https://en.wikipedia.org/wiki)

  - Content and Structure

    The *Brhaj-jataka* is divided into multiple chapters, each focusing on a specific aspect of *horoscopy* (कुंडली). The treatise covers various topics, such as the signs of the zodiac, planets, houses, planetary periods (*dasha systems* - दश प्रणाल्या), and more. It delves into the interpretation of *birth charts* (जन्मपत्रिका), predicting events in an individual's life based on the positions and associations of *celestial bodies* (आकाशपिण्डा) at the *time of birth* (जन्मकाल).

- Planetary Characters and Effects

  One of the notable aspects of the *Brhaj-jataka* is its detailed exposition on the nature and characteristics of planets. Each *planet* (ग्रह) is described with its inherent qualities, and their effects on human lives when placed in different signs or houses.

- Predictive Astrology

  The text emphasizes the predictive aspect of a*strology* - ज्योतिषशास्त्रम्. It offers insights into predicting major life events, such as one's profession, *marital life* - वैवाहिक जीवन, *progeny*- प्रजा, *fortune* - भाग्य, and even potential challenges or misfortunes. There are specific *yogas* (combinations) elucidated that lead to certain results. For instance, combinations that can make a person a ruler, or those that indicate poverty, etc.

- Technical Depth

  *Varāhamihira*'s work is known for its technical depth. The *Brhaj-jataka* provides mathematical calculations for various astrological configurations. It guides astrologers on how to determine planetary strengths, calculate ascendants, and much more.

❖ Laghu-jataka (लघु-जातक): The *Laghu-jataka,* often translated as the "*Small or Minor Horoscopy,*" is another significant work attributed to the ancient Bhartiya astrologer and mathematician *Varāhamihira*. As the name *"Laghu*" (meaning "small" or "brief") suggests, it is a concise version or perhaps a summarization of the more extensive treatise, the *Brhaj-jataka*. Despite its brevity, *the Laghu-jataka* holds its own for its depth and precision in astrological principles. The *Laghu-jataka,* given its concise and straightforward nature, has been a favored text for many in the astrological community. Over the years, it has received various commentaries and interpretations by scholars, further solidifying its position as a foundational text in *Vedic astrology*.

- Content and Focus:

  While the *Brhaj-jataka* is comprehensive and expansive in its exploration of astrological concepts, the *Laghu-jataka,* in contrast, focuses on the core principles and offers a succinct overview of horoscopic astrology. It covers the fundamental aspects of casting a horoscope and the basic interpretative techniques used in Jyotisha.

- Utility

  Due to its concise nature, the *Laghu-jataka* is particularly useful for students and practitioners who need a quick reference or are at the early stages of their astrological studies. It provides an efficient way to understand and recall essential astrological concepts.

- Planetary Interpretations

  Like its more extensive counterpart, the *Laghu-jataka* delves into the nature and effects of planets in various signs and houses. It offers interpretations that can be used to predict specific life events based on the planetary placements in a natal chart.

- Predictive Techniques:

  The text touches upon several predictive techniques, including the use of planetary periods (*dasha systems*) and transits. It provides insights on how to use these techniques in tandem with the natal chart to forecast life events.

卐 **Horoscopy: Marriage**

- Brhad-vivaha-patala (बृहद्-विवाह-पतल): As a dedicated work on marital horoscopy, *Brhad-vivaha-patala* is essential in understanding and interpreting natal charts in the context of marriage. It delves deeply into various factors affecting marital life. This work focuses on the horoscopes pertaining to marriages. It offers guidelines on how to interpret horoscopes to determine marital compatibility and auspicious times for marriage. The treatise provides insights into determining the compatibility of potential partners. This includes analyzing specific houses, planetary positions, and yogas (specific combinations) that influence marriage. It gives astrologers the tools to determine auspicious timings (*muhurtas* - मुहूर्ता) for marriage ceremonies.

- Svalpa-vivaha-patala (स्वाल्प-विवाह-पटल): This is a shorter treatise on marital horoscopy, offering more concise interpretations for marriage-related astrological readings. This text is a more concise version of the *Brhad-vivaha-patala.* It offers straightforward interpretations, focusing on key factors affecting marital compatibility and timings. It's particularly useful for quick references or for those who are at the beginning stages of their astrological studies pertaining to marriage.

卐 **Horoscopy: Journey**

- Brhad-yatra (बृहद-यात्रा) and Yoga-yatra (योगयात्रा): These works discuss the interpretation of horoscopes to determine auspicious times and outcomes for journeys or travels. These twin works delve into the significance of journeys in one's life. They provide astrological insights to determine the best times for travel and the likely outcomes of journeys. Depending on the positions of planets and specific houses in a natal chart, these texts can guide one on when travel would be beneficial or when it might be best to stay put. They also touch upon spiritual journeys or pilgrimages, understanding their significance in one's life chart.

- Svalpa-yatra (स्वल्प-यात्रा): A concise treatise on the horoscopy of journeys. This is the abridged version focusing on the horoscopy of journeys. Like the *Svalpa-vivaha-patala* (स्वल्प-विवाह-पतल), it's designed for quick reference and offers essential insights without delving too deeply into the intricate details.

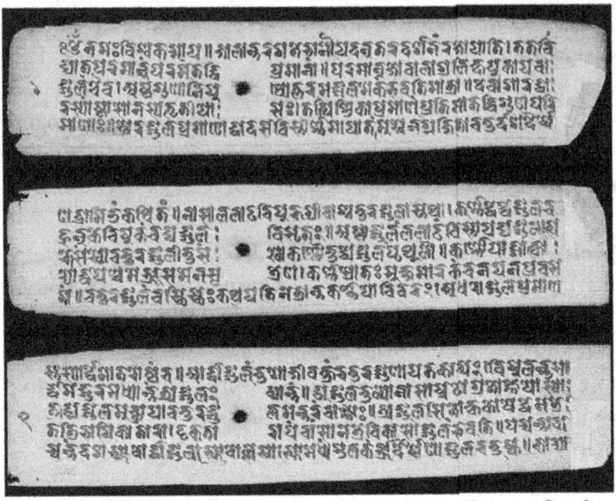

Figure 2.6. Manuscript of the *Brhat-samhita*. (Image Credit: https://en.wikipedia.org/wiki)

## 卐 General Astrology (Samhita)

- Brhat-samhita (बृहत-संहिता): The *Brhat-samhita* is one of the gems of ancient Bhartiya literature (see Figure 2.6). It's a comprehensive treatise, covering a plethora of subjects beyond just astrology. This is an encyclopedic work that touches upon various subjects, including astrology, architecture, agriculture, weather phenomena, gemology, and more. For example, it provides insights into predicting rains, earthquakes, and other natural events based on astrological signs.
  In its *astrological sections* (ज्योतिषखण्डा), it deals with various phenomena, from celestial occurrences to terrestrial events like *rains* (वर्षा), *earthquakes* (भूकम्प), and more. It provides guidelines on predicting these based on astrological configurations. Beyond *astrology*, it touches on architecture (*Vastu Shastra* - वास्तु शास्त्रम्), gemology, agriculture, and even offers omens and their interpretations. This showcases the holistic approach of ancient Bhartiya scholars, intertwining various disciplines.

- Samasa-samhita (समस-संहिता): A shorter version of the *Brhat-samhita*, offering a more concise overview of general astrology topics. A concise counterpart to the B*rhat-samhita*, the *Samasa-samhita* distills the vast knowledge of the former into a more compact form. It focuses primarily on the astrological aspects, offering insights into general astrology without diving too deeply into the interdisciplinary topics of the *Brhat-samhita*.

## 卐 Trigonometry

*Varāhamihira* made some important mathematical discoveries. Among these are certain trigonometric formulae which translated into our present-day notation correspond to

$$\sin x = \cos(\frac{\pi}{2} - x)$$

$$\sin^2 x + \cos^2 x = 1$$

$$\frac{1}{2}(1 - \cos 2x) = \sin^2 x$$

Another important contribution to trigonometry was his sine tables where he improved those of *Aryabhata I* giving more accurate values. It should be emphasised that accuracy was very important for these Bhartiya mathematicians since they were computing sine tables for applications to astronomy and astrology. This motivated much of the improved accuracy they achieved by developing new interpolation methods.

## 卐 Algebra

The *Jaina school of mathematics* investigated rules for computing the number of ways in which $r$ objects can be selected from $n$ objects over the course of many hundreds of years. They gave rules to compute the binomial coefficients $nC_r$ which amount to

$$nC_r = \frac{1}{r!}n(n-1)(n-2)(n-3)\dots(n-r+1)$$

However, *Varāhamihira* attacked the problem of computing $nC_r$ in a rather different way. He wrote the numbers $n$ in a column with $n = 1$ at the bottom. He then put the numbers $r$ in rows with $r = 1$ at the left-hand side. Starting at the bottom left side of the array which corresponds to the values $n = 1, r = 1$, the values of $nC_r$ are found by summing two entries, namely the one directly below the $(n, r)$ position and the one immediately to the left of it. Of course, this table is none other than *Pascal*'s triangle for finding the *binomial coefficients* despite being viewed from a different angle from the way we build it up today.

### 卐 Magic Square

The oldest dateable fourth order magic square in the world is found in an encyclopaedic work written by *Varāhamihira* around 587 CE called *Brhat Samhita*. The magic square is constructed for the purpose of making perfumes using 4 substances selected from 16 different substances. Each cell of the square represents a particular ingredient, while the number in the cell represents the proportion of the associated ingredient, such that the mixture of any four combination of ingredients along the columns, rows, diagonals, and so on, gives the total volume of the mixture to be 18. Although the book is mostly about divination, the magic square is given as a matter of combinatorial design, and no magical properties are attributed to it. The special features of this magic square were commented on by *Bhattotpala* (c. 966 CE).

| | | | |
|---|---|---|---|
| 2 | 3 | 5 | 8 |
| 5 | 8 | 2 | 3 |
| 4 | 1 | 7 | 6 |
| 7 | 6 | 4 | 1 |

The square of *Varāhamihira* as given above has sum of 18. Here the numbers 1 to 8 appear twice in the square. It is a pan-diagonal magic square. Four different magic squares can be obtained by adding 8 to one of the two sets of 1 to 8 sequence. The sequence is selected such that the number 8 is added exactly twice in each row, each column and each of the main diagonals. One of the possible magic squares shown in the right side. This magic square is remarkable in that it is a *90-degree rotation* of a magic square that appears in the 13th century Islamic world as one of the most popular magic squares. *Varāhamihira*'s works contain 35 Sanskritized *Greek* astronomical terms, and he exhibits a good understanding of the Greek astronomy. He praised the *Greeks* (*Yavanas*) for being "*well trained in the sciences*", though impure in ritual order. A *Bṛhat-Saṃhitā* verse, reads:

म्लेच्छा हि यवनास्तेषु सम्यक् शास्त्रमिदं स्थितम् ॥
ऋषिवत् तेऽपि पूज्यन्ते किं पुनर्दैवविद् द्विजः ॥

-adhyāya II, śloka 14

*mlecchā hi yavanās teṣu samyak śāstram idaṃ sthitam,*
*ṛṣivat te'pi pūjyante kiṃ punar daivavid dvijaḥ*

Translated

Meaning:

*"The Yavanas are of low origin. When this science has come to stay with them and when such astrologers are worshipped as sages, how much more should an astrologer of twice-born origin be?"*

**Further Reading**


1. A.M. Shastri 1969, p. 1.
2. A.M. Shastri 1969, p. 15.
3. A.M. Shastri 1969, p. 21.
4. A.M. Shastri 1969, p. 4.
5. A.M. Shastri 1969, pp. 1–3.
6. A.M. Shastri 1991, p. 1.
7. A.M. Shastri 1991, p. 10.
8. A.M. Shastri 1991, p. 121.
9. A.M. Shastri 1991, p. 13.
10. A.M. Shastri 1991, p. 17.
11. A.M. Shastri 1991, p. 18.
12. A.M. Shastri 1991, p. 20.
13. A.M. Shastri 1991, p. 24.
14. A.M. Shastri 1991, p. 248.
15. A.M. Shastri 1991, p. 27.
16. A.M. Shastri 1991, p. 4.
17. A.M. Shastri 1991, p. 5.

18. A.M. Shastri 1991, p. 51.
19. A.M. Shastri 1991, p. 6.
20. A.M. Shastri 1991, p. 8.
21. A.M. Shastri 1991, p. 9.
22. A.M. Shastri 1991, pp. 100–120.
23. A.M. Shastri 1991, pp. 10–11.
24. A.M. Shastri 1991, pp. 120–121.
25. A.M. Shastri 1991, pp. 17–18.
26. A.M. Shastri 1991, pp. 19–20.
27. A.M. Shastri 1991, pp. 3–4.
28. A.M. Shastri 1991, pp. 43–45.
29. A.M. Shastri 1991, pp. 8–9.
30. Abraham Eraly (2014). "Pearls and Pebbles". The First Spring Part 2: Culture in the Golden Age of India. Penguin. ISBN 9789351186465. He belonged, as his name indicates, to a Brahminized family of the Magi, sun-worshipping Zoroastrian priests, and he himself came to be regarded in later times as an incarnation of the sun god.
31. B. Suryanarain Rao, ed. (1986). Sree Varaha Mihira's Brihat Jataka. Motilal Banarsidass. p. 612. ISBN 9788120813953.
32. Brihat Samhita of Varaha Mihira, archive.org, Sanskrit with English translation by Panditabhushana V. Subrahmanya Sastri and Vidwan M. Ramakrishna Bhat. 1946: Bangalore. p. 19.
33. Chaudhuri, Kirti Narayan (1990). Asia Before Europe Economy and Civilisation of the Indian Ocean from the Rise of Islam to 1750. Cambridge University Press. p. 54. ISBN 0521316812.
34. Dániel Balogh 2019, p. 142.
35. Dániel Balogh 2019, p. 142-143.
36. Dániel Balogh 2019, p. 143.
37. Dániel Balogh 2019, pp. 141–142.
38. Datta, Bibhutibhusan; Singh, Awadhesh Narayan (1992). "Magic Squares in India" (PDF). Indian Journal of History of Science. 27 (1): 51–120. Archived from the original (PDF) on 2018-01-17. Retrieved 2018-01-16.
39. Evans, Brian (24 February 2014). The Development of Mathematics Throughout the Centuries: A Brief History in a Cultural Context. John Wiley & Sons. p. 61. ISBN 978-1118853979. Varāhamihira, a mathematician born around 505 CE and died 587 CE, who was also known for innovation with Pascal's triangle.
40. Gopal, Ram (1984). Kālidāsa: His Art and Culture. Concept Publishing Company. p. 15.

41. Hayashi, Takao (1987). "Varahamihira's Pandiagonal Magic Square of the Order Four" (PDF). Historia Mathematica. 14 (2): 159–166. doi:10.1016/0315-0860(87)90019-X.
42. History of Indian Literature. Motilal Banarsidass Publications. 2008. p. 46.
43. J. P. Hogendijk, A. I. Sabra, The Enterprise of Science in Islam: New Perspectives, Published by MIT Press, 2003, ISBN 0-262-19482-1, p. xv.
44. M. Srinivasachariar (1974). History of Classical Sanskrit Literature. Motilal Banarsidass. pp. 94–111. ISBN 9788120802841.
45. M.R. Bhatt 1996, p. 570.
46. M.R. Bhatt 1996, p. 572.
47. M.R. Bhatt 1996, p. 638.
48. McEvilley, Thomas (November 2001). The Shape of Ancient Thought: Comparative Studies in Greek and Indian Philosophies. Allworth Press. p. 385. ISBN 978-1-58115-203-6.
49. Pingree, David (1963). "Astronomy and Astrology in India and Iran". Isis. University of Chicago Press. 54 (2): 229–246. doi:10.1086/349703. JSTOR 228540. S2CID 128083594.
50. Pingree, David (1978). The Yavanajātaka of Sphujidhvaja. Harvard Oriental Series. Vol. 2. pp. 437–438.
51. Sarma, K. V. (2008). "Varahamihira". In Helaine Selin (ed.). Encyclopaedia of the History of Science, Technology, and Medicine in Non-Western Cultures. Dordrecht: Springer Netherlands. pp. 2184–2185. doi:10.1007/978-1-4020-4425-0_9604. ISBN 978-1-4020-4559-2.
52. T.K Puttaswamy (2012). Mathematical Achievements of Pre-modern Indian Mathematicians. p. 141. ISBN 9780123979131.
53. Velandai Gopala Aiyer. The chronology of ancient India: beginning of the Sat Yuga, Dwaper, Treta, and Kali Yuga with date of Mahabharata. Sanjay Prakashan. p. 63.
54. Winternitz, Moriz (1985). History of Indian Literature. Motilal Banarsidass. pp. 45–47. ISBN 978-81-208-0056-4.
55. Winternitz, Moriz (1985). History of Indian Literature. Motilal Banarsidass. pp. 685–697. ISBN 978-81-208-0056-4.
56. A.M. Shastri (1969). India as Seen in the Bṛhatsaṁhitā of Varāhamihira. Motilal Banarsidass. ISBN 9780896842212.
57. A.M. Shastri (1969). India as Seen in the Bṛhatsaṁhitā of Varāhamihira. Motilal Banarsidass. ISBN 9780896842212.

58. A.M. Shastri (1991). Varāhamihira and His Times. Kusumanjali. OCLC 28644897.
59. A.M. Shastri (1991). Varāhamihira and His Times. Kusumanjali. OCLC 28644897.
60. A.M. Shastri 1991, p. 3.
61. D G Dhavale, The Kapitthaka of Varahamihira, in Proceedings of the Symposium on Copernicus and Astronomy, New Delhi, 1973, Indian J. History Sci. 9 (1) (1974), 77-78; 141.
62. D G Dhavale, The Kapitthaka of Varahamihira, in Proceedings of the Symposium on Copernicus and Astronomy, New Delhi, 1973, Indian J. History Sci. 9 (1) (1974), 77-78; 141.
63. Dániel Balogh (2019). Inscriptions of the Aulikaras and Their Associates. De Gruyter. ISBN 9783110649789.
64. Dániel Balogh (2019). Inscriptions of the Aulikaras and Their Associates. De Gruyter. ISBN 9783110649789.
65. G Abraham, Theories for the motion of the Sun in the Pancasiddhantika, Arch. Hist. Exact Sci. 34 (3) (1985), 221-230.
66. G Abraham, Theories for the motion of the Sun in the Pancasiddhantika, Arch. Hist. Exact Sci. 34 (3) (1985), 221-230.
67. G Ifrah, A universal history of numbers: From prehistory to the invention of the computer (London, 1998).
68. G Ifrah, A universal history of numbers: From prehistory to the invention of the computer (London, 1998).
69. Ghayasuddin, Varahamihira, the best Sanskrit source of al-Biruni on Indian Jyotisa, in Proceedings of the Symposium on al-Biruni and Indian Sciences, New Delhi, 1971, Indian J. History Sci. 10 (2) (1975), 139-152;159-160.
70. Ghayasuddin, Varahamihira, the best Sanskrit source of al-Biruni on Indian Jyotisa, in Proceedings of the Symposium on al-Biruni and Indian Sciences, New Delhi, 1971, Indian J. History Sci. 10 (2) (1975), 139-152;159-160.
71. Hayashi, Takao (2008). "Magic Squares in Indian Mathematics". Encyclopaedia of the History of Science, Technology, and Medicine in Non-Western Cultures (2 ed.). Springer. pp. 1252–1259. doi:10.1007/978-1-4020-4425-0_9778. ISBN 978-1-4020-4559-2.
72. https://en.wikipedia.org/wiki/Var%C4%81hamihira
73. K S Shukla, The Pancasiddhantika of Varahamihira. I (errata insert), Ganita 24 (1) (1973), 59-73.
74. K S Shukla, The Pancasiddhantika of Varahamihira. I (errata insert), Ganita 24 (1) (1973), 59-73.

75. K S Shukla, The Pancasiddhantika of Varahamihira. I, in Proceedings of the Symposium on Copernicus and Astronomy, New Delhi, 1973, Indian J. History Sci. 9 (1) (1974) 62-76; 141.
76. K S Shukla, The Pancasiddhantika of Varahamihira. I, in Proceedings of the Symposium on Copernicus and Astronomy, New Delhi, 1973, Indian J. History Sci. 9 (1) (1974) 62-76; 141.
77. K S Shukla, The Panca-siddhantika of Varahamihira. II, Ganita 28 (1977), 99-116.
78. K S Shukla, The Panca-siddhantika of Varahamihira. II, Ganita 28 (1977), 99-116.
79. M. Ramakrishna Bhat, ed. (1996) [1982]. Brhat Samhita of Varahamihira (Second ed.). Motilal Banarsidass. ISBN 9788120810600.
80. M. Ramakrishna Bhat, ed. (1996) [1982]. Brhat Samhita of Varahamihira (Second ed.). Motilal Banarsidass. ISBN 9788120810600.
81. Magic square, https://en.wikipedia.org/wiki/Magic_square.
82. Neugebauer and D Pingree, The Pancasiddhantika of Varahamihira. I, Hist.-Filos. Skr. Danske Vid. Selsk. 6 (1) (1970).
83. Neugebauer and D Pingree, The Pancasiddhantika of Varahamihira. I, Hist.-Filos. Skr. Danske Vid. Selsk. 6 (1) (1970).
84. Neugebauer and D Pingree, The Pancasiddhantika of Varahamihira. II, Hist.-Filos. Skr. Danske Vid. Selsk. 6 (2) (1971).
85. Neugebauer and D Pingree, The Pancasiddhantika of Varahamihira. II, Hist.-Filos. Skr. Danske Vid. Selsk. 6 (2) (1971).
86. Pancasiddhantika, Brihat Jataka, Brihat Samhita and Hora Shastra Various editions in English and Sanskrit. (PDF), http://www.wilbourhall.org/index.html#panca
87. Pingree, David (1969). The Later Pauliśa Siddhānta. Centaurus 14. pp. 172–241.
88. R C Gupta, Varahamihira's calculation of ^C® and the discovery of Pascal's triangle, Ganita Bharati 14 (1-4) (1992), 45-49.
89. R C Gupta, Varahamihira's calculation of ^C® and the discovery of Pascal's triangle, Ganita Bharati 14 (1-4) (1992), 45-49.
90. Sudhakara Dvivedi, following the reading "Kampilyaka", identified Varahamihira's birthplace as present-day Kalpi in Uttar Pradesh. However, this identification is incorrect: the ancient name of Kalpi was Kalapriya, not Kampilyaka.
91. T Hayashi, Varahamihira's pandiagonal magic square of the order four, Historia Math. 14 (2) (1987), 159-166.

92. T Hayashi, Varahamihira's pandiagonal magic square of the order four, Historia Math. 14 (2) (1987), 159-166.
93. T S Kuppanna Sastry, The Vasistha-Paulisa Venus in the Pancasiddhantika of Varahamihira, Indian J. Hist. Sci. 14 (2) (1979), 150-154.
94. T S Kuppanna Sastry, The Vasistha-Paulisa Venus in the Pancasiddhantika of Varahamihira, Indian J. Hist. Sci. 14 (2) (1979), 150-154.
95. The Brihat Jataka (1905) (PDF) – archived from Wayback Machine, https://archive.org/details/brihatjataka00varaiala
96. The Brihat-samhita; complete translation by N. Chidambaram Iyer Online edition with glossary, https://www.wisdomlib.org/hinduism/book/brihat-samhita.
97. Varahamihira, https://mathshistory.st-andrews.ac.uk/Biographies/Varahamihira/.
98. Varahamihira's Pandiagonal Magic Square of the Order Four, https://core.ac.uk/download/pdf/82500954.pdf.

## Virahanka (विरहाङ्क)
## (6th century CE)

Virahanka (विरहाङ्क) was a *prosodist* (गद्यवादक) who is also known for his work on mathematics. He may have lived in the 6th century, but it is also possible that he worked as late as the 8th century. His work on prosody builds on the *Chhanda-sutras* (छन्दसूत्राणि) of *Pingala*, and was the basis for a 12th-century commentary by *Gopala* (गोपाल). *Pingala* had only briefly touched upon *Matra vrutta* (मातृवृत्ताय) while discussing *Arya* (आर्यं) and *Vaitaliya* (वैताल्यं) meters. *Matra vrutta* is more common in P*rakrit* and regional languages. *Virahanka* gave a comprehensive explanation of the *Prastaar* (प्रस्तार) and *Sankhya* (सांख्य) pratyays for matra vrutta in his *Prakrit* work *Vrattajati samuccaya* (वृत्तजातिसमुच्चय:). A more detailed explanation of matra vruttas is given in *Chandonusasana* (चन्दोनुसासन) of Hemacandra (12th cent CE) and the books *Prakrit Paingala* and *Vanibhusana* (वाणीभूषणम्) by *Damodara* – (दामोदर) (15th cent CE).

### 卐 Key Contributions in Mathematics

#### 卐 Matra Prastaar and Virahanka's Sankhyankas

How many poetic meters are possible for a quarter having $n$ matras? Can you enlist all of them?

*Virahanka* had answered these questions. He had developed the *Prastaar* (प्रस्तार प्रत्यय) and *Sankhya* pratyay (सांख्य प्रत्यय) to enlist all the possible meters having $n$ matras.

एष एव प्रस्तारो मात्रावृत्तानां साधितः किन्तु।
मात्रा यत्र न पूर्यति प्रथमं स्पर्शं तत्र देहि॥
– (वृत्तजातिसमुच्चयः ६.२०)

**Virahanka's Actual Problem**

Find number of poetic meters with 8 beats, made of:

- Short syllables, each 1 beat duration
- Long syllables, each 2 beats duration

**Example:** Shardulavikridita (19 syllables, 15 beats)!

L L L S S L S L S S S L L L S L L S L

*ya kun den du tu sha r ha r dha va la ya shubh r vas tra vru la*
*ya vee na va ra dan d man di ta k ra ya shwe t pad ma s na*
*Aa ji chya ja wa lii gha dya l ka sa lay aa he ch mat kaa ri ka*
*De ii the vu ni te ku the a ju n hii na hi ku na tha u ka*

**The algorithm to generate the Matra Prastaar:**

1. The first row consists of all Gs if n is even and an L followed by all Gs if n is odd.
2. Given any row in the prastaar, to generate the next row, scan from the left to identify the first G. Place an L below that. The elements to the right are brought down as they are.
3. The remaining matras to the left are filled in by all Gs, and by placing an L at the beginning, if need be, to keep the total number of matras the same.
4. The last row consists of all Ls.

**Explanation**

The *Matra prastaar* for a quarter having 1 matra has only 1 row which contains only 1 *Laghu* (L).

Table 2.1. *Matra prastaar* for a quarter having 1 *matra*.

| Row No. | Matra 1 |
|---|---|
| 1 | L |

Table 2.2. The matra prastaar will have 2 rows. The first row will have a G and the second row will contain 2 Ls.

Table 2.2. *Matra prastaar* for a quarter having 2 *matras*.

| Row No. | Matra 2 | Matra 1 |
|---|---|---|
| 1 | | G |
| 2 | L | L |

Table 2.3. Since n = 3 is an odd number the first row contains a L at the beginning.

Table 2.3. *Matra prastaar* for a quarter having 3 *matras*.

| Row No. | Matra 3 | Matra 2 | Matra 1 |
|---|---|---|---|
| 1 | | L | G |
| 2 | | G | L |
| 3 | L | L | L |

Table 2.4. Since n = 4 is an even number the first row will contain all Gs.

Table 2.4. *Matra prastaar* for a quarter having 4 *matras*.

| Row No. | Matra 4 | Matra 3 | Matra 2 | Matra 1 |
|---|---|---|---|---|
| 1 | | | G | G |
| 2 | | L | L | G |
| 3 | | L | G | L |
| 4 | | G | L | L |
| 5 | L | L | L | L |

Table 2.5. Since n = 5 is an odd number the first row will contain an L at the beginning.

Table 2.5. *Matra prastaar* for a quarter having 5 *matras*.

| Row No. | Matra 5 | Matra 4 | Matra 3 | Matra 2 | Matra 1 |
|---|---|---|---|---|---|
| 1 | | | L | G | G |
| 2 | | | G | L | G |
| 3 | | L | L | L | G |
| 4 | | | G | G | L |
| 5 | | L | L | G | L |
| 6 | | L | G | L | L |
| 7 | | G | L | L | L |
| 8 | L | L | L | L | L |

Table 2.66. Since n = 6 is an even number the first row will contain all Gs. Now let us closely observe and analyse the prastaar for $n = 6$ matras.

The last column of the first 5 rows (row number 1 to 5) contains all Gs. As we know the value of G is 2. Then the *matra* values of the remaining columns of each of these rows needs to add up to 4. It means that the remaining columns of these rows should be the prastaar for n = 4. Refer Table 2.4, you will see that it exactly matches the prastaar for n = 4.

The last column of row number 6 to 13 contains all Ls. As we know the value of L is 1. Then the *matra* values of the remaining columns of each of these rows needs to add up to 5. It means that the remaining columns of these rows should be the prastaar for n = 5. Refer Table 2.5, you will see that it exactly matches the prastaar for n = 5.

Let $S_n$ represent the number of all the possible poetic meters for a quarter having *n matras* or the number of rows in a *n-matra* prastaar. Then from the above discussion we have:

$$S_6 = S_5 + S_4$$

Table 2.6. *Matra prastaar* for a quarter having 6 *matras*.

| Row No. | Matra 6 | Matra 5 | Matra 4 | Matra 3 | Matra 2 | Matra 1 |
|---|---|---|---|---|---|---|
| 1 | | | | G | G | G |
| 2 | | | L | L | G | G |
| 3 | | | L | G | L | G |
| 4 | | | G | L | L | G |
| 5 | | L | L | L | L | G |
| 6 | | | L | G | G | L |
| 7 | | | G | L | G | L |
| 8 | | L | L | L | G | L |
| 9 | | | G | G | L | L |
| 10 | | L | L | G | L | L |
| 11 | | L | G | L | L | L |
| 12 | | G | L | L | L | L |
| 13 | L | L | L | L | L | L |

Let us consider a *prastaar* for a quarter having *n matras*. The rows in the *prastaar* end either with a G or an L. Now consider those rows that end with a G. Then the *matra* values of the remaining columns of each of these rows needs to add up to $(n-2)$. It means that the remaining columns of these rows should be the *prastaar* for $(n-2)$ matras. Now consider those rows that end with an L. Then the *matra* values of the remaining columns of each of these rows needs to add up to $(n-1)$. It means that the remaining columns of these rows should be the *prastaar* for $(n-1)$ *matras*.

Therefore, in general we can say that:

$$S_n = S_{n-1} + S_{n-2} \; for \; n > 2$$

This was explained by Virahanka in his Sankhya pratyay given below:

> द्वौ द्वौ पूर्वविकल्पौ या मेलयित्वा जायते सङ्ख्या।
> सा उत्तरमात्राणां सङ्ख्याया एष निर्देशः॥
>
> – (वृत्तजातिसमुच्चयः ६.४९)

Then we get the following sequence of *Virahanka's Sankhyankas* (सङ्ख्यांक)

| $n$ | 1 | 2 | 3 | 4 | 5 | 6 | 7 | 8 | 9 | ... |
|---|---|---|---|---|---|---|---|---|---|---|
| $S_n$ | 1 | 2 | 3 | 5 | 8 | 12 | 21 | 34 | 55 | ... |

As you might have already realised what is popularly known as *Fibonacci* sequence is actually the sequence of Virahanka's Sankhyankas.

> Every n-beat rhythm ends with either a 1 or a 2.

> Virahanka Sankhyanka is obtained by adding the previous two numbers.

**Further Reading**


1. Virahanka-fibonacci number, http://new-savanna.blogspot.com/2018/11/virahanka-fibonacci-numbers.html.
2. Virahanka numbers, https://chandrahasblogs.wordpress.com/2021/11/06/virahanka-numbers-part-i/.
3. Virahanka numbers proof, https://mathedu.hbcse.tifr.res.in/wp-content/uploads/2014/01/proof-without-wrords-ak1.pdf.
4. Virahanka numbers, https://www.cse.iitb.ac.in/~cs101/2012.2/resources/VirahankaNumbers.pdf.
5. Singh, Parmanand (1985). "The so-called fibonacci numbers in ancient and medieval India". Historia Mathematica. 12 (3): 229–244. doi:10.1016/0315-0860(85)90021-7. Retrieved 15 October, 2023.
6. The So-called Fibonacci Numbers in Ancient and Medieval India Archived 24 July 2019 at the Wayback Machine by Parmanand Singh, http://www.sfs.uni-tuebingen.de/~dg/sdarticle.pdf
7. Virahanka, https://en.wikipedia.org/wiki/Virahanka.

# Chapter 3

# Early Medieval Period
(521 CE – 1206 CE)

**Brahmagupta (ब्रह्मगुप्त)**
**(598–670 CE)**

Figure 3.1. Brahmagupta (ब्रह्मगुप्त). (Image Credit: https://mathshistory.st-andrews.ac.uk)

Brahmagupta (ब्रह्मगुप्त) was the foremost mathematician of his time (see Figure 3.1). He made advances in *astronomy* (खगोलशास्त्रम्) and most importantly in number systems including algorithms for square roots and the solution of quadratic equations. *Brahmagupta,* whose father was *Jisnugupta* (जिस्नुगुप्त), according to his own statement, was born in 598 CE. Born in *Bhillamāla* (भील्लामाला) in *Gurjaradesa* - गुर्जरादेश (modern Bhinmal in Rajasthan, Bharat) during the reign of the *Chavda dynasty* (चवडा वंश) *ruler* (शासक) *Vyagrahamukha* - व्याग्रहमुख, his ancestors were probably from *Sindh* (*Sindh*). He lived and worked there for a good part of his life.

*Bhillamala* was the capital of the *Gurjaradesa,* the second-largest kingdom of Western Bharat, comprising southern *Rajasthan* and northern *Gujarat* in modern-day Bharat. It was also a centre of learning for mathematics and astronomy. He became an astronomer of the *Brahmapaksha* (ब्रह्मपक्ष) school, one of the four major schools of Bhartiya astronomy during this period. He studied the five traditional *Siddhantas* on Bhartiya astronomy as well as the work of other astronomers including *Aryabhata I, Latadeva, Pradyumna, Varahamihira, Simha, Srisena, Vijayanandin and Vishnuchandra*. In the year 628, at the age of 30, he composed the *Brāhmasphuṭasiddhānta* (ब्राह्मस्फुटसिद्धान्त) ("*improved treatise of Brahma*") which is believed to be a revised version of the received *Siddhanta* of the *Brahmapaksha* school of astronomy.

The book consists of *24 chapters* with *1008 verses* in the *ārya* (आर्य) metre. A good deal of it is astronomy, but it also contains key chapters on mathematics, including algebra, geometry, trigonometry and algorithmics, which are believed to contain new insights due to *Brahmagupta* himself. Brahmagupta is credited with first clear description of the quadratic formula (the solution of the quadratic equation) in his main work, the *Brāhmasphuṭasiddhānta*. In 628 CE, Brahmagupta first described gravity as an attractive force, and used the term "*gurutvākarṣaṇam* (गुरुत्वाकर्षणम्)" in Sanskrit to describe it. Later, *Brahmagupta* moved to *Ujjaini*, Avanti, a major centre for astronomy in central Bharat. At the age of 67, he composed his next well-known work *Khanda-khādyaka* ("edible bite", dated 665), a practical manual of Bhartiya astronomy in the *karana* (*karana*) category meant to be used by students. *Brahmagupta* died in 668 CE, and he is presumed to have died in *Ujjain.*

The earth on all its sides is the same; all people on the earth stand upright, and all heavy things fall down to the earth by a law of nature, for it is the nature of the earth to attract and to keep things, as it is the nature of water to flow ... If a thing wants to go deeper down than the earth, let it try. The earth is the only low thing, and seeds always return to it, in whatever direction you may throw them away, and never rise upwards from the earth.

Brahmagupta in 628 first described gravity as an attractive force

Brahmagupta provided the solution of the quadratic equation.

## 卐 Key Contributions in Mathematics

### 卐 Zero, Positive and Negative Numbers

*Brāhmasphuṭasiddhānta* is one of the first books to provide concrete ideas on *positive numbers, negative numbers, and zero*. This text of mathematical astronomy contains significant mathematical content, including the first good understanding of the role of zero, rules for manipulating both negative and positive numbers, a method for computing square roots, methods of solving linear and quadratic equations, and rules for summing series, *Brahmagupta*'s identity, and Brahmagupta theorem. The book was written completely in verse and does not contain any kind of mathematical notation

*Brahmagupta*'s *Brāhmasphuṭasiddhānta* provides rules for arithmetic manipulations that apply to zero and to negative numbers. The *Brāhmasphuṭasiddhānta* is the earliest known text to treat zero as a number in its own right, rather than as simply a placeholder digit in representing another number as was done by the *Babylonians* or as a symbol for lack of quantity as was done by *Ptolemy* and the *Romans*. In *chapter eighteen* of his *Brāhmasphuṭasiddhānta, Brahmagupta* describes operations on negative numbers.

He first describes addition and subtraction:

> [The sum] of two positives is positives, of two negatives negative; of a positive and a negative [the sum] is their difference; if they are equal it is zero. The sum of a negative and zero is negative, [that] of a positive and zero positives, [and that] of two zeros zero.
> [...]
>
> Brahmasphuṭasiddhānta (18.30.)

> A negative minus zero is negative, a positive [minus zero] is positive; zero [minus zero] is zero. When a positive is to be subtracted from a negative or a negative from a positive, then it is to be added.
>
> Brahmasphuṭasiddhānta (18.32.)

He goes on to describe multiplication:

> The product of a negative and a positive is negative, of two negatives positive, and of positives positive; the product of zero and a negative, of zero and a positive, or of two zeros is zero.[20]
>
> Brahmasphuṭasiddhānta (18.33.)

But his description of division by zero differs from our modern understanding:

> A positive divided by a positive or a negative divided by a negative is positive; a zero divided by zero is zero; a positive divided by a negative is negative; a negative divided by a positive is [also] negative.
>
> Brahmasphuṭasiddhānta (18.34.)

> A negative or a positive divided by zero has that [zero] as its divisor, or zero divided by a negative or a positive [has that negative or positive as its divisor]. The square of a negative or positive is positive; [the square] of zero is zero. That of which [the square] is the square is [its] square root.
>
> Brahmasphuṭasiddhānta (18.35.)

Here *Brahmagupta* states that $\frac{0}{0} = 0$ where $a \neq 0$ he did not commit himself. His rules for arithmetic on negative numbers and zero are quite close to the modern understanding, except that in modern mathematics division by zero is left undefined.

*Brahmagupta*'s understanding of the number systems went far beyond that of others of the period. In the Brahmasphutasiddhanta he defined zero as the result of subtracting a number from itself. He gave some properties as follows:

*"When zero is added to a number or subtracted from a number, the number remains unchanged; and a number multiplied by zero becomes zero."*

He also gives arithmetical rules in terms of fortunes (positive numbers) and debts (negative numbers):

- *A debt minus zero is a debt.*
- *A fortune minus zero is a fortune.*
- *Zero minus zero is a zero.*
- *A debt subtracted from zero is a fortune.*
- *A fortune subtracted from zero is a debt.*
- *The product of zero multiplied by a debt or fortune is zero.*
- *The product of zero multipliedby zero is zero.*
- *The product or quotient of two fortunes is one fortune.*
- *The product or quotient of two debts is one fortune.*
- *The product or quotient of a debt and a fortune is a debt.*
- *The product or quotient of a fortune and a debt is a debt.*

Brahmagupta then tried to extend arithmetic to include division by zero:

- *Positive or negative numbers when divided by zero is a fraction the zero as denominator.*
- *Zero divided by negative or positive numbers is either zero or is expressed as a fraction with zero as numerator and the finite quantity as denominator.*
- *Zero divided by zero is zero.*

### 卐 Gomutrika Method for Multiplication

*Gomutrika* - गोमुत्रिका (sometimes translated as "*like the trajectory of a cow's urine*") is an intriguing title, possibly hinting at a flowing, stepwise progression down and to the right. Here's the breakdown:

**Example:**

Given:

```
 235
×264
------
```

1. Multiply 235 by the units digit (4):

```
 235
× 4
------
 940
```

2. Multiply 235 by the tens digit (6), and shift the result one place to the right:

```
 235
× 6
------
1410
```

3. Multiply 235 by the hundreds digit (2), and shift the result two places to the right:

```
 235
× 2
------
 470
```

4. Now, sum up the three products:

```
  470
 1410
+ 940
-------
62040
```

### 卐 Square Root

Another arithmetical result presented by *Brahmagupta* is his algorithm for computing square roots. This algorithm is discussed in where it is shown to be equivalent to the *Newton-Raphson* iterative formula (in S C Kak, The Brahmagupta algorithm for square rooting, *Ganita Bharati* 11 (1-4) (1989), 27-29.).

## 卐 Linear Equation

*Brahmagupta* gave the solution of the general linear equation in *chapter eighteen* of *Brahmasphuṭasiddhānta,*

*"The difference between rupas, when inverted and divided by the difference of the [coefficients] of the [unknowns], is the unknown in the equation. The rupas are [subtracted on the side] below that from which the square and the unknown are to be subtracted."*

which is a solution for the equation $bx + c = dx + e$ where there *rupas* refers to the constants $c$ and $e$. The solution given is equivalent to

$$x = \frac{(e - c)}{(b - d)}$$

He further gave two equivalent solutions to the general quadratic equation as below.

| |
|---|
| Diminish by the middle [number] the square-root of the rupas multiplied by four times the square and increased by the square of the middle [number]; divide the remainder by twice the square. [The result is] the middle [number].<br>Brahmasphuṭasiddhānta (18.44.) |
| Whatever is the square-root of the rupas multiplied by the square [and] increased by the square of half the unknown, diminished that by half the unknown [and] divide [the remainder] by its square. [The result is] the unknown.[20]<br>Brahmasphuṭasiddhānta (18.45.) |

which are, respectively, solutions for the equation $ax^2 + bx = c$ equivalent to,

$$x = \frac{\pm\sqrt{4ac + b^2} - b}{2a}$$

and

$$x = \frac{\pm\sqrt{ac + \frac{b^2}{4}} - \frac{b}{2}}{a}$$

He went on to solve systems of simultaneous indeterminate equations stating that the desired variable must first be isolated, and then the equation must be divided by the desired variable's coefficient.

In particular, he recommended using "the pulverizer" to solve equations with multiple unknowns.

> Subtract the colors different from the first color. [The remainder] divided by the first [color's coefficient] is the measure of the first. [Terms] two by two [are] considered [when reduced to] similar divisors, [and so on] repeatedly. If there are many [colors], the pulverizer [is to be used].
>
> Brahmasphuṭasiddhānta (18.51.)

Like the algebra of *Diophantus,* the algebra of *Brahmagupta* was syncopated. Addition was indicated by placing the numbers side by side, subtraction by placing a dot over the subtrahend, and division by placing the divisor below the dividend, similar to our notation but without the bar. Multiplication, evolution, and unknown quantities were represented by abbreviations of appropriate terms.

### 卐 Brahmagupta's π

In verse 40, he gives values of $\pi$.

> The diameter and the square of the radius [each] multiplied by 3 are [respectively] the practical circumference and the area [of a circle]. The accurate [values] are the square-roots from the squares of those two multiplied by ten. - Brahmasphuṭasiddhānta (12.40.)

So Brahmagupta uses 3 as a "practical" value of $\pi$ and $\sqrt{10} \approx 3.1622...$ as an "accurate" value of $\pi$, with an error less than 1%.

## 卐 Brahmagupta's Measurements and constructions

In some of the verses before verse 40, Brahmagupta gives constructions of various figures with arbitrary sides. He essentially manipulated right triangles to *produce isosceles triangles, scalene triangles, rectangles, isosceles trapezoids, isosceles trapezoids* with three equal sides, and a scalene *cyclic quadrilateral* (चक्रीय चतुर्भुज).

After giving the value of *pi,* he deals with the geometry of plane figures and solids, such as finding volumes and surface areas (or empty spaces dug out of solids). He finds the volume of rectangular prisms, pyramids, and the frustum of a square pyramid. He further finds the average depth of a series of pits. For the volume of a frustum of a pyramid, he gives the "*pragmatic*" value as the depth times the square of the mean of the edges of the top and bottom faces, and he gives the "*superficial*" volume as the depth times their mean area.

## 卐 Brahmagupta Theorem

In geometry, *Brahmagupta's* theorem states that if a *cyclic quadrilateral is orthodiagonal* (that is, has perpendicular diagonals), then the perpendicular to a side from the point of intersection of the diagonals always bisects the opposite side.

More specifically, let A, B, C and D be four points on a circle such that the lines AC and BD are perpendicular. Denote the intersection of AC and BD by M. Drop the perpendicular from M to the line BC, calling the intersection E. Let F be the intersection of the line EM and the edge AD. Then, the theorem states that F is the midpoint AD.

> The square-root of the sum of the two products of the sides and opposite sides of a non-unequal quadrilateral is the diagonal. The square of the diagonal is diminished by the square of half the sum of the base and the top; the square-root is the perpendicular [altitudes].
>
> \- Brahmasphuṭasiddhānta (12.23.)

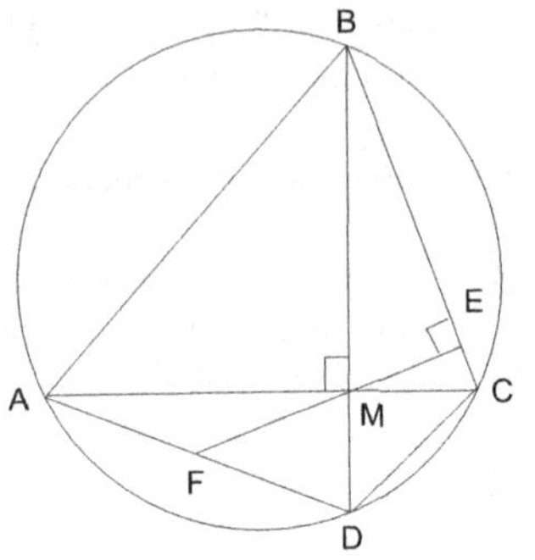


Figure 3.2. $\overline{BM} \perp \overline{AC}$, $\overline{EF} \perp \overline{BC}$, $\Rightarrow \overline{AF} \perp \overline{FD}$, *Brahmagupta*'s theorem states that $\overline{AF} = \overline{FD}$.

He continues to give formulas for the lengths and areas of geometric figures, such as the circumradius of an isosceles trapezoid and a scalene quadrilateral, and the lengths of diagonals in a scalene cyclic quadrilateral. This leads up to *Brahmagupta*'s famous theorem,

> Imaging two triangles within [a cyclic quadrilateral] with unequal sides, the two diagonals are the two bases. Their two segments are separately the upper and lower segments [formed] at the intersection of the diagonals. The two [lower segments] of the two diagonals are two sides in a triangle; the base [of the quadrilateral is the base of the triangle]. Its perpendicular is the lower portion of the [central] perpendicular; the upper portion of the [central] perpendicular is half of the sum of the [sides] perpendiculars diminished by the lower [portion of the central perpendicular].
>
> \- Brahmasphuṭasiddhānta (12.30-31.)

**Proof of the Theorem**

We need to prove that AF = FD. We will prove that both AF and FD are in fact equal to FM (see Figure 3.2).

To prove that AF = FM, first note that the angles FAM and CBM are equal, because they are inscribed angles that intercept the same arc of the circle.

Furthermore, the angles CBM and CME are both complementary to angle BCM (i.e., they add up to 90°), and are therefore equal. Finally, the angles CME and FMA are the same. Hence, AFM is an isosceles triangle, and thus the sides AF and FM are equal.

The proof that FD = FM goes similarly: the angles FDM, BCM, BME and DMF are all equal, so DFM is an isosceles triangle, so FD = FM. It follows that AF = FD, as the theorem claims.

### 卐 Brahmagupta's Sine Table

In *Chapter 2* of his *Brāhmasphuṭasiddhānta,* entitled *Planetary True Longitudes, Brahmagupta* presents a sine table:

> The sines: The Progenitors, twins; Ursa Major, twins, the Vedas; the gods, fires, six; flavors, dice, the gods; the moon, five, the sky, the moon; the moon, arrows, suns [...]
>
> Brahmasphuṭasiddhānta (2.2-5.)

Here *Brahmagupta* uses names of objects to represent the digits of place-value numerals, as was common with numerical data in Sanskrit treatises. Progenitors represents the 14 Progenitors ("*Manu*") in Bhartiya cosmology or 14, "twins" means 2, "*Ursa Major*" represents the seven stars of *Ursa Major* or 7, "*Vedas*" refers to the 4 Vedas or 4, dice represents the number of sides of the traditional die or 6, and so on. This information can be translated into the list of sines, 214, 427, 638, 846, 1051, 1251, 1446, 1635, 1817, 1991, 2156, 2312, 2459, 2594, 2719, 2832, 2933, 3021, 3096, 3159, 3207, 3242, 3263, and 3270, with the radius being 3270 (this numbers represent

$$3270 \sin\frac{\pi n}{48}$$

for $n = 1, 2, 3, \dots 24$.

## 卐 Brahmagupta's Identity

In algebra, *Brahmagupta*'s identity says that, for given $n$, the product of two numbers of the form $(a^2 + nb^2)$ is itself a number of that form. In other words, the set of such numbers is closed under multiplication. Specifically:

$$\begin{aligned}(a^2 + nb^2)(c^2 + nd^2) &= (ac - nbd)^2 + n(ad + bc)^2 \\ &= (ac + nbd)^2 + n(ad - bc)^2\end{aligned}$$

Both equations can be verified by expanding each side of the equation and can be obtained from each other by changing $b$ to $-b$.
This shows that, for any fixed A, the set of all numbers of the form

$$x^2 + Ay^2$$

is closed under multiplication. This identity holds in both the ring of integers and the ring of rational numbers, and more generally in any commutative ring.

For example,

$$(1^2 + 4^2)(2^2 + 7^2) = 26^2 + 15^2 = 30^2 + 1^2$$

The identity is also known as the *Diophantus identity*, as it was first proved by *Diophantus of Alexandria*. It is a special case of Euler's four-square identity, and also of *Lagrange*'s identity.

That identity was rediscovered by *Brahmagupta*, who generalized it and used it in his study of what is now called *Pell*'s equation. His *Brāhmasphuṭasiddhānta* was translated from Sanskrit into Arabic by Mohammad al-Fazari, and was subsequently translated into Latin in 1126. The identity later appeared in *Fibonacci*'s Book of Squares in 1225.

## 卐 Brahmagupta's Discovery to the Solution (Pell's equation)

In its original context, *Brahmagupta* applied his discovery to the solution of what was later called Pell's equation, namely

$$x^2 - Ny^2 = 1.$$

Using the identity in the form

$$(x_1^2 - Ny_1^2)(x_2^2 - Ny_2^2) = (x_1x_2 + Ny_1y_2)^2 - N(x_1y_2 + x_2y_1)^2$$

he was able to "compose" triples $(x_1, y_1, k_1)$ and $(x_2, y_2, k_2)$ that were solutions of
$x^2 - Ny^2 = k$, to generate the new triple

$$(x_1x_2 + Ny_1y_2,\ x_1y_2 + x_2y_1, k_1k_2).$$

Not only did this give a way to generate infinitely many solutions to $x^2 - Ny^2 = 1$ starting with one solution, but also, by dividing such a composition by $k_1k_2$, integer or "nearly integer" solutions could often be obtained. The general method for solving the *Pell equation* given by *Bhaskara* II in 1150, namely the *chakravala* - चक्रवला (cyclic) method, was also based on this identity.

## Brahmagupta's Problem

This problem was given in Bharat by the mathematician *Brahmagupta* in 628 AD in his treatise *Brāhmasphuṭasiddhānt*:

Solve the earlier discovery

$$x^2 - 92y^2 = 1$$

for integers $x, y > 1$

*Brahmagupta* gave the smallest solution as

$$(x, y) = (1151,120)$$

## Brahmagupta's Interpolation Formula

*Brahmagupta*'s interpolation formula is a second-order polynomial interpolation formula developed by Brahmagupta himself. The *Sanskrit couplet* describing the formula can be found in the supplementary part of *Khandakadyaka* a work of *Brahmagupta* completed in 665 CE. The same couplet appears in *Brahmagupta*'s earlier *Dhyana-graha-adhikara* (ध्यान-ग्रह-अधिकार), which was probably written *"near the beginning of the second quarter of the 7th century CE, if not earlier."* *Brahmagupta* was the one of the first to describe and use an interpolation formula using second-order differences.

*Brahmagupta*'s interpolation formula is equivalent to modern-day second-order *Newton–Stirling* interpolation formula.

| x | $x_2$ | ... | $x_r$ | $x_{r+1}$ | $x_n$ |
|---|---|---|---|---|---|
| Differences | $D_1$ | ... | $D_r$ | $D_{r+1}$ | $D_n$ |

Mathematicians prior to *Brahmagupta* used a simple linear interpolation formula. The linear interpolation formula to compute $f(a)$ is

$$f(a) = f_r + tD_r$$

where $t = \frac{(a-x_r)}{h}$

For the computation of $f(a)$, *Brahmagupta* replaces $D_r$ with another expression which gives more accurate values and which amounts to using a second-order interpolation formula.

In *Brahmagupta*'s terminology the difference $D_r$ is the *gatakhanda* (गतखण्ड) meaning past difference or the difference that was crossed over, the difference $D_{r+1}$ is the *bhogyakhanda* (भोग्यखण्ड) which is the difference yet to come. *Vikala* (विकला) is the amount in minutes by which the interval has been covered at the point where we want to interpolate. In the present notations it is $a - x_r$. The new expression which replaces $f_{r+1} - f_r$ is called *sphuta-bhogyakhanda* (स्फूत-भोग्यखण्ड).

This formula was originally stated for the computation of the values of the sine function for which the common interval in the underlying base table was 900 minutes or 15 degrees. So, the reference to 900 is in fact a reference to the common interval *h*.

**Modern Structure**

*Brahmagupta*'s method computation of *shuta-bhogyakhanda* (शुटा-भोग्यखण्ड) can be formulated in modern notation as follows:

$$sphuta - bhogyakhanda = \frac{D_r + D_{r-1}}{2} \pm t\frac{|D_r - D_{r-1}|}{2}$$

The $\pm$ sign is to be taken according to whether $\frac{1}{2}(D_r + D_{r+1})$ is less than or greater than $D_{r+1}$, , or equivalently, according to whether $D_r < D_{r+1}$ or $D_r > D_{r+1}$. *Brahmagupta*'s expression can be put in the following form:

$$sphuta - bhogyakhanda = \frac{D_r + D_{r-1}}{2} + t\frac{|D_r - D_{r-1}|}{2}$$

This correction factor yields the following approximate value for $f(a)$.

$$f(a) = f_r + t \times (sphuta - bhogyakhanda)$$

$$= f_r + t\frac{D_r+D_{r-1}}{2} + t^2\ \frac{D_r+D_{r-1}}{2}$$

This is now known as the *Stirling*'s interpolation formula truncated at the second-order differences. *Brahmagupta* has given a separate formula for the case where the values of the independent variable are not equally spaced.

गत भोग्य खण्डकान्तर दल विकल वधात् शतैर्नवभिराप्तैः ।
तद्युति दलं युतोनं भोग्यादूनाधिकं भोग्यम् ।।

- Dhyana-Graha-Upadesa-Adhyaya, 17; Khandaka Khadyaka, IX, 8

Multiply the vikala by the half the difference of the gatakhanda and the bhogyakhanda and divide the product by 900. Add the result to half the sum of the gatakhanda and the bhogyakhanda if their half-sum is less than the bhogyakhanda, subtract if greater. (The result in each case is sphuta-bhogyakhanda the correct tabular difference.)

Translated using Bhattolpala's (10th century CE) commentary

*Brahmagupta* discovered the *Stirling*'s interpolation formula.

### 卐 Brahmagupta's Formula

*Brahmagupta*'s formula is used to find the area of any *cyclic quadrilateral* (one that can be inscribed in a circle) given the lengths of the sides.

The approximate area is the product of the halves of the sums of the sides and opposite sides of a triangle and a quadrilateral. The accurate [area] is the square root from the product of the halves of the sums of the sides diminished by [each] side of the quadrilateral.

- Brahmasphuṭasiddhānta (12.21.)

*Brahmagupta*'s formula gives the area $K$ of a *cyclic quadrilateral* (a cyclic quadrilateral or inscribed *quadrilateral* (चतुर्भुज) (see Figure 3.3) is a quadrilateral whose vertices all lie on a single circle; this circle is called the circumcircle or circumscribed circle, and the vertices are said to be concyclic) whose sides have lengths $a, b, c, d$ as:

$$K = \sqrt{(s-a)(s-b)(s-c)(s-d)}$$

Where $s$ is the *superperimeter* (the semiperimeter of a polygon is half its perimeter), is defined as below:

$$s = \frac{a+b+c+d}{2}$$

This formula generalizes *Heron*'s formula for the area of a triangle. A triangle may be regarded as a quadrilateral with one side of length zero. From this perspective, as d approaches zero, a cyclic quadrilateral converges into a cyclic triangle (all triangles are cyclic), and *Brahmagupta*'s formula simplifies to *Heron*'s formula.

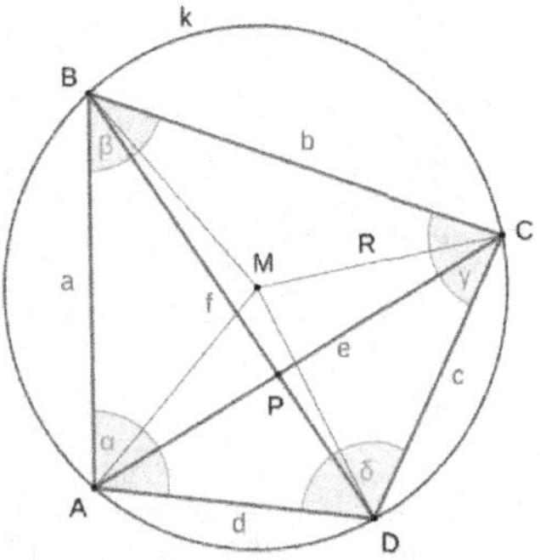


Figure 3.3 A cyclic quadrilateral ABCD. (Image Credit: https://en.wikipedia.org/wiki/)

$$K = \frac{1}{4}\sqrt{(-a+b+c+d)(a-b+c+d)(a+b-c+d)(a+b+c-d)}$$

Another equivalent version is

$$K = \frac{1}{4}\sqrt{(a^2+b^2+c^2+d^2)^2 + 8abcd - 2(a^4+b^4+c^4+d^4)}$$

Its generalized version, *Bretschneider*'s formula, can be used with non-cyclic quadrilateral.

Heron's formula can be thought as a special case of the *Brahmagupta*'s formula for triangles.

## 卐 Rate of Interest

Brahmagupta provided the formulation of rate of interest for financial transactions as below.

*"Five hundred drammas were loaned at an unknown rate of interest, The interest on the money for four months was loaned to another at the same rate of interest and amounted in ten mounths to 78 drammas. Give the rate of interest."*

## 卐 Brahmagupta's Astronomy

*Brahmagupta* directed a great deal of criticism towards the work of rival astronomers, and his *Brāhmasphuṭasiddhānta* displays one of the earliest schisms among Bhartiya mathematicians. The division was primarily about the application of mathematics to the physical world, rather than about the mathematics itself. In *Brahmagupta*'s case, the disagreements stemmed largely from the choice of astronomical parameters and theories. Critiques of rival theories appear throughout the *first ten* astronomical chapters and the eleventh chapter is entirely devoted to criticism of these theories, although no criticisms appear in the *twelfth* and *eighteenth* chapters.

Some of the important contributions made by *Brahmagupta* in astronomy are his methods for calculating the position of heavenly bodies over time (*ephemerides*), their rising and setting, conjunctions, and the calculation of solar and lunar eclipses.

In chapter seven of his *Brāhmasphuṭasiddhānta*, entitled *Lunar Crescent* (चन्द्र अर्धचन्द्र), *Brahmagupta* rebuts the idea that the Moon is farther from the Earth than the Sun. He does this by explaining the illumination of the Moon by the Sun.

- If the moon were above the sun, how would the power of waxing and waning, etc., be produced from calculation of the longitude of the moon? The near half would always be bright.
- In the same way that the half seen by the sun of a pot standing in sunlight is bright, and the unseen half dark, so is [the illumination] of the moon [if it is] beneath the sun.
- The brightness is increased in the direction of the sun. At the end of a bright [i.e. waxing] half-month, the near half is bright and the far half dark. Hence, the elevation of the horns [of the crescent can be derived] from calculation. [...]

He explains that since the Moon is closer to the Earth than the Sun, the degree of the illuminated part of the Moon depends on the relative positions of the Sun and the Moon, and this can be computed from the size of the angle between the two bodies.

Further work exploring the longitudes of the planets, diurnal rotation, lunar and solar eclipses, risings and settings, the moon's crescent and conjunctions of the planets, are discussed in his treatise *Khandakhadyaka*. The treatise contains eight chapters covering such topics as the longitudes of the planets, diurnal rotation, lunar and solar eclipses, risings and settings, the moon's crescent and conjunctions of the planets. The treatise also includes an appendix which in some versions has only one chapter, and in other has three. The treatise was written as a response to *Aryabhata*'s *Ardharatrikapaksa*.

**Further Reading**


1. Alberuni's India. London: Kegan Paul, Trench, Trübner & Co., 1910. Electronic reproduction. Vol. 1 and 2. New York: Columbia University Libraries, 2006. p. 272. Retrieved 3 June 2014.
2. Avari 2013, p. 32.
3. Ayyappappanikkar (1997). Medieval Indian Literature: Surveys and selections. Sahitya Akademi. p. 493. ISBN 978-81-260-0365-5.
4. Bhattacharyya 2011, p. 185.
5. Bhattacharyya 2011, p. 185: "Brahmagupta, one of the most celebrated mathematicians of the East, indeed of the world, was born in the year 598 CE, in the town of Bhillamala during the reign of King Vyaghramukh of the Chapa Dynasty."
6. Bhattacharyya 2011, pp. 185–186.
7. Bose, Mainak Kumar (1988). Late classical India. A. Mukherjee & Co.
8. Bose, Sen & Subbarayappa 1971.
9. Boyer (1991, "China and India" p. 221) "*he was the first one to give a general solution of the linear Diophantine equation ax + by = c, where a, b, and c are integers. [...] It is greatly to the credit of Brahmagupta that he gave all integral solutions of the linear Diophantine equation, whereas Diophantus himself had been satisfied to give one particular solution of an indeterminate equation. Inasmuch as Brahmagupta used some of the same examples as Diophantus, we see again the likelihood of Greek influence in India – or the possibility that they both made use of a common source, possibly from Babylonia. It is interesting to note also that the algebra of Brahmagupta, like that of Diophantus, was syncopated. Addition was indicated by juxtaposition, subtraction by placing a dot over the subtrahend, and division by placing the divisor below the dividend, as in our fractional notation but without the bar. The operations of multiplication and evolution (the taking of roots), as well as unknown quantities, were represented by abbreviations of appropriate words.*"
10. Boyer (1991, p. 220): However, here again Brahmagupta spoiled matters somewhat by asserting that $0 \div 0 = 0$, and on the touchy matter of $a \div 0$, he did not commit himself.
11. Bradley, Michael. The Birth of Mathematics: Ancient Times to 1300, p. 86 (Infobase Publishing 2006).
12. Bradley, Michael. The Birth of Mathematics: Ancient Times to 1300, p. 86 (Infobase Publishing 2006).

13. Brahmagupta; Bhāskara II (1817). Algebra, with Arithmetic and Mensuration, from the Sanscrit of Brahmegupta and Bháscara. Translated by Henry Thomas Colebrooke. John Murray. p. 319.
14. Cai 2023, p. 114; Cooke 1997, p. 208.
15. Coxeter, H. S. M.; Greitzer, S. L.: Geometry Revisited. Washington, DC: Math. Assoc. Amer., p. 59, 1967.
16. Daniel Shanks, Solved and unsolved problems in number theory, p.209, American Mathematical Society, Fourth edition 1993.
17. David Pingree, ed. (1970). Census of the Exact Sciences in Sanskrit Series A. Vol. 1. American Philosophical Society. p. 50.
18. David Pingree, ed. (1970). Census of the Exact Sciences in Sanskrit Series A. Vol. 1. American Philosophical Society. p. 54.
19. Gupta 2008, p. 162.
20. Gupta 2008, p. 163.
21. Henry Thomas Colebrooke. Algebra, with Arithmetic and Mensuration, from the Sanscrit of Brahmegupta and Bháscara, London 1817, p. 339 (online).
22. Hildebrand, Francis Begnaud (1987). Introduction to numerical analysis. Courier Dover Publications. pp. 138–139. ISBN 9780486653631.
23. J. L. Coolidge, "A Historically Interesting Formula for the Area of a Quadrilateral", American Mathematical Monthly, 46 (1939) pp. 345-347.
24. J. L. Coolidge, "A Historically Interesting Formula for the Area of a Quadrilateral", American Mathematical Monthly, 46 (1939) pp. 345-347.
25. Joseph (2000, pp.285–86).
26. Joseph 2000, p. 306.
27. Kaplan, Robert (1999). The Nothing That Is: A Natural History of Zero. London: Allen Lane/The Penguin Press. pp. 68–75. Bibcode:2000tnti.book.
28. Kaplan, Robert (1999). The Nothing that is: A Natural History of Zero. New York: Oxford University Press. pp. 68–75. ISBN 0-19-514237-3.
29. Kitāb al-Jawharatayn al-'atīqatayn al-mā'i'atayn min al-ṣafrā' wa-al-bayḍā': al-dhahab wa-al-fiḍḍah كتاب الجوهرتين العتيقتين المائعتين من الصفراء والبيضاء : الذهب والفضة. Cairo: Maṭba'at Dār al-Kutub wa-al-Wathā'iq al-Qawmīyah bi-al-Qāhirah. 2004. pp. 43–44, 87. OCLC 607846741.

30. Mackenzie, Dana. The Universe in Zero Words: The Story of Mathematics as Told through Equations, p. 61 (Princeton University Press, 2012).
31. Mackenzie, Dana. The Universe in Zero Words: The Story of Mathematics as Told through Equations, p. 61 (Princeton University Press, 2012).
32. Maley, F. Miller; Robbins, David P.; Roskies, Julie (2005). "On the areas of cyclic and semicyclic polygons". Advances in Applied Mathematics. 34 (4): 669–689. arXiv:math/0407300. doi:10.1016/j.aam.2004.09.008. S2CID 119565975.
33. Maley, F. Miller; Robbins, David P.; Roskies, Julie (2005). "On the areas of cyclic and semicyclic polygons". Advances in Applied Mathematics. 34 (4): 669–689. arXiv:math/0407300. doi:10.1016/j.aam.2004.09.008. S2CID 119565975.
34. Marc Chamberland: Single Digits: In Praise of Small Numbers. Princeton University Press, 2015, ISBN 9781400865697, p. 60
35. Meijering, Erik (March 2002). "A Chronology of Interpolation From Ancient Astronomy to Modern Signal and Image Processing". Proceedings of the IEEE. 90 (3): 319–321. doi:10.1109/5.993400.
36. Milne-Thomson, Louis Melville (2000). The Calculus of Finite Differences. AMS Chelsea Publishing. pp. 67–68. ISBN 9780821821077.
37. Pickover, Clifford (2008). Archimedes to Hawking: Laws of Science and the Great Minds Behind Them. Oxford University Press. p. 105. ISBN 978-0-19-979268-9.
38. Pingree, David E. (1970–1994). Pingree's Census of the Exact Sciences in Sanskrit. APS. pp. A4, 256 ff., A5, 239-240 et passim.
39. Plofker (2007, p. 419).
40. Plofker (2007, p. 420).
41. Plofker (2007, p. 423) Here the sums of the squares and cubes of the first n integers are defined in terms of the sum of the n integers itself.
42. Plofker (2007, p. 424) Brahmagupta does not explicitly state that he is discussing only figures inscribed in circles, but it is implied by these rules for computing their circumradius.
43. Plofker (2007, p. 426).
44. Plofker (2007, p. 427) After the geometry of plane figures, Brahmagupta discusses the computation of volumes and surface areas of solids (or empty spaces dug out of solids). His straightforward rules for the volumes of a rectangular prism and pyramid are followed by a more ambiguous one, which may refer to

finding the average depth of a sequence of puts with different depths. The next formula apparently deals with the volume of a frustum of a square pyramid, where the "pragmatic" volume is the depth times the square of the mean of the edges of the top and bottom faces, while the "superficial" volume is the depth times their mean area.
45. Plofker (2007, pp. 418–419).
46. Plofker (2007, pp. 419–420) Brahmagupta discusses the illumination of the moon by the sun, rebutting an idea maintained in scriptures: namely, that the moon is farther from the earth than the sun is. In fact, as he explains, because the moon is closer the extent of the illuminated portion of the moon depends on the relative positions of the moon and the sun, and can be computed from the size of the angular separation $\alpha$ between them.
47. Plofker (2007, pp. 419–420) Brahmagupta's sine table, like much other numerical data in Sanskrit treatises, is encoded mostly in concrete-number notation that uses names of objects to represent the digits of place-value numerals, starting with the least significant. [...].
48. Plofker (2007, pp. 421–427).
49. Plofker (2007, pp. 422) The reader is apparently expected to be familiar with basic arithmetic operations as far as the square-root; Brahmagupta merely notes some points about applying them to fractions. The procedures for finding the cube and cube-root of an integer, however, are described (compared the latter to Aryabhata's very similar formulation). They are followed by rules for five types of combinations: [...].
50. Plofker (2007, pp. 428–434).
51. Raju, C K (2007). Cultural foundations of mathematics: the nature of mathematical proof and the transmission of the calculus from India to Europe in the 16th c. CE. Pearson Education India. pp. 138–140. ISBN 9788131708712.
52. Sachau, Edward C. (1910), Alberuni's India, Volume I, London: Kegan Paul, Trench and Trubner, p. 153 – via archive.org, Brahma-siddhānta, so-called from Brahman, composed by Brahmagupta, the son of Jishnu, from the town of Bhillamāla between Multān and Anhilwāra, 16 yojana from the latter place (?).
53. Sen, Amartya (2005). The Argumentative Indian. Allen Lane. p. 29. ISBN 978-0-7139-9687-6.
54. Stillwell (2004, p. 77).
55. Stillwell (2004, pp. 44–46): In the seventh century CE the Indian mathematician Brahmagupta gave a recurrence relation for

generating solutions of x2 – Dy2 = 1, as we shall see in Chapter 5. The Indians called the Euclidean algorithm the "pulverizer" because it breaks numbers down to smaller and smaller pieces. To obtain a recurrence one has to know that a rectangle proportional to the original eventually recurs, a fact that was rigorously proved only in 1768 by Lagrange.

56. Stillwell (2004, pp. 72–74).
57. Stillwell 2002, p. 76.
58. Stillwell 2002, pp. 72–76.
59. Teresi, Dick (2002). Lost Discoveries: The Ancient Roots of Modern Science. Simon and Schuster. p. 135. ISBN 0-7432-4379-X.
60. Thurston, Hugh (1993). Early Astronomy. New York: Springer-Verlag. ISBN 978-0-387-94107-3.
61. Tiwari, Sarju (1992), Mathematics in History, Culture, Philosophy, and Science: From Ancient Times to Modern Age, Mittal Publications, pp. 91–, ISBN 978-81-7099-404-6.
62. van Bladel, Kevin (28 November 2014), "Eighth Century Indian Astronomy in the Two Cities of Peace", in Asad Q. Ahmed; Benham Sadeghi; Robert G. Hoyland (eds.), Islamic Cultures, Islamic Contexts: Essays in Honor of Professor Patricia Crone, BRILL, pp. 257–294, ISBN 978-90-04-28171-4.
63. Van Brummelen, Glen (2009). The mathematics of the heavens and the earth: the early history of trigonometry. Princeton University Press. p. 329. ISBN 9780691129730. (p.111).
64. Young, M. J. L.; Latham, J. D.; Serjeant, R. B. (2 November 2006), Religion, Learning and Science in the 'Abbasid Period, Cambridge University Press, pp. 302–303, ISBN 978-0-521-02887-5.
65. "Brahmagupta | Indian astronomer". Encyclopedia Britannica.
66. "Brahmagupta-Fibonacci Identity".
67. Al-Biruni's India (c. 1030).
68. Algebra, with Arithmetic and mensuration, from the Sanscrit of Brahmegupta and Bháscara at the Internet Archive, translated by Henry Thomas Colebrooke, https://archive.org/details/algebrawitharith00brahuoft.
69. Avari, Burjor (2013), Islamic Civilization in South Asia: A history of Muslim power and presence in the Indian subcontinent, Routledge, ISBN 978-0-415-58061-8.
70. B Chatterjee, Al-Biruni and Brahmagupta, Indian J. History Sci. 10 (2) (1975), 161-165.
71. B Datta, Brahmagupta, Bull. Calcutta Math. Soc. 22 (1930), 39-51.

72. Bhattacharyya, R. K. (2011), "Brahmagupta: The Ancient Indian Mathematician", in B. S. Yadav; Man Mohan (eds.), Ancient Indian Leaps into Mathematics, Springer Science & Business Media, pp. 185–192, ISBN 978-0-8176-4695-0.
73. Bose, D. M.; Sen, S. N.; Subbarayappa, B. V. (1971), A Concise History of Science in India, New Delhi: Indian National Academy of Science, pp. 95–97, archived from the original on 8 December 2015.
74. Boyer, Carl B. (1991), A History of Mathematics, John Wiley & Sons, Inc, ISBN 0-471-54397-7.
75. Brahmagupta formula, https://en.wikipedia.org/wiki/Brahmagupta%27s_formula.
76. Brahmagupta's Brahma-sphuta-siddhanta edited by Ram Swarup Sharma, Indian Institute of Astronomical and Sanskrit Research, 1966. English introduction, Sanskrit text, Sanskrit and Hindi commentaries (PDF), http://www.wilbourhall.org/index.html#BSS.
77. Brahmasphutasiddhanta at GRETIL (mathematical chapters: 12, 18-20, 21.17-23)
78. Cai, Tianxin (25 July 2023). A Brief History of Mathematics: A Promenade through the Civilizations of Our World. Springer Nature. ISBN 978-3-031-26841-0.
79. Categories: InterpolationIndian mathematicsHistory of mathematics
80. Cooke, Roger (1997), The History of Mathematics: A Brief Course, Wiley-Interscience, ISBN 0-471-18082-3.
81. D Pingree, Biography in Dictionary of Scientific Biography (New York 1970-1990).
82. David Pingree. Census of the Exact Sciences in Sanskrit (CESS). American Philosophical Society. A4, p. 254.
83. E R Suryanarayan, The Brahmagupta polynomials, Fibonacci Quart. 34 (1) (1996), 30-39.
84. G Ifrah, A universal history of numbers : From prehistory to the invention of the computer (London, 1998).
85. G S Bhalla, Brahmagupta's quadrilateral, Math. Comput. Ed. 20 (3) (1986), 191-196.
86. Gupta, R. C. "Second-order interpolation in Indian mathematics upto the fifteenth century". Indian Journal of History of Science. 4 (1 & 2): 86–98.
87. Gupta, Radha Charan (2008), "Brahmagupta", in Selin, Helaine (ed.), Encyclopaedia of the History of Science, Technology, and Medicine in Non-Western Cultures, Springer, pp. 162–163, ISBN 978-1-4020-4559-2.

88. H T Colebrooke, Algebra, with Arithmetic and Mensuration from the Sanscrit of Brahmagupta and Bhaskara (1817).
89. Hess, Albrecht, "A highway from Heron to Brahmagupta", Forum Geometricorum 12 (2012), 191–192.
90. Hess, Albrecht, "A highway from Heron to Brahmagupta", Forum Geometricorum 12 (2012), 191–192.
91. Hockey, Thomas, ed. (2007), "Brahmagupta", Biographical Encyclopedia of Astronomers, Springer Science & Business Media, p. 165, ISBN 978-0387304007
92. http://www.britannica.com/biography/Brahmagupta.
93. Brahmasphuta-siddhanta, https://en.wikipedia.org/wiki/Br%C4%81hmasphu%E1%B9%ADasiddh%C4%81nta
94. Brahmagupta, https://en.wikipedia.org/wiki/Brahmagupta.
95. Brahmagupta formula, https://en.wikipedia.org/wiki/Brahmagupta%27s_formula.
96. Brahmagupta interpolation formula, https://en.wikipedia.org/wiki/Brahmagupta%27s_interpolation_formula.
97. Brahmagupta Fibanacci identity, https://en.wikipedia.org/wiki/Brahmagupta%E2%80%93Fibonacci_identity.
98. Brahmagupta theorem, https://en.wikipedia.org/wiki/Brahmagupta_theorem.
99. Khandakhadyaka, https://en.wikipedia.org/wiki/Khandakhadyaka.
100. Brahmagupta, https://mathshistory.st-andrews.ac.uk/Biographies/Brahmagupta/.
101. J Pottage, The mensuration of quadrilaterals and the generation of Pythagorean triads : a mathematical, heuristical and historical study with special reference to Brahmagupta's rules, Arch. History Exact Sci. 12 (1974), 299-354.
102. Joseph, George G. (2000), The Crest of the Peacock, Princeton University Press, ISBN 0-691-00659-8.
103. Joseph, George G. (2000), The Crest of the Peacock: The Non-European Roots of Mathematics (2nd ed.), Princeton University Press, p. 306, ISBN 978-0-691-00659-8.
104. K Elfering, Die negativen Zahlen und die Rechenregeln mit ihnen bei Brahmagupta, in Mathemata, Boethius Texte Abh. Gesch. Exakt. Wissensch. XII (Wiesbaden, 1985, 83-86.

105.Michael John Bradley (2006). The Birth of Mathematics: Ancient Times to 1300. Publisher Infobase Publishing. ISBN 0816054231. Page 70, 85.
106.O'Connor, John J.; Robertson, Edmund F., "Brahmagupta", MacTutor History of Mathematics Archive, University of St Andrews.
107.O'Leary, De Lacy (2001) [first published 1948], How Greek Science Passed to the Arabs (2nd ed.), Goodword Books, ISBN 8187570245/
108.P K Majumdar, A rationale of Brahmagupta's method of solving ax + c = by, Indian J. Hist. Sci. 16 (2) (1981), 111-117.
109.Plofker, Kim (2007), "Mathematics in India", in Victor Katz (ed.), The Mathematics of Egypt, Mesopotamia, China, India, and Islam: A Sourcebook, Princeton University Press, ISBN 978-0-691-11485-9/
110.R C Gupta, Brahmagupta's formulas for the area and diagonals of a cyclic quadrilateral, Math. Education 8 (1974), B33-B36.
111.R C Gupta, Brahmagupta's rule for the volume of frustum-like solids, Math. Education 6 (1972), B117-B120.
112.R C Gupta, Munisvara's modification of Brahmagupta's rule for second order interpolation, Indian J. Hist. Sci. 14 (1) (1979), 66-72.
113.S C Kak, The Brahmagupta algorithm for square rooting, Ganita Bharati 11 (1-4) (1989), 27-29.
114.S Jha, A critical study on 'Brahmagupta and Mahaviracharya and their contributions in the field of mathematics', Math. Ed. (Siwan) 12 (4) (1978), 66-69.
115.S P Arya, On the Brahmagupta- Bhaskara equation, Math. Ed. 8 (1) (1991), 23-27.
116.S S Prakash Sarasvati, A critical study of Brahmagupta and his works : The most distinguished Indian astronomer and mathematician of the sixth century A.D. (Delhi, 1986).
117.Setsuro Ikeyama (2003). Brāhmasphuṭasiddhānta (Chapter 21) of Brahmagupta with Commentary of Pṛthūdhakasvāmin, critically edited with English translation and notes. INSA.
118.Shashi S. Sharma. Mathematics & Astronomers of Ancient India. Pitambar Publishing. ISBN 9788120914216.
119.Stillwell, John (2002), Mathematics and its history (2nd ed.), Springer, pp. 72–76, ISBN 978-0-387-95336-6
120.Stillwell, John (2004), Mathematics and its History (Second ed.), Springer Science + Business Media Inc., ISBN 0-387-95336-1/
121.T Kusuba, Brahmagupta's sutras on tri- and quadrilaterals, Historia Sci. 21 (1981), 43-55.

122. There are fourteen Progenitors ("Manu") in Indian cosmology; "twins" of course stands for 2; the seven stars of Ursa Major (the "Sages") for 7, the four Vedas, and the four sides of the traditional dice used in gambling, for 6, and so on. Thus Brahmagupta enumerates his first six sine-values as 214, 427, 638, 846, 1051, 1251. (His remaining eighteen sines are 1446, 1635, 1817, 1991, 2156, 2312, 2459, 2594, 2719, 2832, 2933, 3021, 3096, 3159, 3207, 3242, 3263, 3270). The Paitamahasiddhanta, however, specifies an initial sine-value of 225 (although the rest of its sine-table is lost), implying a trigonometric radius of R = 3438 approx= C(')/2π: a tradition followed, as we have seen, by Aryabhata. Nobody knows why Brahmagupta chose instead to normalize these values to R = 3270.

123. Thomas F. Glick; Steven Livesey; Faith Wallis, eds. (2014). Medieval Science, Technology, and Medicine: An Encyclopedia. Routledge. p. 464. ISBN 978-1135459390.

124. Weisstein, Eric W. "Brahmagupta's Problem". MathWorld.

## Bhāskara I (भास्कर I)
## (600–680 CE)

We have very little information about Bhāskara I's life except what can be deduced from his writings. *Bhāskara* I often refers to the *Asmakatantra* instead of the *Aryabhatiya* that he must have been working in a school of mathematicians in *Asmak*a which was probably in the *Nizamabad* District of Andhra Pradesh. If this is correct, and it does seem quite likely, then the school in *Asmaka* would have been a collection of scholars who were followers of *Aryabhata* I and of course these fits in well with the fact that *Bhaskara* I himself was certainly a follower of *Aryabhata I*. He was probably an astronomer who received his astronomical education from his father. This commentary, Āryabhaṭīyabhāṣya (आर्यभटीयभाष्य), written in 629 CE, is among the oldest known prose works in Sanskrit on mathematics and astronomy. He also wrote two astronomical works in the line of *Aryabhat*a's school: the *Mahābhāskarīya* (महाभास्करीय) ("*Great Book of Bhāskara*") and the *Laghubhāskarīya* (लघुभास्करीय) ("*Small Book of Bhāskara*").

First to write numbers in the Hindu–Arabic decimal system with a circle for the zero.

Gave a unique and remarkable rational approximation of the sine function.

### 卐 Key Contributions in Mathematics

### 卐 Sine Approximation Formula

Sine approximation formula is a rational expression in one variable for the computation of the approximate values of the trigonometric sines. This formula is given in his treatise titled *Mahābhāskarīya*. The formula is elegant and simple, and it enables the computation of reasonably accurate values of trigonometric sines without the use of geometry.

(Now) I briefly state the rule (for finding the bhujaphala and the kotiphala, etc.) without making use of the Rsine-differences 225, etc. Subtract the degrees of a bhuja (or koti) from the degrees of a half circle (that is, 180 degrees). Then multiply the remainder by the degrees of the bhuja or koti and put down the result at two places. At one place subtract the result from 40500. By one-fourth of the remainder (thus obtained), divide the result at the other place as multiplied by the anthyaphala (that is, the epicyclic radius). Thus is obtained the entire bahuphala (or, kotiphala) for the sun, moon or the star-planets. So also are obtained the direct and inverse Rsines.

- Mahābhāskarīya (Chapter VII, verses 17-19)

Translation

The sine approximation formula can be expressed using the radian measure of angles as follows:

$$\sin x \approx \frac{16x(\pi - x)}{5\pi^2 - 4x(\pi - x)} \quad for\ 0 \leq x \leq \pi$$

The formula is applicable for values of x° in the range from 0° to 180°. The formula is remarkably accurate in this range. The graphs of sin x and the approximation formula are visually indistinguishable and are nearly identical. The maximum relative error is less than 1.8%.

For a positive integer $n$ this takes the following form:

$$\sin\frac{\pi}{n} \approx \frac{16x(n-1)}{5n^2 - 4n + 4}$$

The formula acquires an even simpler form when expressed in terms of the cosine rather than the sine. Using radian measure for angles from $-\frac{\pi}{2}\ to + \frac{\pi}{2}$ and putting $x = \frac{1}{2}\pi + y$, one gets

$$\cos y \approx \frac{\pi^2 - 4y^2}{\pi^2 + y^2}$$

To express the previous formula with the constant $\tau = 2\pi$, one can use

$$\cos y \approx 1 - \frac{20y^2}{20y^2 + \tau^2}$$

Equivalent forms of *Bhāskara* I's formula have been given by almost all subsequent astronomers and mathematicians of Bharat. For example, *Brahmagupta*'s (598–668 CE) *Brāhmasphuṭasiddhānta* (verses 23–24, chapter XIV) gives the formula in the following form:

$$\mathrm{R} \sin \mathrm{x}^0 \approx \frac{Rx(180 - x)}{10125 - \frac{1}{4}x(180 - x)}$$

卐 **Representation of numbers**

The most important mathematical contribution of *Bhāskara I* concerns the representation of numbers in a positional numeral system. The first positional representations had been known to Bhartiya astronomers approximately 500 years before *Bhāskara's* work. However, these numbers were written not in figures, but in words or allegories and were organized in verses.

For instance, the number *1 was given as moon,* since it exists only once; the *number 2 was represented by wings, twins, or eyes* since they always occur in pairs; the *number 5 was given by the (5) senses*. Similar to our current decimal system, these words were aligned such that each number assigns the factor of the power of ten corresponding to its position, only in reverse order: the higher powers were right from the lower ones.

*Bhāskara*'s numeral system was truly positional, in contrast to word representations, where the same word could represent multiple values (such as 40 or 400). He often explained a number given in his numeral system by stating *ankair api* - अङ्कैर् अपि ("*in figures this reads*"), and then repeating it written with the first nine *Brahmi* numerals, using a small circle for the zero. Contrary to the word system, however, his numerals were written in descending values from left to right, exactly as we do it today. Therefore, since at least 629, the decimal system was definitely known to the Bhartiya scientists. Presumably, *Bhāskara* did not invent it, but he was the first to openly use the Brahmi numerals in a scientific contribution in Sanskrit.

## Further Reading


1. "Bhāskara i | Famous Indian Mathematician and Astronomer". Cuemath. 28 September 2020. Retrieved 3 September 2022.
2. "Bhāskara". Nasa Space Science Data Coordinated Archive. Retrieved 16 September 2017.
3. B. van der Waerden: Erwachende Wissenschaft. Ägyptische, babylonische und griechische Mathematik. Birkäuser-Verlag Basel Stuttgart 1966 p. 90.
4. George Gheverghese Joseph (2009). A passage to infinity: Medieval Indian mathematics from Kerala and its impact. New Delhi: SAGE Publications India Pvt. Ltd. ISBN 978-81-321-0168-0. (p.60).
5. Glen Van Brummelen (2009). The mathematics of the heavens and the earth: the early history of trigonometry. Princeton University Press. ISBN 978-0-691-12973-0. (p.104)
6. Hayashi, Takao (1 July 2019). "Bhāskara I". Encyclopedia Britannica. Retrieved 12 December 2022.
7. Keller (2006a, p. xiii).
8. Keller (2006a, p. xiii) cites [K S Shukla 1976; p. xxv-xxx], and Pingree, Census of the Exact Sciences in Sanskrit, volume 4, p. 297.
9. O'Connor, J. J.; Robertson, E. F. "Bhāskara I – Biography". Maths History. School of Mathematics and Statistics, University of St Andrews, Scotland, UK. Retrieved 5 May 2021.
10. R.C. Gupta (1967). "Bhāskara I's approximation to sine" (PDF). Indian Journal of History of Science. 2 (2). Archived from the original (PDF) on 16 March 2012. Retrieved 20 April 2010.
11. "Bhāskara I". Encyclopedia.com. Complete Dictionary of Scientific Biography. 30 November 2022. Retrieved 12 December 2022.
12. A Mukhopadhyay and M R Adhikari, A step towards incommensurability of $\pi$ and Bhaskara I : An episode of the sixth century AD, Indian J. Hist. Sci. 33 (2) (1998), 119-129.
13. A Mukhopadhyay and M R Adhikari, The concept of cyclic quadrilaterals: its origin and development in India (from the age of Sulba Sutras to Bhaskara I, Indian J. Hist. Sci. 32 (1) (1997), 53-68.
14. Bhaskara_I, https://en.wikipedia.org/wiki/Bh%C4%81skara_I.
15. Bhaskara_I, https://mathshistory.st-andrews.ac.uk/Biographies/Bhaskara_I/.
16. D Pingree, Biography in Dictionary of Scientific Biography (New York 1970-1990).

17. G. Ifrah: The Universal History of Numbers. John Wiley & Sons, New York 2000 ISBN 0-471-39340-1.
18. H.-W. Alten, A. Djafari Naini, M. Folkerts, H. Schlosser, K.-H. Schlote, H. Wußing: 4000 Jahre Algebra. Springer-Verlag Berlin Heidelberg 2003 ISBN 3-540-43554-9, §3.2.1.
19. https://en.wikipedia.org/wiki/Bh%C4%81skara_I%27s_sine_approximation_formula
20. I I Zaidullina, Bhaskara I and his work (Russian), Istor. Metodol. Estestv. Nauk No. 36 (1989), 45-49.
21. J J O'Connor and E F Robertson (November 2000). "Bhāskara I". School of Mathematics and Statistics University of St Andrews, Scotland. Archived from the original on 23 March 2010. Retrieved 22 April 2010.
22. K S Shukla, Hindu mathematics in the seventh century as found in Bhaskara I's commentary on the Aryabhatiya II, Ganita 22 (2) (1971), 61-78.
23. K S Shukla, Hindu mathematics in the seventh century as found in Bhaskara I's commentary on the Aryabhatiya III, Ganita 23 (1) (1972), 57-79
24. K S Shukla, Hindu mathematics in the seventh century as found in Bhaskara I's commentary on the Aryabhatiya IV, Ganita 23 (2) (1972), 41-50.
25. K S Shukla, Hindu mathematics in the seventh century as found in Bhaskara I's commentary on the Aryabhatiya, Ganita 22 (1) (1971), 115-130.
26. K Shankar Shukla, Bhaskara I, Bhaskara I and his works II. Maha-Bhaskariya (Sanskrit) (Lucknow, 1960).
27. K Shankar Shukla, Bhaskara I, Bhaskara I and his works III. Laghu-Bhaskariya (Sanskrit) (Lucknow, 1963).
28. K. S. Shukla. Āryabhaṭīya of Āryabhaṭa, with the commentary of Bhāskara I and Someśvara. Indian National Science Academy (INSA), New- Delhi, 1999.
29. K. S. Shukla. Laghubhāskarīya, Edited and Translated into English, with Explanatory and Critical Notes, and Comments, etc., Department of mathematics and astronomy, Lucknow University, 2012.
30. K. S. Shukla. Mahābhāskarīya, Edited and Translated into English, with Explanatory and Critical Notes, and Comments, etc. Department of mathematics, Lucknow University, 1960.

31. K. Stroethoff, Bhāskara's approximation for the sine, The Mathematics Enthusiast, Vol. 11, No. 3 (2014), 485–492.
32. Keller, Agathe (2006a), Expounding the Mathematical Seed. Vol. 1: The Translation: A Translation of Bhāskara I on the Mathematical Chapter of the Aryabhatiya, Basel, Boston, and Berlin: Birkhäuser Verlag, 172 pages, ISBN 3-7643-7291-5.
33. Keller, Agathe (2006b), Expounding the Mathematical Seed. Vol. 2: The Supplements: A Translation of Bhāskara I on the Mathematical Chapter of the Aryabhatiya, Basel, Boston, and Berlin: Birkhäuser Verlag, 206 pages, ISBN 3-7643-7292-3.
34. M. C. Apaṭe. The Laghubhāskarīya, with the commentary of Parameśvara. Anandāśrama, Sanskrit series no. 128, Poona, 1946.
35. O'Connor, John J.; Robertson, Edmund F., "Bhāskara I", MacTutor History of Mathematics Archive, University of St Andrews.
36. P K Majumdar, A rationale of Bhaskara I's method for solving ax ± c = by, Indian J. Hist. Sci. 13 (1) (1978), 11-17.
37. P K Majumdar, A rationale of Bhatta Govinda's method for solving the equation ax - c = by and a comparative study of the determination of 'Mati' as given by Bhaskara I and Bhatta Govinda, Indian J. Hist. Sci. 18 (2) (1983), 200-205.
38. R C Gupta, Bhaskara I's approximation to sine, Indian J. History Sci. 2 (1967), 121-136.
39. R C Gupta, On derivation of Bhaskara I's formula for the sine, Ganita Bharati 8 (1-4) (1986), 39-41.
40. S. Gottwald, H.-J. Ilgauds, K.-H. Schlote (Hrsg.): Lexikon bedeutender Mathematiker. Verlag Harri Thun, Frankfurt a. M. 1990 ISBN 3-8171-1164-9.
41. T Hayashi, A note on Bhaskara I's rational approximation to sine, Historia Sci. No. 42 (1991), 45-48.
42. T. Hayashi, A note on Bhāskara I's rational approximation to sine, Historia Sci. No. 42 (1991), 45–48.
43. v.harish Mahābhāskarīya of Bhāskarācārya with the Bhāṣya of Govindasvāmin and Supercommentary Siddhāntadīpikā of Parameśvara. Madras Govt. Oriental series, no. cxxx, 1957.

**Śrīdhara (श्रीधर)**
**(between 650–850 CE)**

Figure 3.4 Śrīdhara (श्रीधर). (Image Credit: https://www.ajabgajabfacts.com/wp-content/uploads/2021/09/sridharacharya-biography-in-hindi.jpg)

Śrīdhara (श्रीधर) or Sridharacharya, or Śrīdharācāryya (श्रीधराचार्य) was a *Sanskrit pandit* (संस्कृत पण्डित) and *philosopher* (दार्शनिक) (seem Figure 3.4). Two theories exist concerning his birthplace which are far apart. Some historians give southern Bharat as the place of his birth while other historians believe that Śrīdhara was born in *Bhuriśreṣṭi* (Bhurisriṣṭi or Bhurśuṭ) village in South Rādha at present day Hugli in West Bengal, then undivided Bengal with its Capital at Gaur. His father's name was *Baladevācārya* (बलदेवाचार्य) *or Baladeva Acharya* (बलदेव आचार्य) and his mother's name was *Acchoka Devi* (अच्चोका देवी). His father was a Sanskrit pandit.

*Śrīdhara* is known as the author of two mathematical treatises, namely the *Triśatika* (त्रिसाटिका) (sometimes called the *Pāṭīgaṇitasāra* - पटिगनीतासार) and the Patiganita (पाटिगनीत). However at least three other works have been attributed to him, namely the *Bijaganita* (बीजगणित), *Navasati* (नवसत), *and Brhatpati* (बृहत्पति). Information about these books was given the works of

*Bhaskara II* (writing around 1150), *Makkibhatta* – मक्कीभट्ट (writing in 1377), and *Raghavabhatta* – राघवभट्ट (writing in 1493). There is another mathematical treatise *Ganitapancavimsi* (गणितापञ्चविम्) which some historians believe was written by Śrīdhara

## 卐 Key Contributions in Mathematics

### 卐 Patiganita

The *Patiganita* is written in *verse form* (छन्दस्वरूप). The book begins by giving tables of monetary and metrological units. Following this algorithm are given for carrying out the elementary arithmetical operations, squaring, cubing, and square and cube root extraction, carried out with natural numbers. Through the whole book *Śrīdhara* gives methods to solve problems in terse rules in verse form which was the typical style of Bhartiya texts at this time. All the algorithms to carry out arithmetical operations are presented in this way and no proofs are given. Indeed, there is no suggestion that *Śrīdhara* realised that proofs are in any way necessary. Often after stating a rule *Śrīdhara* gives one or more numerical examples, but he does not give solutions to these examples nor does he even give answers in this work.

After giving the rules for computing with natural numbers, *Śrīdhara* gives rules for operating with rational fractions. He gives a wide variety of applications including problems involving ratios, barter, simple interest, mixtures, purchase and sale, rates of travel, wages, and filling of cisterns. Some of the examples are decidedly non-trivial and one has to consider this as a really advanced work. Other topics covered by the author include the rule for calculating the number of combinations of $n$ things taken m at a time. There are sections of the book devoted to arithmetic and geometric progressions, including progressions with a fractional number of terms, and formulae for the sum of certain finite series are given.

The book ends by giving rules, some of which are only approximate, for the areas of some plane polygons. In fact, the text breaks off at this point but it certainly was not the end of the book which is missing in the only copy of the work which has survived. We do know something of the missing part, however, for the *Patiganitasara* (पटिगनीतासार) or (sometimes called *Trisatika* (300)) is a summary of the *Patiganita* including the missing portion. It was written in *three hundred ślokas* thus his major work *Pāṭīgaṇitasāra* was named *Triśatika*. The book discusses counting of numbers, natural number, zero, measures, multiplication, fraction, division, squares, cubes, rule of three, interest-calculation, joint business or partnership, and mensuration (the main part of geometry concerned with ascertaining sizes, lengths, areas, and volumes).

## 卐 Notable Works

- He gave an exposition on the zero.

  He wrote:

  *"If zero is added to any number, the sum is the same number; if zero is subtracted from any number, the number remains unchanged; if zero is multiplied by any number, the product is zero".*

- In the case of dividing a fraction he has found out the method of multiplying the fraction by the reciprocal of the divisor.
- He wrote on the practical applications of algebra.
- He separated algebra from arithmetic.
- He was the one of the first to give an algorithm for solving quadratic equations (although there is no indication that he considered two solutions).

## 卐 *Śrīdhara*'s Rule for Solving Quadratic Equations

We give details below of *Śrīdhara*'s rule for solving quadratic equations as given by *Bhaskara II*. Unfortunately, as we indicated above, the original is lost and we have to rely on a quotation of *Śrīdhara*'s rule from *Bhaskara II*:

> Multiply both sides of the equation by a known quantity equal to four times the coefficient of the square of the unknown; add to both sides a known quantity equal to the square of the coefficient of the unknown; then take the square root.

To see what this means take

$$ax^2 + bx = c$$

Multiply both sides by $4a$ to get

$$4a^2x^2 + 4abx = 4ac$$

then adding $b^2$ to both sides to get

$$4a^2x^2 + 4abx + b^2 = 4ac + b^2$$

and, taking the square root

$$ax + b = \sqrt{(4ac + b^2)}$$

*Śrīdharācāryya's* formula is also known as the Quadratic formula or *Śrīdharācāryya's* method. *Śrīdharācāryya* Method is used to find solutions to quadratic equations of the general form $ax^2 + bx + c = 0, a \neq 0$ and is given by

$$x = \frac{-b \pm \sqrt{(b^2 - 4ac)}}{2a}$$

## Further Reading


1. O'Connor, J J. "Sridhara". Mac Tutor. School of Mathematics and Statistics University of St Andrews, Scotland. Retrieved September 8, 2021.
2. A I Volodarskii, Mathematical treatise Patiganita by Sridhara (Russian), in 1966 Phys. Math. Sci. in the East (Russian) 'Nauka' (Moscow, 1966), 141-159.
3. A I Volodarskii, Notes on the treatise Patiganita by Sridhara (Russian), in 1966 Phys. Math. Sci. in the East (Russian) 'Nauka' (Moscow, 1966), 182-246.
4. B Datta, On the relation of Mahavira to Sridhara, Isis 17 (1932), 25-33.
5. D Pingree, Biography in Dictionary of Scientific Biography (New York 1970-1990).
6. G G Joseph, The crest of the peacock (London, 1991).
7. Ganitanand, On the date of Sridhara, Ganita Bharati 9 (1-4) (1987), 54-56.
8. K Shankar Shukla, On Sridhara's rational solution of Nx2+1=y^2, Ganita 1 (1950), 1-12.

9. K Shankar Shukla, The Patiganita of Sridharacarya (Lucknow, 1959).
10. O'Connor, John J.; Robertson, Edmund F., "Sridhara", MacTutor History of Mathematics Archive, University of St Andrews.
11. Sharma, Rahul. "शून्य की व्याख्या करने वाला विद्वान : श्रीधराचार्य". ajabgajabfacts.
12. Sridhara, https://en.wikipedia.org/wiki/Sridhara.
13. Sridhara, https://mathshistory.st-andrews.ac.uk/Biographies/Sridhara/.
14. T Hayashi, Sridhara's authorship of the mathematical treatise Ganitapancavimsi, Historia Sci. (2) 4 (3) (1995), 233-250.
15. The Date of Sridharacharya by S. Srikanta Sastri, The Jaina Antiquary (Jan, 1948).

**Lalla (लल्ला)**
**(c. 720–790 CE)**

Lalla (लल्ला) wrote mainly on the application of mathematics to astronomy. *Lalla*'s father was *Trivikrama Bhatta* (त्रिविक्रम भट्ट) and *Trivikrama*'s father, *Lalla*'s paternal grandfather, was named *Samba*. He lived in central Bharat, possibly in the *Lāṭa* region in modern south Gujarat. *Lalla* was known as being one of the leading Bhartiya astronomers of the eighth century. Only two of his works are currently thought to be extant.

## 卐 Key Contributions in Mathematics

### 卐 Astronomy

*Lalla*'s most famous work was entitled *Śiṣyadhīvṛddhidatantra* (शिष्यधिविधिदातन्त्रम्) ("*Treatise which expands the intellect of students*").
This major treatise was in two volumes:

- The *first volume*, on the computation of the positions of the planets, was in thirteen chapters and covered topics such as: mean longitudes of the planets; true longitudes of the planets; the three problems of *diurnal rotation; lunar eclipses; solar eclipses; syzygies; risings and settings*; the shadow of the moon; the moon's crescent; conjunctions of the planets with each other; conjunctions of the planets with the fixed stars; the patas of the moon and sun, and a final chapter in the first volume which forms a conclusion. In *Śiṣyadhīvṛddhidatantra, Lalla* uses Sanskrit numerical symbols.

> ... over the centuries, Sanskrit has lent itself admirably to the rules of prosody and versification. This explains why Indian astronomers [like Lalla] favoured the use of Sanskrit numerical symbols, based on a complex symbolism which was extraordinarily fertile and sophisticated, possessing as it did an almost limitless choice of synonyms.
>
> Translated by Ifrah

Major components of this volume covers following:

- On the Computation of the Positions of the Planets
- On the mean longitudes of the planets.
- On the true longitudes of the planets.
- On the three problems involving diurnal motion.
- On lunar eclipses.
- On solar eclipses.
- On the syzygies.
- On the heliacal settings and risings of the planets.
- On the shadow of the moon.
- On the lunar crescent.
- On planetary conjunctions.
- On conjunctions of the planets with the stars.
- On the pātas of the sun and moon.

❖ The *second volume* was on the sphere. In this volume *Lalla* examined topics such as: graphical representation; the *celestial sphere*; the principle of *mean motion*; the terrestrial *sphere*; motions and stations of the *planets; geography; erroneous knowledge; instruments*; and finally selected problems.

Major components of this part covers following:

- On graphical representations.
- On the construction of the celestial sphere.
- On the principles of mean motion.
- On the terrestrial sphere.
- On the motions and stations of the planets.
- On geography.
- On erroneous knowledge.
- On instruments.
- On certain (selected) problems

Despite writing the most famous treatise giving the views of *Aryabhata* I, *Lalla* did not accept his theory given in the *Aryabhatiya* that the earth rotated. *Lalla* argues in his commentary, like many other Bhartiya *astronomers* before him such as *Varāhamihira* and *Brahmagupta* that if the earth rotated then the speed would have to be such that one would have to ask how do the bees or birds flying in the sky come back to their nests? In fact, *Lalla* misinterpreted some of *Aryabhata I*'s statements about the rotating earth. One has to assume that the idea appeared so impossible to him that he just could not appreciate *Aryabhata I*'s arguments.

## 卐 Perpetual Motion

It is within the *Śiṣyadhīvṛddhidatantra* (the most extensive extant exposition of the views of the *Āryapaksa* – आर्यपक्ष school) that the earliest known description of *perpetual motion* (नित्यगति) is described. *Perpetual motion* is the motion of bodies that continues forever in an unperturbed system. A *perpetual motion* machine is a hypothetical machine that can do work infinitely without an external energy source. This kind of machine is impossible, as it would violate either the first or second law of *thermodynamics*, or both. These laws of *thermodynamics* apply regardless of the size of the system. For example, the motions and rotations of celestial bodies such as planets may appear perpetual, but are actually subject to many processes that slowly dissipate their kinetic energy, such as solar wind, interstellar medium resistance, gravitational radiation and thermal radiation, so they will not keep moving forever.

## 卐 Astrology

The other extant work by *Lalla* is the *Jyotiṣaratnakośa* (ज्योतिषरत्नकोष) ("*Treasury of Jewels*"), a treatise on *catarchic astrology* (कैटार्किक ज्योतिषशास्त्र). This work is one of the earliest known Sanskrit astrological works for determining *auspicious* (शुभं कालः) *and inauspicious times* (अशुभं कालः). No edition of this text has ever been published while the known manuscripts are incomplete.

Astrology at this time was based on astronomical tables and often the horoscopes allow one to identify the tables used. Some *Arabic* horoscopes were based on astronomical tables calculated in Bharat. The most frequently used tables were by *Aryabhata I*. *Lalla* improved on these tables and he produced a set of corrections for the Moon's longitude. One aspect of *Aryabhata I*'s work which *Lalla* did follow was his value of $\pi$.

Lalla uses

$$\pi = \frac{62832}{20000} = 3.1416$$

which is a value correct to the fourth decimal place.

*Lalla* also wrote a *commentary* (भाष्यम्) on *Khandakhadyaka* (खण्डखाद्यक), a work of *Brahmagupta*. *Lalla*'s commentary has not survived but there is another work on astrology by *Lalla* which has survived, namely the *Jyotiṣaratnakośa*. This was a very popular work which was the main one on the subject in Bharat for around 300 years.

**Further Reading**


1. Bracher, Katherine.
2. Plofker (2009, p. 318).
3. Plofker (2009, p. 321).
4. "Lalla." Complete Dictionary of Scientific Biography. Vol. 7. Detroit: Charles Scribner's Sons, 2008. 582-583. Gale Virtual Reference Library. Web. 3 June 2013.
5. B Chatterjee, A glimpse of Aryabhata's theory of rotation of earth, Indian J. History Sci. 9 (1) (1974), 51-55; 141/
6. Bracher, Katherine. The Biographical Encyclopedia Of Astronomers [Electronic Resource] / Edited By K. Bracher ... [Et Al.]. n.p.: New York ; London : Springer, 2006., 2006. UNIV OF OKLAHOMA LIBRARIES's Catalog. Web. 6 June 2013.
7. D Pingree, Biography in Dictionary of Scientific Biography (New York 1970-1990).
8. G Ifrah, A universal history of numbers : From prehistory to the invention of the computer (London, 1998).
9. Lalla, https://en.wikipedia.org/wiki/Lalla/.
10. Perpetual motion, https://en.wikipedia.org/wiki/Perpetual_motion/.
11. K. V. Sarma. "Lalla." Encyclopaedia of the History of Science, Technology, and Medicine in Non-Western Cultures, ed. Helaine Selin, Dordrecht: Kluwer Academic Publishers, 1997, p. 508.
12. Lalla, https://mathshistory.st-andrews.ac.uk/Biographies/Lalla/.
13. M Yano, Aryabhata's possible rebuttal to objections to his theory of the rotation of the Earth, Historia Sci. No. 19 (1980), 101-105.
14. O'Connor, John J.; Robertson, Edmund F., "Lalla", MacTutor History of Mathematics Archive, University of St Andrews.
15. Plofker, Kim (2009), Mathematics in India (book), Princeton University Press, ISBN 978-0-691-12067-6.
16. R C Gupta, Aryabhata I's value of π, Math. Education 7 (1973), B17-B20.

**Virasena (आचार्य विरासेन)**
**(792-853 CE)**

Figure 3.5 Acharya Virasena (आचार्य विरासेन). (Image Credit: https://en.wikipedia.org/wiki/)

Acharya Virasena (आचार्य विरासेन) also known as *Veerasena,* was a *Digambara monk* (दिगम्बर भिक्षु) and belonged to the lineage of *Acharya Kundakunda* (आचार्य कुण्डकुण्ड). He was a Jain philosopher and scholar. He was also known as a famous *orator* (वक्ता) and an accomplished poet. *Virasena* was proficient in *astrology, grammar, logic, mathematics and prosody*. He wrote *Dhavala* (धवला), a commentary on Jain canon *Shatakhandagama* (शतखण्डगाम). He also started the work on *Jayadhavalaa* (जयधवला) which was completed by his disciples. He was among the jewels of *Rashtrakuta king Amoghavarsha* (राष्ट्रकूट राजा अमोघवर्ष).

His lineage started with *Chandrasena* (चन्द्रसेन) who initiated *Aryanandi* (आर्यानन्दि). *Aryanandi* initiated *Virasena* and *Jayasena* (जयसेन). *Virasena* initiated six disciples who were *Dasharayguru* (दशरायगुरु), *Jinasena* (जिनसेन), *Vinayasena* (विनयसेन), *Shripal* (श्रीपाल), *Padmasena* (पद्मसेन) *and Devasena* (देवसेन). *Dasharayguru* and *Jinasena* initiated *Gunabhadra* (गुणभद्र) who later initiated *Lokasena* (लोकसेन). *Vinayasena* initiated *Kumarasena* (कुमारसेन) who started the *Kashtha Sangha* (काष्ठ सङ्घ).

## 卐 Key Contributions in Mathematics

### 卐 Volume of a Frustum

*Virasena* gave the derivation of the volume of a frustum (see Figure 3.6) by a sort of infinite procedure. In geometry, a frustum is the portion of a solid (normally a pyramid or a cone) that lies between two parallel planes cutting the solid.

He worked with the concept of *Ardha Chheda* (अर्ध छेदा):

> The number of times a number could be divided by 2; This coincides with the binary logarithm when applied to powers of two but it is different for other integers, giving the 2-adic order rather than the logarithm.

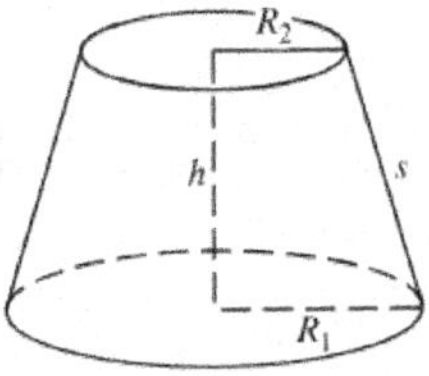


Figure 3.6 Conical frustum.

### 卐 Circumference of a Circle

*Virasena* gave the approximate formula

$$C = 3d + (16d + 16)/113$$

to relate the circumference of a circle, C, to its diameter, d. For large values of d, this gives the approximation

$$\pi \approx \frac{355}{113} = 3.14159292\ldots$$

which is more accurate than the approximation $\pi \approx 3.1416$ given by *Aryabhata* in the *Aryabhatiya*.

## Further Reading


1. Gupta, R. C. (2000), "History of Mathematics in India", in Hoiberg, Dale; Ramchandani, Indu (eds.), Students' Britannica India: Select essays, Popular Prakashan, p. 329.
2. Indranandi. Shrutāvatāra.
3. Jinasena. Ādi Purāṇa.
4. Mishra, V.; Singh, S. L. (February 1997), "First Degree Indeterminate Analysis in Ancient India and its Application by Virasena" (PDF), Indian Journal of History of Science, 32 (2): 127–133, archived from the original (PDF) on 29 November 2014, retrieved 17 September 2014.
5. Nagrajji, Acharya Shri (2003). Agama and Tripitaka: Language and Literature. Concept Publishing Company. p. 530. ISBN 9788170227311.
6. Natubhai Shah 2004, p. 31.
7. Pannalal Jain 1951, pp. 30–31.
8. Satkhandagama: Dhavala (Jivasthana) Satparupana-I (Enunciation of Existence-I) An English Translation of Part 1 of the Dhavala Commentary on the Satkhandagama of Acarya Pushpadanta & Bhutabali Dhavala commentary by Acarya Virasena English tr. by Prof. Nandlal Jain, Ed. by Prof. Ashok Jain ISBN 9788186957479.
9. See, e.g., Shparlinski, Igor (2013), Cryptographic Applications of Analytic Number Theory: Complexity Lower Bounds and Pseudorandomness, Progress in Computer Science and Applied Logic, vol. 22, Birkhäuser, p. 35, ISBN 978-3-0348-8037-4.
10. Jain, Pannalal, ed. (1951), Mahapurana Adipurana of Bhagavata Jinasenacharya, Bharatiya Jnanapitha.
11. Jaini 1991, p. 111.
12. Jaini, Padmanabh S. (1991), Gender and Salvation: Jaina Debates on the Spiritual Liberation of Women, University of California Press, ISBN 0-520-06820-3.
13. Shah, Natubhai (2004) [First published in 1998], Jainism: The World of Conquerors, vol. I, Motilal Banarsidass, ISBN 81-208-1938-1.
14. Singh, A. N. "Mathematics of Dhavala". Lucknow University. Translation of part of the Dhavala.
15. Virsena, https://en.wikipedia.org/wiki/Virasena.

## Prthudakasvami (पृथुदकस्वामी)
## (c. 830 – 890 AD)

Prthudakasvami (पृथुदकस्वामी) was best known for his work on *solving equations*. The solution of a first-degree indeterminate equation by a method called *kuttaka* (or "pulveriser") was given by *Aryabhata I*. This method of finding integer solutions resembles the *continued fraction process* (अंशप्रक्रिया निरन्तरता) and can also be seen as a use of the *Euclidean algorithm*.

### 卐 Key Contributions in Mathematics

### 卐 Algebra

*Brahmagupta* seems to have used a method involving *continued fractions* to find integer solutions of an indeterminate equation of the type

$$ax + c = by$$

*Prthudakasvami*'s commentary on *Brahmagupta*'s work is helpful in showing how "*algebra*", that is the method of calculating with the unknown, was developing in Bharat. Prthudakasvami discussed the *kuttaka* method which he renamed as "*bijagnita* - बिजग्नित्" which means the method of calculating with unknown elements.

To see just how this new idea of algebra was developing in Bharat, we look at the notation which was being used by *Prthudakasvami* in his commentary on *Brahmagupta*'s *Brāhmasphuṭasiddhānta*.

In this commentary *Prthudakasvami* writes the equation

$$10x + 8 = x^2 + 1$$

as:

*yava* 0 *ya* 10 *ru* 8
*yava* 1 *ya* 0 *ru* 1

Here *yava* (यव) is an abbreviation for *yavat* (यवत) *avad* (अवद्) *varga* (वर्ग) which means the "*square of the unknown quantity*", *ya* (य) is an abbreviation for *yavat* havat which means the "*unknown quantity*", and *ru* (रु) is an abbreviation for rupa which means "*constant term*". Hence the top row reads

$$0x^2 + 10x + 8$$

while the second row reads

$$x^2 + 0x + 1$$

The whole equation is therefore

$$0x^2 + 10x + 8 = x^2 + 0x + 1$$

or

$$10x + 8 = x^2 + 1$$

**Further Reading**


1. Chaturveda Prithudaka Swami, https://en.wikipedia.org/wiki/Prithudaka.
2. P K Majumdar, A rationale of Brahmagupta's method of solving ax + c = by, Indian J. Hist. Sci. 16 (2) (1981), 111-117.
3. Prithudaka, https://mathshistory.st-andrews.ac.uk/Biographies/Prthudakasvami/.
4. V Mishra and S L Singh, First degree indeterminate analysis in ancient India and its application by Virasena, Indian J. Hist. Sci. 32 (2) (1997), 127-133.

## Sankara Narayana (शङ्कर नारायण)
## (c. 840 – c. 900 AD)

Sankara Narayana - शङ्कर नारायण (or Śaṅkaranārāyaṇa) was an astronomer and mathematician in the *court of Ravi Kulasekhara* (रवि कुलशेखर के दरबार) (c. 844 – c. 883 AD) of the *Chera Perumal* (चेरा पेरुमाल) kingdom of Kerala. He was a disciple of the astronomer and mathematician Govindasvami. His most famous work was the *Laghubhāskarīyavivaraṇa* (लघुभास्करी विवरण) which was a commentary on the *Laghubhaskariya* of *Bhaskara I* which in turn is based on the work of *Aryabhata I.*

*Laghubhāskarīyavivaraṇa* (Chapter VII), produced in the court of king *Ravi Kulasekhara* at *Kodungallur*, explicitly states that it was composed in Saka Year 791 (= 869/70 AD). It is also mentioned that the year was the 25th regnal year of king *Ravi Kulasekhara*. In the *second verse* of the *Laghubhāskarīyavivaraṇa*, *Narayana* remembers five major precessors in the field of mathematics (*Aryabhata, Varahamihira, Bhaskara I, Govinda and Haridatta*), including his possible master *Govinda* – गोविन्द (c. 800 – c. 860 AD). An unusual feature of the *Laghubhāskarīyavivaraṇa* is the use of katapayadi system of numeration as well as the place-value Sanskrit numerals which *Laghubhāskarīyavivaraṇa* frequently uses.

### 卐 Key Contributions in Mathematics

### 卐 Observatory

The *vivarana* mentions a *royal observatory* - राजवेधशाला (under the charge of *Sankara Narayana*) at *Mahodayapura* (*Kodungallur*). There are references to an instrument called "*Rashi Chakra* - राशि चक्र " marked by a "*Yanthra Valaya* (यन्त्र वलया)" in the *vivarana*. This instrument might be the same as the *Gola Yanthra* (गोला यन्थ्र)/Chakra Yanthra mentioned by famous polymath Aryabhata. The *Chakra Yanthra* (चक्र यन्त्र) was developed further and called *Phalaka Yanthra* (फलक यन्त्र) by *Bhaskara I*.

"Oh [king] Ravi Varma Deva, now deign to tell us quickly, reading off from the armillary sphere installed [at the observatory] in Mahodayapura, duly fitted with all the relevant circles and with the sign (degree-minute) markings, the time of the rising point of the ecliptic (lagna) when the Sun is at 10° in the sign of Capricorn, and also when the Sun is at the end of the sign Libra, which I have noted."

Translation from Harley (1992)

At the directions of *Sankara Narayana,* in every '*katikai* - कतिकै' (= 34 minutes), bells were sounded at different important centres of *Mahodayapura* to announce correct time.

'*katikai* - कतिकै' = 34 minutes

卐 **Solution of the Indeterminate Equation by = ax ± c (a, b, c integers) in integers**

The *Laghubhāskarīyavivaraṇa* was written by *Sankara Narayana* in 869 AD for the author writes in the text that it is written in the *Shaka* year 791 which translates to a date AD by adding 78. It is a text which covers the standard mathematical methods of *Aryabhata* I such as the solution of the indeterminate equation

$$by = ax \pm c$$

(a, b, c integers) in integers which is then applied to astronomical problems.

The standard Bhartiya method involves using the *Euclidean algorithm.* It is called *kuttakara* ("pulveriser") but the term eventually came to have a more general meaning like "*algebra*".

## 卐 Place Value

Perhaps the most unusual feature of the *Laghubhāskarīyavivaraṇa* is the use of *katapayadi numeration* (कटपयादी संख्या) as well as the *place-value* (स्थान-मूल्यम्) *Sanskrit numerals* (संस्कृत संख्या) which *Sankara Narayana* frequently uses. *Sankara Narayana* is the first author known to use *katapayadi numeration* with this name but he did not invent it for it appears to be identical to a system invented earlier which was called *varnasamjna* - वर्णसंज्ञा. The numeration system *varnasamjna* was almost certainly invented by the astronomer *Haridatta* (हरिदत्त), and it was explained by him in a text which many historians believe was written in 684 but this would contradict what *Sankara Narayana* himself writes.

This point is discussed below.

First, we should explain ideas behind *Sankara Narayana*'s *katapayadi numeration*. The system is based on writing numbers using the letters of the *Bhartiya alphabet* (भारतीय वर्णमाला).

> "... the numerical attribution of syllables corresponds to the following rule, according to the regular order of succession of the letters of the Indian alphabet: the first nine letters represent the numbers 1 to 9 while the tenth corresponds to zero; the following nine letters also receive the values 1 to 9 whilst the following letter has the value zero; the next five represent the first five units; and the last eight represent the numbers 1 to 8."
>
> Ifrah (1998)

- Under this system *1 to 5* are represented by four different letters.
- For example, *1* is represented by the letters *ka, ta, pa, ya* which give the system its name (*ka, ta, pa, ya* becomes *katapaya*).
- Then *6, 7, 8* are represented by three letters
- Finally, nine and zero are represented by two letters.
- The system was a spoken one in the sense that *consonants* (विसर्गा) *and vowels* (स्वरा) which are not vocalised have no numerical value. The system is a *place-value system* with zero but one may reasonably ask why such an apparently complicated numeral system might ever

come to be invented. Well, the answer must be that it led to easily remembered mnemonics. In fact, many different "*words*" could represent the same number and this was highly useful for works written in verse as the Bhartiya texts tended to be.

## 卐 Date

Let us return to the interesting point about the *date of Haridatta*. Very unusually for an Bhartiya text, *Sankara Narayana* expresses his thanks to those who have gone before him and developed the ideas about which he is writing. This in itself is not so unusual but the surprise here is that *Sankara Narayana* claims to give the list in chronological order.

His list is:

- *Aryabhata I*
- *Varahamihira*
- *Bhaskara I*
- *Govindasvami*
- *Haridatta*

The chronological order in the list agrees with the dates we have for the first four of these mathematicians. However, putting *Haridatta* after *Govindasvami* would seem an unlikely mistake for *Sankara Narayana* to make if *Haridatta* really did write his text in 684 since *Sankara Narayana* was himself a disciple of *Govindasvami*. If the dating given by *Sankara Narayana* is correct then *katapayadi* numeration had been invented only a few years before he wrote his text.

More information about various dates is as below:

| |
|---|
| *"Angartvambara nanda devamanubhir yate dinanam gane"* |

- *Anga* (अङ्ग) = 6
- *Rtu* (ऋतु) = 6
- *Ambara* (अम्बरा) = 0
- *Nanda* (नन्द) = 9
- *Veda* (वेद) = 4

- *Manu* (मनु) = *14*
- *Order* (आदेश) - *6609414*
- *Reverse Order* (उल्टा क्रम) - *1449066*
- *Kali Date* (काली तिथि) - *3967 years and 86 days = 25 Mithuna* (मिथुन)
- *Kollam Era* (कोल्लम युग) *41 = 870 AD*

> *"Evam Sakabdah punariha candra randhramuni sankhyaya asambhiravagatah"*

*Candra* (चन्द्र) = *1*

- *Randhra* (रन्ध्र) = *9*
- *Muni* (मुनि) = *7*
- *Order* (आदेश) - *197*
- *Reverse Order* (उल्टा क्रम) - *791* (*Saka Year* - सका वर्ष ) = *870 AD*

> *"Angartvambara nanda devamanubhir yate dinanam gane*
> *Graste tigma mayukhamalinitamobhute parahne divi*
> *Prsta praggrahanad dvitiyaghatika grasa pramanam raver*
> *Bharta sri Kulasekharena vilasad velavrtaya bhuva"*
>
> - Sankara Narayana, *Laghubhāskarīyavivaraṇa* Vyakha, Chapter IV (c. 870 AD)

> *"Evam Sakabdah punariha candra randhramuni sankhyaya asambhiravagatah"*.
>
> - Sankara Narayana, *Laghubhāskarīyavivaraṇa* Vyakha, Chapter I (c. 870 AD)

*"Capapravista guru sauri samatva kalam*
*Yamyottaram gamanamantaratah pramanam*
*Acaksvya sarvamavagamya bhatoktamargad*
*Ityuktavan ravirasena nrpabhivandya".*
*"Tada pancavimsati Varsanyatitani devasya"*

- Sankara Narayana, *Laghubhāskarīyavivaraṇa* Vyakha, Chapter VII (c. 870 AD)

- Meeting of *Guru* (=Jupiter) and *Sauri* (=Saturn) in *Capa* (Dhanu) = 25th regnal year of the king = 870 AD

**Further Reading**


1. George Gheverghese Joseph (2009). A Passage to Infinity. New Delhi: SAGE Publications Pvt. Ltd. p. 13. ISBN 978-81-321-0168-0.
2. J.B. harley; David Woodward, eds. (1992). The history of cartography : Volume 2 Book 1 : Cartography in the traditional Islamic and South Asian societies. University of Chicago Press. p. 360. ISBN 0-226-31635-1.
3. Narayanan, M. G. S. Perumāļs of Kerala. Thrissur (Kerala): CosmoBooks, 2013. 390-391 and 408-409.
4. Narayanan, M. G. S. Perumāļs of Kerala. Thrissur (Kerala): CosmoBooks, 2013. 78-79.
5. O'Connor, John J.; Robertson, Edmund F., "Sankara Narayana", MacTutor History of Mathematics Archive, University of St Andrews.
6. S. Venkitasubramonia Iyar; S.Kochukunju Asari, eds. (1949). Laghubhaskariyavivarana. Vol. 162. Trivandrun: TSS.
7. Virendra Nath Sharma (1995). Sawai Jai Singh and His Astronomy. Delhi: Motilal Banarsidass Publishers. ISBN 81-208-1256-X.
8. G Ifrah, A universal history of numbers: From prehistory to the invention of the computer (London, 1998).
9. Sankara, https://mathshistory.st-andrews.ac.uk/Biographies/Sankara/.
10. Narayanan, M. G. S. Perumāļs of Kerala. Thrissur (Kerala): CosmoBooks, 2013. 78-79 and 390-91.

11. P K Majumdar, A rationale of Bhatta Govinda's method for solving the equation ax - c = by and a comparative study of the determination of 'Mati' as given by Bhaskara I and Bhatta Govinda, Indian J. Hist. Sci. 18 (2) (1983), 200-205.
12. Śaṅkaranārāyaṇa, https://en.wikipedia.org/wiki/%C5%9Aa%E1%B9%85karan%C4%81r%C4%81ya%E1%B9%87a.

## Vaṭeśvara (वटेश्वर)
## (born 880 CE)

Vaṭeśvara (वटेश्वर) was a tenth-century mathematician from *Kashmir* (approximately latitude 34° 9′, more specifically *Nagrapathari* (latitude 33° 55′), earlier known as *Nagarapura* a village between Srinagar and Punch) who presented several trigonometric identities. He was the author of *Vaṭeśvara-siddhānta* (वटेश्वर-सिद्धान्त) written in 904 AD, a treatise focusing on astronomy and applied mathematics. The work criticized *Brahmagupta* and defended *Aryabhatta I*. An edition of the *first three chapters* was published in 1962 by *R. S. Sharma* and *Mukund Mishra*. *Al Biruni* referred to the works by *Vaṭeśvara*, particularly the *Karaṇasāra* (करणासार), noting that the author was the son of *Mihdatta* (मिहदत्त) who belonged to *Nagarapura* (also given as *Anandapura* which is now *Vadnagar*). The *Karaṇasāra* uses 821 *Saka era* (899 AD) as a reference year.

It had not been so far possible to know anything regarding this Vaṭeśvara as no book by him had been published or notice anywhere. *M.M.Pt. Sudhakara Dvivedi* wrote in 1892 AD in his *Hanala Tarngini* (हनाला तरङ्गिणी). as follows:

*'Yathā brahmaguptenāryabhaṭādināṃ khaṇḍanaṃ kṛtaṃ tathaiva vaṭeśvareṇa svasiddhānte bahutra brahmaguptakhaṇḍanaṃ kṛtamasti asyaiva "kajanmano'ṣṭau sadalāḥ samā yayuri" tyādinā vartamānasya brahmaṇa āyuḥ sārddhavarṣāṣṭakaṃ gatamiti mataṃ asya siddhāntagrantho mayā sampūrṇo na dṛṣṭaḥ gvāliyaramahārājasyāśritasya śribālajyotirvido gehe'yamastīti śrutvā tatrāsakṛtpatraṃ preṣitaṃ parantvadyāvadhi kimapyuttaraṃ na prāptam'*

"As Brahmagupta refuted Aryabhata and others, in the same way Vaṭeśvara in his siddhanta has refuted Brahmagupta in many ways. It is his opinion that eight years and a half of the age of the current Bramha have already passed. I have not seen the whole of the astronomer Sri Bala attached with the Maharaja of Gwalior, many letters were sent but till now no reply has been received."

Translation

Some sections of *Vaṭeśvara Siddhanta* have been now published by the then *Indian Institute of Astronomical and Sanskrit Research, Delhi* (1930s) and it is possible, as will be shown presently, to identity the author of *Vaṭeśvara Siddhanta* as the author of the book *Vaṭeśvara* mentioned by *Al-biruni.* But it unfortunate the *Pt. Sudhakara Dvivedi* did not indicate where he has seen even fragmentary copies of the book as the present edition is based on only one cope which prepared from one manuscript in the library of the *Lahore University* and the editing leaves much to be desired. According to *Al-biruni, Vaṭeśvara* was the son of *Bhadatta* (? *Mihadatta*). Now, in *Persian script Mahadatta* can be also read as *Mihadatta* in the absence of proper diacritical marks or even as Bhadatta if ma is taken to be ba whch is veru probable. Fortunately, *Vaṭeśvara* has given his parentage in the very *first page* of his *Siddhanta.* He says:

> ***'Brahmāvanīndubudhaśukradivākarāra - jivārkasūnubhagurūn pitarau ca-natvā***
> ***Brāhmaṃ graharkṣagaṇitaṃ mahadattasūnurvakṣye'khilaṃ sphuṭamatīva vaṭeśvaro'ham'***

> "I, Vaṭeśvara, the son of Magadatta, after having bowed to the God, earth, moon, Mercury, Venus, sun, Mars, Jupiter, Saturn, the constellations, the teachers and my parents, give a clear account of the whole of the Bramha mathematics of planets and constellations."
>
> Translation

## Key Contributions in Mathematics

### Yuga

He calculated the *Kalpa* and *Brahma* life time as below:

> ***'Yadyugotthamiha paryayādikaṃ tadgajābhragaganendu tāḍitaṃ kalpajaṃ khakhanakhagrahāhataṃ tadbhavetkamalaviṣṭarāyuṣi'***

$$1\,Kalpa = 108 \times 1\,yuga$$
$$1\,Brahma\,Life = 72000\,Kalpa$$

### 卐 The Ksepa (additive constant) Year of Karansara

Astronomical calculations adopted in the book of *Karanasara* was reported by *Al-biruni*. *Al-biruni* complaints about the bad translation of the book in his possession and then givens the following quotation:

*"He subtracts 821 from the years of the Sakakala. The reminder is the basis. This would be the year 132 for our gauge year. He writes down the number in three different places. He multiplies the first number by 132 degrees. The product gives the number 17424 for our gauge-date. He multiplies the second number by 46 minutes, and gets the product 6072. He multiplies the third number by 34 and gets the number 4488. He divides it by 50 and the quotient represents minute, seconds, etc., viz, 89′ 46″. Then he adds to the sum of degrees in the upper lace 112, changing the seconds to minutes, the minutes to degrees, the degrees to circles. Thus he gets 48 circles 358°41′46″. This is the mean place of the moon when the sun enters Aries."*

*"Further, he divides the degrees of the mean place of the moon by 12. The quotient represents the days. The remainder of the division he multiplies by 60, and adds thereto the minutes of the mean places of the moon."*

*"He divides the sum by 12, and the quotient represents ghatis and minor portions of time. Thus we get 27° 23′ 29″, i.e. adhimasa days. No doubt this number represents the past portion of the adhimasa month, which is at present in the course of formation."*

The author, in regard to the manner in which the measure of the adhimasa is found, makes the following remarks:

*"He divides the lunar number which we have mentioned viz. 132° 44′ 34″ by 12. Thereby he gets as the portio anui 11° 3′ 52″50‴ and the portio mensis 0° 55′ 19″24‴10$^{iv}$. By means of the latter portio he computes the duration of the time in which 30 days sum up as 2 years, 8 months, 16 days, 4 ghatis, 45 chashaka. Then he multiplies the basis by 29 and gets 3828. He adds thereto 20 and divides the sum by 36. The quotient represents the unaratra days, viz.* $106\frac{8}{9}$*."*

*"However, as I have not been able to find the proper explanation of this method, I simply give it as I find it, but I must remark that the amount of unaratra days which corresponds to a single adhimasa month is* $15\frac{78871}{10662}$*."*

- It is clear that the *first paragraph* of the above quotation gives a *method of calculating* the mean longitude of the *moon* when the *sun* just enters the *Aries*.
- The *second gives* the number of *adhimasa* days in the current year.
- The *third gives* the number of *adhimasa* days per annum and per month and then gives the *time after* which an *adhimasa* on the averahe ought to occur/ it also gives the time a method of calculating the *unaratra days*.

It will now be shown how the numerical figures given above can be derives from the astronomical constants given in the astronomical books of *Aryabhata* and his followers. But before doing that it is necessary to point out that due to the copyists the number of adhimasa days is given as $27^0$ 23 *ghatis* and 29 *palas*. It be actually be $29^0$ 53 *ghatis* and 29 *palas*.

| |
|---|
| 1 Ksepa (additive constant) = one revolution and $6\frac{4}{5}$ zodical signs |

## 卐 Revolutions of the Great Bear

Another calculation is obtained when we consider the revolutions of the *Great bear* (महान् ऋक्ष). ***Vaṭeśvara*** says in his book:

| |
|---|
| *'Kamalaviṣṭaravaktrasaroruhasphuṭagirābhihitā muniparyayāḥ*<br>*Ya iha tānapi vacmi yugodbhavān dyucaralabdhavaro bhujago'ṣṭayaḥ'* |
| I, Vatesvara, who have been given a boon by the planets, state the revolutions of the *munis*, as declared clearly by the lotus-like mouth of Bramha, to be one thousand six hundred and ninety-two.<br><br>Translation |

Unfortunately, the commentators have dismissed the phrase meaning 1692 as useless and connected this stanza with the next where the author speaks of the number of revolutions of the apogee of the *sun, Mars and Jupiter*.
Considering the calculations of the position of *Great Bear*, as given in *Karanasara, Al-biruni* says:

*"Subtract 821 from the Saka Kala. The remainder is the basis, i.e., the number of years above 4000 which have elapsed since the beginning of Kaliyuga."*

*"Multiply the basis by 47 and add 68000 to the product. Divide the sum by 10000. The quotient represents the zodiacal signs and fractions of them, i.e. the position of the Great Bear which was sought."*
*"The addition of 68000 prescribed in this rule must be the original position of the Great Bear at the brginning of the basis multiplied by 10000. If we divide 68000 by 10000, we get the quotient* $6\frac{4}{5}$*, i.e. six zodiacal signs and twenty-four degrees of a seventh sing."*

*"It is evident that if we divide the 10000 by 47, the Great Bear has wandered through one zodiacal sign in 212 years, 9 months and 6 days, according to solar time. Accordingly it wanders through one degree of a sign in 7 years, 1 month and 3 days, and through one lunar station in 94 years, 7 months and 20 days."*

Apart from him, *Varahamihira* soke that the *Great Bear* makes 1600 revolutions in *mahayuga*. According to *Aryabhata II*, it makes 1599998 revolutions in 1000 *mahayugas*, i.e., 1599.998 revolutions in a *mahayuga*. According to to the figures of ***Vaṭeśvara*** it will lead to the result that the *Great Bear* will be in the 7$^{th}$ lunar mansion. But *Varahamihra* told its measure as 10$^{th}$ lunar mansion. Thus, these statements can't be reconciled with each other. However, the figure 1692 leads to the figures as stated by *Al-biruni*.

In 432000 years, the *Great Bear* makes according to *Vaṭeśvara* 1692 revolutions. Hence in 10000 years it will pass through $\frac{1692\times12}{432} = 47$ zodical signs. Also, in 4000 years upto *Saka year* 821, it will have passed through $\frac{47\times4}{10} = \frac{94}{5}$ zodical signs, i.e., one revolution and $6\frac{4}{5}$ zodical signs which is the additive constant according to *Al-biruni*.

## 卐 Angular Size of the Earth's Shadow

The method to calculate the angular size of the Earth's shadow is given below:

*"Multiply the bhukti of the moon by 4 and the bhukti of the sun by 13. Divide the difference between the two products by 30 and the quotient is the diameter of the shadow."*

Majority of astronomers, except *Bhaskara I* (only moon), gave this angle in terms of daily angular speeds of the moon and sun.

The diameter of the the shadow at the moon is given by the formula:

$$\begin{aligned} &\mathit{Diameter\ of\ Shadow} \\ &\quad = \mathit{Diameter\ of\ the\ Earth} - \frac{(\mathit{Diameter\ of\ sun} - \mathit{Diameter\ of\ the\ Earth})}{\frac{\mathit{Distance\ of\ the\ sun}}{\mathit{Distance\ of\ the\ moon}}} \end{aligned}$$

Putting down the values, according to Aryabhata, we get the diameter of the shadow as below:

$$\mathit{Shadow} = 1050\ \mathit{yojana} - \frac{4410\ \mathit{yojana} - 1050\ \mathit{yojana}}{\frac{57753336}{4320000}}$$

To get the angle subtended we have to divide the above by the distance of the moon from the earth which is $\frac{216000}{2\pi}$ *yojanas* and then convert it into minutes. It will then be found that the first term is slightly less than $\frac{4}{30}$ of *moon*'s angular speed in minutes and the second term is lightly less the $\frac{14}{30}$ of the angular speed of the *sun*. The angle then can be approximately taken to be that given by *Vaṭeśvara*, the difference between this value and the actual value being only of the order of than $\frac{1}{3200}$ of the angular speed of the *moon*.

Table 3.1 shows various data for comparison obtained or calculated by various mathematicians.

Table 3.1 Comparison of some astronomical constants according to Aryabhata, Brahmagupta, Vatesvara and Suryasiddhanta

| | Values | | | |
|---|---|---|---|---|
| Constants | Aryabhata | Brahmagupta | **Vaṭeśvara** | Suryasiddhanta |
| Revolutions of the constellations | 1582237500 | 1582236450 | 1582237560 | 1582237828 |
| Revolutions of the sun | 432000 | 432000 | 432000 | 432000 |
| Revolutions of the moon | 57753336 | 57753336 | 57753336 | 57753336 |
| Revolutions of the months | 1593336 | 1593336 | 1593336 | 1593336 |
| Revolutions of the days | 25082580 | 25082580 | 25082580 | 25082580 |
| Angular size of the earth's shadow | ... | $\frac{8\ lunar\ motion - 25\ solar\ motion}{60}$ | $\frac{4\ lunar\ motion - 13\ solar\ motion}{30}$ | $\frac{8\ lunar\ motion - 25\ solar\ motion}{60}$ |

## Further Reading


1. Kim Plofker, Mathematics in India, Princeton University Press (2009), p. 326.
2. Kolachana, Aditya; Mahesh, K.; Ramasubramanian, K. (2019), Kolachana, Aditya; Mahesh, K.; Ramasubramanian, K. (eds.), "Hindu astronomer Vaṭeśvara and his works", Studies in Indian Mathematics and Astronomy, Singapore: Springer Singapore, pp. 616–624, doi:10.1007/978-981-13-7326-8_32, ISBN 978-981-13-7325-1, S2CID 194346382, retrieved 30 May 2022.
3. Vaṭeśvara, Vaṭeśvara-siddhānta and Gola of Vaṭeśvara: English translation and commentary, National Commission for the Compilation of History of Sciences in India (1985), p. xxvii.

4. K. S. Shukla, "Ancient Indian Mathematical Astronomy Eleven Centuries ago (Vateswara Siddanta of Vateshwaracharya 880 AD)", Indian Institute of Scientific Heritage (IISH).
5. K. V. Sarma (1997), "Vatesvara", Encyclopaedia of the History of Science, Technology, and Medicine in Non-Western Cultures edited by Helaine Selin, Springer, ISBN 978-0-7923-4066-9.
6. R.N. Rai, Karanasara of Vatesvara, Indian National Science Academy (1970), vol. 6, n. I, p. 34 Archived 2015-06-09 at the Wayback Machine.
7. Vaṭeśvara,
https://en.wikipedia.org/wiki/Va%E1%B9%ADe%C5%9Bvara.
8. Vaṭeśvara-siddhānta,
https://en.wikipedia.org/wiki/Va%E1%B9%ADe%C5%9Bvara-siddh%C4%81nta.

**Mahāvīra (महावीर)**
**(9th century CE)**

Figure 3.7 Mahāvīra (महावीर).

Mahāvīra (महावीर) (or Mahaviracharya - महावीराचार्य, "Mahavira the Teacher") was a 9th-century Jain mathematician possibly born in Mysore, in Bharat (see Figure 3.7). He authored *Gaṇita-sāra-saṅgraha* (गणितासारसंग्रह) (*Ganita Sara Sangraha*) or the *Compendium* on the gist of Mathematics in 850 AD. He was patronised by the *Rashtrakuta king Amoghavarsha* (राष्ट्रकूट राजा अमोघवर्ष). He separated astrology from mathematics. He worked in *Mysore* in southern Bhartiya where he was a member of a school of mathematics. If he was not born in *Mysore* then it is very likely that he was born close to this town in the same region of Bharat. We have essentially no other biographical details although **we can** gain just a little of his personality from the acknowledgement he gives in the introduction to his only known work.

It is the earliest Bhartiya text entirely devoted to mathematics. He expounded on the same subjects on which *Aryabhata* and *Brahmagupta* contended, but he expressed them more clearly. His work is a highly syncopated approach to algebra and the emphasis in much of his text is on developing the techniques necessary to *solve algebraic problems*. He is highly respected among Bhartiya mathematicians, because of his establishment of terminology for concepts such as equilateral, and isosceles triangle; rhombus; circle and semicircle. *Mahāvīra*'s eminence spread throughout *South Bharat* and his books proved inspirational to other mathematicians in Southern Bharat.

The *Gaṇita-sāra-saṅgraha* consisted of nine chapters and included all mathematical knowledge of mid-ninth century Bharat. It provides us with the bulk of knowledge which we have of *Jaina* mathematics and it can be seen as in some sense providing an account of the work of those who developed this mathematics. There were many Bhartiya mathematicians before the time of *Mahāvīra* but, perhaps surprisingly, their work on mathematics is always contained in texts which discuss other topics such as astronomy.

The nine chapters of the *Gaṇita-sāra-saṅgraha* are:

- Terminology
- Arithmetical operations
- Operations involving fractions
- Miscellaneous operations
- Operations involving the rule of three
- Mixed operations
- Operations relating to the calculations of areas
- Operations relating to excavations
- Operations relating to shadows

In the introduction to the work *Mahāvīra* paid tribute to the mathematicians whose work formed the basis of his book. These mathematicians included *Aryabhata I, Bhaskara I,* and *Brahmagupta.*

*Mahāvīra* writes:

> *With the help of the accomplished holy sages, who are worthy to be worshipped by the lords of the world ... I glean from the great ocean of the knowledge of numbers a little of its essence, in the manner in which gems are picked from the sea, gold from the stony rock and the pearl from the oyster shell; and I give out according to the power of my intelligence, the Sara Samgraha, a small work on arithmetic, which is however not small in importance.*

## 卐 Key Contributions in Mathematics

### 卐 Place-Value System

Throughout the work a place-value system with nine numerals is used or sometimes Sanskrit numeral symbols are used. Of interest in *Chapter 1* regarding the development of a place-value number system is *Mahāvīra*'s description of the number 12345654321 which he obtains after a calculation. He describes the number as:

> *... beginning with one which then grows until it reaches six, then decreases in reverse order.*

Notice that this wording makes sense to us using a *place-value system* but would not make sense in other systems. It is a clear indication that *Mahāvīra* is at home with the place-value number system.

### 卐 Rules for Decomposing Fractions

*Mahāvīra*'s *Gaṇita-sāra-saṅgraha* gave systematic rules for expressing a fraction as the sum of unit fractions. This follows the use of unit fractions in Bhartiya mathematics in the *Vedic period,* and the *Śulba Sūtras'* giving an approximation of $\sqrt{2}$ equivalent to below:

$$\sqrt{2} = 1 + \frac{1}{3} + \frac{1}{3.4} - \frac{1}{3.4.34}$$

In the *Gaṇita-sāra-saṅgraha,* the second section of the chapter on arithmetic is named kalā-savarṇa-vyavahāra - काल-सवर्ण-व्यावहार (lit. "*the operation of the reduction of fractions*"). In this, the *bhāgajāti section (verses 55–98)* gives rules for the following:

- To express 1 as the *sum of n unit fractions* (*Gaṇita-sāra-saṅgraha kalāsavarṇa 75, examples in 76*)

*rūpāṃśakarāśīnāṃ rūpādyās triguṇitā harāḥ kramaśaḥ //*
*dvidvitryaṃśābhyastāv ādimacaramau phale rūpe //*

*When the result is one, the denominators of the quantities having one as numerators are [the numbers] beginning with one and multiplied by three, in order. The first and the last are multiplied by two and two-thirds [respectively].*

$$1 = \frac{1}{1.2} + \frac{1}{3} + \frac{1}{3^2} + \cdots + \frac{1}{3^{n-2}} + \frac{1}{\frac{2}{3}3^{n-1}}$$

❖ To express 1 as the *sum of an odd number of unit fractions* (*Gaṇita-sāra-saṅgraha kalāsavarṇa kalāsavarṇa 77*):

$$1 = \frac{1}{2.3.\frac{1}{2}} + \frac{1}{3.4.\frac{1}{2}} + \cdots + \frac{1}{(2n-1).2n.\frac{1}{2}} + \frac{1}{2n.\frac{1}{2}}$$

❖ To express a unit fraction $\frac{1}{q}$ as the *sum of n other fractions* with given numerators $a_1, a_2, \ldots, a_n$ (*Gaṇita-sāra-saṅgraha kalāsavarṇa kalāsavarṇa 78*):

$$\frac{1}{q} = \frac{a_1}{q(q+a_1)} + \frac{a_2}{(q+a_1)(q+a_1+a_2)} + \cdots + \frac{a_{n-1}}{(q+a+\cdots+a_{n-2})(q+a_1+\cdots+a_{n-1})} + \frac{a_n}{a_n(q+a_1+\cdots+a_{n-1})}$$

❖ To express any fraction $\frac{p}{q}$ as a *sum of unit fractions* (*Gaṇita-sāra-saṅgraha kalāsavarṇa kalāsavarṇa 80, examples in 81*):

Choose an integer $i$ such that $\frac{q+i}{p}$ is an integer $r$, then write:

$$\frac{p}{q} = \frac{1}{r} + \frac{i}{r.q}$$

and repeat the process for the second term, recursively.

- To express a unit fraction as the *sum of two other unit fractions* (*Gaṇita-sāra-saṅgraha kalāsavarṇa kalāsavarṇa 85, example in 86*):

$$\frac{1}{n}=\frac{1}{p.n}+\frac{1}{\frac{p.n}{n-1}}$$

where $p$ is to be chosen such that $\frac{p.n}{n-1}$ is an integer (for which $p$ must be a multiple of $n-1$).

$$\frac{1}{a.b}=\frac{1}{a(a+b)}+\frac{1}{b(a+b)}$$

- To express a fraction $\frac{p}{q}$ as the *sum of two other fractions with given numerators a and b* (*Gaṇita-sāra-saṅgraha kalāsavarṇa kalāsavarṇa 87, example in 88*):

$$\frac{p}{q}=\frac{a}{\frac{ai+b}{p}.\frac{q}{i}}+\frac{b}{\frac{ai+b}{p}.\frac{q}{i}.i}$$

Where $i$ is to be chosen such that $p$ divides $ai+b$.

## 卐 Indeterminate Linear Equations

An example of a problem given in the *Gaṇita-sāra-saṅgraha* which leads to indeterminate linear equations is the following:

> *Three merchants find a purse lying in the road. One merchant says "If I keep the purse, I shall have twice as much money as the two of you together". "Give me the purse and I shall have three times as much" said the second merchant. The third merchant said "I shall be much better off than either of you if I keep the purse, I shall have five times as much as the two of you together". How much money is in the purse? How much money does each merchant have?*

If the first merchant has $x$, the second $y$, and $p$ is the amount in the purse then

$$p + x = 2(y + z)$$

$$p + y = 3(x + z)$$

$$p + z = 5(x + y)$$

There is no unique solution but the smallest solution in positive integers is p = 15, x = 1, y = 3, z = 5. Any solution in positive integers is a multiple of this solution as *Mahāvīra* claims.

## 卐 Permutations and Combinations

*Mahāvīra* gave special rules for the use of permutations and combinations which was a topic of special interest in Jaina mathematics.

He also found out the formula for

$$nC_r = \frac{n(n-1)(n-2)\ldots(n-r+1)}{r(r-1)(r-2)\ldots 2.1}$$

## 卐 Approximate Formula for the Area and the Perimeter of an Ellipse

*Mahāvīra* also attempts to solve certain mathematical problems which had not been studied by other Bhartiya mathematicians. For example, he gave an approximate formula for the area and the *perimeter of an ellipse* (दीर्घवृत्तस्य परिधि). The formulas for a conch-like figure have so far been found only in the works of *Mahāvīra* and *Narayana*. It is reasonable to ask what a "*conch-like figure*" is. It is two unequal semicircles (with diameters AB and BC) stuck together along their diameters.

## 卐 Algebra and Geometry

Algebraic identities like

$$a^3 = a(a+b)(a-b) + b^2(a-b) + b^3$$

was established by him for first time in history.

He also described a process for calculating the volume of a sphere and one for calculating the cube root of a number.
He looked at some geometrical results including *right-angled triangles* with rational sides.

As per him, the rule for finding the area of a segment of a circle is given below:

- Gross area: $h.(c+h)/2$
- Neat area: $\frac{1}{4}.c.h.\sqrt{10}$

  where, $c$ denotes the chord and $h$ denotes the height of the segment.

He provided some rules for squaring a number as below:

- $a^2 = a \times a$
- $a^2 = (a+N) \times (a-N) + N^2$ where $N$ is any arbitrary number suitably chosen.
- $(a+b+c+d)^2 = a^2 + b^2 + c^2 + 2a(b+c+\cdots) + 2b(c+\cdots) + \cdots$
- $a^2 = 1+3+5+\cdots$ to $a$ terms.

He asserted that the *square root of a negative number* does not exist.

**Further Reading**


1. Algebra: Sets, Symbols, and the Language of Thought by John Tabak: p.43
2. Census of the Exact Sciences in Sanskrit by David Pingree: page 388
3. Geometry in Ancient and Medieval India by T. A. Sarasvati Amma: page 122
4. Hayashi 2013.
5. Krebs 2004, p. 132.
6. Kusuba 2004, pp. 497–516
7. O'Connor & Robertson 2000.
8. Puttaswamy 2012, p. 231.
9. Selin 2008, p. 1268.
10. Tabak 2009, p. 42.
11. Tabak 2009, p. 43.
12. The Math Book: From Pythagoras to the 57th Dimension, 250 Milestones in the ... by Clifford A. Pickover: page 88

13. A Jain, Mahaviracarya, the man and the mathematician, Acta Cienc. Indica Math. 10 (4) (1984), 275-280.
14. B Datta, On Mahavira's solution of rational triangles and quadrilaterals, Bull. Calcutta Math. Soc. 20 (1932), 267-294.
15. B Datta, On the relation of Mahavira to Sridhara, Isis 17 (1932), 25-33.
16. B S Jain, On the Ganita-Sara-Samgraha of Mahavira (c. 850 A.D.), Indian J. Hist. Sci. 12 (1) (1977), 17-32.
17. Bibhutibhusan Datta and Avadhesh Narayan Singh (1962). History of Hindu Mathematics: A Source Book.
18. D Pingree, Biography in Dictionary of Scientific Biography (New York 1970-1990).
19. E T Bell, Mahavira's Diophantine system, Bull. Calcutta Math. Soc. 38 (1946), 121-122.
20. G G Joseph, The crest of the peacock (London, 1991).
21. G Mazars, Les fractions dans l'Inde ancienne de la civilisation de l'Indus à Mahavira (IXe siècle), in Histoire de fractions, fractions d'histoire (Basel, 1992), 209-218; 405.
22. Hayashi, Takao (2013), "Mahavira", Encyclopædia Britannica
23. J Filliozat, La science indienne antique, in R Taton (ed.), Histoire générale des sciences (Paris, 1957-1964), 159-.
24. Krebs, Robert E. (2004), Groundbreaking Scientific Experiments, Inventions, and Discoveries of the Middle Ages and the Renaissance, Greenwood Publishing Group, ISBN 978-0-313-32433-8
25. Kusuba, Takanori (2004), "Indian Rules for the Decomposition of Fractions", in Charles Burnett; Jan P. Hogendijk; Kim Plofker; et al. (eds.), Studies in the History of the Exact Sciences in Honour of David Pingree, Brill, ISBN 9004132023, ISSN 0169-8729
26. Mahavira, https://en.wikipedia.org/wiki/Mah%C4%81v%C4%ABra_(mathematician)
27. Mahavira, https://mathshistory.st-andrews.ac.uk/Biographies/Mahavira/
28. O'Connor, John J.; Robertson, Edmund F. (2000), "Mahavira", MacTutor History of Mathematics Archive, University of St Andrews
29. Pingree 1970.
30. Pingree, David (1970). "Mahāvīra". Dictionary of Scientific Biography. New York: Charles Scribner's Sons. ISBN 978-0-684-10114-9. (Available, along with many other entries from other encyclopaedias for other Mahāvīra-s, online.)

31. Puttaswamy, T.K (2012), Mathematical Achievements of Pre-modern Indian Mathematicians, Newnes, ISBN 978-0-12-397938-4
32. R C Gupta, Rectification of ellipse from Mahavira to Ramanujan, Ganita Bharati 15 (1-4) (1993), 14-40.
33. Selin, Helaine (2008), Encyclopaedia of the History of Science, Technology, and Medicine in Non-Western Cultures, Springer, Bibcode:2008ehst.book.....S, ISBN 978-1-4020-4559-2
34. T Hayashi, Mahavira's formulas for a conch-like plane figure, Ganita Bharati 14 (1-4) (1992), 1-10.
35. Tabak, John (2009), Algebra: Sets, Symbols, and the Language of Thought, Infobase Publishing, ISBN 978-0-8160-6875-3

## Jayadeva (जयदेव)
## (c. 1000 BCE)

Jayadeva (जयदेव) was a mathematician, who further developed the *cyclic method (Chakravala method* - चक्रवला विधि*)*. He also made significant contributions to *combinatorics*. *Jayadeva*'s works are lost, and he is known only from a *20-verse quotation* in *Udaya-divakara Sundari* - उदय-दिवाकार सुन्दरी (c. 1073), a commentary on *Bhaskara I*'s *Laghubhaskariya*. This means that *Jayadeva* must have lived sometime before 1073, possibly around 1000 CE.

He firther developed *Chakravala method.*

### 卐 Key Contributions in Mathematics

### 卐 Chakravala Method

The *chakravala method* is a *cyclic algorithm* to *solve indeterminate quadratic equations* (अनिर्धारित द्विघात समीकरण), including *Pell's* equation. It is commonly attributed to *Bhāskara II* although some attribute it to *Jayadeva*. *Jayadeva* pointed out that *Brahmagupta*'s approach to solving equations of this type could be *generalized*, and he then described this general method, which was later refined by *Bhāskara II* in his *Bijaganita* (बीजगणिता) treatise.

He called it the *Chakravala method:* chakra meaning "*wheel*" in Sanskrit, a reference to the cyclic nature of the algorithm. This method is also known as the cyclic method and contains traces of mathematical induction.

*Chakra* in Sanskrit means *cycle*. As per popular legend, *Chakravala* indicates a mythical range of mountains which orbits around the Earth like a wall and not limited by light and darkness.

*Brahmagupta* in 628 CE studied indeterminate quadratic equations, including Pell's equation as below:

$$x^2 = Ny^2 + 1$$

for minimum integers $x$ and $y$.

Brahmagupta could solve it for several $N$, but not all.

*Jayadeva* and *Bhāskara II* offered the *first complete solution to the equation,* using the *chakravala method* to find for

$$x^2 = 61y^2 + 1$$

the solution

$$x = 1766319049, y = 226153980$$

In essence, as per *Selenius,*

*"The method represents a best approximation algorithm of minimal length that, owing to several minimization properties, with minimal effort and avoiding large numbers automatically produces the best solutions to the equation. The chakravala method anticipated the European methods by more than a thousand years. But no European performances in the whole field of algebra at a time much later than Bhaskara's, nay nearly equal up to our times, equalled the marvellous complexity and ingenuity of chakravala."*

**Further Reading**


1. "Pell's equation". Maths History. Retrieved 2021-06-14.
2. B. V. Subbarayappa; S. R. N. Murthy, eds. (1988). Scientific Heritage of India. Mythic Society. p. 59.
3. Cajori (1918), p. 197
4. Datta and Singh (1962). History of Hindu Mathematics: A Source Book Parts I and II. Asia Publishing House. pp. 157–160. ISBN 8180903907.
5. Goonatilake, page 127 – 128.
6. Gopal, Madan (1990). K.S. Gautam (ed.). India through the ages. Publication Division, Ministry of Information and Broadcasting, Government of India. p. 79.
7. Helaine Selin, ed. (2008). Encyclopaedia of the History of Science, Technology, and Medicine in Non-Western Cultures. Vol. 1. Springer. p. 200.
8. John Stillwell (2002), Mathematics and its history (2 ed.), Springer, pp. 72–76, ISBN 978-0-387-95336-6.

9. K. V. Sarma (2008). "Jayadeva". In Helaine Selin (ed.). Encyclopaedia of the History of Science, Technology, and Medicine in Non-Western Cultures. Vol. 1 (2nd ed.). Springer. p. 1153. ISBN 978-1-4020-4559-2.
10. Kaye (1919), p. 337.
11. Kumar, page 23.
12. O'Connor, John J.; Robertson, Edmund F., "Pell's equation", MacTutor History of Mathematics Archive, University of St Andrews.
13. Plofker, page 474.
14. The example in this section is given (with notation.
15. The example in this section is given (with notation Q_{n} for k, P_{n} for m, etc.) in: Michael J. Jacobson; Hugh C. Williams (2009), Solving the Pell equation, Springer, p. 31, ISBN 978-0-387-84922-5Florian Cajori (1918), Origin of the Name "Mathematical Induction", The American Mathematical Monthly 25 (5), p. 197-201.
   *"The process of reasoning called "Mathematical Induction" has had several independent origins. It has been traced back to the Swiss Jakob (James) Bernoulli, the Frenchman B. Pascal and P. Fermat, and the Italian F. Maurolycus. [...] By reading a little between the lines one can find traces of mathematical induction still earlier, in the writings of the Hindus and the Greeks, as, for instance, in the "cyclic method" of Bhaskara, and in Euclid's proof that the number of primes is infinite."*
16. Clas-Olaf Selenius, "Kettenbruchtheoretische Erklärung der zyklischen Methode zur Lösung der Bhaskara-Pell-Gleichung", Acta Acad. Abo. Math. Phys. 23 (10) (1963), pp. 1–44.
17. Clas-Olaf Selenius, "Rationale of the chakravala process of Jayadeva and Bhaskara II", Historia Mathematica 2 (1975), pp. 167–184.
18. Edwards, Harold (1977). Fermat's Last Theorem. New York: Springer. ISBN 0-387-90230-9.
19. Florian Cajori (1918), Origin of the Name "Mathematical Induction", The American Mathematical Monthly 25 (5), p. 197-201.
20. G. R. Kaye, "Indian Mathematics", Isis 2:2 (1919), p. 326–356.
21. George Gheverghese Joseph, The Crest of the Peacock: Non-European Roots of Mathematics (1975).
22. g Civilizational Knowledge. Indiana: Indiana University Press. ISBN 0-253-33388-1.
23. Goonatilake, Susantha (1998). Toward a Global Science: Mining Civilizational Knowledge. Indiana: Indiana University Press. pp. 127, 128. ISBN 0-253-33388-1.

24. Gopal, Madan (1990). K.S. Gautam (ed.). India through the ages. Publication Division, Ministry of Information and Broadcasting, Government of India. p. 79.
25. Hoiberg & Ramchandani – Students' Britannica India: Bhaskaracharya II, page 200.
26. Hoiberg, Dale & Ramchandani, Indu (2000). Students' Britannica India. Mumbai: Popular Prakashan. ISBN 0-85229-760-2.
27. Chakravala method, https://en.wikipedia.org/wiki/Chakravala_method.
28. Jayadeva, https://en.wikipedia.org/wiki/Jayadeva_(mathematician).
29. Introduction to Chakravala, https://web.archive.org/web/20110707120031/http://www-groups.dcs.st-and.ac.uk/~history/Miscellaneous/Pearce/Lectures/Ch8_6.html.
30. Kumar, Narendra (2004). Science in Ancient India. Delhi: Anmol Publications Pvt Ltd. ISBN 81-261-2056-8.
31. Ploker, Kim (2007) "Mathematics in India". The Mathematics of Egypt, Mesopotamia, China, India, and Islam: A Sourcebook New Jersey: Princeton University Press. ISBN 0-691-11485-4.

## Daśabala (दशबल)
## (fl. 1055 – 1058)

Daśabala (दशबल) was an astronomer who lived in *Saurashtra* in what is now Rajasthan and wrote several works on calendars, eclipses, and planetary conjunctions. *Daśabala* belonged to a *Buddhist sect* (बौद्ध सम्प्रदाय) and called himself *Bodhisattva* (बोधिसत्त्व). His father's name was *Virocana* (विरोचन) and belonged to a *clan* (कुल:) called *Valabha* (वलभ:). He worked under the patronship of *King Bhoja* (राजा भोज) of the *Paramara dynasty* (परमरा वंश) of Rajasthan.

### 卐 Key Contributions in Mathematics

#### 卐 Overall

All that is known of *Daśabala* is from his two works, the *Cintāmannṇisāraṇikā* - चित्तमणिसारणिका (1055) and the *Karaṇakamalamārtaṇḍa* - करणाकमलमार्तण्डा (1058) which indicate his membership to the *Brahma School of astronomy* (or *Brāhmapakṣa* - ब्रह्मपक्ष, 5th to 12th century). These texts include astronomical calculations in *verse form* (छन्दस्वरूप). They provide means for calendar calculations, determining the positions of the planets, lunar and solar eclipses as well as a *60-year cycle* (६० वर्षचक्रम्) of *Jupiter* - बृहस्पति.

### Further Reading


1. Chattopadhyay, Anjana (2002). "Dasabala". Biographical Dictionary of Indian Scientists: From Ancient to Contemporary. New Delhi: Rupa. p. 344.
2. Dasabala, https://en.wikipedia.org/wiki/Da%C5%9Babala.
3. Pingree, David (1971). "Dasabala". In Gillispie, C.C. (ed.). Dictionary of Scientific Biography. Volume 3. New York: Charles Scribner's Sons. p. 584.

## Śatānanda (शतानन्द)
## (fl. 1068–1099)

Śatānanda (शतानन्द) was an astronomer best known from his manual the *Bhāsvatī* - भास्वती. *Śatānanda* is thought to have lived in *Puri* or *Ujjain* and drew from other texts including the *Sūrya Siddhānta* of *Varāhamihira*. *Śatānanda* was the son of *Sankara* and *Saraswati,* born in *Puri,* Odisha. He may have served in the court of the *Kesari dynasty* (केसरी वंश). Some sources suggest he lived in Ujjain.

### 卐 Key Contributions in Mathematics

### 卐 Chakravala Method

His name literally means the "*joy of hundreds*" in Sanskrit and he used *centesimal numbering* (शताब्दी संख्याकरणम्) in his astronomical manual *Bhāsvatī* published in 1099. The book *has eight short chapters* with *128 verses* which include methods for *preparing almanacs* (पंचांगं सज्जीकरोति). He also includes methods for computing the longitudinal positions of the planets.

Using 528 CE as a reference he computed the *annual precession rate* (वार्षिक पूर्वाग्रह दर) as 1 minute.

The sections include:

- *Tithyādidhruvādhikāra* - तिथिआदिध्रुवाधिकरण (Tithi Dhruva)
- *Grāhadhruvādhikāra* - ग्राहद्रुवाधिकार (Graha Dhruva)
- *Pancāngaspastādhikāra* - पञ्चाङ्गस्पष्टाधिकरण (Calculation of Calendar)
- *Grahaspasfādhikāra* - ग्रहस्पस्फाधिकार (True place of Planets)
- *Triprasnādhikāra* - त्रिप्रश्नाधिकार (Three problems: Time, Place and Direction)
- *Chandragrahanâdhikâra* - चन्द्रग्रहाधिकरण (Lunar Eclipse)
- *Sûryagrahanādhikāra* - सूर्यग्रहणाधिकरण (Solar Eclipse)
- *Parilekhādhikāra* - परिलेखाधिकार (Sketch or graphical presentations of eclipses)

- Several contemporary commentaries of the book were also published. *Aniruddha* (अनिरुद्ध), a son of *Bhavadasa* of Varanasi, wrote a commentary (tika) on *Satananda*'s *Bhāsvatī* in 1495 CE.

**Further Reading**


1. David Pingree, ed. (1970). Census of the Exact Sciences in Sanskrit Series A. Vol. 1. American Philosophical Society. p. 43.Ploker, Kim (2007) "Mathematics in India". The Mathematics of Egypt, Mesopotamia, China, India, and Islam: A Sourcebook New Jersey: Princeton University Press. ISBN 0-691-11485-4.
2. Sarma, K.V. (2007). "Satananda". In Thomas Hockey; Virginia Trimble; Thomas R. Williams; Katherine Bracher; Richard A. Jarrell; Jordan D. Marchéll; F. Jamil Ragep; JoAnn Palmeri; Marvin Bolt (eds.). The Biographical Encyclopedia of Astronomers. p. 1919. doi:10.1007/978-0-387-30400-7_1218.
3. Śatānanda, https://en.wikipedia.org/wiki/%C5%9Aat%C4%81nanda.
4. Panda, Sudhira (2019). "The Bhāsvatī astronomical handbook of Śatānanda". Journal of Astronomical History and Heritage. 22 (3): 536–544.

## Āryabhaṭa II (आर्यभट II)
**(c. 920 – c. 1000)**

Essentially nothing is known of the life of *Āryabhaṭa II* (आर्यभट II). Historians have argued about his date and have come up with many different theories. The most famous work by *Aryabhata II* is the *Mahasiddhanta* (महासिद्धान्त) (or *Arya-siddhanta* - आर्य-सिद्धान्त) which consists of 18 chapters. The manuscripts of his *Maha-Siddhanta* have been discovered from *Gujarat, Rajasthan, Uttar Pradesh,* and *Bengal,* so he probably lived in northern Bharat. The *treatise* is written in Sanskrit verse and the *first twelve chapters* form a treatise on mathematical astronomy covering the usual topics that Bhartiya mathematicians worked on during this period.

The topics included in these twelve chapters are:

- the longitudes of the planets,
- eclipses of the sun and moon,
- the projection of eclipses,
- the lunar crescent,
- the rising and setting of the planets,
- conjunctions of the planets with each other and with the stars.

The *next six chapters* of the book includes topics such as geometry, geography and algebra, which were applied to calculate the longitudes of the planets.

In about *twenty verses* in the treatise, he gives elaborate rules to solve the indeterminate equation:

$$by = ax + c$$

These rules have been applied to a number of different cases such as when $c$ has a positive value, when $c$ has a negative value, when the number of the quotients is an even number, when this number of quotients is an odd number, etc.

## 卐 Key Contributions in Mathematics

### 卐 Algebra

*Āryabhaṭa II* also deduced a method to calculate the *cube root of a number*, but his method was already given by *Āryabhaṭa I*, many years earlier. Bhartiya mathematicians were very keen to give the correct sine tables since they played a vital role to calculate the planetary positions as accurately as possible. *Āryabhaṭa II* played a vital role in it by constructing a sine table, which was accurate up to five decimal places.

### 卐 Sine Table

*Āryabhaṭa II* constructed a sine table correct up to five decimal places when measured in decimal parts of the radius. Bhartiya mathematicians were very interested in giving accurate sine tables since they were used to calculate the planetary positions as accurately as possible.

**Further Reading**


1. David Pingree, ed. (1970). Census of the Exact Sciences in Sanskrit Series A. Vol. 1. American Philosophical Society. p. 53.
2. "Aryabhata II - Biography". Maths History: University of St Andrews. Retrieved 16 July 2022.
3. Āryabhaṭa II, https://en.wikipedia.org/wiki/Aryabhata_II.
4. Āryabhaṭa II, https://mathshistory.st-andrews.ac.uk/Biographies/Aryabhata_II/.
5. B Datta, Two Aryabhatas of al-Biruni, Bull. Calcutta Math. Soc. 17 (1926), 59-74.
6. D Pingree, Biography in Dictionary of Scientific Biography (New York 1970-1990).
7. D Pingree, On the date of the Mahasiddhanta of the second Aryabhata, Ganita Bharati 14 (1-4) (1992), 55-56.
8. O'Connor, John J.; Robertson, Edmund F., "Aryabhata II", MacTutor History of Mathematics Archive, University of St Andrews.
9. S K Jha and V N Jha, Computation of sine-table based on the Mahasiddhanta of Aryabhata II, J. Bihar Math. Soc. 14 (1991), 9-17.

10. T Hayashi, T Kusuba and M Yano, Indian values for $\pi$ derived from Aryabhata's value, Historia Sci. No. 37 (1989), 1-16.
11. V N Jha, Aryabhata II's method for finding cube root of a number, Ganita Bharati 19 (1-4) (1997), 60-68.
12. V N Jha, Indeterminate analysis in the context of the Mahasiddhanta of Aryabhata II, Indian J. Hist. Sci. 29 (4) (1994), 565-578.
13. Vagiswari, A. (2007). "Āryabhaṭa II". In Thomas Hockey; et al. (eds.). The Biographical Encyclopedia of Astronomers. New York: Springer. p. 64. ISBN 978-0-387-31022-0. (PDF version).

## Vijayanandi (विजयानन्दि)
**(c. 940–1010)**

Vijayanandi (विजयानन्दि) (or *Vijayanandin* - विजयानन्दिन्) was the son of *Jayananda* (जयानन्द). He was born into the *Brahman* caste which meant he was from the highest-ranking caste of *Hindu priests*. He was a mathematician and astronomer whose most famous work was the *Karanatilaka* (करणाटिलका). We should note that there was another astronomer named *Vijayanandi* who was mentioned by *Varāhamihira* in one of his works.

Since *Varāhamihira* wrote around 550 and *the Karanatilaka* was written around 966, there must be two astronomers both named "*Vijayanandi*". *Vijayanandi* made contributions to *trigonometry*. The only known information is that he wrote a work called the *Karanatilaka* known only from an Arabic translation, *Ghurrat al-Zijat*, by *al-Biruni*. The *Karanatilaka* has not survived in its original form but we know of the text through an Arabic translation by *al-Biruni*.

It is a work in *fourteen chapters* covering the standard topics of Bhartiya astronomy.

It deals with the topics of:

- units of time measurement
- mean and true longitudes of the sun and moon
- the length of daylight
- mean longitudes of the five planets
- true longitudes of the five planets
- the three problems of *diurnal rotation* (दैनिक परिभ्रमण)
- lunar eclipses, solar eclipses
- the projection of eclipses
- first visibility of the planets
- conjunctions of the planets with each other and with fixed stars
- the moon's crescent
- the *patas* of the moon and sun.

## 卐 Key Contributions in Mathematics

### 卐 Cosmology

The Bhartiyas had a *cosmology* which was based on long periods of time with astronomical events occurring a certain whole number of times within the cycles. This system led to much work on integer solutions of equations and their application to astronomy. In particular there was, according to *Āryabhaṭa I*, a basic period of 4320000 years called a *mahayuga* (महायुग) and it was assumed that the sun, the moon, their apsis and node, and the planets reached perfect conjunctions after this period.

Hence it was assumed that the periods were rational multiples of each other. All the planets and the node and apsis of the moon and sun had to have an integer number of revolutions in the mahayuga. Many Bhartiya astronomers proposed different values for these integral numbers of revolutions. For the number of revolutions of the apsis and node of the moon per mahayuga, *Āryabhaṭa I* proposed 488219 and 232226, respectively. However, *Vijayanandi* changed these numbers to 488211 and 232234. The reasons for giving the new numbers are unclear.

### 卐 Trigonometry

Like other Bhartiya astronomers, *Vijayanandi* made contributions to trigonometry and it appears that his calculation of the periods was computed by using *tables of sines* and *versed sines*. It is significant that accuracy was need in *trigonometric tables* to give accurate astronomical theories and this motivated many of the Bhartiya mathematicians to produce more accurate methods of approximating entries in tables.

**Further Reading**

1. D Pingree, Biography in Dictionary of Scientific Biography (New York 1970-1990).
2. Khan, M.S. (1987). "An Examination of Al-Bīrūnī's Knowledge of Indian Astronomy". International Astronomical Union Colloquium. 91: 139–144. doi:10.1017/S0252921100105962. ISSN 0252-9211.

3. O'Connor, John J.; Robertson, Edmund F., "Vijayanandi", MacTutor History of Mathematics Archive, University of St Andrews.
4. R N Rai, Some observations on Vrddha-vasistha Siddhanta, Indian J. History Sci. 11 (1) (1976), 49-53.
5. Sarton, George; Siegel, Frances (1936). "Forty-Sixth Critical Bibliography of the History and Philosophy of Science and of the History of Civilization (To End of February 1936,--With Special Reference to China and Japan)". Isis. 25 (2): 522–613. doi:10.1086/347115. ISSN 0021-1753.
6. Vijayanandi, https://mathshistory.st-andrews.ac.uk/Biographies/Vijayanandi/.
7. Vijayanandi, https://en.wikipedia.org/wiki/Vijayanandi.

## Śrīpati (श्रीपति)
## (c. 1019 – 1066)

Śrīpati - (श्रीपति) (or Shri-pati - श्री-पति) as an astronomer, astrologer and mathematician. *Śrīpati* was born in *Rohinikhanda* in present-day Maharashtra. His father was *Nagadeva* – नागदेव (sometimes written as *Namadeva* - नमदेव) and *Nagadeva*'s father, *Śrīpati*'s paternal grandfather, was *Kesava* - केशव. *Śrīpati* was a follower of the teaching of *Lalla* writing on astrology, astronomy and mathematics. His mathematical work was undertaken with applications to astronomy in mind, for example a study of spheres. His work on astronomy was undertaken to provide a basis for his astrology. *Śrīpati* was the most prominent mathematicians of the 11th Century.

*Śrīpati* wrote on astrology, astronomy and mathematics. He was mainly focused on *astrology* (ज्योतिषशास्त्रम), and his work on astronomy was aimed at supporting his research on astrology; his work on mathematics, in turn, was aimed at supporting his work on astronomy, such as the study of spheres. *Śrīpati* had introduced one of the main methods of house division in *Jyotiṣa,* known as *Śrīpati Bhāva* (श्रीपति भव) *system* (प्रणाली). *Śrīpati* obtained more fame in astrology than in other areas and it is fair to say that he considered this to be his most important contributions. He wrote the *Jyotisaratnamala* (ज्योतिरत्नमाला) which was an astrology text in twenty chapters based on the *Jyotisaratnakosa* (ज्योतिरत्नकोस) of *Lalla. Śrīpati* wrote a commentary on this work in *Marathi* and it is one of the oldest works to have survived that is written in that language. *Marathi* is the oldest of the regional languages in *Indo-Aryan,* dating from about 1000.

His works are provided below:

- *Dhikotidakarana* (ढिकोतिदकरण) written in 1039, a work of twenty verses on solar and lunar eclipses; Dhruvamanasa written in 1056, a work of *105 verses* on calculating planetary longitudes, eclipses and planetary transits; *Siddhantasekhara* a major work on astronomy in 19 chapters; and *Ganitatilaka* an incomplete arithmetical treatise in 125 verses based on a work by *Śrīpati*
- *Dhīkotida-karana* - धीकोटिदा-करणम् (1039) is 20-verse text that discusses solar and lunar eclipses.

An *udaharana* (1610) by *Ramakrishna Bhatta.*

A *tippana* by *Dinakara,* known from a manuscript copied in 1823.

- *Dhruva-mānasa* - ध्रुव-मानस (1056) is a 105-verse text on calculating planetary longitudes, eclipses and planetary transits.
- *Siddhānta-śekhara* - सिद्धान्त-शेखर (*Siddhanta-shekhara*) is a major work on astronomy in 19 chapters. Some of the notable chapters include:
- Chapter 13: Arithmetic; comprises 55 verses on arithmetic, Measurement, and shadow reckoning.
- Chapter 14: Algebra; discusses various rules of algebra without proof, in verbal form (without algebraic symbols), Contains rules of signs for addition, subtraction, multiplication, division, square, square root, cube and cube root of positive and negative quantities, Contains the rule for solving a quadratic equation
- Chapter 15: Sphere; The work also contains the rules for the solution of simultaneous indeterminate equations of the first degree; these rules are similar to those given by Brahmagupta.

*Gaṇita-tilaka* (गणिता-तिलक) is an incomplete, 125-verse treatise on arithmetics, based on a work by *Shridhara.* Its lost portion possibly consisted of the verses 19-55 of Chapter 13 of *Siddhānta-śekhara.*

*Jyotiṣa-ratna-mālā* (ज्योतिषरत्नमाला) (Jyotisha-ratna-mala) is a 20-chapter text on astrology, based on the *Jyotisha-ratna-kosha* of *Lalla. Śrīpati* wrote a commentary on this work in *Marathi* language: this is one of the oldest surviving Marathi-language works.

*Jātaka-paddhati* (जातक-पद्धति), also known as *Śrīpati-paddhatiḥ* (*Shripati-paddhati*), is an 8-chapter text on astrology.

## 卐 Key Contributions in Mathematics

### 卐 Quadratic Solution

His work on equations in this chapter contains the rule for solving a quadratic equation and, more impressively, he gives the identity:

$$\sqrt{x+\sqrt{y}} = \sqrt{\frac{1}{2}(x+\sqrt{x^2-y})} + \sqrt{\frac{1}{2}(x-\sqrt{x^2-y}}$$

Other mathematics included in *Śrīpati* 's work includes, in particular, rules for the solution of simultaneous indeterminate equations of the first degree that are similar to those given by *Brahmagupta*.

## Overall

The titles of Chapters 13, 14, and 15 of the Siddhantasekhara are Arithmetic, Algebra and On the Sphere. Chapter 13 consists of 55 verses on arithmetic, mensuration, and shadow reckoning. It is probable that the lost portion of the arithmetic treatise Ganitatilaka consisted essentially of verses 19-55 of this chapter. The 37 verses of Chapter 14 on algebra state various rules of algebra without proof. These are given in verbal form without algebraic symbols. In verses 3, 4 and 5 of this chapter Sripati gave the rules of signs for addition, subtraction, multiplication, division, square, square root, cube and cube root of positive and negative quantities.

## Further Reading

1. D Pingree, Biography in Dictionary of Scientific Biography (New York 1970-1990).
2. David Pingree 1970, p. 36.
3. David Pingree 1970, p. 38.
4. David Pingree 1970, p. 50.
5. David Pingree 1970, p. 55.
6. David Pingree 1970, p. 58.
7. David Pingree 1971, p. 25.
8. David Pingree 1976, p. 102.
9. David Pingree 1976, p. 122.
10. David Pingree 1976, p. 168.
11. David Pingree 1976, p. 43.
12. David Pingree 1994, p. 136.
13. David Pingree 1994, p. 209.
14. David Pingree 1994, p. 246.
15. David Pingree 1994, p. 369.
16. David Pingree 1994, p. 394.

17. David Pingree 1994, p. 457.
18. David Pingree 1994, p. 495.
19. David Pingree 1994, p. 503.
20. David Pingree 1994, p. 562.
21. David Pingree 1994, p. 740.
22. O'Connor, John J.; Robertson, Edmund F., "Śrīpati", MacTutor History of Mathematics Archive, University of St Andrews.
23. David Pingree, ed. (1970). Census of the Exact Sciences in Sanskrit Series A. Vol. 1. American Philosophical Society.
24. David Pingree, ed. (1971). Census of the Exact Sciences in Sanskrit Series A. Vol. 2. American Philosophical Society.
25. David Pingree, ed. (1976). Census of the Exact Sciences in Sanskrit Series A. Vol. 3. American Philosophical Society.
26. David Pingree, ed. (1994). Census of the Exact Sciences in Sanskrit Series A. Vol. 5. American Philosophical Society. ISBN 0-87169-213-9.
27. Sripati, https://mathshistory.st-andrews.ac.uk/Biographies/Sripati/.
28. K N Sinha, Algebra of Sripati: an eleventh century Indian mathematician, Ganita Bharati 8 (1-4) (1986), 27-34.
29. K N Sinha, On Sripati's (c. 1039) laws of signs, also his concept of imaginary numbers, Math. Education 10 (1) (1976), B10-B13.
30. K N Sinha, Sripati: an eleventh-century Indian mathematician, Historia Math. 12 (1) (1985), 25-44.
31. K N Sinha, Sripati's 'Ganitatilaka': English translation with introduction, Ganita Bharati 4 (3-4) (1982), 112-133.
32. K N Sinha, Vyaktaganitadhyaya of Sripati's 'Siddhanta'sekhara', Ganita Bharati 10 (1-4) (1988), 40-50.
33. R S Lal and R Prasad, Contributions of Sripati (1039 A.D.) in the solution of first degree simultaneous indeterminate equations, Math. Ed. (Siwan) 20 (3) (1986), 113-118.
34. Ravi P. Agarwal; Syamal K. Sen, eds. (2014). Creators of Mathematical and Computational Sciences. Springer. p. 116. ISBN 9783319108704.
35. Śrīpati, https://en.wikipedia.org/wiki/%C5%9Ar%C4%ABpati.
36. V Arora and V Goel, Life and works of Sripati, Gurukula Kangri Vijnana Patrika Aryabhata 1 (1998), 105-110.

## Hemchandra (हेमचन्द्र)
## (1088–1172/1173 CE)

Figure 3.8 Acharya Hemchandra (आचार्य हेमचन्द्र).

Hemchandra (हेमचन्द्र)'s mother was named *Pahini* (पहिनि) and his father *Chachadev* (चाचदेव) (see Figure 3.8). The name *Hemchandra* was one he took later in life and he was named *Candradeva* (चन्द्रदेव) after he was born. The city of *Dhandhuka* where he was born is about 50 km south west of *Ahmadabad* the capital of Gujarat. *Candradeva*, when still young, was taken to a Jain temple where he became a monk and changed his name to *Somacandra* (सोमचन्द्र). He was instructed in religion, Bhartiya philosophy, the sacred scriptures, logic and grammar. When *Candradeva* was ordained in 1110 into the *Shvetambara* – श्वेताम्बर (White-robed) sect of Jainism and he was given the name *Acharya Hemchandra*.

Gujarat at this time was ruled by the *Solanki dynasty*. Gujarat expanded to its largest extent under this dynasty and learning flourished, particularly in the economic and cultural fields. *King Siddharaja* (राजा सिद्धराज) made excellent use of *Hemchandra*'s great skills, knowledge and learning in ruling Gujarat. *King Kumarapala* (राजा कुमारपाल) succeeded *King Siddharaja* and from 1125 he was advised by Hemchandra. Of course, as a spiritual leader, Hemchandra was in a strong position to see that his ideas were put into practice and indeed this led to Gujarat becoming considerably more advanced in culture and learning. *Hemchandra* was a strong believer in *non-violence* (अहिंसा) and Gujarat flourished in peace for many years. He convinced *King Kumarapala* to bring in laws which not only prohibited violence between people, but also made it illegal to kill animals.

*Hemchandra* convinced the King to make the *Jain religion* the *official religion* (आधिकारिक धर्म) of Gujarat.

Noted as a prodigy by his contemporaries, he gained the title *Kalikālasarvajñ*a (कलिकलासर्वज्ञ), "*the knower of all knowledge in his times*" and father of Gujarati language.

## Father of Gujarati language

He was a prodigious writer, *Hemachandr*a wrote grammars of Sanskrit and Prakrit, *poetry, prosody, lexicons, texts on science and logic* and many branches of Bhartiya philosophy.

His systematic exposition of the Jain path in the *Yogaśāstra* (योगशास्त्रम्) and its auto-commentary is a very influential text in Jain thought (see Figure 3.9). The *Siddha-Hema-Śabdanuśāśana* (सिद्ध-हेम-शब्दानुशासनम्) includes six languages (see Figure 3.10): Sanskrit, the "*standard*" *Prakrit* (virtually Maharashtri Prakrit*), Shauraseni, Magahi, Paiśācī,* the otherwise-unattested *Cūlikāpaiśācī* and *Apabhraṃśa* (virtually *Gurjar Apabhraṃśa*, prevalent in the area of Gujarat and Rajasthan at that time and the precursor of Gujarati language).

He gave a detailed grammar of *Apabhraṃśa* (अपभ्रंश) and also illustrated it with the *folk literature* (लोकसाहित्यम) of the time for better understanding. It is the only known *Apabhraṃśa* grammar. He wrote the grammar in form of rules, with *8 Adhyayas* (Chapters) and "*Tattvaprakaashika Prakash* - तत्त्वप्रकाशिका प्रकाश" with "*maharnav nyaas* - महर्णव न्यास्" in one year. Many copies were made of it, and many schemes were announced for the study of the grammar. Scholars named *Kaakal* and *Kaayasth* took great efforts to teach the grammar.

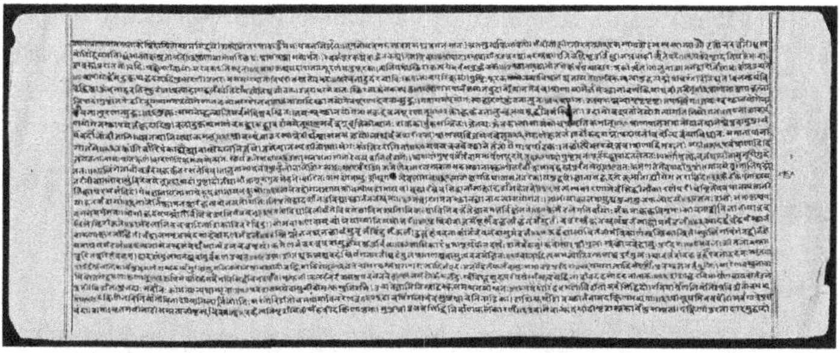

Figure 3.9 A 12th-century manuscript of Hemachandra's *Yogasastra* in Sanskrit. The text is notable for using 1 mm miniaturized Devanagari script. (Image Credit: https://en.wikipedia.org/wiki/Hemachandra)

Figure 3.10. Hemachandra's grammar text *Siddhahemashabdanushasana* in Sanskrit. (Image Credit: https://en.wikipedia.org/wiki/Hemachandra)

To illustrate the grammar, he produced the epic poetry *Dvyashraya Kavya* (द्व्याश्रय काव्य) on the history of *Chaulukya* (चौलुक्य) dynasty. It is an important source of history of region of the time. The epic poem *Trīṣaṣṭiśalākāpuruṣacharitra* (त्रिषष्टिशालाकापुरुषचरित्रम्) or "*Lives of Sixty-Three Great Men*" is a hagiographical treatment of the twenty-four *tirthankaras* (तीर्थङ्करा) and other important persons instrumental in defining the Jain philosophical position, collectively called the "*śalākāpuruṣa* (शलाकापुरुषः)", their asceticism and eventual liberation from the cycle of death and rebirth, as well as the legendary spread of the Jain influence. It still serves as the standard synthesis of source material for the early history of Jainism. The appendix to this work, the *Pariśiṣṭaparvan* (परिशिष्टपर्वणं) or *Sthavirāvalīcarita* (स्थाविरवलीकारिता), contains his own commentary and is in itself a treatise of considerable depth.

His *Kavyanuprakasha* – काव्यानुप्रकाश follows the model of Kashmiri rhetorician Mammata's *Kavya-prakasha* (काव्य-प्रकाश). He quoted other scholars like *Anandavardhana* (आनन्दवर्धनं) and Abhinavagupta (अभिनवगुप्तं) in his works.

His other works are a commentary in rhetoric work *Alankara Chudamani* (अलंकार चूडामणि), *Abhidhana-chintamani* (अभिधान-चिन्तामणि), *Pramana-mimansa* (Pramana-mimansa) (logic - तर्क), *Vitaraga-Stotra* (वितरग-स्तोत्र) (prayers - प्रार्थना).

## 卐 Key Contributions in Mathematics

### 卐 Fibonacci Numbers

His contribution to the *Fibonacci* numbers which was made *fifty years before Fibonacci* wrote *Liber Abaci* with its famous *rabbit* problem.

In a text written about 1150 he looked at the following problem:

*Suppose we assume that lines are composed of syllables which are either short or long. Suppose also that each long syllable takes twice as long to articulate as a short syllable. A line of length n n contains n units where each short syllable is one unit and each long syllable is two units.*

Clearly a line of length $n$ units take the same time to articulate regardless of how it is composed.

*Hemchandra* asks: *How many different combinations of short and long syllables are possible in a line of length n?*

*Hemchandra* then finds the answer explicitly.

Suppose that there are $f(n)$ possibilities for a line of length $n$. The line of length $n$ either ends in a short syllable or in a long syllable. If it is the former than there remains a line of length $n-1$ which can be composed in $f(n-1)$ ways and if the line of length $n$ ends in a long syllable, then there is a line of length $n-2$ remaining which can be composed in $f(n-2)$ ways. Hence, argues Hemachandra,

$$f(n) = f(n-1) + f(n-2)$$

Before we rush to try to change the name of the Fibonacci numbers into *Hemchandra number* (हेमचन्द्र संख्या) it is worth noting that *Gopala* had studied these numbers in about 1135 and Bhartiya mathematicians as early as the 7th century had looked at sequences which are produced by the familiar Fibonacci rule.

### 卐 Syllable

He studied the *rhythms of Sanskrit poetry*. Syllables in Sanskrit are either *long or short.*

Long syllables have twice the length of short syllables.

The question he asked is:

*How many rhythm patterns with a given total length can be formed from short and long syllables?*

For example,

*how many patterns have the length of five short syllables (i.e., five "beats")?*

There are eight: SSSSS, SSSL, SSLS, SLSS, LSSS, SLL, LSL, LLS.
As rhythm patterns, these are xxxxx, xxxx., xxx.x, xx.xx, x.xxx, xx.x., x.xx., x.x.x

## Further Reading


1. "Archived copy" (PDF). Archived from the original (PDF) on 3 March 2019. Retrieved 24 September 2019.
2. "Hemacandra". Jain World. Archived from the original on 29 April 2008. Retrieved 6 May 2008.
3. "Hemachandra | Jaina author | Britannica". www.britannica.com. Retrieved 14 November 2021.
4. Asoke Kumar Majumdar 1956, p. 83.
5. Asoke Kumar Majumdar 1956, p. 97.
6. Datta, Amaresh (1987). Encyclopaedia of Indian Literature: A-Devo. Sahitya Akademi. p. 15. ISBN 978-81-260-1803-1.
7. Datta, Amaresh (1987). Encyclopaedia of Indian Literature: A-Devo. Sahitya Akademi. pp. 15–16. ISBN 978-81-260-1803-1.

8. Datta, Amaresh; various (1 January 2006). The Encyclopaedia Of Indian Literature (A To Devo). Vol. 1. Sahitya Akademi. pp. 15–16. ISBN 978-81-260-1803-1.
9. Doniger 1993, p. 241.
10. Dundas, Paul (2002). The Jains. Psychology Press. pp. 134–135. ISBN 978-0-415-26606-2.
11. Encyclopaedia of Oriental Philosophy. Global Vision Pub House. 2005. p. 278. ISBN 978-81-8220-113-2.
12. Flügel, Peter. "A Short History of Jaina Law": 5.
13. Gujarat (India) (1984). Gazetteers. Directorate of Government Print., Stationery and Publications. p. 183.
14. Hemacandra; R. C. C. Fynes (1998). The Lives of the Jain Elders. Oxford University Press. ISBN 978-0-19-283227-6.
15. Joshi, Dinkar (1 January 2005). Glimpses of Indian Culture. Star Publications. pp. 79–80. ISBN 978-81-7650-190-3.
16. Koshy, Thomas (2001). Fibonacci and Lucas numbers with applications. John Wiley & Sons. ISBN 9780471399698. ... before Fibonacci proposed the problem; they were given by Virahanka (between 600 and 800 AD), Gopala (prior to 1 135 AD), ...
17. Meghani, Jhaverchand (2003). A Noble Heritage: A Collection of Short Stories Based on the Folklore of Saurashtra. Bharatiya Vidya Bhavan. p. xviii.
18. Mehta, Makrand (1 January 1991). Indian Merchants and Entrepreneurs in Historical Perspective: With Special Reference to Shroffs of Gujarat, 17th to 19th Centuries. Academic Foundation. p. 65. ISBN 978-81-7188-017-1.
19. Olle Quarnström (2002), The Yogaśāstra of Hemacandra: A Twelfth Century Handbook of Śvetāmbara Jainism, Harvard University Press, ISBN 978-0674009349.
20. Olle Quarnström, The Yogasastra of Hemacandra: a twelfth century handbook of Svetambara Jainism, 2002, introduction.
21. Pandit, Shankar Pandurang, ed. (1936). The Kumarapalacarita (Prakrta Dvyasraya Kavya) of Hemachandra with commentary of Purnakalashagani. Bombay Sanskrit and Prakrit Series Book LX. P. L. Vaidya (revision) (2 ed.). Poona: The Bhandarkar Oriental Institute. pp. xxiii–xxv.
22. 'Prastaavna', Haim Sanskrit Praveshika by Pandit Shivlal Nemchand Shah, Bhadrankar Prakashan.

23. Singh, Upinder (2008). A History of Ancient and Early Medieval India: From the Stone Age to the 12th Century. Pearson Education India. p. 241. ISBN 978-81-317-1677-9.
24. Tetlow, Philip (2007). The Web's awake: an introduction to the field of Web science and the concept. John Wiley & Sons. ISBN 978-0-470-13794-9. This sequence was first described by the Indian mathematicians Gopala and Hemachandra in 1150, who were investigating the possible ways of exactly packing items of length 1 and 2 into containers. In the West it was first studied by ...
25. Upinder Singh 2016, p. 26.
26. Acharya Hemchandra by Madhya Pradesh Hindi Granth Academy.
27. As per Datta and Jain World, (1088–1173).
28. As per Dundas, (1089–??).
29. As per Gujarat Gazetteers, Volume 18, (1087–1174).
30. As per Indian Merchants and Entrepreneurs, (1089–1173).
31. Asoke Kumar Majumdar (1956). Chaulukyas of Gujarat. Bharatiya Vidya Bhavan. OCLC 4413150.
32. Bibliography of Hemachandra's works, Item 687 Archived 13 November 2021 at the Wayback Machine, Karl Potter, University of Washington.
33. Biography in Encyclopaedia Britannica, http://www.britannica.com/biography/Hemachandra.
34. Cinnaiya, S., Nayara, H. K., & Mathura, R. (2017). Cakita kare Fibonācī. Bengaluru: Pratham Books.
35. Doniger, Wendy, ed. (1993), Purana Perennis: Reciprocity and Transformation in Hindu and Jaina Texts, State University of New York Press, ISBN 0-7914-1381-0
36. Hemachandra, https://en.wikipedia.org/wiki/Hemachandra.
37. Hemachandra, https://mathshistory.st-andrews.ac.uk/Biographies/Hemchandra/.
38. R C C Fynes (trs.), Hemchandra's Lives of Jain Elders (Oxford University Press, Oxford, 1998).
39. S Kak, The Golden Mean and the Physics of Aesthetics (Archive of Physics, physics/0411195, 2004), http://arxiv.org/pdf/physics/0411195.
40. Singh, Upinder (2016), A History of Ancient and Early Medieval India: From the Stone Age to the 12th Century, Pearson Education, ISBN 978-93-325-6996-6.
41. Trishashti Shalaka Purusha Caritra of Hemchandra English translation of books 1-10.

**Bhāskara II (भास्कर II)**
**(c. 1114–1185 CE)**

Figure 3.11 Bhaskara II (भास्कर II). (Image Credit: https://mathshistory.st-andrews.ac.uk/)

Bhāskara is also known as Bhāskara II or as *Bhāskarāchārya* (भास्कराचार्य), this latter name meaning "*Bhaskara the Teacher*" (see Figure 3.11). Since he is known in Bharat as *Bhāskarāchārya* we will refer to him throughout this article by that name. *Bhāskarāchārya*'s father was a *Brahman* named *Mahesvara* (महेश्वर). *Mahesvara* himself was famed as an astrologer. This happened frequently in Bhartiya society with generations of a family being excellent mathematicians and often acting as teachers to other family members.

*Bhāskarāchārya* became head of the astronomical observatory at *Ujjain*, the leading mathematical centre in Bharat at that time. Outstanding mathematicians such as *Varāhamihira* and *Brahmagupta* had worked there and built up a strong school of mathematical astronomy. In many ways *Bhāskarāchārya* represents the peak of mathematical knowledge in the 12th century. He reached an understanding of the number systems and solving equations which was not to be achieved in Europe for several centuries.

A number of works by *Bhāskarāchārya* are known as below:

- *Līlāvatī* - लीलावती (The Beautiful): It is on mathematics. The first section *Līlāvatī* (also known as *pāṭīgaṇita* or *aṅkagaṇita* - अङ्कगणित), named after his daughter, consists of *277 verses*. It covers *calculations, progressions, measurement, permutations,* and other topics.
- *Bījagaṇita* (Seed Counting or Root Extraction): It is on algebra. The *second section Bījagaṇita* (Algebra) has *213 verses*. It discusses zero, infinity, positive and negative numbers, and indeterminate equations including (the now called) *Pell*'s equation, solving it using a *kuṭṭaka* method. In particular, he also solved the

  $$61x^2 + 1 = y^2$$

  case that was to elude Fermat and his European contemporaries centuries later.

- *Grahaganita* - ग्रहगणित: In the third section *Grahagaṇita*, while treating the motion of planets, he considered their instantaneous speeds. He arrived at the approximation: It consists of *451 verses;*

  $$\sin y' - \sin y \approx (y' - y) \cos y \text{ for.}$$

  $y'$ close to $y$, or in modern notation:

  $$\frac{d}{dy} \sin y = \cos y$$

In his words:

| |
|---|
| *bimbārdhasya koṭijyā guṇastrijyāhāraḥ phalaṃ dorjyāyorantaram* |

This result had also been observed earlier by *Muñjalācārya* (मुञ्जलाचार्य) (or *Mañjulācārya* - मञ्जुलाचार्य) *mānasam* (मानसं), in the context of a table of sines.

- *Siddhantasiromani* (सिद्धान्तसिरोमणि) which is in two parts, the first on mathematical astronomy with the second part on the sphere;

- *Vasanabhasya* (वासनाभास्य) of *Mitaksara* (मिताक्षरा) which is *Bhāskarāchārya*'s own commentary on the *Siddhantasiromani;*
- Karanakutuhala (Calculation of Astronomical Wonders) or *Brahmatulya* (ब्रह्मतुल्य) which is a simplified version of the *Siddhantasiromani;* and
- *Vivarana* (विवरण) which is a commentary on the *Shishyadhividdhidatantra* (शिष्यधिविधिदातन्त्रम्) of *Lalla*. It is the first three of these works which are the most interesting, certainly from the point of view of mathematics, and we will concentrate on the contents of these.

> Bhāskara also stated that at its highest point a planet's instantaneous speed is zero.

## 卐 Key Contributions in Mathematics

### 卐 Overall

Some of Bhaskara's contributions to mathematics include the following:

- A proof of the *Pythagorean theorem* by calculating the same area in two different ways and then cancelling out terms to get

$$a^2 + b^2 = c^2.$$

- In Lilavati, solutions of quadratic, cubic and quartic indeterminate equations are explained.
- Solutions of indeterminate quadratic equations (of the type $ax^2 + b = y^2$).
- Integer solutions of linear and quadratic indeterminate equations (*Kuṭṭaka*). The rules he gives are (in effect) the same as those given by the Renaissance European mathematicians of the 17th century.
- A *cyclic Chakravala* method for solving indeterminate equations of the form

$$ax^2 + bx + c = y$$

The solution to this equation was traditionally attributed to *William Brouncker* in 1657, though his method was more difficult than the chakravala method.

- The *first general method* for finding the solutions of the problem

$$x^2 - ny^2 = 1 \text{ (so-called "Pell's equation")}$$

was given by *Bhāskara II.*

- Solutions of *Diophantine equations* of the second order, such as

$$61x^2 + 1 = y^2$$

This very equation was posed as a problem in 1657 by the French mathematician *Pierre de Fermat*, but its solution was unknown in Europe until the time of Euler in the 18th century.

- Solved quadratic equations with more than one unknown, and found negative and irrational solutions.
- Preliminary concept of mathematical analysis.
- Preliminary concept of infinitesimal calculus, along with notable contributions towards integral calculus.
- Conceived differential calculus, after discovering an approximation of the derivative and differential coefficient.
- Stated *Rolle*'s theorem, a special case of one of the most important theorems in analysis, the mean value theorem. Traces of the general mean value theorem are also found in his works.
- Calculated the derivatives of trigonometric functions and formulae.
- In *Siddhanta-Śiromaṇi, Bhāskara* developed spherical trigonometry along with a number of other trigonometric results.

## 卐 Algebra

His *Bījaganita* ("Algebra") was a work in *twelve chapters*. It was the first text to recognize that a positive number has two square roots (a positive and negative square root). His work *Bījaganita* is effectively a treatise on algebra and contains the following topics:

- Positive and negative numbers.
- The 'unknown' (includes determining unknown quantities).
- Determining unknown quantities.

- Surds (includes evaluating surds and their square roots).
- Kuṭṭaka (for solving indeterminate equations and Diophantine equations).
- Simple equations (indeterminate of second, third and fourth degree).
- Simple equations with more than one unknown.
- Indeterminate quadratic equations, of the type:

$$ax^2 + b = y^2$$

- Solutions of indeterminate equations of the second, third and fourth degree.
- Quadratic equations.
- Quadratic equations with more than one unknown.
- Operations with products of several unknowns.

*Bhāskara* derived a *chakravala method* for solving indeterminate quadratic equations of the form

$$ax^2 + bx + c = 7$$

*Bhāskara*'s method for finding the solutions of the problem

$$Nx^2 + 1 = y^2$$

is of considerable importance for N = 8, 11, 32, 61 and 67.

- When N = 61 he found the solutions $x = 226153980$ and $y = 1776319049$.

- When N = 67 he found the solutions $x = 5967, y = 48842$.

He studied many Diophantine problems.

Given that he was building on the knowledge and understanding of *Brahmagupta* it is not surprising that *Bhāskarāchārya* understood about zero and negative numbers. However, his understanding went further even than that of *Brahmagupta*.

To give some examples before we examine his work in a little more detail, we note that he knew that $x^2 = 9$ had two solutions.

He also gave the formula:

$$\sqrt{a \pm \sqrt{b}} = \sqrt{\frac{a + \sqrt{a^2 - b}}{2}} \pm \sqrt{\frac{a - \sqrt{a^2 - b}}{2}}$$

In *Bījagaṇita, Bhāskarāchārya* attempted to improve on *Brahmagupta's* attempt to divide by zero (and his own description in *Līlāvatī*) when he wrote:

> *A quantity divided by zero becomes a fraction the denominator of which is zero. This fraction is termed an infinite quantity. In this quantity consisting of that which has zero for its divisor, there is no alteration, though many may be inserted or extracted; as no change takes place in the infinite and immutable God when worlds are created or destroyed, though numerous orders of beings are absorbed or put forth.*

So *Bhāskarāchārya* tried to solve the problem by writing

$$\frac{n}{0} = \infty$$

At first sight we might be tempted to believe that *Bhāskarāchārya* has it correct, but of course he does not.
If this were true then 0 times $\infty$ must be equal to every number n, so all numbers are equal. The Bhartiya mathematicians could not bring themselves to the point of admitting that one could not divide by zero.

Equations leading to more than one solution are given by *Bhāskarāchārya*:

> *Example: Inside a forest, a number of apes equal to the square of one-eighth of the total apes in the pack are playing noisy games. The remaining twelve apes, who are of a more serious disposition, are on a nearby hill and irritated by the shrieks coming from the forest. What is the total number of apes in the pack?*

The problem leads to a quadratic equation and *Bhāskarāchārya* says that the two solutions, namely 16 and 48, are equally admissible.

The *kuttaka* method to solve indeterminate equations is applied to equations with three unknowns. The problem is to find integer solutions to an equation of the form

$$ax + by + cz = d$$

An example he gives is:

> *Example: The horses belonging to four men are 5, 3, 6 and 8. The camels belonging to the same men are 2, 7, 4 and 1. The mules belonging to them are 8, 2, 1 and 3 and the oxen are 7, 1, 2 and 1. all four men have equal fortunes. Tell me quickly the price of each horse, camel, mule and ox.*

Of course, such problems do not have a unique solution as *Bhāskarāchārya* is fully aware.

He finds one solution, which is the minimum, namely horses 85, camels 76, mules 31 and oxen 4.

*Bhāskarāchārya*'s conclusion to the *Bījagaṇita* is fascinating for the insight it gives us into the mind of this great mathematician:

> *A morsel of tuition conveys knowledge to a comprehensive mind; and having reached it, expands of its own impulse, as oil poured upon water, as a secret entrusted to the vile, as alms bestowed upon the worthy, however little, so does knowledge infused into a wise mind spread by intrinsic force.*

> *It is apparent to men of clear understanding, that the rule of three terms constitutes arithmetic and sagacity constitutes algebra. Accordingly I have said ... The rule of*

*three terms is arithmetic; spotless understanding is algebra. What is there unknown to the intelligent? Therefore for the dull alone it is set forth.*

*Bhāskarāchārya* is finding integer solution to

$$195x = 221y + 65$$

He obtains the solutions $(x, y) = (6,5)$ or $(23,20)$ or $(40,35)$ and so on.

In the *final chapter* on combinations *Bhāskarāchārya* considers the following problem.

Let an $n$-digit number be represented in the usual decimal form as;

$$d_1 d_2 d_3 \dots d_n \quad (*)$$

where each digit satisfies $1 \le d_j \le 9, j = 1,2,3, \dots, n$. Then *Bhāskarāchārya*'s problem is to find the total number of numbers of the form (*) that satisfy

$$d_1 + d_2 + d_3 \dots + d_n = S$$

In his conclusion to *Līlāvatī*, *Bhāskarāchārya* writes:

*Joy and happiness is indeed ever increasing in this world for those who have Lilavati clasped to their throats, decorated as the members are with neat reduction of fractions, multiplication and involution, pure and perfect as are the solutions, and tasteful as is the speech which is exemplified.*

The *Bījagaṇita* is a work in *twelve chapters.*

The topics are:

- positive and negative numbers
- zero
- the unknown
- surds

- the kuttaka
- indeterminate quadratic equations
- simple equations
- quadratic equations
- equations with more than one unknown
- quadratic equations with more than one unknown; operations with products of several unknowns; and
- the author and his work.

Having explained how to do arithmetic with negative numbers, *Bhāskarāchārya* gives problems to test the abilities of the reader on calculating with negative and affirmative quantities:

> *Example: Tell quickly the result of the numbers three and four, negative or affirmative, taken together; that is, affirmative and negative, or both negative or both affirmative, as separate instances; if thou know the addition of affirmative and negative quantities.*

Negative numbers are denoted by placing a dot above them:

> *The characters, denoting the quantities known and unknown, should be first written to indicate them generally; and those, which become negative should be then marked with a dot over them.*
>
> *Example: Subtracting two from three, affirmative from affirmative, and negative from negative, or the contrary, tell me quickly the result ...*

In *Bījagaṇita, Bhāskarāchārya* attempted to improve on *Brahmagupta*'s attempt to divide by zero (and his own description in *Līlāvatī*) when he wrote:

> *A quantity divided by zero becomes a fraction the denominator of which is zero. This fraction is termed an infinite quantity. In this quantity consisting of that which has zero for its divisor, there is no alteration, though many may be inserted or extracted; as no change takes place in the infinite and immutable God when*

*worlds are created or destroyed, though numerous orders of beings are absorbed or put forth.*

## Trigonometry

The *Siddhānta Shiromani* (written in 1150) demonstrates *Bhāskara*'s knowledge of trigonometry, including the sine table and relationships between different trigonometric functions.
He also developed spherical trigonometry, along with other interesting trigonometrical results.
In particular *Bhāskara* seemed more interested in trigonometry for its own sake than his predecessors who saw it only as a tool for calculation.
Among the many interesting results given by *Bhāskara*, results found in his works include computation of *sines of angles* of 18 and 36 degrees, and the now well-known formulae for:

$$sin(a+b) = \sin a \; cob \; b + \cos a \sin b$$

and

$$\sin(a-b) = \sin a \;\; cosb - \cos a \;\; sinb$$

## Calculus

His work, the *Siddhānta Shiromani,* is an astronomical treatise and contains many theories not found in earlier works. Preliminary concepts of infinitesimal calculus and mathematical analysis, along with a number of results in trigonometry, differential calculus and integral calculus that are found in the work are of particular interest.
Evidence suggests *Bhāskara* was acquainted with some ideas of differential calculus.
*Bhāskara* also goes deeper into the '*differential calculus*' and suggests the differential coefficient vanishes at an extremum value of the function, indicating knowledge of the concept of '*infinitesimals*'.

There is evidence of an early form of *Rolle*'s theorem in his work. The modern formulation of *Rolle*'s theorem states that if

$$f(a) = f(b) = 0,$$

then

$$f'(x) = 0 \text{ for some } x \text{ with } a < x < b$$

In this astronomical work he gave one procedure that look like a precursor to infinitesimal methods.

In terms that is if

$$x \approx y$$

Then

$$\sin(y) - \sin(x) \approx (y - x)\cos(y)$$

that is a derivative of sine although he did not develop the notion on derivative.

*Bhāskara* uses this result to work out the position angle of the ecliptic, a quantity required for accurately predicting the time of an eclipse.

In computing the instantaneous motion of a planet, the time interval between successive positions of the planets was no greater than a *truti*, or a $\frac{1}{33750}$ of a second, and his measure of velocity was expressed in this infinitesimal unit of time.

He was aware that when a variable attains the maximum value, its differential vanishes.

He also showed that when a planet is at its farthest from the earth, or at its closest, the equation of the centre (measure of how far a planet is from the position in which it is predicted to be, by assuming it is to move uniformly) vanishes. He therefore concluded that for some intermediate position the differential of the equation of the centre is equal to zero. In this result, there are traces of the general mean value theorem, one of the most important theorems in analysis, which today is usually derived from *Rolle*'s theorem. The mean value formula for inverse interpolation of the sine was later founded by *Parameshvara* in the 15th century in the *Lilavati Bhasya,* a commentary on *Bhāskara's Līlāvatī.*

## 卐 Astronomy

Using an astronomical model developed by *Brahmagupta* in the 7th century, *Bhāskara* accurately defined many astronomical quantities, including, for example, the length of the sidereal year, the time that is required for the Earth to orbit the Sun, as approximately 365.2588 days which is the same as in *Suryasiddhanta*. The modern accepted measurement is 365.25636 days, a difference of 3.5 minutes.

His mathematical astronomy text *Siddhanta Shiromani* is written in two parts: (i) the first part on mathematical astronomy and (ii) the second part on the sphere.

The twelve chapters of the first part cover topics such as:

- Mean longitudes of the planets.
- True longitudes of the planets.
- The three problems of diurnal rotation (it is an astronomical term referring to the apparent daily motion of stars around the Earth, or more precisely around the two celestial poles. It is caused by the Earth's rotation on its axis, so every star apparently moves on a circle, that is called the diurnal circle.)
- Syzygies.
- Lunar eclipses.
- Solar eclipses.
- Latitudes of the planets.
- Sunrise equation
- The Moon's crescent.
- Conjunctions of the planets with each other.
- Conjunctions of the planets with the fixed stars.
- The paths of the Sun and Moon.

The second part contains thirteen chapters on the sphere. It covers topics such as:

- Praise of study of the sphere.
- Nature of the sphere.
- Cosmography and geography.
- Planetary mean motion.
- Eccentric epicyclic model of the planets.
- The armillary sphere.
- Spherical trigonometry.
- Ellipse calculations.
- First visibilities of the planets.
- Calculating the lunar crescent.

- Astronomical instruments.
- The seasons.
- Problems of astronomical calculations.

> A particle of tuition conveys science to a comprehensive mind: and having reached it, expands of its own impulse. As oil poured on water, as a secret entrusted to the vile, as alms bestowed upon the worthy, however little, so does science infused into a wise mind spread by intrinsic force.
>
> \- Līlāvatī

**Further Reading**


1. "Anand Narayanan". IIST.
2. "Great Indian Mathematician - Bhaskaracharya". Indiavideo.dot.org. 22 September 2015. Archived from the original on 12 December 2021.
3. "Pavuluri Mallana", Wikipedia, 2021-04-04, retrieved 2023-01-04
4. "Pearls of Wisdom". The Indian Express. Retrieved September 1, 2012.
5. "The Great Bharatiya Mathematician Bhaskaracharya ll". The Times of India. ISSN 0971-8257. Retrieved 24 May 2023.
6. 50 Timeless Scientists von K.Krishna Murty.
7. Algebra with Arithmetic and Mensuration from the Sanscrit of Brahmegupta and Bhascara by Henry Colebrooke, Scholiasts of Bhascara p., xxvii.
8. B.I.S.M. quarterly, Poona, Vol. 63, No. 1, 1984, pp 14-22.
9. Bhaskara NASA 16 September 2017.
10. Bhaskaracharya (1936). Lilavati Ganitamu (in Telugu). Translated by Pidaparti Krishnamurti Sastry. Vizianagaram: Srividya Press. Retrieved 7 October 2022.
11. Bhau Daji (1865). "Brief Notes on the Age and Authenticity of the Works of Aryabhata, Varahamihira, Brahmagupta, Bhattotpala and Bhaskaracharya". Journal of the Royal Asiatic Society of Great Britain and Ireland. pp. 392–406.
12. Burton 2011, p. 106
13. Chopra 1982, pp. 52–54.
14. Colebrooke 1817.

15. Cooke 1997, pp. 213–215.
16. David Pingree, ed. (1970). Census of the Exact Sciences in Sanskrit Series A. Vol. 1. American Philosophical Society. p. 45.
17. Eves 1990, p. 228
18. IERS EOP PC Useful constants. An SI day or mean solar day equals 86400 SI seconds. From the mean longitude referred to the mean ecliptic and the equinox J2000 given in Simon, J. L., et al., "Numerical Expressions for Precession Formulae and Mean Elements for the Moon and the Planets" Astronomy and Astrophysics 282 (1994), 663–683.
19. Indian Journal of History of Science, Volume 35, National Institute of Sciences of India, 2000, p. 77.
20. K. V. Ramesh; S. P. Tewari; M. J. Sharma, eds. (1990). Dr. G. S. Gai Felicitation Volume. Agam Kala Prakashan. p. 119. OCLC 464078172.
21. M. S. Mate; G. T. Kulkarni, eds. (1974). Studies in Indology and Medieval History: Prof. G. H. Khare Felicitation Volume. Joshi & Lokhande Prakashan. p. 42-47. OCLC 4136967.
22. Mathematical Achievements of Pre-modern Indian Mathematicians von T.K Puttaswamy.
23. Mathematics in India by Kim Plofker, Princeton University Press, 2009, p. 182.
24. Mazur 2005, pp. 19–20.
25. Pingree 1970, p. 299.
26. Plofker 2007, p. 477.
27. Plofker 2009, p. 71.
28. Poulose 1991, p. 79.
29. Poulose 1991, p. 79.
30. Proceedings, Indian History Congress, Volume 40, Indian History Congress, 1979, p. 71.
31. S. Balachandra Rao (13 July 2014), ನವ ಜನ್ಮಶತಾಬ್ದಿಯ ಗಣಿತರ್ಷಿ ಭಾಸ್ಕರಾಚಾರ್ಯ, Vijayavani, p. 17.
32. Sahni 2019, p. 50.
33. Scientist (13 July 2014), ನವ ಜನ್ಮಶತಾಬ್ದಿಯ ಗಣಿತರ್ಷಿ ಭಾಸ್ಕರಾಚಾರ್ಯ, Vijayavani, p. 21.
34. Selin 2008, pp. 269–273.
35. Shukla 1984, pp. 95–104.
36. Stillwell 2002, p. 74.
37. Students& Britannica India. 1. A to C by Indu Ramchandani.

38. T. A. Saraswathi (2017). "Bhaskaracharya". Cultural Leaders of India - Scientists. Publications Division Ministry of Information & Broadcasting. ISBN 9788123024851.
39. The Illustrated Weekly of India, Volume 95. Bennett, Coleman & Company, Limited, at the Times of India Press. 1974. p. 30. Deshasthas have contributed to mathematics and literature as well as to the cultural and religious heritage of India. Bhaskaracharaya was one of the greatest mathematicians of ancient India.
40. Verses 128, 129 in Bijaganita Plofker 2007, pp. 476–477.
41. White 1978, pp. 52–53.
42. गणिती (Marathi term meaning Mathematicians) by Achyut Godbole and Dr. Thakurdesai, Manovikas, First Edition 23, December 2013. p. 34.
43. A A Krishnaswami Ayyangar, Remarks on Bhaskara's approximation to the sine of an angle, Math. Student 18 (1950), 12.
44. B Chaudhary and P Jha, Studies of Bhaskara's works in Mithila, Ganita Bharati 12 (1-2) (1990), 27-32.
45. B Datta, The two Bhaskaras, Indian Historical Quarterly 6 (1930), 727-736.
46. Bhaskaracharya's Lilavathi - Book Review, http://nithin.nagaraj.googlepages.com/lilavathi.pdf.
47. Bhāskara-II, https://en.wikipedia.org/wiki/Bh%C4%81skara_II.
48. Bhaskara-II, https://mathshistory.st-andrews.ac.uk/Biographies/Bhaskara_II/.
49. Biography in Encyclopaedia Britannica.
50. Burton, David M. (2011), The History of Mathematics: An Introduction (7th ed.), McGraw Hill, ISBN 978-0-07-338315-6.
51. Chopra, Pran Nath (1982), Religions and communities of India, Vision Books, ISBN 978-0-85692-081-3.
52. C-O Selenius, Kettenbruch theoretische Erklarung der zyklischen Methode zur Losung der Bhaskara-Pell-Gleichung, Acta Acad. Abo. Math. Phys. 23 (10) (1963).
53. C-O Selenius, Rationale of the chakravala process of Jayadeva and Bhaskara II, Historia Math. 2 (1975), 167-184.
54. Colebrooke, Henry T. (1817), Arithmetic and mensuration of Brahmegupta and Bhaskara.
55. Cooke, Roger (1997), "The Mathematics of the Hindus", The History of Mathematics: A Brief Course, Wiley-Interscience, pp. 213–215, ISBN 0-471-18082-3.

56. D A Somayaji, Bhaskara's calculations of the gnomon's shadow, Math. Student 18 (1950), 1-8.
57. D Pingree, Biography in Dictionary of Scientific Biography (New York 1970-1990).
58. Eves, Howard (1990), An Introduction to the History of Mathematics (6th ed.), Saunders College Publishing, ISBN 978-0-03-029558-4.
59. G G Joseph, The crest of the peacock (London, 1991).
60. G Ifrah, A universal history of numbers : From prehistory to the invention of the computer (London, 1998).
61. George Gheverghese Joseph. The Crest of the Peacock: Non-European Roots of Mathematics, 2nd Edition. Penguin Books, 2000.
62. Goonatilake, Susantha (1999), Toward a global science: mining civilizational knowledge, Indiana University Press, ISBN 978-0-253-21182-8.
63. BHaskata-II, http://www.britannica.com/biography/Bhaskara-II.
64. Ian Pearce. Bhaskaracharya II at the MacTutor archive. St Andrews University, 2002.
65. K S Patwardhan, S A Naimpally and S L Singh, Lilavati of Bhaskaracarya (Delhi 2001).
66. Khagol Mandal, https://web.archive.org/web/20061031120945/http://www.khagolmandal.com/index.php?option=com_content&task=view&id=45&Itemid=45&limit=1&limitstart=0.
67. Līlāvatī, https://en.wikipedia.org/wiki/L%C4%ABl%C4%81vat%C4%AB.
68. M G Inamdar, A formula of Bhaskara for the chord of a circle leading to a formula for evaluating sin a∞, Math. Student 18 (1950), 9-11.
69. Mazur, Joseph (2005), Euclid in the Rainforest, Plume, ISBN 978-0-452-28783-9.
70. O'Connor, John J.; Robertson, Edmund F., "Bhāskara II", MacTutor History of Mathematics Archive, University of St Andrews University of St Andrews, 2000.
71. Pingree, David (1970–1980). "Bhāskara II". Dictionary of Scientific Biography. Vol. 2. New York: Charles Scribner's Sons. pp. 115–120. ISBN 978-0-684-10114-9.
72. Pingree, David Edwin (1970), Census of the Exact Sciences in Sanskrit, vol. 146, American Philosophical Society, ISBN 9780871691460.
73. Plofker 2009, p. 71.

74. Plofker, Kim (2007), "Mathematics in India", in Katz, Victor J. (ed.), The Mathematics of Egypt, Mesopotamia, China, India, and Islam: A Sourcebook, Princeton University Press, ISBN 9780691114859.
75. Plofker, Kim (2009), Mathematics in India, Princeton University Press, ISBN 9780691120676.
76. Poulose, K. G. (1991), K. G. Poulose (ed.), Scientific heritage of India, mathematics, Ravivarma Samskr̥ta granthāvali, vol. 22, Govt. Sanskrit College (Tripunithura, India)
77. R C Gupta, Bhaskara II's derivation for the surface of a sphere, Math. Education 7 (1973), A49-A52.
78. R C Gupta, The last combinatorial problem in Bhaskara's Lilavati, Ganita Bharati 18 (1-4) (1996), 14-20.
79. R Calinger (ed.), Classics of Mathematics (New Jersey, 1995).
80. S A Naimpally, Approximate Formula For The Length of a Chord, Ganita Bharati 9 (1987) 57-58.
81. S P Arya, On the Brahmagupta- Bhaskara equation, Math. Ed. 8 (1) (1991), 23-27.
82. S R Sinha, Bhaskara's Lilavati, Bull. Allahabad Univ. Math. Assoc. 15 (1951), 9-16.
83. Sahni, Madhu (2019), Pedagogy Of Mathematics, Vikas Publishing House, ISBN 978-9353383275.
84. Sarkār, Benoy Kumar (1918), Hindu achievements in exact science: a study in the history of scientific development, Longmans, Green and co.
85. Seal, Sir Brajendranath (1915), The positive sciences of the ancient Hindus, Longmans, Green and co.
86. Selin, Helaine, ed. (2008), "Astronomical Instruments in India", Encyclopaedia of the History of Science, Technology, and Medicine in Non-Western Cultures (2nd edition), Springer Verlag Ny, ISBN 978-1-4020-4559-2.
87. Selin, Helaine; D'Ambrosio, Ubiratan, eds. (2001), "Mathematics across cultures: the history of non-western mathematics", Science Across Cultures, Springer, 2, ISBN 978-1-4020-0260-1.
88. Shukla, Kripa Shankar (1984), "Use of Calculus in Hindu Mathematics", Indian Journal of History of Science, 19: 95–104.
89. Stillwell, John (2002), Mathematics and its history, Undergraduate Texts in Mathematics, Springer, ISBN 978-0-387-95336-6.
90. V Madhukar Mallayya and K Jha, Bhaskara's concept of numeration in decuple proportions - earliest reference in Vedas with Yaska's 'Nirukta' throwing light on the notion of succession in enumeration :

an anticipation of Peano's axioms, Ganita-Bharati 17 (1-4) (1995), 85-91.

91. V Madhukar Mallayya, Arithmetic operation of division with special reference to Bhaskara II's Lilavati and its commentaries, Indian J. Hist. Sci. 32 (4) (1997), 315-324.
92. V Madhukar Mallayya, Pancarasikadau - the Indian golden rule compound (with special reference to Bhaskara's Lilavati and its commentaries, Gurukula Kangri Vijnana Patrika Aryabhata 1 (1998), 29-50.
93. V Madhukar Mallayya, Various methods of squaring with special reference to the Lilavati of Bhaskara II and the commentary Kriyakramakari of Sankara and Narayana, Ganita Sandesh 11 (1) (1997), 31-36.
94. Victor J. Katz, ed. (10 August 2021). The Mathematics of Egypt, Mesopotamia, China, India, and Islam: A Sourcebook. Princeton University press. p. 447. ISBN 978-0691114859.
95. W. W. Rouse Ball. A Short Account of the History of Mathematics, 4th Edition. Dover Publications, 1960.
96. White, Lynn Townsend (1978), "Tibet, India, and Malaya as Sources of Western Medieval Technology", Medieval religion and technology: collected essays, University of California Press, ISBN 978-0-520-03566-9.

# Chapter 4

# Late Medieval Period
(1206 CE - 1526 CE)

## Mahendra Sūri (महेन्द्र सूरि)
## (c. 1340 – 1400)

Mahendra Sūri (महेन्द्र सूरि) is the 14th century Jain astronomer who wrote the *Yantraraja* (यन्त्रराज), the first Bhartiya treatise on the astrolabe. He was trained by *Madana Sūri* (मदन सूरि), and was teacher to *Malayendu Sūri* (मलयेन्दु सूरि). Jainism had a strong influence on mathematics particularly in the last couple of centuries BC. By the time of *Mahendra Suri,* however, Jainism had lost support as a national religion and was much less vigorous.

The *Delhi Sultanate* was established around 1200 and from that time on Muslim culture flourished in Bharat. The ideas of Islamic astronomy began to appear in works in the Sanskrit language and it is the Islamic ideas on the astrolabe which *Mahendra Suri* wrote on in his famous text. It is clear from the various references in the text and also from the particular values that *Mahendra Suri* uses for the angle of the ecliptic etc. that his work is based on Islamic rather than traditional Bhartiya astronomy works.

*Mahendra Sūri* was patronized by the Tughluq ruler of Delhi, *Firūz Shāh* (r. 1351–1388), who evinced keen interest in astronomy. *Firūz Shāh* had earlier caused the *Bṛhatsaṃhitā* (बृहत्संहिता) of *Varāhamihira* to be translated into Persian. At the sultan's instance, *Mahendra Sūri* studied the astrolabe and introduced it to the Sanskrit audience in 1370 in his *Yantrarāja*. Its circulation was largely, if not wholly, confined to astronomers who worked within the Islamic and Ptolemaic traditions.

He was the first to write a Sanskrit treatise - *Yantrarāja* on the astrolabe.

### 卐 Key Contributions in Mathematics

#### 卐 Astronomy

*Mahendra Suri's* fame rests on the work *Yantrarāja,* which introduced the astrolabe to the Bhartiya astronomer. An astrolabe is an astronomical instrument dating back to ancient times. It serves as a star chart and physical model of visible heavenly bodies. Its various functions also make it an elaborate inclinometer and an analog calculation device capable of working out several kinds of problems in astronomy.

Yantrarāja was the first astrolabe in Bharat.

The *Yantrarāja* is best described as an astrolabe user's manual. It explains how this *king* (*raja* – राजा) of instruments (*yantra* - यन्त्रम्) is to be constructed and commissioned for purposes of *observation*. The *saumya yantra* - सौम्य यन्त्र (northern instrument) projected from the *South Pole* (दक्षिण ध्रुव) and the *yāmya yantra* - याम्य यन्त्र (southern instrument) projected from the *North Pole* (उत्तर ध्रुव) are discussed separately, followed by a description of the *phaṇīndra yantra* - फणीन्द्र यन्त्र (the serpentine instrument), which combines both. A detailed discussion on the application of the astrolabe is found in *Malayendu Sūri*'s commentary on the *Yantrarāja*.

This commentary is also interesting because it provided, for the first time to Bhartiya astronomers, tables for ready reference, which helped simplify calculations. The commentator gave latitudes of 75 cities. *Malayendu* made a list of latitudes for 32 stars, which were identified as relevant for purposes of calculations in Bharat.

**Further Reading**


1. Glick; et al., eds. (2005). Medieval Science, Technology, and Medicine: An Encyclopedia. Routledge. p. 464. ISBN 0-415-96930-1. the Jain astronomer Mahendra Suri (fl. 1370)...wrote the first Indian treatise on the astrolabe, called the Yantraraja (1370).
2. Padmakara Dvivedi (ed)., Dvivedi, Sudhakara. (1933). Gaṇaka Taraṅgiṇī or Lives of Hindu Astronomers. Benaras: Jyotish Prakash Press.
3. Pingree, David. (1981). Census of the Exact Sciences in Sanskrit, Series A, Vol 4. Philadelphia: American Philosophical Society.
4. Raikwa, K.K. (ed). (1936). Yantrarāja of Mahendraguru with the Commentary of Malayendu Sūri. Bombay: Nirnaya Sagar Press.
5. Astrolabe, https://en.wikipedia.org/wiki/Astrolabe.

6. D Pingree, Biography in Dictionary of Scientific Biography (New York 1970-1990).
7. D Pingree, Islamic astronomy in Sanskrit, J. Hist. Arabic Sci. 2 (2) (1978), 315-330; 425.
8. Devadevan, Manu V. "Academia website". Academia.
9. K. V. Sarma (2008), "Mahendra Suri", Encyclopaedia of the History of Science, Technology, and Medicine in Non-Western Cultures (2nd edition) edited by Helaine Selin, Springer, ISBN 978-1-4020-4559-2.
10. Mahendra_Suri, https://en.wikipedia.org/wiki/Mahendra_S%C5%ABri.
11. Mahendra_Suri, https://mathshistory.st-andrews.ac.uk/Biographies/Mahendra_Suri/.
12. R C Gupta, Indian values of the sinus totus, Indian J. Hist. Sci. 13 (2) (1978), 125-143.
13. Y Ohashi, Early history of the astrolabe in India, Indian J. Hist. Sci. 32 (3) (1997), 199-295.

## Nārāyaṇa Paṇḍita (नारायण पण्डित)
## (1325–1400)

Nārāyaṇa Paṇḍita (नारायण पण्डित) was a great mathematician. *Plofker* writes that his texts were the most significant Sanskrit mathematics treatises after those of *Bhāskara II*, other than the *Kerala school*. He wrote the *Ganita Kaumudi* - गणिता कौमुदी (lit "*Moonlight of mathematics*") in 1356 about mathematical operations. The work anticipated many developments in combinatorics.

*Narayana Pandit* wrote two works, an arithmetical treatise called (i) *Ganita Kaumudi* and (ii) an algebraic treatise called *Bijaganita Vatamsa* (बिजगणिता वातमसा). *Narayanan* is also thought to be the author of an elaborate commentary of *Bhāskara II*'s Līlāvatī, titled *Karmapradipika* - कर्मप्रदीपिका (or *Karma-Paddhati*).

Although the *Karmapradipika* contains little original work, it contains seven different methods for squaring numbers, a contribution that is wholly original to the author, as well as contributions to algebra and *magic squares* (जादू वर्ग).

### 卐 Key Contributions in Mathematics

#### 卐 Narayana's cows

It is an integer sequence which *Narayana* described as the *number of cows present each year*, starting from one cow in the first year, where every cow has one baby cow each year starting in its fourth year of life.

The first few terms of the sequence are as follows: 1, 1, 1, 2, 3, 4, 6, 9, 13, 19, …

*Narayana's cows'* sequence: $a(0) = a(1) = a(2) = 1$; hereafter

$$a(n) = a(n-1) + a(n-3)$$

This comment covers a family of sequences which satisfy a recurrence of the form

$$a(n) = a(n-1) + a(n-m), \text{ with } a(n) = 1 \; for \; n = 0 \ldots m-1.$$

The generating function is

$$\frac{1}{1 - x - x^m}$$

Also,

$$a(n) = \sum_{i=0..\lfloor(\frac{n}{m})\rfloor} binomial(n - (m-1) * i, i)\,.$$

This family of binomial summations or recurrences gives the number of ways to cover (without overlapping) a linear lattice of n sites with molecules that are m sites wide.

The family $a(n) = a(n-1) + a(n-m)$ with $a(n) = 1$ for $n = 0..m-1$ can be generated by considering the sums (A102547):

| | | | | | | | | | | | | |
|---|---|---|---|---|---|---|---|---|---|---|---|---|
| 1 | 1 | 1 | 1 | 1 | 1 | 1 | 1 | 1 | 1 | 1 | 1 | 1 |
| | | | 1 | 2 | 3 | 4 | 5 | 6 | 7 | 8 | 9 | 10 |
| | | | | | | 1 | 3 | 6 | 10 | 15 | 21 | 28 |
| | | | | | | | | | 1 | 4 | 10 | 20 |
| | | | | | | | | | | | | 1 |
| 1 | 1 | 1 | 2 | 3 | 4 | 6 | 9 | 13 | 19 | 28 | 41 | 60 |

with (in this case 3) leading zeros added to each row.

Here are a few notable interpretations and applications:

- Compositions of Numbers: It represents the number of compositions of *n into parts 1 and* 3. For example, there are 6 ways to partition 5, namely 11111, 2111, 1211, 1121, 1112, 212.

  Equivalently, number of compositions (ordered partitions) of n-1 into parts 1 and 2 with no two 2's adjacent.

  E.g., there are 6 such ways to partition 5, namely 11111, 2111, 1211, 1121, 1112, 212, so a(6) = 6.

- Special Start: This sequence could have begun 1,0,0,1,1,1,2,3,4,6,9,... but that would spoil many nice properties.

- The $a(n+2)$ is the number of n-bit 0-1 sequences that avoid both 00 and 010.

- The $a(n-4)$ is the number of n-bit sequences that start and end with 0 but avoid both 00 and 010.
  For n >= 6, such a sequence necessarily starts 011 and ends 110; deleting these 6 bits is a bijection to the preceding item.

- Also, number of compositions of n+1 into parts congruent to 1 mod m. Here m=3, etc.

- a(n-3) is the top left entry of the n-th power of any of the 3 x 3 matrices as below:

$$\begin{bmatrix} 0 & 1 & 0 \\ 0 & 1 & 1 \\ 1 & 0 & 0 \end{bmatrix} \quad \begin{bmatrix} 0 & 0 & 1 \\ 1 & 1 & 0 \\ 0 & 1 & 0 \end{bmatrix} \quad \begin{bmatrix} 0 & 1 & 0 \\ 0 & 0 & 1 \\ 1 & 0 & 1 \end{bmatrix} \quad \begin{bmatrix} 0 & 0 & 1 \\ 1 & 0 & 0 \\ 0 & 1 & 1 \end{bmatrix}$$

- The compositions of $n$ in which each natural number is colored by one of p different colors are called p-colored compositions of n.

  For

$$n >= 3, 2 * a(n-3)$$

  equals the number of 2-colored compositions of $n$ *with all parts* $>=$ 3, such that no adjacent parts have the same color.

- Rabbit Population: It can be seen as representing the number of pairs of rabbits existing at period $n$ generated by one pair, where all pairs become fertile after 3 periods and generate a new pair in all subsequent periods.

- Binary Words: It can also represent the number of binary words of certain lengths with specific constraints, such as avoiding consecutive 0's of certain lengths or certain patterns like 00 and 010.

- Matrix Powers: The sequence can be derived from the top-left entry of the powers of specific matrices. For instance, it is the top-left entry of the $n$-th power of the matrix $[1,0,1; 1,0,0; 0,1,0]$.

- Tilings and Arrangements: It is related to the number of ways to tile rectangles with specific tiles or the arrangement of tatami mats in rooms satisfying certain conditions.

- Kangaroo Hops: A unique interpretation involves a kangaroo jumping through points on an integer interval, making hops restricted to $\{-1,1,2\}$.

- Testing Strategies: The sequence can also be related to optimal testing strategies in a problem involving finding a positive sample with cost considerations.

## Supergolden Ratio

In mathematics, two quantities are in the supergolden ratio if their quotient equals the unique real solution to the equation

$$x^3 = x^2 + 1$$

This solution is commonly denoted by $\Psi$. The name supergolden ratio results of an analogy with the golden ratio $\Psi$ which is the positive root of the equation

$$x^2 = x + 1$$

Using formulas for the cubic equation, one can show that

$$\Psi = \frac{1}{3}\left(1 + \sqrt[3]{\frac{29 + 3\sqrt{93}}{2}} + \sqrt[3]{\frac{29 - 3\sqrt{93}}{2}}\right)$$

or,

using the hyperbolic cosine,

$$\Psi = \frac{2}{3}\cosh\left(\frac{1}{3}cpsh^{-1}\left(\frac{29}{2}\right)\right) + \frac{1}{3}$$

The decimal expansion of this number begins as 1.4655712318767680266567312...

## 卐 Overall

*Narayana*'s other major works contain a variety of mathematical developments, including a rule to calculate approximate values of square roots, investigations into the second order indeterminate equation

$$nq^2 + 1 = p^2$$

solutions of indeterminate higher-order equations, mathematical operations with zero, several geometrical rules, methods of integer factorization, and a discussion of magic squares and similar figures.

Evidence also exists that *Narayana* made minor contributions to the ideas of differential calculus found in Bhāskara *II*'s work. Narayana has also made contributions to the topic of cyclic quadrilaterals. *Narayana* is also credited with developing a method for systematic generation of all permutations of a given sequence.

### Further Reading


1. Ian G. Pearce (2002). Mathematicians of Kerala Archived 2008-12-19 at the Wayback Machine. MacTutor History of Mathematics archive. University of St Andrews.
2. J. J. O'Connor and E. F. Robertson (2000). Narayana Archived 2008-01-24 at the Wayback Machine, MacTutor History of Mathematics archive.
3. Kim Plofker (2009), Mathematics in India: 500 BCE–1800 CE, Princeton, NJ: Princeton University Press, ISBN 978-0-691-12067-6
4. Kusuba, Takanori (2004), "Indian Rules for the Decomposition of Fractions", in Charles Burnett; Jan P. Hogendijk; Kim Plofker; et al. (eds.), Studies in the History of the Exact Sciences in Honour of David Pingree, Brill, p. 497, ISBN 9004132023, ISSN 0169-8729
5. "Narayana - Biography". Maths History. Retrieved 3 October 2022.

6. Crilly, Tony (1994). "A Supergolden Rectangle". The Mathematical Gazette. 78 (483): 320–325. doi:10.2307/3620208. JSTOR 3620208. S2CID 125782726.
7. Supergolden ratio, https://en.wikipedia.org/wiki/Supergolden_ratio.
8. OEIS, https://oeis.org/A000930 in in OEIS.
9. Narayana Pandita, https://en.wikipedia.org/wiki/Narayana_Pandita_(mathematician).

## Madhava of Sangamagrama (संगमग्रामस्य माधव)
## (c. 1340 – c. 1425)

Madhava (माधव) of Sangamagramma was born near *Cochin* on the coast in the Kerala state in southwestern Bharat. It is only due to research into *Keralese* mathematics over the last twenty-five years that the remarkable contributions of *Madhava* have come to light. All the mathematical writings of *Madhava* have been lost, although some of his texts on astronomy have survived. However, his brilliant work in mathematics has been largely discovered by the reports of other Keralese mathematicians such as *Nilakantha* who lived about 100 years later.

In a manuscript preserved in the *Oriental Institute*, Baroda, *Madhava* has been referred to as:

> *"Mādhavan vēṇvārōhādīnām karttā ... Mādhavan Ilaññippaḷḷi Emprān"*

It has been noted that the epithet '*Emprān*' refers to the *Emprāntiri community* (एम्प्रान्तिरि समुदाय), to which *Madhava* might have belonged to.

*Mādhava* of Sangamagrāma was an Bhartiya mathematician and astronomer who is considered as the founder of the Kerala school of astronomy and mathematics. One of the greatest astronomers of the late Middle Ages, *Madhava* made pioneering contributions to the study of infinite series, calculus, trigonometry, geometry, and algebra.

He was the first to use *infinite series* (अनन्त श्रृङ्खला) approximations for a range of trigonometric functions, which has been called the

"*decisive step onward from the finite procedures of ancient mathematics to treat their limit-passage to infinity*".

The term "*Ilaññippaḷḷi*" has been identified as a reference to the residence of *Mādhava*. This is corroborated by *Mādhava* himself. In his short work on the moon's positions titled *Veṇvāroha*, (वेण्वारोह) *Mādhava* says that he was born in a house named *bakuḷādhiṣṭhita . . . vihāra.*

This is clearly Sanskrit for *Ilaññippaḷḷi*. Ilaññi is the Malayalam name of the evergreen tree *Mimusops elengi* and the Sanskrit name for the same is *Bakuḷa. Palli* is a term for village. The Sanskrit house name *bakuḷādhiṣṭhita . . . vihāra* has also been interpreted as a reference to the Malayalam house name *Iraññi ninna ppaḷḷi* and some historians have tried to identify it with one of two currently existing houses with names *Iriññanavaḷḷi* and *Iriññārapaḷḷi* both of which are located near Irinjalakuda town in central Kerala. This identification is far-fetched because both names have neither phonetic similarity nor semantic equivalence to the word "*Ilaññippaḷḷi*".

Most of the writers of astronomical and mathematical works who lived after *Madhava*'s period have referred to *Madhava* as "*Sangamagrama Madhava*" and as such it is important that the real import of the word "*Sangamagrama* - संगमग्राम" be made clear. There is a small town on the southern banks of the *Nila River* (नील नदी), around 10 kilometers upstream from *Tirunavaya,* called *Kūḍallūr.* The exact literal Sanskrit translation of this place name is *Samgamagram*: kūṭal in Malayalam means a confluence (which in Sanskrit is samgama) and ūr means a village (which in Sanskrit is grama). Also, the place is at the confluence of the *Nila River* and its most important tributary, namely, the *Kunti river* (कुन्ति नदी). Incidentally there is still existing a *Nambudiri* (Malayali Brahmin) family by name Kūtallūr Mana a few kilometers away from the *Kudallur* village. The family has its origins in *Kudallu*r village itself. For many generations this family hosted a great *Gurukulam* specialising in *Vedanga* (वेदङ्गः).

That the only available manuscript of *Sphuṭacandrāpti* (स्फुटचन्द्राप्तिः), a book authored by *Madhava,* was obtained from the manuscript collection of *Kūtallūr Mana* might strengthen the conjecture that *Madhava* might have had some association with *Kūtallūr Mana*.

*K. V. Sarma* has identified *Madhava* as the author of the following works:

- *Golavada* (गोलवाद)
- *Madhyamanayanaprakara* (मध्यमानयनप्रकार)
- *Mahajyanayanaprakara* - महाज्यानयनप्रकार (Method of Computing Great Sines)
- *Lagnaprakarana* (लग्नप्रकरण)
- *Venvaroha* (वेण्वारोह)
- *Sphuṭacandrāpti* (स्फुटचन्द्राप्ति)
- *Aganita-grahacara* (अगणित-ग्रहचार)

- *Chandravakyani* (चन्द्रवाक्यानि) (*Table of Moon-mnemonics* - चन्द्र-स्मृतिविज्ञानस्य सारणी)

## 卐 Key Contributions in Mathematics

### 卐 Infinite Series

Among his many contributions, he discovered infinite series for the trigonometric functions of sine, cosine, arctangent, and many methods for calculating the circumference of a circle. One of *Madhava*'s series is known from the text *Yuktibhāṣā* (युक्तिभाषा), which contains the derivation and proof of the power series for inverse tangent, discovered by *Madhava*.

In the text, *Jyeṣṭhadeva* (ज्येष्ठदेव) describes the series in the following manner:

> The first term is the product of the given sine and radius of the desired arc divided by the cosine of the arc. The succeeding terms are obtained by a process of iteration when the first term is repeatedly multiplied by the square of the sine and divided by the square of the cosine. All the terms are then divided by the odd numbers 1, 3, 5, .... The arc is obtained by adding and subtracting respectively the terms of odd rank and those of even rank. It is laid down that the sine of the arc or that of its complement whichever is the smaller should be taken here as the given sine. Otherwise the terms obtained by this above iteration will not tend to the vanishing magnitude.

This is a remarkable passage describing *Madhava*'s series, but remember that even this passage by *Jyesthadeva* was written more than 100 years before *James Gregory* rediscovered this series expansion. Perhaps we should write down in modern symbols exactly what the series is that *Madhava* has found.

The first thing to note is that the Bhartiya meaning for sine of θ would be written in our notation as $r\sin\theta$ and the Bhartiya cosine of would be $r\cos\theta$ in our notation, where r is the radius.

Thus, the series is:

$$r\theta = \frac{r\ sin\theta}{\cos\theta} - \left(\frac{1}{3}\right) r \frac{(sin\theta)^3}{(\cos\theta)^3} + \left(\frac{1}{5}\right) r \frac{(sin\theta)^5}{(\cos\theta)^5} - \cdots$$

or, equivalently (putting tan $= \frac{sin}{cos}$) and cancelling r gives

$$\theta = \tan\theta - \frac{\tan^3\theta}{3} + \frac{\tan^5\theta}{5} - \frac{\tan^7\theta}{7} + \cdots$$

where $n = 0, 1, 2, \ldots$

Now this series is *Gregory*'s series:

$$\tan^{-1}\theta = \theta - \frac{1}{3}\theta^3 + \frac{1}{5}\theta^5 - \cdots$$

Now *Madhava* put $q = \frac{\pi}{4}$ into his series to obtain:

$$\frac{\pi}{4} = 1 - \frac{1}{3} + \frac{1}{5} - \cdots$$

None of *Madhava*'s works, containing any of the series expressions attributed to him, have survived. These series expressions are found in the writings of the followers of Madhava in the Kerala school. At many places these authors have clearly stated that these are "*as told by Madhava*".

Thus, the enunciations of the various series found in Tantrasamgraha and its commentaries can be safely assumed to be in "*Madhava's own words* - माधवस्य स्ववचनम् ".

The translations of the relevant verses as given in the *Yuktidipika* (युक्तिदिपिका) commentary of *Tantrasamgraha* (also known as *Tantrasamgraha-vyakhya*) by *Sankara Variar* (circa. 1500 - 1560 CE) are reproduced below. These are then rendered in current mathematical notations.

*Madhava* also gave a table of almost accurate values of *half-sine chords* for *twenty-four arcs* drawn at equal intervals in a quarter of a given circle. It is thought that the way that he found these highly accurate tables was to use the equivalent of the series expansions as follows.

Madhava's sine series is stated in verses 2.440 and 2.441 in Yukti-dipika commentary (Tantrasamgraha-vyakhya) by Sankara Variar. A translation of the verses follows.

*Multiply the arc by the square of the arc, and take the result of repeating that (any number of times). Divide by the squares of the successive even numbers (such that current is multiplied by previous) increased by that number and multiplied by the square of the radius. Place the arc and the successive results so obtained one below the other, and subtract each from the one above. These together give the jiva [sine], as collected together in the verse beginning with "vidvan" etc.*

卐 **Madhava's sine (jya or jiva) Series**

$$\sin\theta = q - \frac{\theta^3}{3!} + \frac{\theta^5}{5!} - \frac{\theta^7}{7!} + \cdots = \sum_{k=0}^{\infty} \frac{(-1)^k}{(2k+1)!}\theta^{2k+1}$$

卐 **Madhava's cosine (kotijya) Series**

Madhava's cosine series is stated in verses 2.442 and 2.443 in Yukti-dipika commentary (Tantrasamgraha-vyakhya) by Sankara Variar. A translation of the verses follows.

*Multiply the square of the arc by the unit (i.e. the radius) and take the result of repeating that (any number of times). Divide (each of the above numerators) by the square of the successive even numbers decreased by that number and multiplied by the square of the radius. But the first term is (now)(the one which is) divided by twice the radius. Place the successive results so obtained one below the other and subtract each from the one above. These together give the śara as collected together in the verse beginning with stena, stri, etc.*

$$\cos\theta = 1 - \frac{\theta^2}{2!} + \frac{\theta^4}{4!} - \frac{\theta^6}{6!} + \cdots = \sum_{k=0}^{\infty} \frac{(-1)^k}{(2k)!}\theta^{2k}$$

## 卐 Madhava's arctangent Series

(Tantrasamgraha-vyakhya) by Sankara Variar. A translation of the verses is given below. Jyesthadeva has also given a description of this series in Yuktibhasa.

*Now, by just the same argument, the determination of the arc of a desired sine can be (made). That is as follows: The first result is the product of the desired sine and the radius divided by the cosine of the arc. When one has made the square of the sine the multiplier and the square of the cosine the divisor, now a group of results is to be determined from the (previous) results beginning from the first. When these are divided in order by the odd numbers 1, 3, and so forth, and when one has subtracted the sum of the even(-numbered) results from the sum of the odd (ones), that should be the arc. Here the smaller of the sine and cosine is required to be considered as the desired (sine). Otherwise, there would be no termination of results even if repeatedly (computed).*

*By means of the same argument, the circumference can be computed in another way too. That is as (follows): The first result should by the square root of the square of the diameter multiplied by twelve. From then on, the result should be divided by three (in) each successive (case). When these are divided in order by the odd numbers, beginning with 1, and when one has subtracted the (even) results from the sum of the odd, (that) should be the circumference.*

$$\arctan \mathrm{x} = x - \frac{x^3}{3!} + \frac{x^5}{5!} - \frac{x^7}{7!} + \cdots = \sum_{k=0}^{\infty} \frac{(-1)^k}{(2k+1)!} x^{2k+1}, where\ |x| \leq 1$$

All three series were later independently discovered in 17th century Europe. The series for sine and cosine were rediscovered by *Isaac Newton* in 1669, and the series for arctangent was rediscovered by *James Gregory* in 1671 and *Gottfried Leibniz* in 1673, and is conventionally called *Gregory*'s series.

The specific value

$$\arctan 1 = \frac{1}{4}\pi$$

can be used to calculate the circle constant $\pi$, and the arctangent series for 1 is conventionally called *Leibniz*'s series.

In recognition of *Madhava*'s priority, in recent literature these series are sometimes called the *Madhava–Newton series, Madhava–Gregory series,* or *Madhava–Leibniz* series.

No surviving works of *Madhava* contain explicit statements regarding the expressions which are now referred to as *Madhava* series. However, in the writing of later Kerala school mathematicians *Nilakantha Somayaji* and Jyeshthadeva one can find unambiguous attributions of these series to *Madhava*. These later works also include proofs and commentary which suggest how *Madhava* may have arrived at the series.

## 卐 Madhava's Sine Table

*Madhava*'s sine table is the table of trigonometric sines constructed by *Madhava* himself. The table lists the trigonometric sines of the twenty-four angles from 3.75° to 90° in steps of 3.75° (1/24 of a right angle, 90°). The table is encoded in the letters of the *Devanagari* alphabet using the *Katapayadi system* (कटपयादी व्यवस्था), giving entries the appearance of the verses of a poem in Sanskrit (see Figure 4.1). *Madhava*'s original work containing the sine table has not been found. The table is reproduced in the *Aryabhatiyabhashya* (आर्यभटियाभाष्य) of *Nilakantha Somayaji* (नीलकण्ठ सोमयाजी) and also in the *Yuktidipika* (युक्तिदीपिका) /*Laghuvivrti* (लघुविवृत्ति) commentary of *Tantrasamgraha* by *Sankara Variar* (circa. 1500-1560). The Figure below gives *Madhava*'s sine table in *Devanagari* as reproduced in Cultural foundations of mathematics by *C.K. Raju*. The first twelve lines constitute the entries in the table. The last word in the thirteenth line indicates that these are "*as told by Madhava*".

श्रेष्ठं नाम वरिष्ठानां हिमाद्रिर्वेदभावनः।
तपनो भानुसूक्तज्ञो मध्यमं विद्धि दोहनम्॥
धिगाज्यो नाशनं कष्टं छत्रभोगाशयाम्बिका।
म्रिगाहारो नरेशोऽयं वीरो रणजयोत्सुकः॥
मूलं विशुद्धं नाळस्य गानेषु विरळा नराः।
अशुद्धिगुप्ता चोरश्रीः शङ्कुकर्णो नगेश्वरः॥
तनुजो गर्भजो मित्रं श्रीमानत्र सुखी सखे! ।
शशी रात्रौ हिमाहारो वेगज्ञः पथि सिन्धुरः॥
छायालयो गजो नीलो निर्मलो नास्ति सत्कुले।
रात्रौ दर्पणमभ्राङ्गं नागस्तुङ्गनखो बली॥
धीरो युवा कथालोलः पूज्यो नारीजनैर्भगः।
कन्यागारे नागवल्ली देवो विश्वस्थली भृगुः॥
तत्परादिकलान्तास्तु महाज्या माधवोदिताः।
स्वस्वपूर्वविशुद्धे तु शिष्टास्तत्खण्डमौर्विकाः॥ २.९.५ ॥

Figure 4.1. *Madhava*'s sine table in Devanagari.

To understand the meaning of the values tabulated by *Madhava,* consider some angle whose measure is A.

Consider a circle of unit radius and center O. Let the arc PQ of the circle subtend an angle A at the center O. Drop the perpendicular QR from Q to OP; then the length of the line segment RQ is the value of the trigonometric sine of the angle A. Let PS be an arc of the circle whose length is equal to the length of the segment RQ. For various angles A, *Madhava*'s table gives the measures of the corresponding angles ∠POS in arcminutes, arcseconds and sixtieths of an arcsecond (see Figure 4.2).

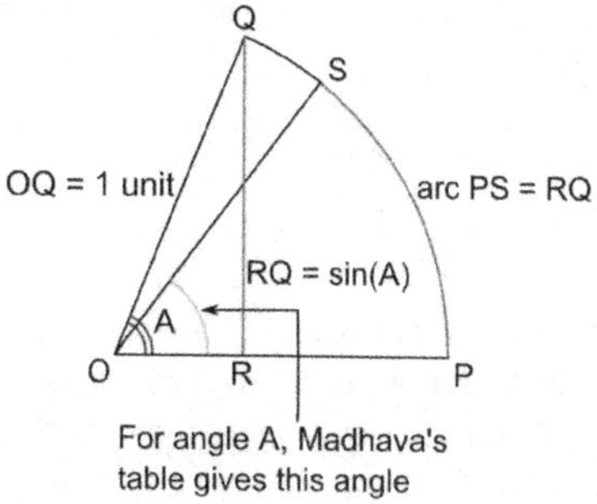


Figure 4.2. Diagram explaining the meaning of the values in *Madhava*'s table.

> multiply 0.3826834 radians by 180/π to convert to 21.92614 degrees, which is
> 1315 arcminutes 34 arcseconds 07 sixtieths of an arcsecond, abbreviated 13153407.

In the *Katapayadi system* the digits are written in the reverse order, so that the entry corresponding to 22.50° is 70435131.

**Derivation of Trigonometric sines From Madhava's Table**

For an angle whose measure is A, let

$$\angle POS = m \quad arcminutes, s \quad arcseconds, t \quad sixtieths\ of\ an\ arcsecond$$

Then

$$\sin(A) = \angle POS$$
$$= length\ of\ arc\ PS$$
$$= \angle POS\ in\ radius$$
$$= \frac{\pi}{180 \times 60}(m + \frac{s}{60} + \frac{t}{60 \times 60})$$

Each of the lines in the table specifies eight digits. Let the digits corresponding to angle A (read from left to right) be

$$d_1, d_2, d_3, d_4, d_5, d_6, d_7, d_8$$

Then according to the rules of the *Katapayadi system* of Kerala mathematicians we have

$$m = d_8 \times 1000 + d_7 \times 100 + d_6 \times 10 + d_5$$
$$s = d_4 \times 10 + d_3$$
$$t = d_2 \times 10 + d_1$$

The table below (see Table 4.1) has entries for the twenty-four angles from 3.75° to 90° in steps of 3.75° (first column). The second column contains the values tabulated by *Madhava* in Devanagari in the form he used.

These are taken from Malayalam Commentary of Karanapaddhati by P.K. Koru and are slightly different from the table given in Cultural foundations of mathematics by C.K. Raju.

The third column contains ISO 15919 transliterations of the lines in the second column. The numbers in the second column are decoded into arcminutes, arcseconds, and sixtieths of an arcsecond in Arabic numerals in the fourth column. The values of the trigonometric sines derived from the numbers specified in *Madhava*'s table (see Table 4.1) are listed in the fifth column, using *Madhava*'s value of 3.1415926535922 for $\pi$. The actual values of the sines to 8 places are listed in the sixth column.

Table 4.1 *Madhava*'s Sine Table.

| Angle A degree | Sin A in Madhava's notation | | | Madhava's sin A | Accurate sin A |
|---|---|---|---|---|---|
| | in Devanagari script using Madhava's Katapayadi system | ISO 15919 transliteration | Decoded angle in minutes' seconds'' sixtieths of second''' | | |
| 3.75 | श्रेष्ठं नाम वरिष्ठानां | śreṣṭhaṁ nāma variṣṭhānāṁ | 224' 50'' 22''' | 0.065403 | 0.065403 |
| 7.5 | हिमाद्रिर्वेदभावनः | himādrirvēdabhāvanaḥ | 448' 42'' 58''' | 0.130526 | 0.130526 |
| 11.25 | तपनो भानु सूक्तज्ञो | tapanō bhānu sūktajñō | 670' 40'' 16''' | 0.19509 | 0.19509 |
| 15 | मध्यमं विद्धि दोहनं | maddhyamaṁ viddhi dōhanaṁ | 889' 45'' 15''' | 0.258819 | 0.258819 |
| 18.75 | धिगाज्यो नाशनं कष्टं | dhigājyō nāśanaṁ kaṣṭaṁ | 1105' 01'' 39''' | 0.321439 | 0.321439 |
| 22.5 | छन्नभोगाशयाम्बिका | channabhōgāśayāmbikā | 1315' 34'' 07''' | 0.382683 | 0.382683 |
| 26.25 | मृगाहारो नरेशोयं | mrgāhārō narēśōyaṁ | 1520' 28'' 35''' | 0.442289 | 0.442289 |
| 30 | वीरो रणजयोत्सुकः | vīrō raṇajayōtsukaḥ | 1718' 52'' 24''' | 0.5 | 0.5 |
| 33.75 | मूलं विशुद्धं नाळस्य | mūlaṁ viśuddhaṁ nāḷasya | 1909' 54'' 35''' | 0.55557 | 0.55557 |
| 37.5 | गानेषु विरळा नराः | gāneṣu viraḷā narāḥ | 2092' 46'' 03''' | 0.608761 | 0.608761 |
| 41.25 | अशुद्धिगुप्ता चोरश्रीः | aśuddhiguptā cōraśrīḥ | 2266' 39'' 50''' | 0.659346 | 0.659346 |
| 45 | शङ्कुकर्णो नगेश्वरः | śaṅkukarṇō nageśvaraḥ | 2430' 51'' 15''' | 0.707107 | 0.707107 |
| 48.75 | तनुजो गर्भजो मित्रं | tanujō garbhajō mitraṃ | 2584' 38'' 06''' | 0.75184 | 0.75184 |
| 52.5 | श्रीमानत्र सुखी सखे | śrīmānatra sukhī sakhē | 2727' 20'' 52''' | 0.793353 | 0.793353 |
| 56.25 | शशी रात्रौ हिमाहारौ | śaśī rātrou himāhārou | 2858' 22'' 55''' | 0.83147 | 0.83147 |
| 60 | वेगज्ञः पथि सिन्धुरः | vēgajñaḥ pathi sindhuraḥ | 2977' 10'' 34''' | 0.866025 | 0.866025 |
| 63.25 | छाया लयो गजो नीलो | chāya layō gajō nīlō | 3083' 13'' 17''' | 0.896873 | 0.896873 |

| | | | | | |
|---|---|---|---|---|---|
| 67.5 | निर्मलो नास्ति सत्कुले | nirmalō nāsti satkulē | 3176' 03" 50"' | 0.92388 | 0.92388 |
| 71.25 | रात्रौ दर्पणमभ्राङ्गं | rātrou darpaṇamabhrāṅgaṁ | 3255' 18" 22"' | 0.94693 | 0.94693 |
| 75 | नागस्तुङ्ग नखो बली | nāgastuṅga nakhō balī | 3320' 36" 30"' | 0.965926 | 0.965926 |
| 78.75 | धीरो युवा कथालोल: | dhīrō yuvā kathālōlaḥ | 3371' 41" 29"' | 0.980785 | 0.980785 |
| 82.5 | पूज्यो नारीजनैर्भग: | pūjyō nārījanairbhagaḥ | 3408' 20" 11"' | 0.991445 | 0.991445 |
| 86.25 | कन्यागारे नागवल्ली | kanyāgārē nāgavallī | 3430' 23" 11"' | 0.997859 | 0.997859 |
| 90 | देवो विश्वस्थली भृगु: | devō viśvasthalī bhr̥guḥ | 3437' 44" 48"' | 1 | 1 |

**Example**

*Madhava*'s table lists the following digits corresponding to the angle 45.00°:

$d_1, d_2, d_3, d_4, d_5, d_6, d_7, d_8$

5 1 1 5 0 3 4 2

$$m = 2 \times 1000 + 4 \times 100 + 3 \times 10 + 0 \; arcminutes$$
$$= 2430 \; arcminutes$$

$$s = 5 \times 10 + 1 \; arcseconds$$
$$= 51 \; arcseconds$$

$$t = 1 \times 10 + 5 \; sixtieths \; of \; an \; arcsecond$$
$$= 15 \; sixtieth \; of \; an \; arcsecond$$

The value of the trigonometric sine of 45.00° as given in Madhava's table is:

$$\sin 45^0 = \frac{\pi}{180 \times 60}(2430 + \frac{51}{60} + \frac{15}{60 \times 60})$$

Substituting the value of $\pi$ computed by *Madhava* in the above expression, sin 45° is found to be 0.70710681. This is accurate to 6 decimal places.

## Madhava's Value of pi

To complete the numerical computations the value of pi ($\pi$) must be known. It is appropriate to use the value of $\pi$ computed by Madhava himself (which is accurate to 11 decimal places). Nilakantha Somayaji has given this value of $\pi$ in his Āryabhaṭīya-Bhashya as follows:

सङ्गमग्रामजो माधवः पुनरत्यासन्नां परिधिसंख्यामुक्तवान् -

विबुधनेत्रगजाहिहुताशनत्रिगुणवेदभवारणबाहवः ।
नवनिखर्वमिते वृतिविस्तरे परिधिमानमिदं जगदुर्बुधाः ॥

A transliteration of the last two lines:

*vibudha-netra-gaja-ahi-hutāśana*
*tri-guṇa-veda-bha-vāraṇa-bāhavaḥ*
*nava-nikharva-mite vr̥tivistare*
*paridhi-mānam idaṁ jagadur budhāḥ*

The various words indicate certain numbers encoded in a scheme known as the bhūtasaṃkhyā system. The meaning of the words and the numbers encoded by them (beginning with the units place) are detailed in the following translation of the verse:

*"Gods (vibudha : 33), eyes (netra : 2), elephants (gaja : 8), snakes (ahi : 8), fires (hutāśana : 3), three (tri : 3), qualities (guṇa : 3), vedas (veda : 4), nakṣatras (bha : 27), elephants (vāraṇa : 8), and arms (bāhavaḥ : 2) - the wise say that this is the measure of the circumference when the diameter of a circle is nava-nikharva (900,000,000,000)."*

So, the translation of the poem using the bhūtasaṃkhyā system will simply read: *"2827433388233 is, as the wise say, the circumference of a circle whose diameter is nava-nikharva (900,000,000,000)".*

That is, divide 2827433388233 (the number from the first two lines of the poem in reverse order) by nava-nikharva (900,000,000,000) to get the value of $\pi$. This calculation yields the value $\pi = 3.1415926535922$. This is the value of $\pi$ used by Madhava in his further calculations and is accurate to 11 decimal places.

(Tantrasamgraha-vyakhya) by Sankara Variar. A translation of the verses is given below. Jyesthadeva has also given a description of this series in continuation to earlier stated in Yuktibhasa.

*By means of the same argument, the circumference can be computed in another way too. That is as (follows): The first result should by the square root of the square of the diameter multiplied by twelve. From then on, the result should be divided by three (in) each successive (case). When these are divided in order by the odd numbers, beginning with 1, and when one has subtracted the (even) results from the sum of the odd, (that) should be the circumference.*

**Another Formula for the Circumference of a Circle**

The second part of the quoted text specifies another formula for the computation of the circumference $c$ of a circle having diameter $d$. This is as follows:

$$c = \sqrt{12d^2} - \frac{\sqrt{12d^2}}{3 \times 3} + \frac{\sqrt{12d^2}}{3^2 \times 5} - \frac{\sqrt{12d^2}}{3^3 \times 7} + \cdots$$

Since $c = \pi d$ this can be reformulated as a formula to computer $\pi$ as follows:

$$\pi = \sqrt{12}(1 - \frac{1}{3 \times 3} + \frac{1}{3^2 \times 5} - \frac{1}{3^3 \times 7} + \cdots)$$

We know that *Madhava* obtained an approximation for $\pi$ correct to 11 decimal places when he gave

$$\pi = 3.14159265359$$

which can be obtained from the last of *Madhava*'s series above by taking 21 terms.

## Madhava's Correction Term

*Madhava*'s correction term is a mathematical that can be used to give a better approximation to the value of the mathematical constant $\pi$ (pi) than the partial sum approximation obtained by truncating the *Madhava-Leibniz* infinite series for $\pi$.

The *Madhava-Leibniz* infinite series for $\pi$ is given below:

$$\frac{\pi}{4} = 1 - \frac{1}{3} + \frac{1}{5} - \cdots$$

Taking the partial sum of the first $n$ terms we have the following approximation to $\pi$

$$\frac{\pi}{4} = 1 - \frac{1}{3} + \frac{1}{5} - \frac{1}{7} + \cdots + (-1)^{n-1}\frac{1}{2n-1}$$

Denoting the *Madhava* correction term by $F(n)$, we have the following better approximation to $\pi$:

$$\frac{\pi}{4} \approx 1 - \frac{1}{3} + \frac{1}{5} - \frac{1}{7} + \cdots + (-1)^{n-1}\frac{1}{2n-1} + (-1)^n F(n)$$

Three different expressions have been attributed to madhava as possible values of $F(n)$, namely

$$F_1(n) = \frac{1}{4n}$$

$$F_2(n) = \frac{n}{4n^2+1}$$

$$F_3(n) = \frac{(n^2+1)}{4n^3+5n}$$

In the extant writings of the mathematicians of the Kerala school there are some indications regarding how the correction terms $F_1(n)$ *and* $F_2(n)$ have been obtained, but there is no indication on how the expression $F_3(n)$ has been obtained. This has led to a lot of speculative work on how the formulas might have been derived.

The expression for $F_2(n)$ *and* $F_3(n)$ are given expliciety in the *Yuktibhasha*, a major treatise on mathematics and astronomy authored by the Bhartiya astronomer *Jyesthadeva* of the Kerala school of mathematics around 1530, but that for $F_1(n)$ appears there only as a step in the argument leading to the derivation of $F_2(n)$.

The *Yuktidipika-Laghuvivrthi* commentary of *Tantrasangraha*, a treatise written by *Nilakantha Somayaji* an astronomer/mathematician belonging to the Kerala school of astronomy and mathematics and completed in 1501, presents the second correction term in the following verses (*Chapter 2: Verses 271 - 274*):

"व्यासे वारिधिनिहते रूपहृते[2] व्याससागराभिहते ।
त्रिशरादिविषमसंख्याभक्तमृणं स्वं पृथक् क्रमात् कुर्यात् ॥ २७१ ॥

यत्संखयायात्र हरणे कृते निवृत्ता हृतिस्तु जामितया ।
तस्या उर्ध्वगताया समसङ्ख्या तद्दलं गुणोऽन्ते स्यात् ॥ २७२ ॥

तद्वर्गो रूपयुतो हारो व्यासाब्धिघाततः प्राग्वत् ।

ताभ्यामाप्तं स्वमृणे कृते धने क्षेप एव करणीयः ॥ २७३ ॥

लब्धः परिधिः सूक्ष्मो बहुकृत्वो हरणतोऽतिसूक्ष्मः स्यात्" ॥ २७४ ॥

To the diameter multiplied by 4 alternately add and subtract in order the diameter multiplied by 4 and divided separately by the odd numbers 3, 5, etc. That odd number at which this process ends, four times the diameter should be multiplied by the next even number, halved and [then] divided by one added to that [even] number squared. The result is to be added or subtracted according as the last term was subtracted or added. This gives the circumference more accurately than would be obtained by going on with that process.

Translation

In modern notations this can be stated as follows (where $d$ is the diameter of the circle):

$$\text{Circumference} = 4d - \frac{4d}{3} + \frac{4d}{5} - \cdots \pm \frac{4d}{p} \mp \frac{4d(p+1)/2}{1+(p+1)^2}$$

If we set $p = 2n - 1$, then last term in the term in right hand side of the above equation reduces to $4dF_2(n)$. The same commentary also gives the correction term $F_3(n)$ in the following verses (Chapter 2: Verses 295 - 296):

"एभ्यः सूक्ष्मतरोऽन्यो विलिख्यते कश्चनापि संस्कारः ।
अन्ते समसंख्यादलवर्गः सैको गुणः स एव पुनः ॥ २९५ ।
युगगुणितो रूपयुतः समसंख्यादलहृतो भवेद् हारः ।
त्रिशरादिविषमसंख्याहरणात् परमेतदेव वा कार्यम्" । २९६ ॥

A subtler method, with another correction. [Retain] the first procedure involving division of four times the diameter by the odd numbers, 3, 5, etc. [But] then add or subtract it [four times the diameter] multiplied by one added to the next even number halved and squared, and divided by one added to four times the preceding multiplier [with this] multiplied by the even number halved.

Translation

In modern notations this can be stated as follows:

$$\text{Circumference} = 4d - \frac{4d}{3} + \frac{4d}{5} - \cdots \pm \frac{4d}{p} \mp \frac{4dm}{(1+4m)+(p+1)^2}$$

where the multiplier $m = 1 + \left(\frac{p+1}{2}\right)^2$. If we set $p = 2n - 1$, the last term in the right-hand side of the above equation reduces to $4dF_3(n)$.

**Further Reading**


1. "An overview of Indian mathematics". Indian Maths. School of Mathematics and Statistics University of St Andrews, Scotland. Retrieved 7 July 2006.
2. "Indians predated Newton 'discovery' by 250 years". press release, University of Manchester. 13 August 2007. Archived from the original on 21 March 2008. Retrieved 5 September 2007.
3. "Jyesthardeva". Biography of Jyesthadeva. School of Mathematics and Statistics University of St Andrews, Scotland. Retrieved 7 July 2006.
4. "Neither Newton nor Leibniz – The Pre-History of Calculus and Celestial Mechanics in Medieval Kerala". MAT 314. Canisius College. Archived from the original on 6 August 2006. Retrieved 9 July 2006.
5. "Science and Mathematics in India". South Asian History. India Resources. Archived from the original on 17 October 2012. Retrieved 6 May 2020.
6. "Science and technology in free India" (PDF). Government of Kerala—Kerala Call, September 2004. Prof. C.G.Ramachandran Nair. Archived from the original (PDF) on 21 August 2006. Retrieved 9 July 2006.
7. "The Kerala School, European Mathematics and Navigation". Indian Mathemematics. D.P. Agrawal – Infinity Foundation. Retrieved 9 July 2006.
8. "The Yuktibhasa Calculus Text" (PDF). The Pre-History of Calculus and Celestial Mechanics in Medieval Kerala. Dr Sarada Rajeev. Retrieved 9 July 2006.
9. A.P. Jushkevich (1961). Geschichte der Mathematik im Mittelalter (German translation, Leipzig, 1964, of the Russian original, Moscow, 1961). Moscow.
10. Bag 1976.
11. C. K. Raju (2001). "Computers, mathematics education, and the alternative epistemology of the calculus in the Yuktibhāṣā" (PDF). Philosophy East & West. 51 (3): 325–362. doi:10.1353/pew.2001.0045. S2CID 170341845. Retrieved 11 February 2020.
12. C. K. Raju (2007). History of Science, Philosophy and Culture in Indian Civilization General Editor D. P. Chattopadhyaya Volume X Part 4. Cultural Foundations of Mathematics: The Nature of Mathematical Proof and the Transmission of the Calculus from India to Europe in the 16th c. CE. New Delhi: Centre for Studies in

Civilizations and Dorling Kindersley (India) Pvt Ltd. pp. 173–174. ISBN 978-81-317-0871-2.
13. Charles Whish (1834), "On the Hindu Quadrature of the circle and the infinite series of the proportion of the circumference to the diameter exhibited in the four Sastras, the Tantra Sahgraham, Yucti Bhasha, Carana Padhati and Sadratnamala", Transactions of the Royal Asiatic Society of Great Britain and Ireland, 3 (3): 509–523, doi:10.1017/S0950473700001221, JSTOR 25581775
14. D F Almeida, J K John and A Zadorozhnyy (2001). "Keralese mathematics: its possible transmission to Europe and the consequential educational implications". Journal of Natural Geometry. 20 (1): 77–104.
15. David Edwin Pingree (1981). Census of the exact sciences in Sanskrit. Vol. 4. Philadelphia: American Philosophical Society. pp. 414–415.
16. Divakaran, P. P. (2007). "The First Textbook of Calculus: "Yuktibhāṣā"". Journal of Indian Philosophy. 35 (5/6): 417–443. doi:10.1007/s10781-007-9029-1. ISSN 0022-1791. JSTOR 23497280. S2CID 170254981.
17. For a review of Ganita-yukti-bhasa (Rationales in Mathematical Astronomy) of Jyesthadeva by Mathematical Association of America see : Homer S. White (17 July 2009). "Ganita-Yukti-Bhāsā (Rationales in Mathematical Astronomy) of Jyesthadeva". The Mathematical Association of America. Retrieved 28 May 2022.
18. For example: Gupta 1973; Joseph 2011, p. 428.
19. For example: Gupta 1992.
20. For more details on contents see Kinokuniya DataBase: "Ganita-yukti-bhasa (Rationales in Mathematical Astronomy) of Jyesthadeva". Retrieved 1 May 2010.
21. George E. Andrews, Richard Askey, Ranjan Roy (1999). Special Functions. Cambridge University Press. p. 58. ISBN 0-521-78988-5.
22. George Ghevarghese Joseph (2009). A Passage to Infinity Medieval Indian Mathematics from Kerala and Its Impact. New Delhi: SAGE Publications India Pvt Ltd. pp. 132–133. ISBN 978-81-321-0168-0.
23. George Gheverghese Joseph (2000). The crest of the peacock. Internet Archive. Princeton University Press. ISBN 978-0-691-00659-8.
24. Gold, D.; Pingree, D. (1991), "A hitherto unknown Sanskrit work concerning Madhava's derivation of the power series for sine and cosine", Historia Scientiarum, 42: 49–65.
25. Gregory received a letter from John Collins including Newton's sine and cosine series in late 1670. He discovered the general Taylor series

and sent a now-famous letter back to Collins in 1671 including several specific series including the arctangent. See Roy 1990.
26. Gupta 1973.
27. Gupta, R. C. (1992). "On the remainder term in the Madhava-Leibniz's series". Ganita Bharati. 14 (1–4): 68–71.
28. Ian G. Pearce (2002). Madhava of Sangamagramma. MacTutor History of Mathematics archive. University of St Andrews.
29. J. J. O'Connor and E. F. Robertson (2000). "Madhava of Sangamagramma". MacTutor History of Mathematics archive. School of Mathematics and Statistics, University of St Andrews, Scotland. Archived from the original on 14 May 2006. Retrieved 8 September 2007.
30. Joseph, George Gheverghese (October 2010) [1991]. The Crest of the Peacock: Non-European Roots of Mathematics (3rd ed.). Princeton University Press. ISBN 978-0-691-13526-7.
31. K V Sarma (1972). A History of the Kerala School of Hindu Astronomy. Hoshiarpur.
32. K V Sarma; S Hariharan (1991). "Yuktibhāṣā of Jyeṣṭhadeva: A book on rationales in Indian Mathematics and Astronomy: An analytic appraisal" (PDF). Indian Journal of History of Science. 26 (2). Archived from the original (PDF) on 28 September 2006. Retrieved 9 July 2006.
33. K. Chandra Hari (2003). "Computation of the true moon by Madhva of Sangamagrama". Indian Journal of History of Science. 38 (3): 231–253. Retrieved 27 January 2010.
34. K. V. Sarma (1972). "A History of the Kerala School of Hindu Astronomy (in perspective). Hoshiarpur: Vishveshvaranand Institute of Sanskrit & Indological Studies, Panjab University. p. 51.
35. K. V. Sarma (1973). Computation of the True Moon by Madhava of sangamagrama. Hoshiarpur: Vishveshvaranand Institute of Sanskrit and Indological Studies, Panjab University. p. 12.
36. K. V. Sarma (1973). Sputachandrapti: Computation of the True Moon by Madhava of Sangamagrama. Hoshiarpur, Punjab: Vishveshvaranand Institute of Sanskrit and Indological Studies, Panjab University. p. 8.
37. K. V. Sarma with explanatory notes in English by K. Ramasubrahmanyan, M. D. Srinivas, M. S. Sriram (2008). Ganita-Yukti-Bhasha of Jyeshthadeva. Volume I - Mathematics. New Delhi: Hindustan Book Agency. pp. 201–207. ISBN 978-81-85931-81-4.

38. K. V. Sarma; S. Hariharan (eds.). "A book on rationales in Indian Mathematics and Astronomy—An analytic appraisal" (PDF). Yuktibhāṣā of Jyeṣṭhadeva. Archived from the original (PDF) on 28 September 2006. Retrieved 9 July 2006.
39. K.V. Sarma (2004). Ganita Yuktibhasa (Volume III). Shimla: Indian Institute of Advanced Study. ISBN 81-7986-052-3.
40. Katz, Victor J. (1 June 1995). "Ideas of Calculus in Islam and India". Mathematics Magazine. 68 (3): 163–174. doi:10.1080/0025570X.1995.11996307. ISSN 0025-570X.
41. Madhava extended Archimedes' work on the geometric Method of Exhaustion to measure areas and numbers such as π, with arbitrary accuracy and error limits, to an algebraic infinite series with a completely separate error term. C T Rajagopal and M S Rangachari (1986). "On medieval Keralese mathematics". Archive for History of Exact Sciences. 35 (2): 91–99. doi:10.1007/BF00357622. S2CID 121678430.
42. Newton (1669) De analysi per aequationes numero terminorum infinitas was circulated as a manuscript but not published until 1711.
43. O'Connor, John J.; Robertson, Edmund F. (2000). "Madhava of Sangamagramma". MacTutor History of Mathematics archive.
44. P. P. Divakaran (2018). The Mathematics of India: Concepts, Methods, Connections. Cochin: Springer - Hindustan Book Agency. pp. 282–290. ISBN 978-981-13-1773-6.
45. Publisher's (Indian Institute of Advanced Study) web page on the book:"Ganita Yuktibhasa by K.V. Sarma". Archived from the original on 17 March 2010. Retrieved 1 May 2010.
46. Purananuru 229.
47. Puthumana Somayaji. Karanapaddhati (with a commentary in Malayalam by P.K. Koru). Cherpu, Kerala, India: Astro Printing and Publishing Company. (Published in 1953).
48. R C Gupta (1973). "The Madhava-Gregory series". Math. Education. 7: B67–B70.
49. R. C. Gupta (1975). "Madhava's and other medieval Indian values of pi". Math. Education. 9 (3): B45–B48.
50. Raju 2007, p. 119.
51. Raju 2007, p. 231.
52. Raju 2007, pp. 114–120.
53. Ranjan Roy (2011). Sources in the Development of Mathematics Infinite Series and Products from the Fifteenth to the Twenty-first

Century. New York: Cambridge University Press. p. 5. ISBN 978-0-521-11470-7.

54. Roy, Ranjan (1990). "The Discovery of the Series Formula for π by Leibniz, Gregory and Nilakantha" (PDF). Mathematics Magazine. 63 (5): 291–306. doi:10.2307/2690896. JSTOR 2690896. Archived from the original (PDF) on 24 February 2012.
55. Sarma 1972.
56. Sarma, K. V. (1977). Contributions to the study of Kerala school of Hindu astronomy and mathematics. Hoshiarpur: V V R I.
57. Sarma, K.V. (2009). Ganita Yuktibhasa (in Malayalam and English). Vol. III. Indian Institute of Advanced Study, Shimla, India. ISBN 978-81-7986-052-6. Archived from the original on 17 March 2010. Retrieved 16 December 2009.
58. Sayahna Foundation (20 November 2020). "Yukthibhasha digital edition" (PDF).
59. See publishers's (Springer's) web page on the book: Ganita-Yukti-Bhasa (Rationales in Mathematical Astronomy) of Jyesthadeva. ISBN 9781848820722. Retrieved 29 April 2010.
60. T. Hayashi, T. Kusuba and M. Yano. "The correction of the Madhava series for the circumference of a circle", Centaurus 33 (pages 149–174). 1990.
61. The 13-digit accurate value of π, 3.1415926535898, can be reached using the infinite series expansion of π/4 (the first sequence) by going up to n = 76.
62. Yuktibhâsâ, Part I (ed) with notes by Ramavarma (Maru) Thampuran and A. R. Akhileswara Aiyer, Magalodayam Ltd., Trichur, Kerala, 1123 Malayalam Era, 1948 CE.
63. A K Bag, Madhava's sine and cosine series, Indian J. History Sci. 11 (1) (1976), 54-57.
64. Anderson, Marlow; Katz, Victor; Wilson, Robin, eds. (2004). Sherlock Holmes in Babylon and Other Tales of Mathematical History. Mathematical Association of America. pp. 107–174. doi:10.4169/j.ctt13x0n0r.
65. Bag, A.K. (1976). "Madhava's sine and cosine series" (PDF). Indian Journal of History of Science. Indian National Academy of Science. 11 (1): 54–57. Archived from the original (PDF) on 5 July 2015. Retrieved 21 August 2016.
66. Bressoud, David (2002). "Was calculus invented in India?". College Mathematics Journal. 33 (1): 2–13.

doi:10.1080/07468342.2002.11921911. Reprinted in Anderson & al. 2004, pp. 131–137

67. C T Rajagopal and M S Rangachari, On an untapped source of medieval Keralese mathematics, Arch. History Exact Sci. 18 (1978), 89-102.
68. C T Rajagopal and M S Rangachari, On medieval Keralese mathematics, Arch. History Exact Sci. 35 (1986), 91-99.
69. C. T. Rajagopal & M. S. Rangachari (1978). "On an Untapped Source of Medieval Keralese Mathematics". Archive for History of Exact Sciences. 18 (2): 89–102. doi:10.1007/BF00348142. S2CID 51861422.
70. C. T. Rajagopal and M. S. Rangachari (1986). "On Medieval Kerala Mathematics". Archive for History of Exact Sciences. 35 (2): 91–99. doi:10.1007/BF00357622. JSTOR 41133779. S2CID 121678430.
71. D Gold and D Pingree, A hitherto unknown Sanskrit work concerning Madhava's derivation of the power series for sine and cosine, Historia Sci. No. 42 (1991), 49-65.
72. Filali, Mahmoud (2012). "Harmonic analysis and applications". Kybernetes. 41: 129–144. doi:10.1108/03684921211213160.
73. For a thorough discussion of the computation of Madhava's sine table with historical references: C.K. Raju (2007). Cultural foundations of mathematics: The nature of mathematical proof and the transmission of calculus from India to Europe in the 16 thc. CE. History of Philosophy, Science and Culture in Indian Civilization. Vol. X Part 4. Delhi: Centre for Studies in Civilizations. pp. 114–123.
74. For an account of Madhava's computation of the sine table see: Van Brummelen, Glen (2009). The mathematics of the heavens and the earth: the early history of trigonometry. Princeton: Princeton University Press. pp. 113–120. ISBN 978-0-691-12973-0.
75. G G Joseph, The crest of the peacock (London, 1991).
76. Gold, David; Pingree, David (1991). "A Hitherto Unknown Sanskrit Work concerning Mādhava's Derivation of the Power Series for Sine and Cosine". Historia Scientiarum. 42: 49–65.
77. Gregory–Leibniz–Nilakantha formula: Gawrońska, Natalia; Słota, Damian; Wituła, Roman; Zielonka, Adam (2013). "Some generalizations of Gregory's power series and their applications" (PDF). Journal of Applied Mathematics and Computational Mechanics. 12 (3).
78. Gupta 1987; Katz 1995; Roy 2021, Ch. 1. Power Series in Fifteenth-Century Kerala, pp. 1–22

79. Gupta, Radha Charan (1973). "The Mādhava-Gregory series". The Mathematics Education. 7. B: 67–70.
80. Gupta, Radha Charan (1975). "Mādhava's and other medieval Indian values of Pi". The Mathematics Education. 9. B: 45–48.
81. Gupta, Radha Charan (1976). "Mādhava's power series computation of the sine". Gaṇita. 27 (1–2): 19–24.
82. Gupta, Radha Charan (1987). "South Indian Achievements in Medieval Mathematics". Gaṇita Bhāratī. 9: 15–40. Reprinted in Ramasubramanian, K., ed. (2019). Gaṇitānanda: Selected Works of Radha Charan Gupta on History of Mathematics. Springer. pp. 417–442. doi:10.1007/978-981-13-1229-8_40.
83. Gupta, Radha Charan (1991). "The Mādhava–Gregory series for tan−1x". Indian Journal of Mathematics Education. 11 (3): 107–110.
84. Gupta, Radha Charan (1992). "On the remainder term in the Mādhava–Leibniz series". Gaṇita Bhāratī. 14 (1–4): 68–71.
85. Hayashi, Takao; Kusuba, Takanori; Yano, Michio (1990). "The correction of the Madhava series for the circumference of a circle". Centaurus. 33 (2): 149–174. doi:10.1111/j.1600-0498.1990.tb00725.x.
86. Horvath, Miklos (1983). "On the Leibnizian quadrature of the circle" (PDF). Annales Universitatis Scientiarum Budapestiensis (Sectio Computatorica). 4: 75–83.
87. Joseph, George Gheverghese (2011) [1st ed. 1991]. The Crest of the Peacock: Non-European Roots of Mathematics (3rd ed.). Princeton University Press. ISBN 978-0-691-13526-7.
88. K V Sarma, A History of the Kerala School of Hindu Astronomy (Hoshiarpur, 1972).
89. Katz, Victor J. (1995). "Ideas of Calculus in Islam and India" (PDF). Mathematics Magazine. 68 (3): 163–174. doi:10.1080/0025570X.1995.11996307. JSTOR 2691411. Reprinted in Anderson & al. 2004, pp. 122–130.
90. Katz, Victor J., ed. (2007). "Chapter 4: Mathematics in India IV. Kerala School". The mathematics of Egypt, Mesopotemia, China, India and Islam: A source book. Princeton University Press. pp. 480–495. ISBN 978-0-691-11485-9.
91. Leibniz later included the series for sine and cosine in Leibniz (1676) De quadratura arithmetica circuli ellipseos et hyperbola cujus corollarium est trigonometria sine tabulis, which was only finally published in 1993. However, he had been sent Newton's sine and cosine series by Henry Oldenburg in 1675 and did not claim to have discovered them. See: Levrie, Paul (2011). "Lost and Found: An

Unpublished ζ(2)-Proof". Mathematical Intelligencer. 33: 29–32. doi:10.1007/s00283-010-9179-y.
92. Madhava correction term, https://en.wikipedia.org/wiki/Madhava%27s_correction_term.
93. Madhava series, https://en.wikipedia.org/wiki/Madhava_series.
94. Madhava, https://en.wikipedia.org/wiki/Madhava_of_Sangamagrama.
95. Madhava, https://mathshistory.st-andrews.ac.uk/Biographies/Madhava/.
96. Madhava' sine table, https://en.wikipedia.org/wiki/Madhava%27s_sine_table.
97. Madhava–Gregory–Leibniz series: Benko, David; Molokach, John (2013). "The Basel Problem as a Rearrangement of Series". College Mathematics Journal. 44 (3): 171–176. doi:10.4169/college.math.j.44.3.171.
98. Madhava–Leibniz–Gregory series: Danesi, Marcel (2021). "1. Discovery of π and Its Manifestations". Pi (π) in Nature, Art, and Culture. Brill. pp. 1–30. doi:10.1163/9789004433397_002.
99. Nilakantha–Gregory series: Campbell, Paul J. (2004). "Borwein, Jonathan, and David Bailey, Mathematics by Experiment". Reviews. Mathematics Magazine. 77 (2): 163. doi:10.1080/0025570X.2004.11953245.
100. P. Rajasekhar (June 2011). "Derivation of remainder term for the Series expansion of π as depicted in Yukthibhasa and its modern Interpretation". Bulletin of Kerala Mathematics Association. 8 (1): 17–39.
101. Plofker, Kim (2005). "Relations between approximations to the sine in Kerala mathematics". In Emch, Gérard G.; Sridharan, R.; Srinivas, M. D. (eds.). Contributions to the History of Indian Mathematics. Gurgaon: Hindustan Book Agency. pp. 135–152. doi:10.1007/978-93-86279-25-5_6.
102. Plofker, Kim (2009). Mathematics in India. Princeton: Princeton University Press. pp. 217–254. ISBN 978-0-691-12067-6.
103. Pouvreau, David (2003). Trigonométrie et "développements en séries" en Inde médiévale (in French). IREM de Toulouse.
104. Pouvreau, David (2015). "Sur l'accélération de la convergence de la série de Madhava-Leibniz". Quadrature (in French). 97: 17–25.
105. Probst, Siegmund (2015). "Leibniz as reader and second inventor: The cases of Barrow and Mengoli". In Goethe, N.; Beeley, P.; Rabouin, D.

(eds.). G.W. Leibniz, Interrelations between Mathematics and Philosophy. Springer. pp. 111–134. doi:10.1007/978-94-017-9664-4_6.
106. R C Gupta, Madhava's and other medieval Indian values of pi, Math. Education 9 (3) (1975), B45-B48.
107. R C Gupta, Madhava's power series computation of the sine, Ganita 27 (1-2) (1976), 19-24.
108. R C Gupta, Madhava's rule for finding angle between the ecliptic and the horizon and Aryabhata's knowledge of it, in History of oriental astronomy, New Delhi, 1985 (Cambridge, 1987), 197-202.
109. R C Gupta, On the remainder term in the Madhava-Leibniz's series, Ganita Bharati 14 (1-4) (1992), 68-71.
110. R C Gupta, The Madhava-Gregory series, Math. Education 7 (1973), B67-B70.
111. Raju, Chandrakant K. (2007). Cultural Foundations of Mathematics: The Nature of Mathematical Proof and the Transmission of the Calculus from India to Europe in the 16th c. CE. History of Science, Philosophy and Culture in Indian Civilization. Vol. X Pt. 4. New Delhi: Pearson Longman. ISBN 81-317-0871-3.
112. Ranjan Roy (13 June 2011). "Power Series in Fifteenth-Century Kerala", except from Sources in the Development of Mathematics: Infinite Series and Products from the Fifteenth to the Twenty-first Century. Cambridge University Press. ISBN 978-0-521-11470-7.
113. Roy 2021, Ch. 8. De Analysi per Aequationes Infinitas, pp. 165–185.
114. Roy, Ranjan (1990). "The Discovery of the Series Formula for $\pi$ by Leibniz, Gregory and Nilakantha" (PDF). Mathematics Magazine. 63 (5): 291–306. doi:10.1080/0025570X.1990.11977541. Reprinted in Anderson & al. 2004, pp. 111–121
115. Roy, Ranjan (2021) [1st ed. 2011]. Series and Products in the Development of Mathematics. Vol. 1 (2nd ed.). Cambridge University Press.
116. Sarma, K.V.; Ramasubramanian, K.; Srinivas, M.D.; Sriram, M.S. (2008). Ganita-Yukti-Bhasa (Rationales in Mathematical Astronomy) of Jyesthadeva. Sources and Studies in the History of Mathematics and Physical Sciences. Vol. I–II (1st ed.). Springer (jointly with Hindustan Book Agency, New Delhi). pp. LXVIII, 1084. Bibcode:2008rma..book.....S. ISBN 978-1-84882-072-2. Retrieved 17 December 2009.
117. Sarma, Krishna Venkateswara (1972). "2. Anticipation of modern mathematical discoveries by Kerala astronomers" (PDF). A History of

the Kerala School of Hindu Astronomy. Hoshiarpur: Vishveshvaranand Institute. pp. 11–28.

118. T Hayashi, T Kusuba and M Yano, The correction of the Madhava series for the circumference of a circle, Centaurus 33 (2-3) (1990), 149-174.

119. The Aryabhatiam of Aryabhattacharya with the Bhashya of Nilakantha Somasutvan, Part1-Gaṇitapāda, Edited by K. Sambasiva Sastri, Trivandrum Sanskrit Series No.101. p. 55. https://archive.org/details/Trivandrum_Sanskrit_Series_TSS http://www.sanskritebooks.org/2013/02/trivandrum-sanskrit-series-anantasayana-samskrita-granthavali/.

120. Van Brummelen, Glen (2009). "§3.8. Taylor Series for Trigonometric Functions in Madhava's Kerala School". The Mathematics of the Heavens and the Earth: The Early History of Trigonometry. Princeton University Press. pp. 113–120. doi:10.2307/j.ctv1pzk6f0.7. ISBN 978-0-691-12973-0.

121. Whish, Charles M. (1834). "XXXIII. On the Hindú Quadrature of the Circle, and the infinite Series of the proportion of the circumference to the diameter exhibited in the four S'ástras, the Tantra Sangraham, the Yucti Bháshá, Carana Padhati, and Sadratnamáka". Transactions of the Royal Asiatic Society. 3 (3): 509–523. doi:10.1017/S0950473700001221. JSTOR 25581775.

122. Young, Paul Thomas (2022). "From Madhava–Leibniz to Lehmer's Limit". American Mathematical Monthly. 129 (6): 524–538. doi:10.1080/00029890.2022.2051405.

123. Yuktibhasya, https://en.wikipedia.org/wiki/Yuktibh%C4%81%E1%B9%A3%C4%81.

## Vatasseri Parameshvara Nambudiri (वातसेरी परमेश्वर नम्बुदिरी) (c. 1380 – 1460)

Vatasseri Parameshvara Nambudiri (वातसेरी परमेश्वर नम्बुदिरी) was a major a mathematician and astronomer of the Kerala school of astronomy and mathematics founded by *Madhava of Sangamagrama*. He was also an astrologer. *Parameshvara* (परमेश्वर) was a proponent of observational astronomy in medieval Bharat and he himself had made a *series of eclipse observations* (ग्रहणनिरीक्षणानाम् श्रृङ्खला) to verify the accuracy of the computational methods then in use.

Based on his eclipse observations, *Parameshvara* proposed several corrections to the astronomical parameters which had been in use since the times of *Aryabhata*. The computational scheme based on the revised set of parameters has come to be known as the *Drgganita* (दृग्गनीता) *or Drig system* (दृग् प्रणाली). *Parameshvara* was also a prolific writer on matters relating to astronomy. At least 25 manuscripts have been identified as being authored by *Parameshvara*. *Parameshvara* was a Hindu of *Bhrgugotra* (भृगु गोत्र) following the *Ashvalayanasutra* (अश्वलायनसूत्रम्) of the *Rigveda* (ऋग्वेद). *Parameshvara*'s family name (*Illam*) was *Vatasseri* and his family resided in the village of *Alathiyur* (Sanskritised as *Asvatthagrama*) in *Tirur*, Kerala. *Alathiyur* or *Alattur* (10°51′N, 76°08′E) is situated on the northern bank of the river Nila (river Bharathappuzha) at its mouth in Kerala.

He was a grandson of a disciple of *Govinda Bhattathiri* (1237–1295 CE), a legendary figure in the astrological traditions of Kerala.

*Parameshvara* studied under teachers *Rudra* (रुद्र) *and Narayana* (नारायण), and also under *Madhava of Sangamagrama*. *Damodara* (दामोदर), another prominent member of the Kerala school, was his son and also his pupil. *Parameshvara* was also a teacher of *Nilakantha Somayaji* (1444–1544) the author of the celebrated *Tantrasamgraha* (तन्त्रसंग्रह).

**Works by Parameshvara**

The following works of *Parameshvara* are well-known. A complete list of all manuscripts attributed to Parameshvara is available in Pingree.

- *Bhatadipika* (भटदीपिका) – Commentary on Āryabhaṭīya of Āryabhaṭa I
- *Karmadipika* (कर्मदीपिका) – Commentary on Mahabhaskariya of Bhaskara I
- *Paramesvari* (परमेश्वरी) – Commentary on Laghubhaskariya of Bhaskara I
- *Sidhantadipika* (सिद्धान्तदीपिका) – Commentary on Mahabhaskariyabhashya of Govindasvāmi
- *Vivarana* (विवरण) – Commentary on Surya Siddhanta and Lilāvati
- *Drgganita* (दृग्गनिता) – Description of the Drig system (composed in 1431 CE)
- *Goladipika* (गोलादिपिका) – Spherical geometry and astronomy (composed in 1443 CE)
- *Grahanamandana* (ग्रहनमन्दन) – Computation of eclipses (Its epoch is 15 July 1411 CE.)
- *Grahanavyakhyadipika* (ग्रहनव्याख्यादिपिका) – On the rationale of the theory of eclipses
- *Vakyakarana* (वाक्यकरण) – Methods for the derivation of several astronomical tables

## Key Contributions in Mathematics

### Eclipse Observation

He made a series of *eclipse observations* (ग्रहणनिरीक्षणम्) over *five deacades* (पञ्च दशकानि). Constantly attempted to compare these with the theoretically computed positions of the planets. He revised planetary parameters based on his observations. Sun and moon, mean as well as true longitudes, of both *Paramesvara* and his disciple *Nilakantha* precisely agree with those determined using the most modern algorithms.

He was a prolific write on adtronomty as well as astrology and is best kwnoedn for his work *Drgganita,* which was formulated after taking into account the astronomical observations for a period of 55 years. A ddetailed account of his original works and commenteries are available in the *Drgganita* critically edited by *K. V. Sharma*.

His date can be gleaned from the dates of two textes - *Drgganita* being *Saka* 1353 (1431 AD) and *Goladipika II* inscribed with *Saka* 1364 (1443 AD).

He has made a mention of the important eclipses that have taken place during 1398 to 1431 AD in a long discussion at the end of the chapter of Eclipses in his commentary Siddhantadipika on the *Mahabhaskariya-bhasya* (महाभास्करी-भाष्यः) of *Govindasvamin* (गोविन्दस्वामिन्).

He has produced three works exclusively dealing with eclipse, viz.,

- *Grahanastaka* consisting of only 10 verses
- *Grahanamandana* consisting with 100 verses written in Kanara that deals with computational aspects of eclipses
- *Grahananyayadipika* of 85 verses presents the underlying theory of eclipses.

Both these works do not contain any information on the eclipses oberserved by him. The account on eclipses is available in his commentary Siddhantadpika on the *Mahabhaskariya-bhasya* of *Govindasvamin* at the end of the chapter on eclipses.

The *Grahanamandana* makes use of the *Kalikhanda* (कालिखण्ड) of 1648157 days, which *K. V. Sharma* has indicated, as corresponding to the *Kali* (कलि) year 4512, kataka 17, i.e.., 15th July 1410 AD.
Modern computation suggests that the epoch actually is 17th July 1411 AD.
We can have some idea of the accuracy of computations in this manual by comparing the mean longitudes accurately determined by the Acharya himself for the sun, moon, moon's apogee and node:

*Kalidina* = 1648157 days: 17 July 1411 AD, Friday, Ujjain sunrise 05:25:36, JD (TDT) = 2236622.5201.

## Ayanamsa of Paramesvara

It can be understood from *Nilkantha*'s remarks in *Jyotirmimamsa* that *Paramesvaracarya* had precisely fixed the *ayanamsa* (अयनंसा) to be $15^0$ in *Kali* year 4536 AD or AD 1435. If we adopt this, the *ayanamsa* for AD 1411 turns out to be $14^0 36'$.

| Planets | Paramesvara's Sideral $\lambda$ | Col. 2+$14^0 49'$ | Col. 2+$14^0 36'$ | Modern $\lambda$ | Modern True $\lambda$ | True $\lambda$ of Paramesvara's |
|---|---|---|---|---|---|---|
| Sun | $108^0 15' 07''$ | $123^0 04' 07''$ | $122^0 51' 07''$ | $122^0 42' 07''$ | $121^0 43' 45''$ | $121^0 44' 40''$ |

| | | | | | | |
|---|---|---|---|---|---|---|
| Moon | 60°17′1″ | 75°06′01″ | 74°53′01″ | 74°27′43″ | 75°43 | 75°49′45″ |
| Apogee | 65°23′18″ | 80°12′18″ | 79°59′18″ | 79°52′36″ | 75°45′22″ | - |
| Node | 331°41′04″ | 346°30′04″ | 346°17′04″ | 346°04′12″ | - | - |

By comparing columns 3 and 4 with column 5 it appears that the mean Sun of *Paramesvara* and modern $\lambda$ differed by 9 minutes of arc when constrasted using his own *ayanamasa*. But the true Sun of P*aramesvara* is precisely equal to the modern true Sun, a fact that reflects the ingenuity of *Paramesvara*. Column 3 suggests that the tradition value of ayanamasa was in surplus by nearly 13 minutes (after counting for the difference in equation of center) and *Paramesvara* rightly and precisely fixed the *siddhantic ayanamasa* to be 15° in AD 1435. It is possible that at the time of writing *Grahanamandana* in 1411 AD, *Paramesvara* probably had only the traditional value of 14°49′ at had and under such circumstances the sayana longitudes of *Paramesvara* would have been in error by a quarter degree.

As regards the accuracy of the results of this treatise on eclipses, *Paramesvara* himself had made the following observation in the concluding part:

| |
|---|
| *iti pūrvaśāstrayuktyā siddhamidaṃ grahaṇakarma nirdiṣṭam,*<br>*kālo'nena ca siddhaḥ kadācid iha bhidyate svalpam /* |
| Thus has been enunciated the computation of eclipses according to principles derived from the ancient texts. The times (of contact etc.) as obtained from this (calculation) may, at times, differ slightly (from observation).<br><br>Translation |

Modern astronomical data of solar eclipses recorded by *Paramesvara* are shown in Table 4.2 and Table 4.3.

Table 4.2 Modern astronomical data of solar eclipses recorded by *Paramesvara.*

| Location: $10^{0}51'$N, $76^{0}43'$E (Allattur) UT+05:00 | | | | | | | | |
|---|---|---|---|---|---|---|---|---|
| Date | Eclipse | Visibility | Beginning | Maximum | Magnitude | Moon/ Sun | Time | Sun's altitude |
| 9/11/1398 | Annular | Partial | 08:18 | 09:56 | 0.906 | 0.938 | 11:57 | $60^{0}$ |
| 19/10/1408 | Hybrid | Partial | 15:27 | 16:36 | 0.551 | 1.001 | 17:37 | $0^{0}.9$ |
| 1/2/1413 | Total | Partial | 07:39 | 08:47 | 0.675 | 1.045 | 10:04 | $47^{0}.4$ |
| 23/1/1422 | Total | Partial | 07:11 | 07:43 | 0.141 | 1.035 | 08:18 | $23^{0}.5$ |
| 8/8/1431 | Annular | Partial | 06:41 | 07:56 | 0.674 | 0.946 | 09:26 | $49^{0}.7$ |
| 2/2/1432 | Partial | Partial | 07:59 | 08:25 | 0.070 | 1.014 | 08:53 | $32^{0}.3$ |
| 10/11/1425 | - | No eclipse | | | | | | |
| 19/8/1430 | - | No eclipse | | | | | | |
| 10/11/1425 | Partial | Partial | 15:24 | 15:59 | 0.087 | 0.946 | 16:32 | $11^{0}.9$ |
| 19/8/1430 | Partial | Partial | 06:15 | 07:18 | 0.077 | 0.935 | 07:53 | $29^{0}.5$ |

Table 4.3 Observational data of *Paramesvara* [$10^{0}51'$N, $76^{0}43'$E].

| | | | Beginning | | | | Modern Data | |
|---|---|---|---|---|---|---|---|---|
| No. | Kalidina | Date | Gnomon shadow | Time | End | Duration (Paramesvara) | Beginning | End |
| 1 | 1643524 | 9/11/1398 | 11 | 08:13 | 11:49 | 3hour36minutes (9 nadikas) | 08:18 | 11:57 |
| 2 | 1647156 | 19/10/1408 | - | - | 17:31 | - | 15:27 | 17:37 |
| 3 | 1648722 | 1/2/1413 | 24 | 07:45 | - | - | 07:39 | 10:04 |
| 4 | 1652000 | 23/1/1422 | 35 | 07:15 | - | - | 07:11 | 08:18 |
| 5 | 16555484 | 8/8/1431 | - | - | - | - | 06:41 | 09:26 |
| 6 | 1655662 | 2/2/1432 | 15 | 08:05 | - | - | 07:59 | 08:53 |
| 7 | 1653387 | 10/11/1425 | No eclipse | - | - | - | No eclipse | - |
| 8 | 1655130 | 19/8/1430 | No eclipse | - | - | - | No eclipse | - |

A discussion on *Paramesvara* shall remain incomplete without a reference to his magnum opus - *Drgganita.*

Acoording to the information available in the text (*sloka 46*) the composition took place in *Saka* 1353 corresponding to AD 1431.

The epoch as such will be the expiry of the *Kali* year 4532: *Kalidina* = 1655352.34: JD[TDT] = 2243817.54490972; Thursday, March 29, 1431, 05:55 LMT [Ujjain Sunrise].

Discovered drek-ganita, a mode of astronomy based on observations.

## 卐Algebra and Geometry

One of *Parameshvara*'s more significant contributions was his mean value type formula for the inverse interpolation of the sine.
He was the first mathematician to give a formula for the radius of the circle circumscribing a *cyclic quadrilateral* (चक्रीय चतुर्भुज).

With the sides of the cyclic quadrilateral being a, b, c, and d, the radius R of the circumscribed circle is:

$$R = \sqrt{\frac{(ab+cd)(ac+bd)(ad+bc)}{(-a+b+c+d)(a-b+c+d)(a+b-c+d)(a+b+c-d)}}$$

The expression is sometimes attributed to *Lhuilier* (1782), 350 years later.

### Further Reading


1. A.K. Bag (May 1980). "Indian literature on mathematics during 1400–1800 AD" (PDF). Indian Journal of History of Science. 15 (1): 79–93. Archived from the original (PDF) on 9 March 2012.
2. Radha Charan Gupta [1977] "Parameshvara's rule for the circumradius of a cyclic quadrilateral", Historia Mathematica 4: 67–74
3. Radha Charan Gupta [1979] "A mean-value-type formula for inverse interpolation of the sine", Ganita 30 (1–2): 78—82.
4. Bhaskara, Mahabhaskariyam: With Parameshvara's commentary called Karmadipika (Poona, 1945).

5. David Edwin Pingree (1981). Census of the exact sciences in Sanskrit. A. Vol. 4. American Philosophical Society. pp. 187–192. ISBN 978-0-87169-213-9.
6. David Pingree, Biography in Dictionary of Scientific Biography (New York 1970–1990).
7. K. Chandra Hari (2003). "Eclipse observations of Parameshvara, the 14th–15th-century astronomer of Kerala" (PDF). Indian Journal of History of Science. 38 (1): 43–57. Archived from the original (PDF) on 16 March 2012. Retrieved 28 January 2010.
8. K. K. Raja (1963) "Astronomy and mathematics in Kerala", Brahmavidya 27; 136–143.
9. K. Shankar Shukla (1957) The Surya-siddhanta with the commentary of Parameshvara (Lucknow).
10. K. V. Sarma (2008), "Paramesvara", Encyclopaedia of the History of Science, Technology, and Medicine in Non-Western Cultures (2nd edition) edited by Helaine Selin, Springer, ISBN 978-1-4020-4559-2.
11. Kim Plofker (1996) "An example of the secant method of iterative approximation in a fifteenth-century Sanskrit text", Historia Mathematica 23 (3): 246–256.
12. Munjala, Laghumanasam: with commentary by Parameshvara (Poona, 1944).
13. Parameshvara Nambudiri, https://en.wikipedia.org/wiki/Parameshvara_Nambudiri.
14. T.A. Sarasvati Amma (1979) Geometry in ancient and medieval India, (Delhi).

## Gaṅgeśa Upādhyāya (गङ्गेश उपाध्याय)
## (first half of the 14th century)

Gaṅgeśa (गङ्गेश) or Gaṅgeśa Upādhyāya was a philosopher, logician and mathematician from the kingdom of *Mithila*. He established the *Navya-Nyāya* - नव्य-न्याय ("*New Logic*") school. His *Tattvachintāmaṇi* - तत्त्वचिन्तामणि (*The Jewel of Thought on the Nature of Things*), also known as *Pramāṇacintāmaṇi* - प्रमाणचिन्तामणि (*The Jewel of Thought on the Means of Valid Knowledge*), is the basic text for all later developments. The logicians of this school were primarily interested in defining their terms and concepts related to *non-binary logical* (अ द्विचक्रीय तार्किक) categories.

*Gaṅgeśa* was born at *Karion* village on the banks of the *Kamala River*, 19 km south-east of *Darbhanga* in a *Brahmin* family. According to tradition, he was illiterate in his early years but later he acquired the knowledge of logic as a *boon* (वरः) from the *goddess Kali* (देवी काली).

### 卐 Key Contributions in Mathematics

### 卐 The Tattvacintāmaṇi

The text is divided into *four khaṇḍas* (books):

- *Pratyakṣakhaṇḍa* – प्रत्यक्षखण्ड (book on perception)
- *Anumānakhaṇḍa* - अनुमानखण्ड (book on inference)
- *Upamānakhaṇḍa* - उपमानाखण्ड (book on comparison) and
- *Śabdakhaṇḍa* - शब्दखण्ड (book on verbal testimony).

The first book opens with a *salutation to Shiva* (शिवाय नमस्कार). The *Tattvacintāmaṇi* is a systematic account of *epistemology, logic, and the philosophy of grammar*. Other subjects, such as the *proofs of God* (ईश्वरस्य प्रमाणानि), are treated incidentally.

*Gaṅgeśa* refers to his own teachings as the *New Nyaya*. This term *New Nyaya* is not to be understood as implying any great originality in theory on Gangesa's part, but rather originality in method. His work differs from the oldest *Nyaya* in that he accepts many tenets of the *Vaisesika* (वैसेसिक) school, and in his arrangement of *Nyaya* teachings under four headings rather than under the 16 subjects (*padartha*) of the *Old Nyaya*.

The newness of *Gaṅgeśa*'s method is newness of style and of organization. He is far more precise, more careful to define his terms, than were his predecessors; these virtues of his work are responsible for the fact that perhaps half of *Navya-Nyaya* literature is based either directly on the *Tattvacintāmaṇi* or on a commentary on the *Tattvacintāmaṇi*.

Large sections of the treatise have not yet been translated into English or any other Bhartiya languages. Broadly, T*attvachintāmaṇi* is divided into four books dealing respectively with perception (*pratyaksha*), inference (*anumāna*), comparison (*upamāna*) and verbal testimony (*sabda*).

According to *Nyaya doctrines* (न्यायसिद्धान्ताः), these are the four means for deriving valid knowledge. The following references provide sources where one can find detailed accounts of the contents of *Tattvachintāmaṇi*.

Tattvachintāmaṇi has attracted many commentaries. *Vardhamana Mahopadhyaya,* a son of *Gaṅgeśa,* has himself written a commentary on *Tattvachintāmaṇi. The History of Logic* gives brief accounts of as many as *48 commentaries*.

He established *Navya- Nyāya darśana.*

### 卐 Navya-Nyāya ("New Logic") school

The *Navya-Nyāya* or Neo-Logical *darśana* (view, system, or school) of Bhartiya logic and Bhartiya philosophy was founded in the 13th century CE by the philosopher *Gangeśa Upādhyāya* and continued by *Raghunatha Siromani* of *Nabadwipa* (नबद्वीपः) in Bengal. It was a development of the classical *Nyāya darśana* (न्याय दर्शन). Other influences on *Navya-Nyāya* were the work of earlier philosophers *Vācaspati Miśra* - वाचस्पति मिश्र (900–980 CE) and *Udayana* - उदयन (late 10th century). It remained active in Bharat through to the 18th century.

*Gangeśa*'s book *Tattvacintāmaṇi* was written partly in response to *Śrīharśa*'s *Khandanakhandakhādya* (खण्डनखण्डखाद्य), a defence of *Advaita Vedānta* (अद्वैत वेदान्त), which had offered a set of thorough criticisms of *Nyāya* theories of thought and language. In his book, Gangeśa both addressed some of those criticisms and – more important – critically examined the *Nyāya darśana* itself. He held that, while *Śrīharśa* had failed to successfully challenge the *Nyāya realist ontology* (न्याय यथार्थवादी अस्तित्वशास्त्र), his and *Gangeśa*'s own criticisms brought out a need to improve and refine the logical and linguistic tools of *Nyāya* thought, to make them more rigorous and precise.

*Tattvacintāmani* dealt with all the important aspects of Bhartiya philosophy, logic, set theory, and especially epistemology, which *Gangeśa* examined rigorously, developing and improving the *Nyāya* scheme, and offering examples. The results, especially his analysis of cognition, were taken up and used by other darśanas.

*Navya-Nyāya* developed a sophisticated language and conceptual scheme that allowed it to raise, analyse, and solve problems in logic and epistemology.

It systematised all the *Nyāya* concepts into four main categories which are:

- perception (*pratyakşa*),
- inference (anumāna),
- comparison or similarity (upamāna), and
- testimony (sound or word; śabda).

Great stalwarts like *Basudev Sarvabhauma, Raghunath Shiromani, Jagadish Tarkalankar, Gadadhar Bhattacharya,* and *Mathuranatha Tarkavagisha* have contributed further in the development of the subject. *Professor John Vattanky* has contributed significantly to the modern understanding of *Navya-Nyāya.*

**Further Reading**

1. Banerjee, Suresh Chandra. "Bhattacharya, Gadadhar". Banglapedia.
2. BHATTACHARYYA, SIBAJIBAN. "GADĀDHARA BHAṬṬĀCĀRYA'S "VIṢAYATĀVĀDA" (Continued)". Journal of Indian Philosophy. 14 (3): 217-302 (86 pages).
3. BHATTACHARYYA, SIBAJIBAN. "GADĀDHARA BHAṬṬĀCĀRYA'S "VIṢAYATĀVĀDA"". Journal of Indian Philosophy. 14 (2): 109-193 (85 pages).
4. BHATTACHARYYA, SIBAJIBAN. "Some Features of Navya-Nyāya Logic". Philosophy East and West. 24 (3): 329-342 (14 pages).
5. BHATTACHARYYA, SIBAJIBAN. "Some Features of the Technical Language of Navya-Nyāya". Philosophy East and West. 40 (2): 129-149 (21 pages).

6. Can logic & Critical reasoning remove suffering permanently? - Gaṅgeśa & his Tattva-cintā-maṇi, retrieved 8 November 2021.
7. Ingalls, Daniel Henry Holmes (1951). Materials for the Study of Navya-nyāya Logic. Delhi: Motilal Banarsidass. pp. 5–6. ISBN 9788120803848.
8. Karl H. Potter, Sibajiban Bhattacharyya (1993). Volume 6 of Encyclopedia of Indian philosophies: Indian philosophical analysis Nyaya-Vaiśesika from Gangeśa to Raghunatha Śiromani. Motilal Banarsidass. p. 85. ISBN 9788120803077.
9. Ray, Kanailal. "Jagadish Tarkalankar". Banglapedia.
10. Ray, Kanailal. "Raghunath Shiromani". Banglapedia.
11. Satis Chandra Vidyabhusana (1920). A History of Indian Logic: Ancient, Mediaeval and Modern Schools. Delhi: Motilal Banarsidas. p. 405. ISBN 9788120805651.
12. Satis Chandra Vidyabhusana (1920). A History of Indian Logic: Ancient, Mediaeval and Modern Schools. Delhi: Motilal Banarsidas. p. 7. ISBN 9788120805651.
13. Satis Chandra Vidyabhusana (1920). A History of Indian Logic: Ancient, Medieval and Modern Schools. Delhi: Motilal Banarsidas. p. 454. ISBN 9788120805651.
14. Satis Chandra Vidyabhusana. A History of Indian Logic. Motilal Banarsidass. pp. 454–487.
15. Vattanky, John (1984). Gaṅgeśa's Philosophy of God: Analysis, Text, Transl. and Interpretation of Iśvaravāda Section of Gaṅgeśa's Tattvacintāmaṇi with a Study on the Development of Nyāya Theism. Madras: Adyar Libr. and Research Centre.
16. Vidyabhusana, Satis Chandra (1920). A History of Indian Logic: Ancient, Mediaeval and Modern Schools. Delhi: Motilal Banarsidass. p. 407. ISBN 9788120805651.
17. Vidyabhusana, Satis Chandra (1920). A History of Indian Logic: Ancient, Mediaeval and Modern Schools. Delhi: Motilal Banarsidass. pp. 405–6. ISBN 9788120805651.
18. BHATTACHARYYA, SIBAJIBAN. "GADĀDHARA BHAṬṬĀCĀRYA'S 'VIṢAYATĀVĀDA.'" Journal of Indian Philosophy 14, no. 2 (1986): 109–93.
19. BHATTACHARYYA, SIBAJIBAN. "GADĀDHARA BHAṬṬĀCĀRYA'S 'VIṢAYATĀVĀDA' (Continued)." Journal of Indian Philosophy 14, no. 3 (1986): 217–302.
20. Bhattacharyya, Sibajiban. Some Features of Navya-Nyāya Logic. Philosophy East and West 24, no. 3 (1974): 329–42.

21. Bhattacharyya, Sibajiban. Some Features of the Technical Language of Navya-Nyāya. Philosophy East and West 40, no. 2 (1990): 129–49.
22. Bimal Krishna Matilal, The Navya-Nyaya Doctrine of Negation: The Semantics and Ontology of Negative Statements in Navya-Nyaya Philosophy (Harvard University Press, 1968) ISBN 0-674-60650-7.
23. Gaṅgeśa Upādhyāya, https://en.wikipedia.org/wiki/Ga%E1%B9%85ge%C5%9Ba.
24. Gangesa, Classical Indian philosophy of induction: the Nyāya viewpoint, (Translator: Kisor Kumar Chakrabarti), ISBN 9780739147054, OCLC 665834163.
25. Gangesa, Tattva-cintā-maṇi, ("Jewel"), translated by Stephen Phillips, Jewel of Reflection on the Truth about Epistemology. 3 volumes, London: Bloomsbury, 2020.
26. Gaṅgeśa, https://en.wikipedia.org/wiki/Ga%E1%B9%85ge%C5%9Ba.
27. Navya nyaya, https://en.wikipedia.org/wiki/Navya-Ny%C4%81ya.
28. J. N. Mohanty, Classical Indian Philosophy (Rowman & Littlefield, 2000) ISBN 0-8476-8933-6.
29. Karl H. Potter, Sibajiban Bhattacharyya (1993). Volume 6 of Encyclopedia of Indian philosophies: Indian philosophical analysis Nyaya-Vaiśesika from Gangeśa to Raghunatha Śiromani. Motilal Banarsidass. pp. 85–311. ISBN 9788120808942.
30. Sarvepalli Radhakrishnan, et al. [edd], History of Philosophy Eastern and Western: Volume One (George Allen & Unwin, 1952).
31. Satis Chandra Vidyabhusana. A History of Indian Logic. Motilal Banarsidass. pp. 405–453.
32. Tattvachintamani, https://en.wikipedia.org/wiki/Tattvachintamani.
33. Vattanky, John, Nyāyapañcānana B. Viśvanātha, Nyāyapañcānana B. Viśvanātha, and Dinakarabhaṭṭa. Nyāya Philosophy of Language: Analysis, Text, Translation and Interpretation of Upamāna and Śabda Sections of Kārikāvalī, Muktāvalī and Dinakarī. (Delhi: Sri Satguru Publications, 1995).
34. Vattanky, John. A System of Indian Logic: The Nyana Theory of Inference. (London: Routledge, 2015).
35. Vattanky, John. Development of Nyāya theism. (New Delhi: Intercultural Publications, 1993).

**Nilakantha Somayaji (नीलकण्ठ सोमयाजी)**
**(c. 14 June 1444 – 1545)**

Figure 4.3. Nilakantha Somayaji (नीलकण्ठ सोमयाजी).

Keḷallur Nilakantha Somayaji - केाल्लूर नीलकण्ठ सोमायजी (also referred to as *Keḷallur Comatiri*) was a major mathematician and astronomer of the Kerala school of astronomy and mathematics (see Figure 4.3). One of his most influential works was the comprehensive astronomical treatise *Tantrasamgraha* (तन्त्रसंग्रहः) completed in 1501.

He had also composed an elaborate commentary on *Aryabhatiya* called the *Aryabhatiya Bhasya* (आर्यभटिया भास्य). In this *Bhasya, Nilakantha* had discussed *infinite series* expansions of trigonometric functions and problems of algebra and spherical geometry.

*Grahapariksakrama* is a manual on making observations in astronomy based on instruments of the time.

Known popularly as *Kelallur Chomaathiri,* he is considered an equal to *Vatasseri Parameshwaran Nambudiri*. *Nilakantha* was born into a *Brahmin* family which came from South Malabar in Kerala. *Nilakantha Somayaji* was one of the very few authors of the scholarly traditions of Bharat who had cared to record details about his own life and times.

**His Major Works**

The following is a brief description of the works by *Nilakantha Somayaji* dealing with astronomy and mathematics.

- *Tantrasamgraha* (तन्त्रसंग्रहः)
- *Golasara* (गोलासार): Description of basic astronomical elements and procedures
- *Sidhhantadarpana* (सिद्धान्तदर्पण): A short work in 32 slokas enunciating the astronomical constants with reference to the Kalpa and specifying his views on astronomical concepts and topics.
- *Candrachayaganita* (चन्द्रचायगनीता): A work in 32 verses on the methods for the calculation of time from the measurement of the shadow of the gnomon cast by the moon and vice versa.
- *Aryabhatiya-bhashya* (आर्यभटिया-भाष्य): Elaborate commentary on Aryabhatiya.
- *Sidhhantadarpana-vyakhya* (सिद्धान्तदर्पण-व्याख्य): Commentary on his own Siddhantadarapana.
- *Chandrachhayaganita-vyakhya* (चन्द्रछायागणिता-व्याख्य): Commentary on his own Chandrachhayaganita.
- *Sundaraja-prasnottara* (सुन्दरराज-प्रस्नोत्तर): Nilakantha's answers to questions posed by Sundaraja, a Tamil Nadu-based astronomer.
- *Grahanadi-grantha* (ग्रहणादि-ग्रन्थ): Rationale of the necessity of correcting old astronomical constants by observations.
- *Grahapariksakrama* (ग्रहपरिक्रमाम्): Description of the principles and methods for verifying astronomical computations by regular observations.
- *Jyotirmimamsa* (ज्योतिर्मिमसा): Analysis of astronomy

In one of his works titled *Siddhanta-star* (सिद्धान्त-तारकम्) and also in his own commentary on *Siddhanta-darpana* (सिद्धान्तदर्पणम्), *Nilakantha Somayaji* has stated that he was born on *Kali*-day 1,660,181 which works out to 14 June 1444 CE.

A contemporary reference to *Nilakantha Somayaji* in a *Malayalam* work on astrology implies that Somayaji lived to a ripe old age even to become a centenarian. *Sankara Variar*, a pupil of *Nilakantha Somayaji*, in his commentary on T*antrasamgraha* titled *Tantrasamgraha-vyakhya* (तन्त्रसंग्रह-व्याख्यात), points out that the *first and last verses* of *Tantrasamgraha* contain chronograms specifying the *Kali*-days of the commencement (1,680,548) and of completion (1,680,553) of *Somayaji*'s magnum opus Tantrasamgraha. Both these days occur in 1500 CE.

In *Aryabhatiya-bhashya, Nilakantha Somayaji* has stated that he was the son ofi and he had a brother named *Sankara. Somayaji* has further stated that he was a *Bhatta* (भट्टः) belonging to the *Gargya gotra* (गार्ग्य गोत्र) and was a follower of *Asvalayana-sutra* (अश्वलायन-सूत्रम्) of *Rigveda.* References in his own *Laghuramayana* (लघुरामायणम्) indicate that *Nilakantha Somayaji* was a member of the Kelallur family (Sanskritised as *Kerala-sad-grama*) residing at *Kundagrama,* now known as *Trikkandiyur* in modern Tirur, Kerala. His wife was named *Arya* and he had two sons *Rama* and *Dakshinamurti.*

*Nilakantha Somayaji* studied vedanta and some aspects of astronomy under one Ravi. However, it was *Damodara,* son of *Kerala-drgganita* (केरल-द्रगनिता) author *Paramesvara,* who initiated him into the science of astronomy and instructed him in the basic principles of mathematical computations. The great Malayalam poet *Thunchaththu Ramanujan Ezhuthachan* (थुंचथ्थु रामानुजन एझुठाचन) is said to have been a student of *Nilakantha Somayaji.*

The epithet *Somayaji is* a title assigned to or assumed by a *Namputiri* (नम्पुतिरि) who has performed the vedic ritual of Somayajna. So, it could be surmised that *Nilakantha Somayaji* had also performed a *Somayajna* ritual and assumed the title of a *Somayaji* in later life. In colloquial Malayalam usage the word *Somayaji* has been corrupted to *Comatiri.*

## 卐 Key Contributions in Mathematics

### 卐 Polymath

*Nilakantha*'s writings substantiate his knowledge of several branches of Bhartiya philosophy and culture. It is said that he could refer to a *Mimamsa* (मीमामसा) authority to establish his view-point in a debate and with equal felicity apply a grammatical dictum to the same purpose. In his writings he refers to a *Mimamsa authority* (मीमामसा अधिकारः), quotes extensively from *Pingala*'s *chandas-sutra,* scriptures, *Dharmasastras* (धर्मशास्त्र), *Bhagavata* (भागवत) and *Vishnupurana* (विष्णुपुराण) also. *Sundararaja,* a contemporary Tamil astronomer, refers to *Nilakantha* as *sad-darshani-parangata* (सद्-दर्शनी-परङ्गता), one who had mastered the six systems of Bhartiya philosophy.

## 卐 Astronomy

In his *Tantrasangraha, Nilakantha* revised *Aryabhata*'s model for the planets *Mercury* and *Venus*. His equation of the centre for these planets remained the most accurate until the time of *Johannes Kepler* in the 17th century. In his Aryabhatiyabhasya, a commentary on *Aryabhata*'s *Aryabhatiya, Nilakantha* developed a computational system for a partially heliocentric planetary model in which *Mercury, Venus, Mars, Jupiter* and *Saturn* orbit *the Sun,* which in turn orbits the Earth, similar to the *Tychonic system* later proposed by *Tycho Brahe* in the late 16th century. Most astronomers of the Kerala school who followed him accepted this planetary model.

## 卐 Tantrasamgraha

The treatise was completed in 1501 CE. It consists of 432 verses in Sanskrit divided *into eight chapters*.

Tantrasamgraha had spawned a few commentaries:

- Tantrasamgraha-vyakhya of anonymous authorship and
- Yuktibhāṣā authored by Jyeshtadeva in about 1550 CE.

*Tantrasangraha,* together with its commentaries, bring forth the depths of the mathematical accomplishments the Kerala school of astronomy and mathematics, in particular the achievements of the remarkable mathematician of the school Sangamagrama Madhava. In his *Tantrasangraha, Nilakantha* revised *Aryabhata*'s model for the planets Mercury and Venus.

A brief account of the contents of Tantrasamgraha is presented below.

A descriptive account of the contents is available in *Bharatheeya Vijnana/Sastra Dhara*. Full details of the contents are available in an edition of *Tantrasamgraha* published in the *Indian Journal of History of Science*.

- Chapter 1 (*Madhyama-prakaranam* - मध्यम-प्रकरणम्): The purpose of the astronomical computation, civil and sidereal day measurements, lunar month, solar month, intercalary month, revolutions of the planets, theory of intercalation, planetary revolution in circular orbits, computation of kali days, mathematical operations like addition, subtraction, multiplication, division, squaring and determining square root, fractions, positive and negative numbers, computation of mean planets, correction for longitude, longitudinal time, positions of the planets at the beginning of Kali era, planetary apogees in degrees. (40 slokas)

- Chapter 2 (*Sphuta-prakaranam* - स्फुत-प्रकरणम् (On true planets)): Computation of risings, and arcs, construction of a circle of diameter equal to the side of a given square, computation of the circumference without the use of square and roots, sum of series, sum of the series of natural numbers, of squares of numbers, of cubes of numbers, processes relating to Rsines and arcs, computation of the arc of a given Rsine, computation of the circumference of a circle, derivation of Rsines for given Rversed sine and arc, computation of Rsine and arcs, accurate computation of the 24 ordained Rsines, sectional Rsines and Rsine differences, sum of Rsine differences, summation of Rsine differences, computation of the arc of an Rsine according to Madhava, computation of Rsine and Rversed sine at desired point without the aid of the ordained Rsines, rules relating to triangles, rules relating to cyclic quadrilaterals, rules relating to the hypotenuse of a quadrilateral, computation of the diameter from the area of the cyclic quadrilateral, surface area of a sphere, computation of the desired Rsine, the ascensional difference, sun's daily motion in minutes of arc, application of ascensional difference to true planets, measure of day and night on applying ascensional difference, conversion of the arc of Rsine of the ascensional difference, etc. (59 slokas)
- Chapter 3 (*Chhaya-prakaranam* - छाया-प्रकरणम् (Treatise on shadow)): Deals with various problems related with the sun's position on the celestial sphere, including the relationships of its expressions in the three systems of coordinates, namely ecliptic, equatorial and horizontal coordinates. (116 slokas)
- Chapter 4 (*Chandragrahana-prakaranam* - चन्द्राग्रहण-प्रकरणम् (Treatise on the lunar eclipse)): Diameter of the Earth's shadow in minutes, Moon's latitude and Moon's rate of motion, probability of an eclipse, total eclipse and rationale of the explanation given for total eclipse, half duration and first and last contacts, points of contacts and points of release in eclipse, and their method of calculation, visibility of the contact in the eclipse at sunrise and sunset, contingency of the

invisibility of an eclipse, possibility of the deflection, deflection due to latitude and that due to declination. (53 slokas)

- Chapter 5 (*Ravigrahana-prakaranam* - रविग्रहन-प्रकरणम् (Treatise on the solar eclipse)): Possibility of a solar eclipse, minutes of parallax in latitude of the sun, minutes of parallax in latitude of the moon,. maximum measure of the eclipse, middle of the eclipse, time of first contact and last contact, half duration and times of submergence and emergence, reduction to observation of computed eclipse, mid eclipse, non prediction of an eclipse. (63 slokas)
- Chapter 6 (*Vyatipata-prakaranam* - Vyatipata-prakaranam (On vyatipata)): Deals with the complete deviation of the longitudes of the sun and the moon. (24 slokas)
- Chapter 7 (*Drikkarma-prakaranam* - द्दक्कर्म-प्रकरणम् (On visibility computation)): Discusses the rising and setting of the moon and planets. (15 slokas)
- Chapter 8 (*Sringonnati-prakaranam* - श्रृङ्गोन्नति-प्रकरणम् (On elevation of the lunar cusps)): Examines the size of the part of the moon which is illuminated by the sun and gives a graphical representation of it. (40 slokas)

In astronomy, the spherical triangle formed by the zenith, the celestial north pole and the Sun is called the astronomical triangle. Its sides and two of its angles are important astronomical quantities.

The sides are $90° - \varphi$ where $\varphi$ is the observer's terrestrial latitude, $90° - \delta$ where $\delta$ is the Sun's declination and $90°$ - a where a is the Sun's altitude above the horizon. The important angles are the angle at the zenith which is the Sun's azimuth and the angle at the north pole which is the Sun's hour angle. The problem is to compute two of these elements when the other three elements are specified.

There are precisely ten different possibilities and *Tantrasamgraha* contains discussions of all these possibilities with complete solutions one by one in one place.

The terrestrial latitude of an observer's position is equal to the zenith distance of the Sun at noon on the equinoctial day. The effect of solar parallax on zenith distance was known to Bhartiya astronomers right from *Aryabhata*.

*Tantrasamgraha* gives the magnitude of this correction and also a correction due to the finite size of the Sun.

| |
|---|
| *Nilakantha Somayaji* first discussed the effect of *solar parallax* on the *observer's latitude.* |

## 卐 Jyotirmimamsa

In astronomy, *Jyotirmimamsa* (analysis of astronomy) is a treatise on the methodology of astronomical studies authored by *Nilakantha Somayaji*. This book stresses the necessity and importance of astronomical observations to obtain correct parameters for computations and to develop more and more accurate theories. It even discounts the role of revealed wisdom and divine intuitions in studying astronomical phenomena. *Jyotirmimamsa* is sometimes cited as proof to establish that modern methodologies of scientific investigations were known to ancient and medieval Bhartiyas. *Neelkantha Somayaji* insisted that computational results should tally with that of observations and astronomical parameters and constants should be revised periodically. To come to more precise conclusions, *Neelkantha Somayaji* have discussions with the astronomer and mathematicians of other schools.

The nature of the astronomical and mathematical work, the divine intuition, the experimental details of the science, corrections to the planetary parameters, reasons for the corrections for the planetary revolutions, *Vedic authority* (वैदिकाधिकारः) for inference in astronomy, relative accuracy of different systems, and correction through eclipses, true motion, position, etc., of planets are some of the topics discussed in Jyotirmimamsa.

The following is an outline of the various topics discussed in *Jyotirmimamsa*:

- The necessity of revising the astronomical constants at regular intervals for correcting the parameters connected with astronomy.
- The meaning of devatha prasada, which is manifested in the intuition of the astronomers, as a prerequisite for obtaining the accurate values and the correct approach in the astronomy.
- The experimental determinations of the astronomical constants and the tools used for these determinations.

- The need and the importance of conducting experiments in astronomical studies.
- Significance of the astronomical books and the base of collecting data from those books.
- Application of corrections, called bija correction, for the astronomical figures.
- The corrections known as the *Bhatta* correction.
- Justification of the changes made by the astronomer Lallacharya in his book *Sishyaddhi vruddhi tantra* (शिष्यद्धि वृद्धि तन्त्रम्).
- Reasons for difference in the mean planets though the revolutions are identical.
- Vedic authority for inference as a means to derive the number of planetary revolutions
- Different systems of astronomy.
- The numbers of planetary revolutions enunciated by *Sripati* and *Brahmagupta*.
- The numbers of revolutions of planets, apogees and nodes, number of days, in a *Kalpa* (in astronomy, a period of 14×72×4320000 years).
- Zero points of the planets at the commencement of Kali Yuga, corrections to planetary revolutions.
- Relative accuracy of the different astronomical systems.
- Mean planets according to the Sunrise system of Aryabhatiya, mean planets, Moon's apogee and Moon's node according to Siddhantasekhara.
- Application of the values in the astronomical calculations, astronomical corrections given on the bases of the eclipses.
- Eclipses observed by *Parameswaracharya*.
- Method of corrections given by other astronomers.
- Demonstration of the validity of those corrections through eclipses, precision of equinoxes, calculation and the correction, correction of the periphery of the manda epicycle, discussion on the precision of the equinoxes, corrections due to the precession of the equinoxes.
- Sine table for *praanakalaantara* (प्रणकलान्तर), sine table for the ascensional differences.
- Derivation of the 36 Rsines.
- Graphic proof for the relation of the sides and hypotenuse.
- Reason for the reduction of the minutes of arc of the planetary orbital to the visible celestial sphere.

**Further Reading**


1. "Nilakantha - Biography". Maths History. Retrieved 11 January 2022.
2. *K. Ramasubramanian & M.S. Sriram (2003). "Corrections to the terrestrial latitude in Tantrasamgraha" (PDF). Indian Journal of History of Science. 38 (2): 129–144. Archived from the original (PDF) on 9 March 2012. Retrieved 18 January 2010.
3. A.K. Bag (1980). "Indian literature on mathematics during 1400 - 1800 AD" (PDF). Indian Journal of History of Science. 15 (1): 79–93. Archived from the original (PDF) on 9 March 2012. Retrieved 30 January 2010.
4. C.M. Whish (1835). "On the Hindu quadrature of the circle and the infinite series of the proportion of the circumference to the diameter exhibited in the four Sastras, the Tantra Sahgraham, Yucti Bhasha, Carana Padhati and Sadratnamala". Transactions of the Royal Asiatic Society of Great Britain and Ireland. III (iii): 509–23.
5. George G. Joseph (2000). The Crest of the Peacock: Non-European Roots of Mathematics, p. 408. Princeton University Press.
6. Glen van Brummelen (2009). The mathematics of the heavens and the earth: The early history of trigonometry. Princeton University Press. pp. 128–129. ISBN 9780691129730.
7. Gopalakrishnan, N. (2004). Baharatheeya Vijnana / Saastra Dhaara [Handbbok of Ancient Indian Scientific Books] (PDF). Heritage Publication Series. Vol. 78. Thiruvanannthapuram, India: Indian Institute of Scientific Heritage. pp. 18–20. Retrieved 12 January 2010.
8. J J O'Connor; E F Robertson (November 2000). "Nilakantha Somayaji". School of Mathematics and Statistics University of St Andrews, Scotland. Retrieved 17 January 2010.
9. K. Ramasubramanian, M. D. Srinivas, M. S. Sriram (1994). "Modification of the earlier Indian planetary theory by the Kerala astronomers (c. 1500 AD) and the implied heliocentric picture of planetary motion", Current Science 66, p. 784-790.
10. K.V. Sarma (editor). "Tantrasamgraha with English translation" (PDF) (in Sanskrit and English). Translated by V.S. Narasimhan. Indian National Academy of Science. p. 48. Archived from the original (PDF) on 9 March 2012. Retrieved 17 January 2010.
11. M. S. Sriram; K. Ramasubramanian & M. D. Srinivas (2002). 500 years of Tantrasangraha — A landmark in the history of astronomy. Shimla: Inter-University Center, Indian Institute of Advanced Study. p. 185. ISBN 81-7986-009-4."Indian Institute of Advanced Study".

Archived from the original on 16 January 2010. Retrieved 18 January 2010.

12. M.S. Sriram (25 July 2000). "Meeting reports: Five hundred years of Tantrasangraha - A landmark in the history of astronomy" (PDF). Current Science. 79 (2): 150–151. Retrieved 1 February 2010.
13. Open Library Reference: Nīlakaṇṭha Somayājī (1985). Tantrasaṅgrahaḥ gaṇitam : savyākhyaḥ. OL 2697994M. Retrieved 18 January 2010.
14. P. Vinod Bhattathiripad; K.D. Nambudripad (3 May 2007). "Yaagam (Yajnam)". Namboothiri Websites Trust. Retrieved 4 February 2010.
15. Radaha Charan Gupta. "Solution of the astronomical triangle as found in Tantrsasamgraha (A.D.1500)" (PDF). Indian Journal of History of Science. Indian National Academy of Science. 9 (1). Archived from the original (PDF) on 9 March 2012. Retrieved 18 January 2010.
16. Roddam Narasimha (2009). "The Chequered Histories of Epistemology of Science". In Bharati Ray (ed.). Different types of history. project of History of Science, Philosophy, and Culture in Indian Civilisation. Vol. XIV Part 4. Delhi: Centre for Studies in Civilisation. pp. 96–97.
17. Tantrasamgraha, ed. K.V. Sarma, trans. V. S. Narasimhan in the Indian Journal of History of Science, issue starting Vol. 33, No. 1 of March 1998
18. Jyotirmimamsa, https://en.wikipedia.org/wiki/Jyotirmimamsa.
19. K. V. Sarma (2008) "Nilakantha Somayaji", Encyclopaedia of the History of Science, Technology, and Medicine in Non-Western Cultures (2nd edition) edited by Helaine Selin, Springer, ISBN 978-1-4020-4559-2.
20. K.D. Swaminathan (1972). "Jyotisa in Kerala" (PDF). Indian Journal of History of Science. 8 (1 & 2): 28–36. Retrieved 3 February 2010.
21. K.V. Sarma (ed.). "Tantrasamgraha with English translation" (PDF) (in Sanskrit and English). Translated by V.S. Narasimhan. Indian National Academy of Science. p. 48. Archived from the original (PDF) on 9 March 2012. Retrieved 17 January 2010.
22. Nilakantha Somayaji, https://en.wikipedia.org/wiki/Nilakantha_Somayaji.
23. R.C. Gupta. "Second order interpolation in Indian mathematics up to the fifteenth century" (PDF). Indian Journal of History of Science. 4 (1 & 2): 87–98. Archived from the original (PDF) on 9 March 2012.

24. Ramasubramanian, K (1998). "Model of planetary motion in the works of Kerala astronomers". Bulletin of the Astronomical Society of India. 26 (11–31): 11. Bibcode:1998BASI...26...11R.
25. Ranjan Roy (December 1990). "The discovery of the series formula for $\pi$ by Leibnitz, Gregory and Nilakantha". Mathematics Magazine. Mathematical Association of America. 63 (5): 291–306. doi:10.2307/2690896. JSTOR 2690896. Retrieved 6 September 2016.
26. Shailesh A Shirali (May 1997). "Nilakantha, Euler and pi". Resonance: 28–43. Retrieved 6 September 2016.
27. Somayaji, Nilakantha (1977). Sarma, K.V. (ed.). Jyotirmimamsa [Investigations on astronomical theories] (1st ed.). Hoshiarpur: Vishveshvaranand Vishva Bandhu Institute of Sanskrit and Indological Studies, Panjab University.
28. Tantrasamgraha, https://en.wikipedia.org/wiki/Tantrasamgraha.

## Keshava of Nandigrama
## (fl. 1496–1507)

Keśava (केशवः) was an astrologer and astronomer (*jyotishi* (ज्योतिषी) *or daivajña* (दैवज्ञा)) from *Nandigrama* in present-day western Bharat. *Keśava* flourished around 1496–1507. He was a son of *Kamala-kara* of *Kaushika gotra* (कौशिका गोत्रम्), and a pupil of *Vaijanatha* - वैजनाथः (or *Vaidyanatha* - वैद्यनाथः). His wife's name was *Lakshmi* (लक्ष्मी).

He had three sons, who were famous *jyotishis*:

- *Ananta* (fl. 1534): wrote *Kalanirnayavabodha* (कलनिर्णय अवबोध) and a commentary (1534 CE) on Varaha-mihira's *Laghu-jataka*.
- *Ganesha* (born 1507): wrote a number of works during 1522–1554; his great-grandson - also called Ganesha - wrote *Shiromani-prakasha* (शिरोमणि-प्रकाशः).
- *Rama* (fl. 1525/1550): his son *Nrsimha* (born 1548) wrote *Graha-kaumudi* (ग्रह-कौमुदी), *Kheta-muktavali* (खेत-मुक्तावली), *Graha-dasha-phala* (ग्रह-दश-फलः), *Graha-dipika* (ग्रह-दीपिका), *Varsha-phala-dipika* (वर्षा-फल-दीपिका), *Harsa-kaumudi* - हरसा-कौमुदी (a commentary on *Ganesha*'s *Graha-laghava* - ग्रह-लघवः), and *Hillaja-dipika* - हिलाज-दीपिका.

### 卐 Key Contributions in Mathematics

### 卐 Astronomy

*Muhurtadipika* (मुहूर्तदीपिका) by his son *Gaṇeśa* lists several works written by *Keśava*. Sometimes, *Gaṇeśa*'s works are also attributed to *Keśava*.

Works written by *Keśava* include:

- *Graha-kautuka* (1496 CE)
- *A commentary on 'Graha-kautuka*
- *Graha-siddhi*
- *Tithi-siddhi*
- *Graha-chalana*
- *Ganita-dipika*

- *Jataka-paddhati* (जातक-पद्धति), also known as *Keshava-paddhati; Brhat-keshavi* is an enlarged version of this text - it is an extremely concise text containing only 42 verses, and was very popular as a handbook on mathematical calculations essential for *jataka*.
- A commentary on *Jataka-paddhati*.
- *Tajika-paddhati* (ताजिका-पद्धति), also known as *Varsha-phala-paddhati* (वर्षा-फल-पद्धति) or *Tajika-keshavi* (ताजिका-केशवि) - a work on *Tajika* (Arabic-Persian astrology), it contains 26 verses.
- *Siddhanta-vasana* (सिद्धान्त-वासना).
- *Kayasthachara-paddhati* (कायस्थचर-पद्धति).
- *Kundastaka-lakshana* (कुण्डष्टक-लक्षणम्).

The following works of *Keśava* survive in form of manuscripts, several of which are incomplete:

- *Graha-kautuka* (over 10 manuscripts), and possibly the commentary on it; over 10 manuscripts survive.
- *Jataka-paddhati* and *Brhat-keshavi* (over 200 manuscripts), and the commentary on it (over 25 manuscripts).
- *Tajika-paddhati* (around 50 manuscripts).
- *Muhurtatattva,* containing two parts - Muhurta-khanda and Samhita-khanda (around 100 manuscripts).
- *Sudhiranjani* (2 manuscripts), a *karana* (concise exposition of astronomy) and apparently an appendix to *Varsha-paddhati.*

*Gaṇeśa*'s *Graha-laghava* or *Siddhanta-rahasya* was apparently based on his father's *Graha-kautuka.*

## 卐 Commentaries

Besides *Keśava* himself, several later authors have written commentaries (*tika* - टीका) on his works:

- *Jataka-paddhti*
- *Keshava himself*
- *Vishva-natha* (1618) *at Kashi*
- *Praudha-manorama by Diva-kara* (1626) *at Kashi*
- *Vasana-bhashya by Dharmeshvara* (c. 1600–1650) *in Malava*
- *Jataka-kaustubha by Narayana* (1678) *at Kashi*
- *Guru-dasa* (1824*) at Jalandhara*

- *several modern editors*
- *Tajika-paddhati*
- *Mallari* (fl. 1612)
- *Vishva-natha* (fl. 1612/1630)
- *Muhurta-tattva*
- *Muhurta-dipika by Keshava's son Ganesha (born 1507)*
- *Vishva-natha* (fl. 1612/1630)

**Further Reading**


1. David Pingree, ed. (1971). Census of the Exact Sciences in Sanskrit Series A. Vol. 2. American Philosophical Society. pp. 65–66.
2. David Pingree 1981, p. 126.
3. David Pingree 1981, p. 36.
4. David Pingree 1981, p. 92.
5. David Pingree 1981, p. 93.
6. David Pingree 1981, p. 98.
7. David Pingree, ed. (1970). Census of the Exact Sciences in Sanskrit Series A. Vol. 1. American Philosophical Society. p. 40.
8. David Pingree, ed. (1976). Census of the Exact Sciences in Sanskrit Series A. Vol. 3. American Philosophical Society. p. 202.
9. K. Ramakalyani (2019). "Gaṇeśa Daivajña's upapattis for some rules in the Līlāvatī". In K. Ramasubramanian; Takao Hayashi; Clemency Montelle (eds.). Bhāskara-prabhā: Sources and Studies in the History of Mathematics and Physical Sciences. Sources and Studies in the History of Mathematics and Physical Sciences. Springer. p. 102. doi:10.1007/978-981-13-6034-3_5. ISBN 9789811360343.
10. Keshava, https://en.wikipedia.org/wiki/Keshava_of_Nandigrama.

## Raghunātha Śiromaṇi (रघुनाथ शिरोमणि)
## (c. 1477–1547)

Raghunātha Śiromaṇi (रघुनाथ शिरोमणि) was a philosopher and logician. He was the head (The Chancellor) of the Ancient *Mithila University* (मिथिला विश्वविद्यालय) also known as *Mithila Vidyapeeth*. He was born in a *brahmin* family at *Nabadwip* in present-day *Nadia* district of West Bengal state. He was the grandson of *Śulapāṇi* - शूलपाणी (c. 14th century CE), a noted writer on *Smṛti* (स्मृति) from his mother's side. He was a pupil of *Vāsudeva Sārvabhauma* (वासुदेव सर्वभौम). He brought the new school of *Nyaya, Navya Nyāya*, representing the final development of Bhartiya formal logic, to its zenith of analytic power.

*Raghunātha's* analysis of relations revealed the true nature of number, inseparable from the abstraction of natural phenomena, and his studies of metaphysics dealt with the negation or nonexistence of a complex reality. His most famous work in logic was the Tattvacintāmaṇidīdhiti, a commentary on the *Tattvacintāmaṇi* of *Gangeśa Upādhyāya*, founder of the Navya Nyāya school.

### 卐 Key Contributions in Mathematics

### 卐 Formal Logic

Formal logic is also known as symbolic logic and is widely used in mathematical logic. It uses a formal approach to study reasoning: it replaces concrete expressions with abstract symbols to examine the logical form of arguments independent of their concrete content. In this sense, it is topic-neutral since it is only concerned with the abstract structure of arguments and not with their concrete content. Formal logic provides tools and techniques for precise argumentation. *Navya Nyāya,* with its emphasis on detailed argumentative methodologies, benefited immensely from this precision, allowing scholars to articulate their arguments with clarity and rigor.

Formal logic offers an analytical framework to dissect complex philosophical issues. *Navya Nyāya* philosophers adopted this structured approach to break down intricate subjects into smaller, more manageable components, making it easier to discuss, debate, and understand.

The New *Nyaya* school developed a rich technical vocabulary that was used to express logical and epistemological concepts.

This vocabulary, rooted in formal logic, facilitated more intricate discussions and debates among scholars. Formal logic provided a foundation for *Navya Nyāya* thinkers to delve into epistemological issues, such as the nature of knowledge, sources of valid knowledge (*pramāṇas*), and the criteria for truth. This led to sophisticated analyses of perception, inference, comparison, and verbal testimony.

## Further Reading


1. "The Sunday Tribune - Spectrum". www.tribuneindia.com. Retrieved 6 February 2022.
2. Language: From I-dentity to My-dentity, the contemporary deployment of a new category, svatva (endowment, possessed-ness, entitlement, my-ness), introduced by Raghunatha.
3. Raghunatha Siromani, https://en.wikipedia.org/wiki/Raghunatha_Siromani.
4. Raghunatha: A Name of Negatives, descriptive information of Raghunatha with some controversial issues (his connection with Mahaprabhu Shri Chaitanya) and bibliography.
5. Vidyabhusana, Satis Chandra (2006) [1920]. A History of Indian Logic: Ancient, Mediaeval, and Modern Schools. Delhi: Motilal Banarsidass. p. 463. ISBN 81-208-0565-8.

# Chapter 5

# Early Modern to Late Modern Period
(1527 CE – 1900 CE)

## Gaṇeśa Daivajna (गणेश दैवज्ञः)
## (born 1507, fl. 1520-1554)

Gaṇeśa Daivajna (गणेश दैवज्ञः) was a sixteenth century astronomer, astrologer, and mathematician from western Bharat who wrote books on methods to *predict eclipses, planetary conjunctions, positions, and make calculations for calendars*. His most major work was the *Grahalaghava* (ग्रहलाघवः) which was included *ephemeris* (क्षणिकम्) and calendar calculations.

*Gaṇeśa* was born in *Nandigrama* (see also *Golagrama*) where his father *Keśava* was a Brahmin astronomer. His mother's name has been noted as *Lakshmi* and he spent his entire life at Nandigrama. The location of *Nandigrama* has been suggested by some as being in Gujarat but more careful study of his work places it in *Nandgaon* in present day Maharashtra. He wrote several works including *Grahalaghava* (ग्रहलाघवः), *Siddhantarahasya* (सिद्धान्तरहस्य), *Buddhivilāsinī* (बुद्धिविलासिनी), *and Laghutithicintamani* (लघुतिथिचिन्तामणि). His work *Buddhivilāsinī* (c. 1546) includes commentaries on the mathematics of *Bhaskara*'s *Līlāvatī*. His book *Muhūrtadipikā* (मुहूर्तदीपिका) includes a commentary on his father *Keśava*'s work *Muhūrtatattva* (मुहूर्ततत्त्वम्). *Gaṇeśa*'s grandfather *Kamalakara* was also an astrologer as were his brothers *Ananta* and *Rama*. Works by *Keśava* include *Grahakautuka* (1496) on the calculation of eclipses, *Jatakapaddhati* for the production of horoscopes, and the T*ajikapaddhati*, which covered Islamic thoughts on astrology. Figure 5.1 shows the illustration of *Ganesa*'s dissection of a circle into triangles.

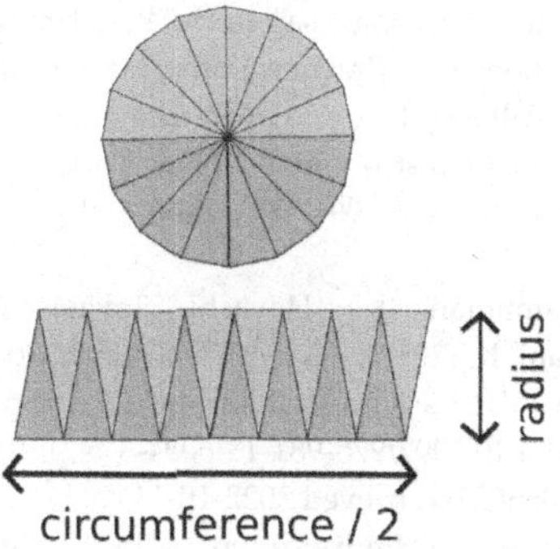


Figure 5.1. Illustration of *Ganesa*'s dissection of a circle into triangles (using 16 sides in this case). (Image Credit: https://en.wikipedia.org/wiki/)

## 卐 Key Contributions in Mathematics

### 卐 Geometry

One of his demonstrations is of the area of a circle calculated by dissection of the circle into a regular polygon. He started with a *polygon with 12 sides* and doubled it through 24, 48, 96, 192, to 384 and came up with an approximation for

$$\pi = \frac{3927}{1250} = 3.1416$$

**Further Reading**


1. Sarma, S.R. (2010). "Nandigrama of Ganesa Daivajna". Indian Journal of History of Science. 45 (4): 569–574.
2. Colebrooke, H.T. (1873). Miscellaneous Essays. Volume III. London: Trubner and Co. p. 406.
3. Joseph, George Gheverghese (2016). Indian Mathematics: Engaging With The World From Ancient To Modern Times. World Scientific. p. 295.
4. Pingree, David (1973). "Kesava". In Gillispie, C.C. (ed.). Dictionary of Scientific Biography. Volume 7. New York: Charles Scribner's Sons. pp. 314–316.
5. Plofker, Kim (2014). "Ganesa". In Hockey, Thomas; Trimble, Virginia; Williams, Thomas R.; Bracher, Katherine; Jarrell, Richard A.; Marché, Jordan D.; Palmeri, JoAnn; Green, Daniel W. E. (eds.). Biographical Encyclopedia of Astronomers. New York: Springer. pp. 781–782. doi:10.1007/978-1-4419-9917-7. ISBN 978-1-4419-9916-0. S2CID 242158697.
6. Ramasubramanian, K.; Hayashi, Takao; Montelle, Clemency; Ramakalyani, K. (2019), "Gaṇeśa Daivajña's upapattis for some rules in the Līlāvatī", Bhāskara-prabhā, Singapore: Springer Singapore, pp. 102–114, doi:10.1007/978-981-13-6034-3_5, ISBN 978-981-13-6033-6, S2CID 146066627, retrieved 2022-10-14
7. Grahalaghava (Hindi)Edition by Sudhakar Dwivedi, https://archive.org/details/in.ernet.dli.2015.448248.

8. Ganesha, https://en.wikipedia.org/wiki/Ga%E1%B9%87e%C5%9Ba_Daivajna.
9. Jataka Alankara, https://archive.org/details/jatakaalankaraofganeshadaivajnatippanig aneshdattphatka_202003_396_n.

## Achyutha Pisharadi (अच्युत पिशरादि)
## (c. 1550 – 7 July 1621)

Achyuta Pisharodi (अच्युत पिशरादि) was a Sanskrit grammarian, astrologer, astronomer and mathematician who studied under *Jyeṣṭhadeva* and was a member of *Madhava of Sangamagrama*'s Kerala school of astronomy and mathematics. He was born in 1550 at *Thrikkandiyur* (aka Kundapura), Tirur, Kerala, Bharat. He is remembered mainly for his part in the composition of his student *Melpathur Narayana Bhattathiri*'s (मेलपथुर नारायण भट्टथिरी) devotional poem, *Narayaneeyam* (नारायणीयम्). *Pisharati* is known to have scolded and provoked an errant *Narayana* to take up the *Brahmin's* duties of prayer and religious practices. He accepted *Narayana* as his student. Later when Pisharati was struck with paralysis (or rheumatism by another account), *Narayana,* unable to bear the pain of his dear guru, by way of *Gurudakshina* (गुरुदक्षिणा) took the disease upon himself.

As a result, *Pisharati* is said to have been cured, but no medicine could cure *Narayana*. As a last resort, *Narayana* went to *Guruvayur* and requested *Thunchaththu Ramanujan Ezhuthachan* (थुंचथ्थु रामानुजन एझुठाचन), a great devotee of *Guruvayoorappan* (गुरुवायोरप्पन), to suggest a remedy for his disease. *Ramajunan Ezhuthachan* advised him to compose a poetical work on the *Avatars* - अवताराः (incarnations) of *Lord Vishnu* (भगवान् विष्णुः) beginning with that of *Matsya* - मत्स्यः (Fish). Narayana composed beautiful slokas in praise of *Lord Guruvayurappan* and recited them before the deity. He was soon cured of his disease. The book of *slokas* (श्लोकाः) written by *Narayana* were named *Narayaneeyam*. The day on which *Narayana* dedicated his *Narayaneeyam* to *Sri Guruvayurappan* is celebrated as "*Narayaneeyam Dinam* - नारायणीयम् दिनम् " every year at *Guruvayur* - गुरुवायुर.

### 卐 Key Contributions in Mathematics

#### 卐 Overall

He discovered the techniques of *'the reduction of the ecliptic'*. He authored *Sphuṭanirṇaya* (स्फुटानिर्णाय), *Rāśigolasphuṭānīti* - राशिगोलस्फुटानीति (raasi meaning zodiac, gola meaning sphere and neeti roughly meaning rule),

*Karanottama* (1593) and a *four-chapter* treatise *Uparagakriyakrama* (उपरागक्रियाक्रमः) on lunar and solar eclipses.

Other key works are mentioned below:

- *Praveśaka*: An introduction to Sanskrit grammar.
- *Karaṇottama*: Astronomical work dealing with the computation of the mean and true longitudes of the planets, with eclipses, and with the vyatūpātas of the sun and moon.
- *Uparāgakriyākrama*: Treatise on lunar and solar eclipses.
- *Sphuṭanirṇaya*: Astronomical text.
- *Chāyāṣṭaka* (चायष्टक): Astronomical text.
- *Uparāgaviṃśati* (उपरागविंशति): Manual on the computation of eclipses.
- *Rāśigolasphuṭānīti*: Work concerned with the reduction of the moon's true longitude in its own orbit to the ecliptic.
- *Veṇvārohavyākhyā* (वेङ्वारोहव्याख्या): Malayalam commentary on the *Veṇvāroha* of *Mādhava of Saṅgamagrāma* written at the request of the *Azhvanchery Thambrakkal.*
- *Horāsāroccaya* (होरासरोच्चया): An adaptation of the *Jātakapaddhati* of *Śrīpati.*

**Further Reading**


1. "Astronomy and Mathematics in Kerala" in Brahmavidyā, 27 (1963), 158–162.
2. Achyutha Pisharadi, https://en.wikipedia.org/wiki/Achyutha_Pisharadi.
3. David Pingree. "Acyuta Piṣāraṭi". Dictionary of Scientific Biography.
4. K. Kunjunni Raja. The Contribution of Kerala to Sanskrit Literature (Madras, 1958), pp. 122–125.
5. K. V. Sarma (2008), "Acyuta Pisarati", Encyclopaedia of the History of Science, Technology, and Medicine in Non-Western Cultures (2nd edition) edited by Helaine Selin, p. 19, Springer, ISBN 978-1-4020-4559-2.
6. S. Venkitasubramonia Iyer. "Acyuta Piṣāroṭi; His Date and Works" in JOR Madras', 22 (1952–1953), 40–46.

## Dinakara (दिनाकारः)
**(fl. 1550)**

Dinakara (दिनाकारः) was an astronomer who produced a treatise *Candrārkī* - चन्द्रार्की (consisting of 33 verses) on the preparation of Bhartiya calendars (*pancanga* - पंचङ्गः) based on the movement of the sun and moon. The treatise exists in as many as *150 manuscript* versions as well as with commentaries, some of which have been interpreted, translated and compared with modern approaches. *Dinakara* is noted as the son of *Rāmeśvara* and greatgrandson of *Dunda* from the *Moḍhajñāti* family and at the time of writing his treatise was residing at *Bārejya* (which has been assumed to be *Bariya* in *Rewa Kantha*, Gujarat).

### 卐 Key Contributions in Mathematics

### 卐 Overall

His major works were *astronomical tables* (खगोलीय सारणी) made for *Śaka* 1500 (Sunday, April 9, 1578 CE) and Śaka 1505 (AD 1583). His treatise *Kheṭakasiddhi* - खेतकसिद्धि (1578) contains rules for determining the positions of the five planets based on *Brahmatulya* (ब्रह्मतुल्यः) of *Bhāskara II*. The *Candrārkī* (चन्द्रार्की) has tables on solar and lunar movement.
His third work *Tithisāraṇī* - तिथिसारणी (or *Dinakarasāraṇī* - दिनकारसारणी) (1583) covers the making of pancangas, the calculation of days, tithis, nakṣatras, and yogas, and was based partly on the earlier work *Mahādevī of Mahādeva*. *Dinakara*'s work was an influence for *Haridatta II*'s *Jagadbhūṣaṇa* - जगद्भूमिः (1638).
*Achalajit of Muraripupura* in Gurjara wrote another Candrārkī in 1655, by converting *Dinakara*'s work from the system followed by the *brahma-paksha* (ब्रह्मपक्षः) school to that of the *saura-paksha* (सौरापक्षः) school.

### Further Reading

1. David Pingree, ed. (1981). Census of the Exact Sciences in Sanskrit Series A. Vol. 4. American Philosophical Society. p. 12.

2. Ikeyama, Setsuro; Chapront-Touzé, Michelle; Joeveer, Mihkel; Durham, Ian T.; Martres, M. J.; Rust, David M.; Lévy, Jacques; Débarbat, Suzanne; Bònoli, Fabrizio (2007). "Dinakara". In Hockey, Thomas; Trimble, Virginia; Williams, Thomas R.; Bracher, Katherine (eds.). The Biographical Encyclopedia of Astronomers. New York, NY: Springer New York. pp. 299–299. doi:10.1007/978-0-387-30400-7_361. ISBN 978-0-387-31022-0.
3. Kolachana, A.; Montelle, C.; Dhammaloka, J.; Melnad, K.; Mahesh, K.; Vyas, P. (2018). "The Candrārkī of Dinakara: A Text Related to Solar and Lunar Tables". Journal for the History of Astronomy. 49 (3): 306–344. doi:10.1177/0021828618787556.
4. Candrarki MS, https://archive.org/details/chandrarki-CGV-PSS-0132-1274/.
5. Dinakara, https://en.wikipedia.org/wiki/Dinakara.
6. Kolachana, Aditya; Montelle, Clemency; Dhammaloka, J.; Melnad, K.; Mahesh, K.; Vyas, P.; Ramasubramanian, K.; Sriram, M.S.; Pai, V. (2018). "A Critical Edition of the Candrārkī of Dinakara: A Text Concerning Solar and Lunar Tables". History of Science in South Asia. 6: 127–161. doi:10.18732/hssa.v6i0.35.

## Mathurānātha Śarman (मथुरानाथ शर्मन्)
## (fl. 1609)

Mathurānātha Śarman (मथुरानाथ शर्मन्) was a *calendrist* (पञ्चाङ्गकारः) and *astronomer* (खगोलशास्त्रज्ञः) who lived in Bengal and belonged to the *Saurapakṣa tradition* (सौरपक्ष परम्परा) of Bhartiya astronomy.

### 卐 Key Contributions in Mathematics

### 卐 Overall

He is known from his work the *Ravisiddhāntamañjarī* (रविसिद्धान्तमञ्जरी) or *Sūryasiddhāntamañjarī* (सूर्यसिद्धान्तमञ्जरी), with planetary longitude tables and the means to compute solar eclipses. He may also have been the author of some other works including the *Pañcaṅgaratna* (पञ्चाङ्गरत्नम्) and the *Praśnaratnāṅkura* (प्रश्नरत्नाङ्कुर) or *Samayāmṛta* (समयमृतः).

### Further Reading


1. Ikeyama, Setsuro (2007), Hockey, Thomas; Trimble, Virginia; Williams, Thomas R.; Bracher, Katherine (eds.), "Mathurānātha Śarman", The Biographical Encyclopedia of Astronomers, New York, NY: Springer New York, pp. 745–745, doi:10.1007/978-0-387-30400-7_910, ISBN 978-0-387-31022-0, retrieved 2023-03-21
2. Mathurānātha Śarman (1911). Ravisiddhāntamañjarī. Bibliotheca Indica, no. 198, edited by Viśvambhara Jyotisārnava. Calcutta: Asiatic Society.
3. Mathurānātha Śarman, https://en.wikipedia.org/wiki/Mathuranatha_Sarman.

## Citrabhānu (चित्रभानुः)
**(fl. 16th century)**

Citrabhānu (चित्रभानुः) was a mathematician of the Kerala school and a student of *Nilakantha Somayaji*. He was a *Nambudiri brahmin* (नम्बुदिरी ब्राह्मण) from the town of *Covvaram* near present day Trissur. He is noted for a *karaṇa* (करणम्), a concise astronomical manual, dated to 1530, an algebraic treatise, and a commentary on a poetic text. *Nilakantha* and he were both teachers of *Shankara Variyar* (शंकर वरियार).

### 卐 Key Contributions in Mathematics

#### 卐 Integer Solution of Diophantine Euquations

He gave integer solutions to *21 types of systems* of two simultaneous *Diophantine equations* in two unknowns.

These types are all the possible pairs of equations of the following seven forms:

$$x + y = a$$

$$x - y = b$$

$$xy = c$$

$$x^2 + y^2 = d$$

$$x^2 - y^2 = e$$

$$x^3 + y^3 = f$$

$$x^3 - y^3 = g$$

For each case, *Chitrabhanu* gave an explanation and justification of his rule as well as an example. Some of his explanations are algebraic, while others are geometric.

**Further Reading**


1. Citrabhānu, https://en.wikipedia.org/wiki/Chitrabhanu_(mathematician).
2. Hayashi, Takao; Kusuba, Takanori (1998), "Twenty-one algebraic normal forms of Citrabhānu", Historia Mathematica, 25 (1): 1–21, doi:10.1006/hmat.1997.2171, MR 1613702.
3. Joseph, George Gheverghese (10 December 2009). A Passage to Infinity: Medieval Indian Mathematics from Kerala and Its Impact. ISBN 9788132104810.
4. Joseph, George Gheverghese (2009), A Passage to Infinity: Medieval Indian Mathematics from Kerala and Its Impact, SAGE Publications India, p. 21, ISBN 9788132104810.
5. Plofker, Kim (2009). Mathematics in India. Princeton: Princeton University Press. pp. 220, 319, 323. ISBN 9780691120676.

## Jyeṣṭhadeva (ज्येष्ठदेवः)
## (c. 1500 – c. 1575)

Jyeṣṭhadeva (ज्येष्ठदेवः) was an astronomer-mathematician of the Kerala school of astronomy and mathematics founded by *Madhava of Sangamagrama*. He is best known as the author of Yuktibhāṣā, a commentary in Malayalam of *Tantrasamgraha* by *Nilakantha Somayaji*.

In *Yuktibhāṣā* (युक्तिभाषा), *Jyeṣṭhadeva* had given complete proofs and rationale of the statements in *Tantrasamgraha*. This was unusual for traditional Bhartiya mathematicians of the time. The *Yuktibhāṣā* is now believed to contain the essential elements of the calculus and one of the earliest treatises on the subject. *Jyeṣṭhadeva* also authored *Drkkarana* (दृक्-करणम्) a treatise on astronomical observations. There are a few references to *Jyeṣṭhadeva* scattered across several old manuscripts. From these manuscripts, one can deduce a few bare facts about the life of *Jyeṣṭhadeva*. He was a Nambudiri belonging to the *Parangngottu* (पराङ्गोट्टु) family (Sanskrtised as *Parakroda*) born about the year 1500 CE.

He was a pupil of *Damodara* and a younger contemporary of *Nilakantha Somayaji*. *Achyuta Pisharati* was a pupil of *Jyeṣṭhadeva*. In the concluding verse of his work titled *Uparagakriyakrama* (उपरागक्रियाक्रमः), completed in 1592, *Achyuta Pisharati* has referred to *Jyeṣṭhadeva* as his aged benign teacher.

From a few references in *Drkkarana*, a work believed to be of *Jyeṣṭhadeva*, one may conclude that *Jyeṣṭhadeva* lived up to about 1610 CE.

*Parangngottu*, the family house of *Jyeṣṭhadeva*, still exists in the vicinity of *Trikkandiyur* and *Alathiyur*. There are also several legends connected with members of *Parangngottu* family.

Little is known about the mathematical traditions in Kerala prior to *Madhava of Sangamagrama*. *Vatasseri Paramesvara* was a direct disciple of *Madhava*. *Damodara* was a son of *Paramesvara*. *Nilakantha Somayaji* and *Jyeṣṭhadeva* were pupils of *Damodara*. *Jyeṣṭhadeva*'s pupil was *Achyuta Pisharati* and *Melpathur Narayana Bhattathiri* was *Achyuta Pisharati*'s students.

## 卐 Key Contributions in Mathematics

### 卐 Manual of Proofs

*Jyeṣṭhadeva* is known to have composed only two works, namely, *Yuktibhāṣā* (also known as *Gaṇita-yukti-bhāṣā* - गणिता-युक्ति-भाषा or *Gaṇitanyāyasaṅgraha* - गणितान्यायसङ्ग्रहः) and *Drkkarana*. The former is commentary with rationales of T*antrasamgraha* of *Nilakantha Somayaji* and the latter is a treatise on astronomical computations.

The *Jyeṣṭhadeva*, written in Malayalam, is a consolidation of the discoveries by *Madhava of Sangamagrama, Nilakantha Somayaji, Parameshvara, Jyeshtadeva, Achyuta Pisharati*, and other astronomer-mathematicians of the Kerala school. It also exists in a Sanskrit version, with unclear author and date, composed as a rough translation of the Malayalam original.

Three factors make *Yuktibhāṣā* unique in the history of the development of mathematical thinking in the Bhartiya subcontinent:

- It is composed in the spoken language of the local people, namely, the Malayalam language. This is in contrast to the centuries-old Bhartiya tradition of composing scholarly works in the Sanskrit language which was the language of the learned.
- The work is in prose, again in contrast to the prevailing style of writing even technical manuals in verse. All the other notable works of the Kerala school are in verse.
- Most importantly, *Yuktibhāṣā* was composed intentionally as a manual of proofs. The very purpose of writing the book was to record in full detail the rationales of the various results discovered by mathematicians-astronomers of the Kerala school, especially of *Nilakantha Somayaji.* This book is proof enough to establish that the concept of proof was not unknown to Bhartiya mathematical traditions.

Figure 5.2. Front and back cover of the Palm-leaf manuscripts of the *Yuktibhasa*, composed by *Jyesthadeva* in 1530, translated by C. M. Whish in 1826. (Image Credit: https://en.wikipedia.org/wiki/)

The work contains proofs and derivations of the theorems that it presents. Modern historians used to assert, based on the works of Bhartiya mathematics that first became available, that early Bhartiya scholars in astronomy and computation lacked in proofs, but *Yuktibhāṣā* demonstrates otherwise (see Figure 5.2 and Figure 5.3).

### 卐 Infinite Series

Some of its important topics include the infinite series expansions of functions; power series, including of $\pi$ and $\pi/4$; trigonometric series of sine, cosine, and arctangent; Taylor series, including second and third order approximations of sine and cosine; radii, diameters and circumferences; and tests of convergence.

*Yuktibhāṣā* mainly gives rationale for the results in *Nilakantha*'s *TantraSamgraha*. It is considered an early text to give some ideas of calculus like *Taylor* and infinity series, predating Newton and *Leibniz* by two centuries. Explanation of sine rules can be presented as per Figure 5.4.

The treatise was largely unnoticed outside Bharat, as it was written in the local language of Malayalam. In modern times, due to wider international cooperation in mathematics, the wider world has taken notice of the work. For example, both Oxford University and the Royal Society of Great Britain have given attribution to pioneering mathematical theorems of Bhartiya origin that predate their Western counterparts.

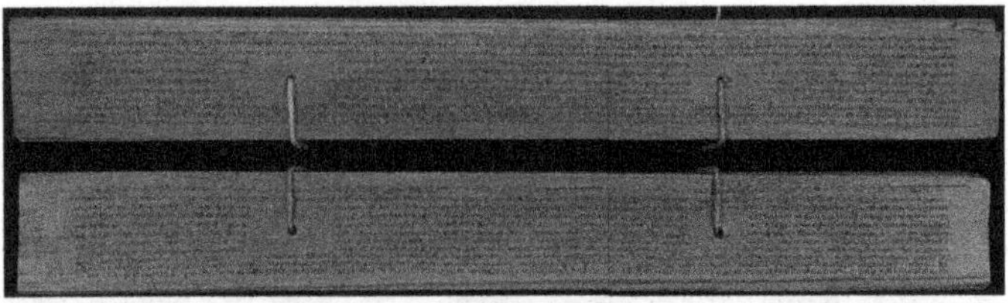

Figure 5.3. Pages from the *Yuktibhasa*. (Image Credit: https://en.wikipedia.org/wiki/)

## 卐 Astronomy

*Chapters eight* to *seventeen* deal with subjects of astronomy: planetary orbits, celestial spheres, ascension, declination, directions and shadows, spherical triangles, ellipses, and parallax correction. The planetary theory described in the book is similar to that later adopted by Danish astronomer *Tycho Brahe*.

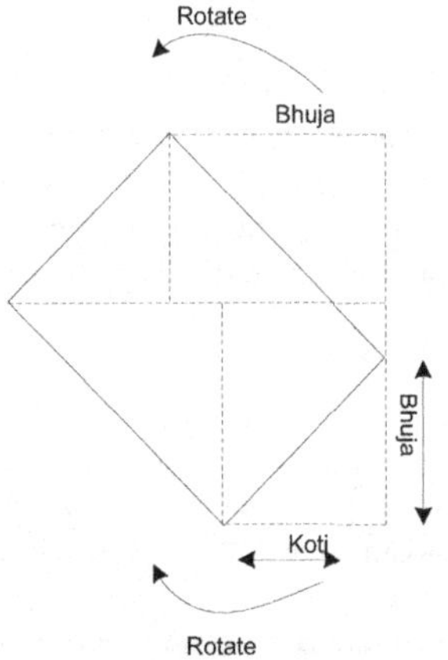


Figure 5.4. Explanation of the sine rule in *Yuktibhāṣā*.

The topics covered in the eight chapters are computation of mean and true longitudes of planets, Earth and celestial spheres, fifteen problems relating to ascension, declination, longitude, etc., determination of time, place, direction, etc., from gnomonic shadow, eclipses, *Vyatipata* - व्यातिपातः (when the sun and moon have the same declination), visibility correction for planets and phases of the moon.

Specifically,

- *Grahagati* (ग्रहात्): Planetary motion, bhagola: sphere of the zodiac, madhyagraha: mean planets, sūryasphuṭa: true sun, grahasphuṭa: true planets.
- *Bhū-vāyu-bhagola* (भू-वायु-भागोला): Spheres of the earth, atmosphere, and asterisms, ayanacalana: precession of the equinoxes.
- *Pañcadaśa-praśna* (पंचदश-प्राश्ना): Fifteen problems relating to spherical triangles.
- *Dig-jñāna* (खनन-ज्ञानम्): orientation, chāyā-gaṇita: shadow computations, lagna: rising point of the ecliptic, nati-lambana: parallaxes of latitude and longitude.
- *Grahaṇa*: Eclipse.
- *Vyatīpāta*
- Visibility correction of planets.
- Moon's cusps and phases of the moon.

卐 **Overall**

This subject treated in the mathematics part of the *Yuktibhāṣā* can be divided into seven chapters:

- *Parikarma* (पारिकर्म): logistics (the eight mathematical operations).
- *Daśapraśna* (दशप्राष्ण): ten problems involving logistics.
- *Bhinnagaṇita* (भिन्नगणिता): arithmetic of fractions.
- *Trairāśika* (त्रैराशिका): rule of three.
- *Kuṭṭakāra* (कुट्टकार): pulverisation (linear indeterminate equations).
- *Paridhi-vyāsa* (परिधि-व्यास): relation between circumference and diameter: infinite series and approximations for the ratio of the circumference and diameter of a circle.
- *Jyānayana* (ज्ञानायन): derivation of Rsines: infinite series and approximations for sines.

The *first four chapters* of the contain elementary mathematics, such as division, the Pythagorean theorem, square roots, etc. Novel ideas are not discussed until the *sixth chapter* on *circumference of a circle* (वृत्तस्य परिधिः). *Yuktibhāṣā* contains a derivation and proof for the power series of inverse tangent, discovered by *Madhava*. In the text, *Jyesthadeva* describes *Madhava*'s series in the following manner:

The first term is the product of the given sine and radius of the desired arc divided by the cosine of the arc. The succeeding terms are obtained by a process of iteration when the first term is repeatedly multiplied by the square of the sine and divided by the square of the cosine. All the terms are then divided by the odd numbers 1, 3, 5, .... The arc is obtained by adding and subtracting respectively the terms of odd rank and those of even rank. It is laid down that the sine of the arc or that of its complement whichever is the smaller should be taken here as the given sine. Otherwise the terms obtained by this above iteration will not tend to the vanishing magnitude.

Translation

The *Yuktibhāṣā* contains many elementary and complex mathematical topics, including:

- Proofs for the expansion of the sine and cosine functions.
- The sum and difference formulae for sine and cosine.
- Integer solutions of systems of linear equations (solved using a system known as *kuttakaram*).
- Geometric derivations of series.
- Early statements of Taylor series for some functions.
- Tests of convergence for sums.

**Further Reading**


1. K.V. Sarma (1991). "Yuktibhāṣā of Jyeṣṭhadeva: A book of rationales in Indin mathematics and astronomy – an analytical appraisal" (PDF). Indian Journal of History of Science. 26 (2): 185–207. Archived from the original (PDF) on 12 September 2014. Retrieved 28 January 2010.
2. "An overview of Indian mathematics". Indian Maths. School of Mathematics and Statistics University of St Andrews, Scotland. Retrieved 7 July 2006.

3. "Jyesthardeva". Biography of Jyesthadeva. School of Mathematics and Statistics University of St Andrews, Scotland. Retrieved 7 July 2006.
4. "Science and Mathematics in India". South Asian History. India Resources. Archived from the original on 17 October 2012. Retrieved 6 May 2020.
5. "The Kerala School, European Mathematics and Navigation". Indian Mathemematics. D.P. Agrawal – Infinity Foundation. Retrieved 9 July 2006.
6. "The Yuktibhasa Calculus Text" (PDF). The Pre-History of Calculus and Celestial Mechanics in Medieval Kerala. Dr Sarada Rajeev. Retrieved 9 July 2006.
7. C. K. Raju (2001). "Computers, mathematics education, and the alternative epistemology of the calculus in the Yuktibhāṣā" (PDF). Philosophy East & West. 51 (3): 325–362. doi:10.1353/pew.2001.0045. S2CID 170341845. Retrieved 11 February 2020.
8. Charles Whish (1834), "On the Hindu Quadrature of the circle and the infinite series of the proportion of the circumference to the diameter exhibited in the four Sastras, the Tantra Sahgraham, Yucti Bhasha, Carana Padhati and Sadratnamala", Transactions of the Royal Asiatic Society of Great Britain and Ireland, 3 (3): 509–523, doi:10.1017/S0950473700001221, JSTOR 25581775
9. Divakaran, P. P. (2007). "The First Textbook of Calculus: "Yuktibhāṣā"". Journal of Indian Philosophy. 35 (5/6): 417–443. doi:10.1007/s10781-007-9029-1. ISSN 0022-1791. JSTOR 23497280. S2CID 170254981.
10. For a review of Ganita-yukti-bhasa (Rationales in Mathematical Astronomy) of Jyesthadeva by Mathematical Association of America see: Homer S. White (17 July 2009). "Ganita-Yukti-Bhāsā (Rationales in Mathematical Astronomy) of Jyesthadeva". The Mathematical Association of America. Retrieved 28 May 2022.
11. George Gheverghese Joseph (2000). The crest of the peacock. Internet Archive. Princeton University Press. ISBN 978-0-691-00659-8.
12. J J O'Connor; E F Robertson (November 2000). "Jyesthadeva". School of Mathematics and Statistics University of St Andrews, Scotland. Retrieved 28 January 2010.
13. K V Sarma; S Hariharan (1991). "Yuktibhāṣā of Jyeṣṭhadeva: A book on rationales in Indian Mathematics and Astronomy: An analytic appraisal" (PDF). Indian Journal of History of Science. 26 (2).

Archived from the original (PDF) on 28 September 2006. Retrieved 9 July 2006.
14. K. V. Sarma (1972). A history of the Kerala school of Hindu astronomy (in perspective). Hoshiarpur, Panjab University: Vishveshvaranand Institute of Sanskrit & Indological Studies. p. 59. Bibcode:1972hksh.book.....S.
15. K.V. Sarma (2004). Ganita Yuktibhasa (Volume III). Shimla: Indian Institute of Advanced Study. ISBN 81-7986-052-3.
16. P.P. Divakaran, P. P. (December 2007). "The First Textbook of Calculus: Yuktibhāṣā". Journal of Indian Philosophy. Springer Netherlands. 35 (5–6): 417–443. doi:10.1007/s10781-007-9029-1. ISSN 0022-1791. S2CID 170254981.
17. Publisher's (Indian Institute of Advanced Study) web page on the book:"Ganita Yuktibhasa by K.V. Sarma". Archived from the original on 17 March 2010. Retrieved 1 May 2010.
18. Sarma, K.V. (2009). Ganita Yuktibhasa (in Malayalam and English). Vol. III. Indian Institute of Advanced Study, Shimla, India. ISBN 978-81-7986-052-6. Archived from the original on 17 March 2010. Retrieved 16 December 2009.
19. Sayahna Foundation (20 November 2020). "Yukthibhasha digital edition" (PDF), https://books.sayahna.org/ml/pdf/yukthibhasha.pdf
20. Yuktibhâsâ, Part I (ed) with notes by Ramavarma (Maru) Thampuran and A. R. Akhileswara Aiyer, Magalodayam Ltd., Trichur, Kerala, 1123 Malayalam Era, 1948 CE.
21. Biography of Jyesthadeva – School of Mathematics and Statistics University of St Andrews, Scotland, https://mathshistory.st-andrews.ac.uk/Biographies/Jyesthadeva/
22. Details on the English translation of Yuktibhāṣā by K. V. Sarma: Sarma, K.V., Ramasubramanian, K., Srinivas, M.D., Sriram, M.S. (2008). Ganita-Yukti-Bhasa (Rationales in Mathematical Astronomy) of Jyeṣṭhadeva: Volume I: Mathematics, Volume II: Astronomy. Sources and Studies in the History of Mathematics and Physical Sciences. Springer jointly with Hindustan Book Agency, New Delhi, India. ISBN 978-1-84882-072-2. (This is a critical translation of the original Malayalam text by K.V. Sarma with explanatory notes by K. Ramasubramanian, M.D. Srinivas and M.S. Sriram.)
23. For a modern explanation of Jyeṣṭhadeva's proof of the power series expansion of the arctangent function: Victor J. Katz (2009). "12". A history of mathematics: An introduction (3 ed.). Addison Wesley. pp. 450–455. ISBN 978-0-321-38700-4.

24. For a review of the English translation of Yuktibhāṣā: Homer S. White (17 July 2009). "Ganita-Yukti-Bhāsā (Rationales in Mathematical Astronomy) of Jyesthadeva". MAA Reviews. The Mathematical Association of America. Retrieved 30 January 2010.[dead link]
25. George Gheverghese Joseph (2000). The Crest of the Peacock: The Non-European Roots of Mathematics. Princeton University Press. pp. 416. ISBN 978-0-691-00659-8.
26. Jyesthardeva, https://en.wikipedia.org/wiki/Jye%E1%B9%A3%E1%B9%ADhadeva
27. K. V. Sarma (1972). A history of the Kerala school of Hindu astronomy (in perspective). Vishveshvaranand Indological series. Vol. 55. Vishveshvaranand Institute of Sanskrit & Indological Studies, Hoshiarpur, Panjab University. Bibcode:1972hksh.book.....S.
28. K.V. Sarma. "Tradition of Aryabhatiya in Kerala : Revision of planetary parameters" (PDF). Indian Journal of History of Science. 12 (2): 194–199. Archived from the original (PDF) on 29 November 2014. Retrieved 30 January 2010.
29. Plofker, Kim (2009). "7 The school of Madhava in Kerala". Mathematics in India. Princeton University Press. pp. 219–254. ISBN 9780691120676.
30. R.C. Gupta (1973). "Addition and subtraction theorems for the sine and the cosine functions in medieval india" (PDF). Indian Journal of History of Science. 9 (2): 164–177. Archived from the original (PDF) on 29 November 2014.
31. Sarma, K.V.; Ramasubramanian, K.; Srinivas, M.D.; Sriram, M.S. (2008). Ganita-Yukti-Bhasa (Rationales in Mathematical Astronomy) of Jyesthadeva. Sources and Studies in the History of Mathematics and Physical Sciences. Vol. I–II (1st ed.). Springer (jointly with Hindustan Book Agency, New Delhi). pp. LXVIII, 1084. Bibcode:2008rma..book.....S. ISBN 978-1-84882-072-2. Retrieved 17 December 2009.
32. Yuktibhāṣā, https://en.wikipedia.org/wiki/Yuktibh%C4%81%E1%B9%A3%C4%81.

## Kamalakara (कमलाकारः)
## (1616 – 1700)

Kamalakara (कमलाकारः) was an astronomer and mathematician, came from a learned family of scholars from *Golagrama,* a village situated in Maharashtra State near *Partha-puri* (Pathari) on the northern bank of the river *Godāvarī.* His father was *Nrsimha* who was born in 1586. Two of *Kamalakara*'s three brothers were also astronomer and mathematicians: *Divakara,* who was the eldest of the brothers born in 1606, and *Ranganatha* who was youngest. Kamalākara learnt astronomy from his elder brother *Divākara,* who compiled five works on astronomy. His family later moved to *Vārāṇasī. Kamalākara*'s major work, "*Siddhāntatattvaviveka* - सिद्धान्ततत्वविवेक ", was compiled in Varanasi at about 1658 and has been published by *Sudhakar Dwivedi* in the *Vārāṇasī series.*

This work consists of 13 chapters in 3,024 verses.

It deals with the topics of:

- Units of time measurement.
- Mean motions of the planets.
- True longitudes of the planets.
- The three problems of diurnal rotation.
- Diameters and distances of the planets.
- The earth's shadow.
- The moon's crescent.
- Risings and settings.
- Syzygies.
- Lunar eclipses, solar eclipses.
- Planetary transits across the sun's disk.
- The patas of the moon and sun.
- The "great problems".
- Along a conclusion.

His other works include *Śeṣavāsanā* (शेषवासना) and *Sauravāsanā* (सौरवासना). *Kamalākara* was bitterly opposed to *Munishvara,* the author of *Siddhāntasārvabhauma*. It is wrongly believed by some moderners that *Kamalākara* discovered the idea that the pole star we see at present is not exactly at the pole.

But this idea was first expressed in *Brahmananda Purana* and *Matsya Purana* by sage *Veda Vyaasa*:

| |
|---|
| "उत्तअनापदा-पुत्रो-सौ मेधिभूतो ध्रुवो द्वे ।।<br>स हि भ्रामण भ्तामयते नित्यं चन्द्रादित्यौ ग्रहैः सह ।।<br><br>*"uttAnapAda-putro-asau meDhibhooto dhruvo divi ।।<br>sa hi bhraman bhtaamayate nityam chandraadityau grahaiH saha ।। ".* |
| The meaning of this expression is "Uttanapada's son Dhruva is fixed like a pole in the Heaven, but it is moving itself and is making all the planets together with Sun and Moon move.<br><br>Translation |

### Key Contributions in Mathematics

### Overall

He combined traditional Bhartiya astronomy with Aristotelian physics and Ptolemaic astronomy as presented by Islamic scientists. In the *third chapter* of the *Siddhanta-tattva-viveka, Kamalākara* used the addition and subtraction theorems for the sine and the cosine to give trigonometric formulae for the sines and cosines of double, triple, quadruple and quintuple angles. In particular he gives formulae for $sin(A/2)$ and $sin(A/4)$ in terms of $sin(A)$ and iterative formulae for $sin(A/3)$ and $sin(A/5)$. He has assumed a value of 60 units for the radius of the Earth and gives values for sines at 1° intervals. *Kamalākara* also gives a table for finding the right ascension of a planet from its longitude.

### Further Reading


1. "Biography of Kamalakara". Archived from the original on 31 March 2012.
2. Achar 2007.

3. D Pingree, Biography in Dictionary of Scientific Biography (New York 1970–1990)
4. A K Bag, "Indian literature on mathematics during 1400–1800 A.D.", Indian J. Hist. Sci. 15 (1) (1980), 79–93.
5. A N Singh, "Hindu trigonometry", Proc. Benares Math. Soc. 1 (1939), 77–92.
6. Achar, Narahari (2007). "Kamalākara". In Thomas Hockey; et al. (eds.). The Biographical Encyclopedia of Astronomers. New York: Springer. p. 609. ISBN 978-0-387-31022-0. (PDF version)
7. Dadhigrama, village in India associated with astronomers and mathematicians
8. David Pingree, "Islamic astronomy in Sanskrit", J. Hist. Arabic Sci. 2 (2) (1978), 315–330; 425.
9. Dvivedi, Sudhakar (1935). The Siddhantatattvaviveka of Kamalakara. Benares.
10. G G Joseph (1991). "Mathematics in India". The Crest of the Peacock: Non-European Roots of Mathematics. London.
11. Golagrama, village in India associated with astronomers and mathematicians
12. Jambusagaranagara, region in India associated with astronomers and mathematicians
13. Kamalakara, https://en.wikipedia.org/wiki/Kamalakara.
14. O'Connor, John J.; Robertson, Edmund F., "Kamalakara", MacTutor History of Mathematics Archive, University of St Andrews
15. Pingree, David Erwin (1970). Census of the Exact Sciences in Sanskrit, Volume 1. American Philosophical Society. pp. 21, 386. ISBN 978-0-87169-081-4.
16. Radha Charan Gupta, "Addition and subtraction theorems for the sine and the cosine in medieval India", Indian J. History Sci. 9 (2) (1974), 164–177.
17. Radha Charan Gupta, "Kamalakara's mathematics and construction of Kundas", Ganita Bharati 20 (1-4) (1998), 8–24.
18. Radha Charan Gupta, "Sines and cosines of multiple arcs as given by Kamalakara", Indian J. History Sci. 9 (2) (1974), 143–150.
19. Radha Charan Gupta, "Sines of sub-multiple arcs as found in the Siddhanta-tattva-viveka", Ranchi Univ. Math. J. 5 (1974), 21–27.

## Jagannātha Samrāṭ (जगन्नाथ सम्राट्)
## (c. 600 BCE)

Paṇḍita Jagannātha Samrāṭ (पाण्डित जगन्नाथ सम्राट्) was an astronomer and mathematician who served in the court of *Jai Singh II* of Amber, and was also his *guru* (गुरु). *Jagannātha,* whose father's name was *Gaṇeśa,* and grandfather's *Viṭṭhala* was from a Vedic family originally from Maharashtra. At the suggestion of *Jai Singh,* he learned Arabic and Persian, in order to study Islamic astronomy. Having become proficient in these languages, he translated texts in these languages into Sanskrit.

These translations include:

- *Rekhā-gaṇita* (रेखागणित), a translation of Euclid's Elements made from Nasir al-Din al-Tusi's Arabic recension of the same. For this work, he had to coin more than a hundred Sanskrit mathematical terms.
- *Siddhānta-sāra-kaustubha* (सिद्धान्तसारकौस्तुभा), a translation of Ptolemy's Almagest from Nasir al-Din al-Tusi's Arabic version.

### 卐 Key Contributions in Mathematics

### 卐 Overall

His original works include:

- *Siddhānta-samrāṭ* (सिद्धान्तसम्राट्), which describes astronomical instruments, their design and construction, and observations. It also describes the use of these observations in correcting parameters and preparing almanacs. It mentions how Jai Singh, who earlier used astronomical instruments (such as the astrolabe) made of metal, later switched to huge outdoor observatories (such as the *Jantar Mantar*), as they were more precise; also, they were made of stone and mortar rather than brick, to diminish the effects of wear-and-tear and climate.
- *Yantra-prakāra* (यन्त्र-प्राकारः), which describes astronomical instruments, measurements, computations, etc. in more detail, and also observations made by him.

- *Jagannātha* held that when theory and observation differed, observation was the true *pramāṇa* (प्रमाण:) and overruled theory. While he used and described a number of astronomical instruments, telescopes were not one of them.

**Further Reading**


1. Achar, Narahari (2007). "Jagannātha Samrāṭ". In Thomas Hockey; et al. (eds.). The Biographical Encyclopedia of Astronomers. New York: Springer. p. 584. ISBN 978-0-387-31022-0. (PDF version, Google Books).
2. Harilal Harshadarai Dhruva. "The Rekhaganita or Geometry in Sanskrit", pp. 35 ff. Bombay: Bombay Sanskrit Series, no. LXI, 1901.
3. Jagannatha Samrat, https://en.wikipedia.org/wiki/Jagannatha_Samrat.
4. K. V. Sarma (2008), "Jagannātha Samrāṭ", in Helaine Selin (ed.), Encyclopaedia of the History of Science, Technology, and Medicine in Non-Western Cultures, Dordrecht: Kluwer Academic Publishers (published 1997), pp. 460–61, Bibcode:2008ehst.book.
5. The Rekhaganita Sanskrit text with English introduction. Two volumes. (PDF), http://www.wilbourhall.org/index.html#rekha.

## Sankara Varman (शङ्कर वर्मन्)
## (1774–1839)

Sankara Varman (शङ्कर वर्मन्) was an astronomer-mathematician belonging to the Kerala school of astronomy and mathematics. He is best known as the author of *Sadratnamala* (सद्रत्नामाला), a treatise on astronomy and mathematics, composed in 1819. *Sankara Varman* is considered as the last notable figure in the long line of illustrious astronomers and mathematicians in the Kerala school of astronomy and mathematics beginning with *Madhava of Sangamagrama*. Sadratnamala was composed in the traditional style followed by members of the Kerala school at a time when Bharat had been introduced to the western style of mathematics and of writing books in mathematics. One of *Varman*'s contributions to mathematics was his computation of the value of the mathematical constant $\pi$ correct to *17 decimal places*.

*Varman* was born as a younger prince in the principality of *Katattanad*, in North Malabar, Kerala, in the year 1774. He had two elder brothers, the eldest being *Raja Udaya varma*, the ruler of the principality, and second one being *Rama Varma*, the crown prince.

Local people referred to *Varman* as *Appu Thampuran* (अप्पु थम्पुराण). *Katattanad* principality was under the suzerainty of the *Zamorin* of Calicut. The family deity of *Sankara Varman* was *Goddess Parvati* (देवी पार्वती) installed in the temple of *Lokamalayarkavu* (लोकमलायरकवु) and he was also a devotee of *Lord Krishna* (भगवान् श्रीकृष्ण).

*Varman* completed the composition of *Sadratnamala* in 1819. He also wrote a Malayalam commentary on *Sadratnamala.* Two versions of the text of Sadratnamala and their commentaries have been discovered. A critical examination of the manuscripts suggests that both are written by Varman, one being a revision of the other.

During the invasion of *Malabar* by *Tipu Sultan/Hyder Ali* during 1766 - 1781, many people including members of royal families fled Malabar and took shelter in Travancore. This brought the principality of Katattanad also into close contact with the Travancore Royal Family. *Varman* was especially attached to *Maharaja Swati Tirunal* (1813–47).

*Varman* was an astute astrologer. There is a legend that he predicted the time of his own death as being in 1839, the year in which he died.

## 卐 Key Contributions in Mathematics

### 卐 Overall

*Sadratnamala* is an astronomical-mathematical treatise in Sanskrit written by *Sankara Varman,* an astronomer-mathematician of the Kerala school of mathematics, in 1819 (see Figure 5.5). Even though the book has been written at a time when western mathematics and astronomy had been introduced in Bharat, it is composed purely in the traditional style followed by the mathematicians of the Kerala school. *Sankara Varman* has also written a detailed commentary on the book in Malayalam. *Sadratnamala* is one of the books cited in C. M. Whish's paper on the achievements of the Kerala school of mathematics. This paper published in the Transactions of the *Royal Asiatic Society* of Great Britain and Ireland in 1834, was the first ever attempt to bring the accomplishments of Keralese mathematicians to the attention of Western mathematical scholarship.

Whish wrote in his paper thus:

*"The author of Sadratnamalah is SANCARA VARMA, the younger brother of the present Raja of Cadattanada near Tellicherry, a very intelligent man and acute mathematician. This work, which is a complete system of Hindu astronomy, is comprehended in two hundred and eleven verses of different measures, and abounds with fluxional forms and series, to be found in no work of foreign or other Indian countries."*

**Synopsis of the book**

The book contains 212 verses divided into six chapters, called *prakarana-s* (प्राकरण-स):

- Chapter 1: Gives the names of numerals; defines the eight operations of addition, subtraction, multiplication, division, squaring, extracting square root, cubing, and extracting cube root.

- Chapter 2: Lists the different measures, namely, the measures of time, angles, lunar days, planets and stars, almanacs, length, grain weight, money and the directions.
- Chapter 3: Defines the rule of three and syllabic enumeration; explains methods for the computation of the elements of the almanac, namely, mean and true sun, moon and *planets, lunar day, yoga and karana*; gives methods for determining the time elapsed after sunrise and after sunset.
- Chapter 4: Deals with arcs and sines and its application in astronomical measurements and computations.
- Chapter 5: Deals with computations relating to the *shadow, eclipse, vyatipata*, retrograde motion of the planets and apses of the moon.
- Chapter 6: Explains the necessity of periodic revision of astronomical constants; gives a full description of *parahita-karana* (परहित-करणम्).

शङ्करवर्ममहाराजविरचिता

## सद्रत्नमाला

### I. अथ प्रथमं परिकर्माष्टकप्रकरणम्

**[मङ्गलाचरणम्]**

श्रीपार्वत्याश्च लोकावनिधरसरिदाराम[1]सान्निध्यवत्याः
पादाम्भोजं गुरूणामपि सततमनुस्मृत्य नत्वारुणादीन् ।
ज्योतिश्शास्त्राब्धिकीर्णां विशद[2]गणित[3]सारोक्तिसद्रत्नमालां
संगृह्यैनां[4] लिखामः पिपठिषुजनसन्धारणा[5]स्वल्पयत्नाम् ॥ 1 ॥

निरुपाधिकृपानिधीनुपासे क्षितिपीयूषभुजो महानुभावान् ।
यदनुग्रहतो विधूतदोषः परिपूर्णाखिलमङ्गलो भवेयम् ॥ 2 ॥

**[ग्रन्थनिर्माणे प्रोत्साहनम्]**

श्रीपोळ्ळतिरिवंशमौक्तिकमणेः श्रीकेरलालङ्कृते-[1]
रार्यस्योदयवर्मणः शुभमतेः[2] श्रीभैमिभूमीपतेः ।
श्रीमत्सोदररामवर्मयुवराजार्याज्ञया तन्यते
तन्त्रं शङ्करवर्मणेदमखिलज्योतिर्विदां[3] प्रीतये ॥ 3 ॥[4]

Figure 5.5. Opening verses of Sadratnamala (in Devanagari). (Image Credit: https://en.wikipedia.org/wiki/)

**Further Reading**


1. C.M. Whish (1834), "On the Hindu Quadrature of the circle and the infinite series of the proportion of the circumference to the diameter exhibited in the four Sastras, the Tantra Sahgraham, Yucti Bhasha, Carana Padhati and Sadratnamala", Transactions of the Royal Asiatic Society of Great Britain and Ireland, Royal Asiatic Society of Great Britain and Ireland, 3 (3): 509–523, doi:10.1017/S0950473700001221, JSTOR 25581775
2. Charles Whish (1834), "On the Hindu Quadrature of the circle and the infinite series of the proportion of the circumference to the diameter exhibited in the four Sastras, the Tantra Sahgraham, Yucti Bhasha, Carana Padhati and Sadratnamala", Transactions of the Royal Asiatic Society of Great Britain and Ireland, Royal Asiatic Society of Great Britain and Ireland, 3 (3): 509–523, doi:10.1017/S0950473700001221, JSTOR 25581775
3. J J O'Connor; E F Robertson (November 2000). "An overview of Indian mathematics". School of Mathematics and Statistics, University of St Andrews Scotland. Retrieved 19 December 2009.
4. Sadratnamala, https://en.wikipedia.org/wiki/Sadratnamala.
5. Sankara Varman, https://en.wikipedia.org/wiki/Sankara_Varman.
6. Sarma, K.V. (2001). "Sadratnamala of Sankara Varman" (PDF). Indian Journal of History of Science. Indian National Academy of Science, New Delhi. 36 (3–4 (Supplement)): 1–58. Archived from the original (PDF) on 2 April 2015. Retrieved 17 December 2009.
7. Sarma, K.V. (2001). "Sadratnamala of Sankara Varman" (PDF). Indian Journal of History of Science. Indian National Academy of Science, New Delhi. 36 (3–4 (Supplement)): 1–58. Archived from the original (PDF) on 2 April 2015. Retrieved 17 December 2009.

**Radhanath Sikdar (राधानाथ सिकदार)**
**(5 October 1813 – 17 May 1870)**

Figure 5.6. Radhanath Sikdar. (Image Credit: https://en.wikipedia.org/wiki/)

Radhanath Sikdar (राधानाथ सिकदार) was a mathematician who is best known for calculating the height of *Mount Everest* (see Figure 5.6). He was the first person to calculate the height of *Mount Everest*, in 1852. In 1831, George Everest, the Surveyor General of India, was in the pursuit of a mathematician who had specialised in Spherical Trigonometry, so that they could be a part of the *Great Trigonometric Survey (GTS)*. In 1832, under the leadership of Everest, the longitudinal series of the "*triangle*" survey was completed from Sironj in Central Bharat to Calcutta in Bengal.

While still working on mapping Calcutta, Bengal, Everest had begun his search for a mathematician, and soon enough, John Tytler, a professor of Mathematics at the Hindu College, now known as the *Presidency College*, recommended his 19-year-old pupil, *Radhanath Sikdar*.

*Radhanath*, a student of the college since 1824, was one of the first two Bhartiyas to read *Isaac Newton's Principia* and by 1832; he had studied *Euclid's Elements, Thomas Jephson's Fluxion* and Analytical Geometry and Astronomy by Windhouse. Taking inspiration from these prestigious papers, he devised a new method to draw a common tangent to two circles, when he was just a teenager. There was little doubt about *Radhanath*'s proficiency in his subject, and he secured the job at the GTS on 19 December 1831 as a "*computer*" at a salary of thirty rupees per month. Soon he was sent to *Sironj* near *Dehradun*. Even as seven other Bengali '*computers*' worked alongside him, *Radhanath* soon showed his superior skills in mathematics and became Everest's favourite colleague. So much so, that he once stopped his transfer to another department. *Radhanath*'s job was to carry geodetic surveys—the study of the earth's geometric shape orientation in space and gravitational field. He did not just use the established methods but invented his own to accurately measure these factors. In 1851, *Radhanath* was promoted to the position of Chief Computer and transferred to Calcutta. Here, he was also a superintendent for the Meteorological Department. It appears that while Everest and Waugh both extolled him for his exceptional mathematical abilities, his relations with the colonial administration were far from cordial. Two specific instances are on record.

In 1851 a Survey Manual (eds. *Capt. H. L. Thullier and Capt. F. Smyth*) was published by the Survey Department. The preface to the Manual mentions that the more technical and mathematical chapters of the Manual were written by *Babu Radhanath Sikdar*. The Manual proved to be immensely useful to surveyors. However, the *third edition* (1875) did not contain that preface, so that Sikdar's memorable contribution was de-recognized. The incident was condemned by a section of British surveyors. The paper Friend of Bharat in 1876 called it *'robbery of the dead'*.

It is also on record that *Sikdar* was fined a sum of 200 rupees in 1843 for having vehemently protested against the unlawful exploitation of survey department workers by the Magistrate Vansittart. The incident was reported in detail in The Bengal Spectator edited by Ramgopal Ghosh.

*Sikdar* had retired from service in 1862, and was later appointed as teacher of mathematics at the General Assembly's Institution (now Scottish Church College).

| |
|---|
| He measured the mount Everest's height as *8839 m*. |

## 卐 Key Contributions in Mathematics

### 卐 Trigonometry and Everest

*Radhanath* started measuring the height of mountains. The brilliant mathematician, who had perhaps never seen Mount Everest, discovered in 1852 that *Kangchenjunga,* which was considered to be the tallest in the world, wasn't really so. Compiling data about Mount Everest from six observations, he eventually came to the conclusion that it was the tallest in the world.

It was during the computations of the northeastern observations that *Radhanath* had calculated the height of *Peak XV at exactly 29,000 ft (8839 m),* but Waugh added an arbitrary two feet because he was afraid that the *Sikdar*'s figure would be considered a rounded number rather than an accurate one. He officially announced this finding in March 1856, and this remained the height of Mount Everest till an Bhartiya survey re-calculated it to be 29,029 ft or 8848 m in 1955.

Technological advancements, data from the thousands of climbers, and the discovery of different routes to the summit all have led to a more accurate calculation of the height of Mount Everest—a peak that grows at the rate of 4 mm every year and whose summit is slowly moving northeastwards each passing year.

In 1854, Sikdar along with his Derozian friend *Peary Chand Mitra* (पियरी चन्द मित्र) started the Bengali journal Masik Patrika, for education and empowerment of women. He used to write in a simple and uncluttered style that was rather atypical for the age.

**Further Reading**


1. "Call to rename Mt Everest after Radhanath Sikdar, who measured its height". Hindustan Times.
2. "How tall will Mount Everest get before it stops growing?". BBC Online.
3. "Radhanath Sikdar: A past we must preserve". The Telegraph (India).
4. "When did modern science emerge in India?". The Hindu.

5. For details, see Ashish Lahiri, Radhanath Sikdar: Beyond the Peak, Boi-Chitra, Kolkata, 2010.
6. Hasnat, Abul. "Shikder, Radhanath". Banglapedia. Retrieved 26 July 2015.
7. Staff List: General Assembly's Institution (1856-1907) in 175th Year Commemoration Volume. Scottish Church College, April 2008. page 569
8. "The man who 'discovered' Everest". 20 October 2003. Retrieved 14 January 2021.
9. Bagal, Jogesh Chandra Unabingsha Shatabdir Bangla, 1941
10. Chaudhury, Ajana A Brief Historical Review of Early Weather Work in Kolkata, Psyche and Society, Kolkata, May 2009
11. Lahiri, Ashish Radhanath Sikdar and Colonial Science, Sahitya Samsad, Kolkata, 2017
12. Nath, Sankar Kumar Radhanath Sikdar: Taththeyer Aaloye, Chirayata, Kolkata-73, October 2012
13. Radhanath Sikdar, https://en.wikipedia.org/wiki/Radhanath_Sikdar.

**Ramchundra (रामचन्द्र)**
**(1821–1880)**

Figure 5.7. Ramchundra (रामचन्द्र). (Image Credit: https://mathshistory.st-andrews.ac.uk/)

Ramchundra (Ramachandra Lal) (रामचन्द्र लाल) was a mathematician. *Ramchundra*'s father, Rai *Sunder Lal Mathur*, was employed in the Bhartiya revenue department and, at the time of *Ramchundra*'s birth, he had been posted to *Panipat* which is a city in north west of Bharat about 70 km from Delhi (see Figure 5.7). In fact, the family came from Delhi and they were of the *Kayastha caste* (कायस्थ गोत्रम्), that is a Hindu caste of scribes who worked in the Bhartiya bureaucracy. *Ramchundra*'s early education was at home, a standard for Bharat at that time, but he also attended school in Panipat from the age of six. The family moved back to Delhi but *Ramchundra*'s father *Sunder Lal* died in 1831 leaving the family in a very difficult financial position. *Ramchundra*'s mother had to care for six young boys.

Ramchundra married a deaf and dumb girl in 1832 when he was eleven years old. This is a very early age for, at this time most Bhartiya girls married around the age of fourteen, but for boys the age for marriage was around eighteen. This was an arranged wedding (the girl was the daughter of a wealthy businessman and came with a sizeable dowry) and *Ramchundra* only learnt after the wedding that his wife was deaf and dumb. That he married at this young age is almost certainly so that the family could benefit from the dowry.

In 1833, *Ramchundra* entered the English Government School in Delhi. He spent six years at this school but he learnt little mathematics there. In fact, it was towards the end of these six years that he became interested in the topic and studied mathematics books at home. However, he did well at school and was awarded minor scholarships which helped out financially. Of course, having to care for and support an invalid wife made things even harder and he found it difficult to purchase the mathematics texts he was so keen to read. He left the English Government School in 1839 and for two years he was employed as a writer. This not only meant he earned money to support his wife, but it also allowed him to help support the education of his five brothers. Delhi College was set up by the *East India Company* in 1824 based on a college for literature, science and art that had been founded by the citizens of Delhi in 1792 on the site of an old Madrasa outside the Ajmeri gate. The College incorporated the English Government School which *Ramchundra* attended. *Delhi College*, originally intended to promote Oriental learning, soon added the teaching of astronomy and mathematics based on European methods.

In 1841 *Ramchundra* competed for a scholarship and was awarded a senior scholarship of thirty rupees per month. He held this scholarship for three years, enabling him to complete his education, and then, in 1844, he was he was appointed as a teacher of European science in the Oriental department of Delhi College. We note that Delhi College today is known as Zakir Husain Delhi College and has been part of the University of Delhi since 1925.

*Ramchundra* began a programme of translating and compiling works into Urdu. He produced Sari-ul-Fahm, aimed at teaching algebra, trigonometry and other mathematical topics up to the differential and integral calculus.

Ramchundra writes:

*"These translations were introduced into the Oriental department as class books; so that in two or three years many students in the Arabic and Persian departments were, to a certain extent, acquainted with English science: and the doctrines of the ancient philosophy, taught through the medium of Arabic, were cast into the shade before the more reasonable and experimental theories of modern science."*

*Ramchundra,* along with the advanced students in the English and Oriental departments of the Delhi College, founded a Society of the Diffusion of Knowledge in 1847. They published a monthly paper which was designed to follow the same style as the Spectator:

*"... in which not only were the dogmas of the Mohamedan and Hindu philosophy exposed, but also many of the Hindu superstitions and idolatries were openly attacked."*

As one might expect, this led to attacks from Hindus and even his friends.

In another monthly publication, founded in September 1847 and published by Delhi College, *Ramchundra* contributed articles on a wide variety of topics including: A Description of the Diving Bell, by which Sunken Materials may be retrieved from the Sea; A Discussion of the Mistakes that Hindu Learned Men have made in *Various Sciences in the Shastras;* On Astronomy; On the work of *Sir Isaac Newton;* A discussion of the Relationship of the Human Mind and Body; On Demosthenes; On Confucius; and A biography of the *Safavid Shah Abbas.* He also discussed the latest scientific developments in agriculture and gave summaries of new books on popular science.

In 1850 *Ramchundra* published at Calcutta, at his own expense, the *book A treatise on problems of maxima and minima solved by algebra* (see Figure 5.8).

He writes:

*"For the last four or five years I was desirous of solving almost all problems of Maxima and Minima by the principles of Algebra, and not by those of the Differential Calculus. ... I flatter myself with the hope that my labours will be of some use to those Mathematical students who are not advanced in their study of the Differential Calculus, and that the lovers of science, both in India and Europe, will give support to my undertaking."*

# A TREATISE

ON

# PROBLEMS OF MAXIMA AND MINIMA,

## SOLVED BY ALGEBRA.

BY RAMCHUNDRA,

TEACHER OF SCIENCE, DELHI COLLEGE.

"The problems which relate to the Maxima and Minima, or the greatest or least values of variable quantities, are among the most interesting in the Mathematics; they are connected with the highest attainments of wisdom and the greatest exertions of power; and seem like so many immoveable columns erected in the infinity of space, to mark the eternal boundary which separates the regions of possibility and impossibility from one another."

2ND DISS. ENCY. BRIT.

CALCUTTA:

PRINTED BY P. S. D'ROZARIO AND CO., TANK SQUARE.

1850.

Figure 5.8. Problems of Maxima and Minima: Solved by Algebra. (Image Credit: https://www.google.co.in/books/edition/A_Treatise_on_Problems_of_Maxima_and_Min/lO27y59Yxv4C?hl=en)

However, when *Augustus De Morgan* received a copy of the book in 1850, he had a very different opinion of it. He *reprinted* the work in London in 1859 and wrote an Editor's Preface to the book which begins:

*"In the year 1850 my friend the late J E Drinkwater-Bethune forwarded to England a number of copies of a work on Maxima and Minima, by Ramchundra, teacher of science, Delhi College, with directions to present copies to various persons, and among others to myself. On examining this work I saw in it, not merely merit worthy of encouragement, but merit of a peculiar kind, the encouragement of which, as it appeared to me, was likely to promote native effort towards the restoration of the native mind in India."*

A review of this 1859 reprint containing *De Morgan's Editor's* Preface also appeared in The Calcutta Review.

After working as a translator in the English camp for a few months, he was appointed as headmaster of the *Thomason Civil Engineering College* at Roorkee in January 1858. He held this position for eight months before becoming the headmaster of a newly founded Delhi District School towards the end of 1858. We note that by this time the rebellion was defeated and Delhi was again under the control of the British. *Ramchundra*'s health was not good at this stage and he retired from his position as headmaster on grounds of ill health in 1866 when 45 years of age. However, he continued working in the service of the *Maharaja of Patiala,* a state in the Punjab in north west Bharat. In 1870 he became director of education in Patiala. His health deteriorated rapidly, however, and he devoted much of his time in his last years to Christian missionary work.

## 卐 Key Contributions in Mathematics

### 卐 Maxima and Minima by Algebra

In the first chapter of his book, he solved the problems of *Maxima and Minima* by the principles of Algebra and not by those of the *Differential Calculus.* All those problems which brought out equations of the second degree were of course easily solved by the method of imaginary roots given in some works on Algebra, particularly in *Wood's Algebra* by *Lund,* and the *Encyclopaedia Metropolitana.* But even these problems in several cases required particular artifices, without which it was impossible for me to solve them. All these problems are solved in the first chapter of this little work. Besides the method of imaginary roots, he had given another, quite independent of Imaginary quantities, quantities which to many beginners of Mathematics, appear somewhat mysterious and unintelligible. This latter method he ventured to call a new method, because in all mathematical works which he had access to, he had never seen a single problem of Maxima or Minima solved by it, though it is used to reduce an affected quadratic to a pure one in a great many works on Algebra.

All the problems solved in the second chapter of his book brings out cubic equations, the solution of which on the condition of *Maximum or Minimum*, required a new method, which he could not find, though he searched for it in several works enumerated hereafter. He then resolved to find out a method, and in intervals of leisure during three years he continually thought on the subject, and at last found it out. This is a method which appears extremely simple and easy, though it baffled all my endeavours for the space of three years. He called it new, for he did not find it in any book he looked into.

The third and fourth chapters, and the supplement contain problems and general solutions of particular equations of the fourth, fifth, and the sixth degree, together with those problems in which two or three variable quantities enter. The methods used in these parts of the work, though more difficult and intricate than that used in the second chapter, were easily discovered.

This work contains about 130 problems taken chiefly from the following works:

- Simpson's Fluxions,
- Hall's Differential Calculus,
- Gregory's Examples,
- Connel's Differential Calculus,
- Walton's Differential Calculus,
- Ritchie's Differential Calculus,
- Young's Differential Calculus,
- Encyclopedia Britannica,
- Hirsch's Geometry,
- works on Mixed Mathematics, &c.

Besides the problems solved here, many more may be solved by the methods given in this treatise. He had also given definitions, formulae, and propositions necessary for the study of this work in the Introduction.

**Further Reading**


1. A De Morgan, Editor's Preface, in A Treatise of Problems on Maxima And Minima, Solved by Algebra (W H Allen & Co, 1859).
2. C Musès, De Morgan's Ramanujan: An Incident in Recovering our Endangered Cultural Memory of Mathematics, Mathematical Intelligencer 20 (3) (1998), 47-51.
3. C Simmons, Augustus De Morgan behind the scenes, in M Pitici (ed.), The Best Writing on Mathematics 2012 (Princeton University Press, 2012), 186-198.
4. C Simmons, Augustus De Morgan behind the scenes, The College Mathematics Journal 42 (1) (2011), 33-40.
5. C. Muses, "De Morgan's Ramanujan". Mathematical Intelligencer, vol. 20, no. 3, 1998, pp. 47–51.
6. D Raina, Ramchundra's Treatise Through the Unsentimentalised Light of Mathematics or the Mathematical Foundation of a Cultural Project, Historia Mathematica 19 (1992), 371-384.
7. De Morgan's Preface to Ramchundra's book (https://mathshistory.st-andrews.ac.uk/Extras/De_Morgan_1859_Preface/.
8. Dhruv Raina, "Ramchundra's Treatise Through the Unsentimentalised Light of Mathematics or the Mathematical Foundation of a Cultural Project", Historia Mathematica, 19, 1992, 371 384.
9. P J Nahin, Chases and Escapes: The Mathematics of Pursuit and Evasion (Princeton University Press, 2012).
10. R D Paul, Ramchundra Mathematical Genius 1821-1880, in Triumphs of His grace: lives of eight Indian Christian laymen of the early days of Protestant Christianity in India: every one of whom was a triumph of His grace (Christian Literature Society, 1967), 186-199.
11. R Gupta, Infinity versus Ramchundra, The Hindu Business Line. http://www.thehindubusinessline.com/blink/explore/infinity-versus-ramchundra/article6940652.ece.
12. Ramchandra's Treatise of Problems on Maxima And Minima (https://mathshistory.st-andrews.ac.uk/Extras/Ramchundra_Preface/.
13. Ramchundra (1859), A Treatise on Problems of Maxima and Minima: Solved by Algebra, London: Wm. H. Allen and Co.
14. Ramchundra, A treatise on problems of maxima and minima solved by algebra. Revised from the 1859 original, edited, rewritten and with an introduction by M S Rangachari. With an introductory note by M

Mahadevan (The Association of Mathematics Teachers of India, Chennai, 2007).
15. Ramchundra, https://en.wikipedia.org/wiki/Ramchundra.
16. Ramchundra, https://mathshistory.st-andrews.ac.uk/Biographies/Ramchundra/.
17. Ramchundra, Preface, A Treatise of Problems on Maxima and Minima, Solved by Algebra (W H Allen & Co, 1859).
18. Ramchundra's autobiographical notes (https://mathshistory.st-andrews.ac.uk/Extras/Ramchundra_autobiography/.
19. Review: A treatise on problems of maxima and minima solved by algebra, by Ramchundra, The Athenaeum 1656 (23 July 1859), 117-118.
20. Review: A treatise on problems of maxima and minima solved by algebra, by Ramchundra, The Calcutta Review 14 (July-December 1850), xxxii-xxxiii.
21. Review: A treatise on problems of maxima and minima solved by algebra, by Ramchundra, The Colonial Church Chronicle and Missionary Journal 13 (1860), 354-357.
22. Review: A treatise on problems of maxima and minima solved by algebra, by Ramchundra, The Mathematical Monthly (November 1859), 66-71.
23. Review: A treatise on problems of maxima and minima solved by algebra, by Ramchundra, The Saturday Review of Politics, Literature, Science and Art 7-8 (12 November 1859), 586-587.
24. S Irfan Habib and D Raina, The Introduction of Modern Science into India: A Study of Ramchundra, Educationist and Mathematician, Annals of Science 46 (1989), 597-610.
25. S.Irfan Habib and Dhruv Raina, "The Introduction of Modern Science into India: A Study of Ramchundra, Educationist and Mathematician", Annals of Science, 46, (1989), 597 610; also Habib and Raina, "Vaijnanik Soch ko Samarpit", Sancha, June July 1988, 76 83.
26. Scientists of India, Speaking for Science. Master Ramchandra, Science Reporter 51 (5) (May 2014), 56-57.
27. The Calcutta Review of Ramchundra's 1850 book (https://mathshistory.st-andrews.ac.uk/Extras/Calcutta_Review_1850/.
28. The Calcutta Review of Ramchundra's 1859 reprint (https://mathshistory.st-andrews.ac.uk/Extras/Calcutta_Review_1859/.

## Jadav Chandra Chakravarti (जादवचन्द्र चक्रवर्ती)
## (1855 – 26 November 1920)

Professor Jadav Chandra Chakravarti (जादवचन्द्र चक्रवर्ती) was a prominent Bengali mathematician who was born in 1855 at the village of *Tatulia* in *Kamarkhanda,* just few miles from Sirajganj city of the then British Raj (now Bangladesh) to *Krishna Chandra Chakravarti* and *Durga Sundari.* He moved to Kolkata for higher education and obtained his M.A. degree in mathematics from Presidency College at the University of Calcutta in 1882. While he was a student of Presidency, he taught physics and chemistry at St. Paul's Cathedral Mission College to pay for his university degree.

*Chakravarti* started his career as a mathematics teacher at Kolkata City College. Then he joined at Aligarh Muslim College on 1 January 1888 at 200 rupees salary. In 1905, his salary was increased to 300 rupees. One of the famous students of his was Ziauddin Ahmad. He retired from Aligarh College on 28 February 1916.

In 1901, he came back to his home town Sirajganj and founded a school for local children. After his retirement in 1916, he was also elected as the Chairman of Sirajganj Municipality. Before joining to Aligarh Univ. He was a high-level British officer, in the princely state of Cooch Behar. He was instrumental in fixing the match between *Maharajah Nripendra Narayan of Cooch Behar* and *Sunity Devi* D/O *Sri Keshub Chandra Sen,* the founder of the new dispensation. Jadav Babu served cooch behar for some years.

In 1905. Morison remarked in his work "*The History of M.A.O College, Aligarh*" that *Prof. Chakravai* came from Calcutta at a time when the politicians of Bengal were being very roughly handled at Aligarh and it is too much hope that remarks were not occasioanly dropped which must have been painful to him. But besides providng himself an admirable teacher, a most loyal colleague, he has shown a sympathetic interest in the progress of the institution and the cause which it was founded to promote; the esteem and confidence which he now commands among Muhammadan trustees were most conspicuously shown by his appointment as *Registrar of the College* in 1899 – a post which he continued to hold at much inconvenience to himself but to the great advantage of the institution.

*Prof. Chakravarti* died at the age of 69 on 26 November 1920 at his Kolkata residence.

## 卐 Key Contributions in Mathematics

### 卐 Arithmetic and Algebra

He was famous for his two books named *Arithmetic* and *Algebra*. His first book *Arithmetic* was published in 1890 and his second book *Algebra* was published in 1912.

The main work for which he is known is the treatise "*Arithmetic*". When it appeared, the firstv edition of January 1880 was part of a movement breaking with older ways of teaching Arithmetic.

Thus, first significant mathetical text famous in old as well as modern times has been well acknowledged by the teachers and learned of mathematics. The book has been reprinted more than *100 times* which probably is a world record and bear testimony to a super class work. Further, the first edition has been paid the compliment of continuous use as a text since it first appeared.

His books were translated into a number of languages including *Bengali, Urdu, Hindi, Assamese and Nepali*.

Some interesting features of the book on *Arithmetic* are problmes relating to clocks, races and games of skill, calendar, *Shubhankar*'s rule, limits of error etc. which differ in many ways from any existing text book on the subject.

Certainly, a glance at the contents of *Profesor Ckakravati*'s *Arithmetic* may well emphasize to the attentive reader that a few topics of *Grogorian method* for the calculation of a leap year, calculation of different Eras in different countries of the world like *Cristian Era*, the *Hejira*, the *Skakhabda*, the *Samvat* and the *Bangabda*, calculation of a day of any given date etc., instead if being relatively isolated, are in fact related to just about everything; from life science to social science.

Some very interesting examples of the book are:

- What day will be 6th April 2001 A.D.?
- Prove that the last day of a century will be either Sunday, Monday, Wednesday or Friday.
- Determine the number of years between 16 BC and 1800 AD.
- Calculate the number of days in the 20th century
- Hezrat Muhammad left Mecca for Madina on 20th September 622 AD (the Hejira Era begings from this date). You want to know what as the day?

One can find it out by the general rule given in *Professor Ckavarti*'s book. The list of such beautiful examples is far too broad for one to attempt any meaningful complete survey.

**Further Reading**


1. গণিত জাদুকরের গল্প. Risingbd.com (in Bengali). Retrieved 28 March 2019. (https://www.risingbd.com/art-literature-news/79034)
2. "Babu Jadav Chandra Chakravarti by DR. MA Pathan". Aligarh Muslim University. Retrieved 28 March 2019. (https://www.researchgate.net/publication/315788591)
3. Chakravarti, Jadav Chandra (June 2010). Arithmetic by Jadav Chandra Chakravarti. ISBN 978-1175032164.
4. Jadav Chandra Chakravarti, https://en.wikipedia.org/wiki/Jadav_Chandra_Chakravarti.

**Pathani Samanta (पठानि समन्ता)**
**(1835–1904)**

Figure 5.9. Pathani Samanta (पठानि समन्ता). (Image Credit: https://en.wikipedia.org/wiki/)

Pathani Samanta - पठानि समन्ता (alias Chandrasekhar - चन्द्रशेखर) better known as *Mahamahopadhyaya Chandrasekhara Singha Harichandana Mahapatra Samanta* (महामहोपाध्याय चन्द्रशेखरसिंह हरिश्चन्दन महापात्र समन्त), was an astronomer, mathematician and scholar who measured the distance from the Earth to the Sun with a bamboo pipe, and traditional instruments (see Figure 5.9). He was born on 13 December 1835 in *Purnimanta Pousha Krishna Ashtami* (पूर्णिमन्त पौष कृष्ण अष्टमी), and died on 11 June 1904 in *Purnimanta Adhika Jyeshtha Krishna Trayodashi* (पूर्णिमन्त अधिकारी ज्येष्ठ कृष्ण त्रयोदशी).

*Samanta* married *Sita Devi*, the daughter of *King Anugul*, in 1857 in a rather dramatic way after the bride's family rejected the alliance on the couple's wedding day because *Samanta* didn't look princely enough, according to his family history, which was written by his grandson *Raghunath Singh Samanta* and published in the book "*Pathani Samanta Jeebani Darpana*".

He reportedly won over the bride's family at the wedding with his faultless *sloka recitation.* Odisha has kept his legacy relevant by displaying his work in the state museum, naming the planetarium in *Bhubaneswar* after him; and dedicating educational institutions, scholarships, and amateur astronomy clubs to his memory. Astronomers and astrophysicists both in Bharat and beyond have praised his work, earning him the moniker "*Bharatn Tycho.*"

However, the general public is mostly unaware of this brilliant astronomer who observed the universe with only the naked eye, as well as of the incredible scientific advances he accomplished with only a few pieces of bamboo and wood and the sheer force of his brilliance. He deserves to be celebrated just like *Aryabhatta, Bhaskara,* and others - probably as the *last torch bearer of the Indian traditional astronomy.*

His research and observations were compiled into a book called *Siddhanta Darpana,* which was written in both Sanskrit and the Odia script. He earned the *Mahamahopadhyaya* Award in 1893, for his usage of traditional instruments for astronomical observations. *Mahamahopadhyaya* is an honorific title given to prestigious scholars by the Government of India. Prior to 1947, the title was bestowed by the British Raj, and before them, by the kings of ancient Bharat. In ancient Bharat, a scholar that wrote works based on topics related to the shastras was granted the title *Mahopadhyaya.* The title *Mahamahopadhyaya* was bestowed on the best amongst the *Mahopadhyaya* scholars.

*Samanta* was born in the princely state of *Khandpara,* in the *Nayagarh* district of the Bhartiya state of Odisha. He was the son of *Samanta Syamabandhu Singha* and *Bishnumali Devi.* He was born into a royal family.

The legend depicted on the walls of the *Pathani Samanta Planetarium* in *Bhubaneswar* states that he was born to a royal couple the loss of many children, leaving them yearning for a healthy child. Hence, soon after his birth, he was given away in adoption to a Muslim fakir to ward off the evil eye, a belief that was strongly prevalent at the time. In remembrance of the fakir and to ward off bad omens, the couple nicknamed their son 'P*athani*'.

He went on to study Sanskrit, and later researched traditional Bhartiya astronomy.

During his youth time, *Samanta* measured the length of the shadows throughout the day by using bamboo and wood to create measuring instruments, which he called mana yantra. He also measured time by using his version of a sundial.

*Samanta* was a self-taught astronomer and learned by reading the books available at the Royal Library until age 15. During his research, Samanta designed many of his instruments by using everyday materials such as wooden sticks and bamboo. After studying mathematics and traditional astronomy he used his knowledge to match predictions made by ancient Bhartiya mathematicians and astronomers such as *Aryabhata, Varahamihira, and Brahmagupta.*

He carried out research in measurements using only a bamboo pipe and two wooden sticks. His findings were recorded in his book titled Siddhanta Darpana and were mentioned in the European and American press in 1899. *Samanta*'s calculations were eventually used in the preparation of almanacs in Odisha. The treatises *Samanta* was referring to had only clues to the observational devices used, so he decided to make his own measuring instruments made of locally available bamboo and wood. They used basic geometry and trigonometry to calculate distance, height, and time. There are many local tales of Samanta measuring the height at which birds fly, finding the height of trees, and persons using the length of shadows and calculating the distance and height of mountains from his fixed location using an instrument he invented called mana yantra. He used his own versions of the *sundial* and *imprsundialater* clocks to measure time.

He was the only Bhartiya astronomer who discovered all three irregularities of the moon independently of European astronomers, which were unknown to ancient Bhartiya astronomers. He continued to teach and attracted pupils worldwide despite his persistent health problems and insomnia. On June 11, 1904, he died suddenly from fever and infection.

## 卐 Key Contributions in Mathematics

### 卐 Astronomical Instruments Design

Armed with the devices, *Chandrasekhara* spent nights after nights in the all-absorbing game of *star gazing* (तारा प्रेक्षमाणः). At the age of twenty-three, *Chandrasekhara* started to record his observations systematically and compiled these findings in a masterpiece treatise written in Sanskrit on palm leaves called *Siddhanta Darpana* (सिद्धान्त दर्पण) which he completed at the age of only thirty-four. But it took 30 years to get published in *Devnagari* script from Kolkata. It contains 2500 number of slokas of which 2284 are compiled by himself and the rest are due to other scholars. This book contains numerous instances of astronomical methods of determining the position and motion of planets, mathematical treatment of spherical astronomy, instrumentation techniques, improvements over earlier measurements, theories and models.

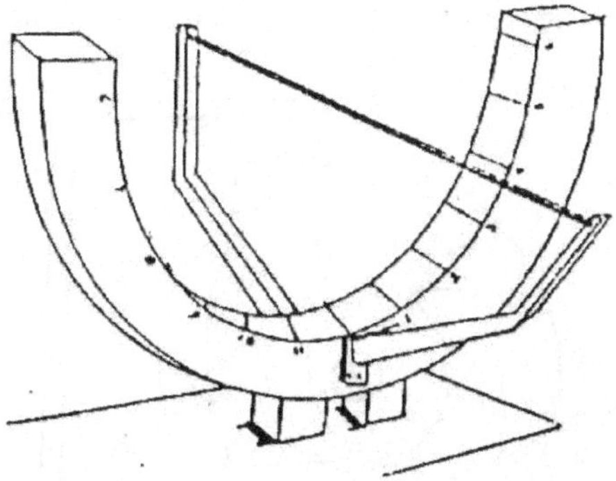

Figure 5.10. *Chapa Yantra*. (Image Credit: Odisha Review)

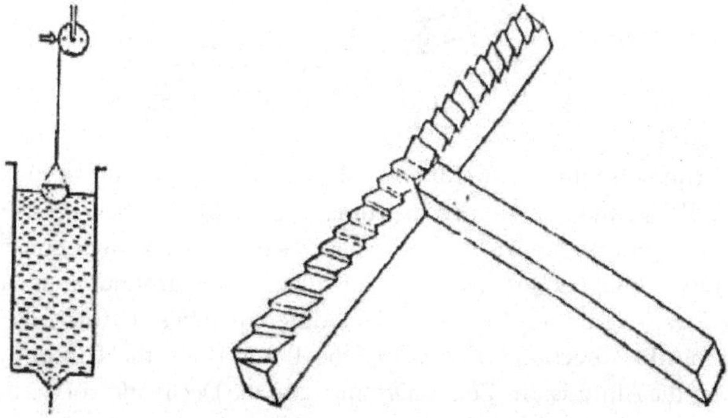

Figure 5.11. *Swayambaha Yantra* (left), *Mana Yantra* (right). (Image Credit: Odisha Review)

His instruments can be broadly classified into three categories, i.e., instruments for measuring *Time* (कालः), *Versatile instruments* (बहुमुखी वाद्ययन्त्राणि) and *Armillary Sphere*. In the context of the highly sophisticated instruments in use these days for astronomical measurements (capable of millionths part of a degree in accuracy), it will be interesting to have an idea of the instruments he was actually using. Instruments for measuring time include sun-dials like *Chapa Yantra* (चापा यन्त्र) (see Figure 5.10), *Chakra Yantra* (चक्र यन्त्र) (see Figure 5.13 (right)), *Golardha Yantra* (गोलर्धा यन्त्र), and of course, a Water clock called *Swayambaha Yantra* (स्वयम्बहा यन्त्र) (see Figure 5.11 (left)). *Chakra Yantra* which measures time for an entire day consists of a graduated disc with a staff fixed at the centre. *Chapa Yantra*, basically a semi-circular dial with a pointer aligned at the Pole-star, measures time for half of a day. *Golardha Yantra* (see Figure 5.13 (left)) is a hemi-spherical sun-dial.

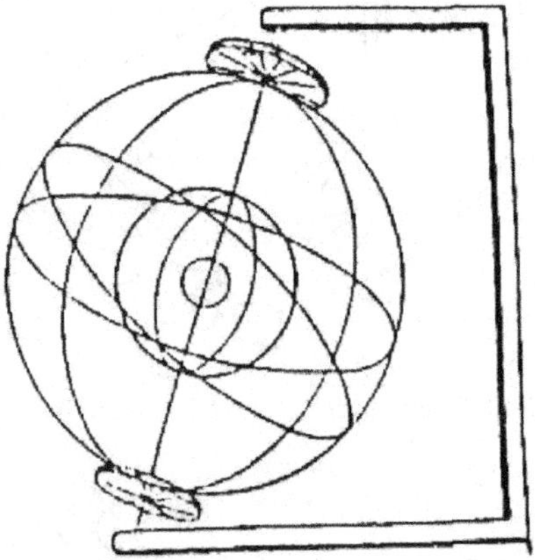

Figure 5.12. *Gola Yantra*. (Image Credit: Odisha Review)

The instruments which were for versatile use primarily include a stick called *Shanku* (शङ्कु) and a 'T' shaped instrument called *Mana Yantra* (मान यन्त्र) (see Figure 5.11 (right)). *Shanku*, or *Gnomon* as it is popularly known, consists of a stick of measured height fixed vertically on a levelled ground. By measuring its shadow length cast by the Sun at different times, it was possible to determine the directions of a place, the Local Time, the Latitude of the observer, the Altitude, the Zenith Distance and the Declination of the Sun and its position along the *Zodiac belt*. However, his most favourite instrument was the Mana Yantra which readily measures angles in the sky as well as on the ground. It consists of a wooden staff to which is attached a cross-piece in the form of a 'T'.

The cross-piece is notched into stairs and pierced with holes to indicate the angle subtended by the distant object at the free end of the staff. With this instrument in hand, he is said to have measured the height of the *Saptasajya hill* with a fair degree of accuracy once upon a request made by the then British Commissioner of *Cuttack Mr. Cooks.* With *Mana Yantra,* he also measured the height of the *Mahendragiri hill* of *Manjusha* in Parikuda upon a request made by the *king of Manjusha* and his value was verified by the then Madras Government. Both *Shanku* and *Mana Yantra* can also be used to determine the height and distance of a hill, tree, lamp-post or cloud. Of course, for such a task (where both the Height and Distance of the object are unknown), separate measurements from two different distances have to be taken. Armillary Sphere or *Gola Yantra* (see Figure 5.12) was a very common device prevalent among classical Bhartiya astronomers and is basically a three-dimensional model of the celestial sphere.

It was used as a demonstration kit for showing to students' various great circles used in astronomy as well as for measuring the Latitude and Longitude of planets. In this mode, it was able to show the direct and retrograde motion of the planets along the *zodiac circle* (राशिचक्रम्).

*Samanta Chandrasekhara,* at the first hand, attempted to improve the existing calendar system that regulates the daily rituals of the Hindu society. This was very important, as the native Hindu almanac computed from the rules given by the *old Siddhantas* (पुरातन सिद्धान्ताः) were falling into serious errors and no two almanacs agreed in their computations. Samanta *Chandrasekhara* has, therefore, painstakingly re-determined the elements of the *old Siddhantas* aided by his practical observations. The ephemeris computed from his elements proved to be more accurate than any other almanacs available at that time, and hence, provided a standard source of reference in the otherwise chaotic situation prevailing at that time. Not surprisingly, his calendar system was accepted in many learned circles including the temple authorities of *Puri* for performing various day-to-day rituals of the deity.

He has prescribed several corrections to be applied to the mean motion of the planets so as to precisely determine their positions in the sky. The *ephemeride* computed from his elements were in close agreement with the Nautical Almanac followed in Europe at that time. While the Bengali almanac may be in error by as much as 40, the corresponding error in *Siddhanta Darpana* (सिद्धान्त दर्पण) was restricted to only half a degree.

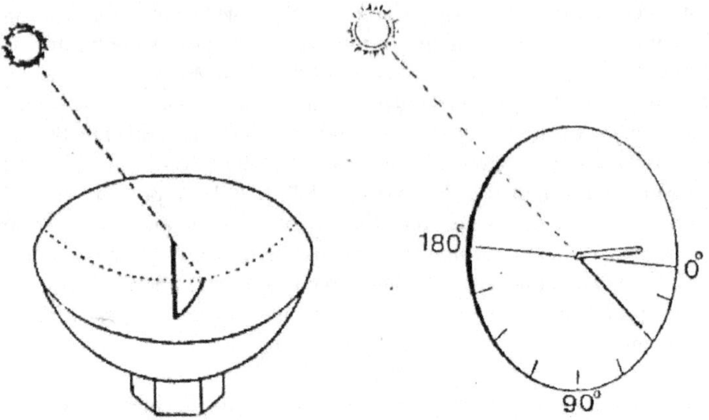


Figure 5.13. *Golardha Yantra* (left), *Chakra Yantra* (right). (Image Credit: Odisha Review)

*Chandrasekhara* has discovered some original corrections to be applied to the mean motion of the Moon to account for its three important anomalies, i.e., *Evection, Variation and Annual Equation*. Although some of these anomalies were known to independent observers at various *tomes, Chandrasekhara* was the only Bhartiya astronomer who has detected and measured all these three values.

Another important contribution of *Chandrasekhara* was the improvement of the values of *Parallax for the Sun* and the Moon which were so important for the calculation and prediction of eclipses. To determine the parallax of the Sun and the Moon, their respective distances from the Earth have to be expressed in terms of *Earth radii. Chandrasekhara* has suggested an innovative experiment with a coin so as to determine the ratio of the Distance to the Diameter for the Sun and the Moon.

**Further Reading**

1. "About I Rewa Prasad Dwivedi". Retrieved 8 October 2015, http://rewaprasaddwivedi.net/about/.
2. "Galagali Ramacharya I Karnataka Samskrit University". www.ksu.ac.in. Retrieved 19 October 2016.

3. "Samanta Chandrasekhar". Archived from the original on February 22, 2014.
4. "Sri Sri Anandamayi Ma's Devotees". Anandamayi Ma. Retrieved 26 September 2014.
5. "Srouta Saiva Siddhanta", https://books.google.com/books?id=Y0FfADSV6o4C.
6. Belvalkar, S.K. (1943). "Our Cordial Greetings to the Ganganath Jha Research Institute". Annals of the Bhandarkar Oriental Research Institute. 24 (3/4): 224–226. JSTOR 41688502.
7. Cultural contours of India, p37.
8. Dutt, Kartik Chandra, ed. (1999). Who's who of Indian Writers, 1999: A-M. New Delhi: Sahitya Akademi. p. 495. ISBN 978-81-260-0873-5.
9. Gulerī, Candradhara Śarmā (1987). Gulerī racanāvalī (in Hindi). Kitābaghara, https://books.google.com/books?id=-YIhAAAAMAAJ.
10. Katti, Madhuri. "Chandrasekhar Samanta: India's Eye in the Sky". Live History India. Retrieved 28 September 2020.
11. Katti, Madhuri. "Chandrasekhar Samanta: India's Eye in the Sky". Live History India. Retrieved 28 September 2020.
12. Panda, Bipin Bihari (2000). "PATHANI SAMANTA AND HIS THEORY OF PLANETARY MOTION" (PDF).
13. Sharma, Dasharatha (1970) Lectures on Rajput history and culture p1 Motilal Banarsidass, Delhi.
14. The contribution of Kerala to Sanskrit Literature; K. Kunjunni Raja; University of Madras 1980; page 257.
15. ମହାମହୋପାଧ୍ୟାୟ ଚନ୍ଦ୍ରଶେଖର ସିଂହ ହରିଚନ୍ଦନ ମହାପାତ୍ର ସାମନ୍ତ; Odia pronunciation: [mɔhamɔho:padʰjajɒ t͡ʃɔndrose:kʰɔr siŋho horit͡ʃɔndo:n mɔhapatro: samo:nto:].
16. Mrmlonline, http://www.mrmlonline.com/ap_eds_hathibhai_jamnagar_mahamahopadhyaya_shastri_swami_tirtha_trivikrama.html.
17. Worldcat identity, http://www.worldcat.org/identities/lccn-n81141805/.
18. Lalye, P. G. (1 January 2002). Mallinātha. Sahitya Akademi. ISBN 9788126012381.
19. Mahamahopadhyaya, https://en.wikipedia.org/wiki/Mahamahopadhyaya.
20. Naik, P. C.; Satpathy, L. (1998). "Samanta Chandra Sekhar: The great naked eye astronomer". Bulletin of the Astronomical Society of India. 26: 33–49. Bibcode:1998BASI...26...33N.

21. Nikunja Bihari Sahu, The Legacy of Samanta Chandrasekhara, Odisha Review, January 2012, pp. 31-34. (UNESCO Kalinga prize Special).
https://magazines.odisha.gov.in/Orissareview/2012/Jan/engpdf/31-34.pdf
22. Pathani_Samanta, https://en.wikipedia.org/wiki/Pathani_Samanta.
23. Satpathy, L., ed. (2003). "Samanta Chandra Sekhar and his Contributions to Ancient Indian Astronomy.". Ancient Indian Astronomy and Contributions of Samanta Chandra Sekhar. ISBN 9788173194320.
24. Sriram, M. S.; Ramasubramanian, K.; Srinivas, M. D. (2002). "Samanta Chandra Sekhara and His Treatise Siddhantadarpana.". 500 years of Tantrasangraha: a landmark in the history of astronomy. p. 127. ISBN 9788179860090.
25. ପଠାଣି ସାମନ୍ତ; Odia pronunciation: [pɔntʰani samoːntoː]

**Ashutosh Mukherjee (आशुतोष मुखर्जी)**
**(4 April 1921 – 3 April 1923)**

Figure 5.14. Ashutosh Mukherjee (आशुतोष मुखर्जी). (Image Credit: https://en.wikipedia.org/wiki/)

Sir Ashutosh Mukherjee (आशुतोष मुखर्जी) was a prolific Bengali educator, jurist, barrister and mathematician (see Figure 5.14). He was the first student to be awarded a dual degree (MA in Mathematics and MSc in physics) from Calcutta University. Perhaps the most emphatic figure of Indian education, he was a man of great personality, high self-respect, courage and towering administrative ability. The second Indian Vice-Chancellor of the *University of Calcutta* for four consecutive two-year terms (1906–1914) and a fifth two-year term (1921–23), Mukherjee was responsible for the foundation of the Bengal Technical Institute in 1906, which was later known as *Jadavpur University* and the *University College of Science* (Rajabazar Science College) of the *Calcutta University* in 1914.

Mukherjee also played a vital role in the founding of the *University College of Law* popularly known as *Hazra Law College*. The *Calcutta Mathematical Society* was also founded by Mukherjee in 1908 and he served as the president of the Society from 1908 to 1923. He was also the president of the inaugural session of the *Indian Science Congress* in 1914 held at the Rajabazar Science College, which he founded. The *Ashutosh College* was also founded under his stewardship in 1916, when he was Vice-chancellor of *University of Calcutta.* He is often called "*Banglar Bagh*" ('The Bengal Tiger') for his high self-esteem, courage and academic integrity. According to historian D. R. Bhandarkar, the epithet '*Vikramaditya* - विक्रमादित्य ' is also ascribed to *Sir Ashutosh Mukherjee.*

*Ashutosh Mukherjee* was born on 29 June 1864 at Bowbazar, Kolkata in a *Brahmin family* to *Jagattarini Devi* and *Ganga Prasad Mukhopadhyaya,* a well-known doctor who founded the South Sub Urban School in Calcutta. Among his ancestors were several distinguished Sanskrit scholars, including *Pandit Ramchandra Tarkalankar,* a professor of nyaya who had been appointed by *Warren Hastings* to that chair at the Sanskrit College in Kolkata. Brought up in an atmosphere of science and literature at home, young *Ashutosh* went to the *Sisu Vidayalaya* at *Chakraberia,* Bhowanipore and showed an early aptitude for mathematics. When he was young, he met *Ishwar Chandra Vidyasagar* who was a major influence on him. He was a student of *Madhusudan Das.*

*Mukherjee* married *Jogamaya Devi Bhattacharyya* in 1885. The couple had seven children, *Kamala* (born 1895), *Rama Prasad* (1896-1983), *Syama Prasad* (1901-1953), *Uma Prasad* (1902-1997), *Amala* (born 1905), *Bama Prasad* (born 1906) and *Ramala* (born 1908). His son *Syama Prasad Mookerjee,* the most notable of his children, founded the *Bharatiya Jana Sangh,* the direct precursor to the modern *Bharatiya Janata Party. Rama Prasad* became a judge in the *High Court of Calcutta* while *Uma Prasad* became famed as a Himalayan trekker and a travel writer - being awarded the *Sahitya Akademi Award* for his travelogue *Manimahesh.*

*Sir Ashutosh Mukherjee* was born into a Hindu Brahmin family in Kolkata, West Bengal. His ancestral town was Jirat in Hooghly District, West Bengal. He was a very meritorious student and he came to Kolkata to study in Medical College with the help of the wealthy people of Jirat. Later he settled down in Bhawanipore area of Kolkata.

In November 1879, at the age of fifteen, *Mukherjee* passed the entrance exam of the Calcutta University in which he stood second and received a first-grade scholarship. In the year 1880, he took admission at the *Presidency College* (now *Presidency University*) in Kolkata where he met *P. C. Ray* and *Narendranath Dutta* who would later become famous as *Swami Vivekananda.* Later that year, though only a first-year undergraduate, he published his first mathematical paper, on a new proof of the 25th proposition of *Euclid's first book.*

In 1883, *Mukherjee* topped the BA examination at *Calcutta University* to complete a postgraduate degree in mathematics. In 1883, *S.N. Banerjee* wrote an article in the newspaper Bengalee against the orders of the Calcutta High Court and he was arrested in contempt of court. Protests and hartals erupted across Bengal and other cities, led by a group of students headed by Mukherjee at Calcutta high court. In 1884, he won the Harishchandra Prize for academic achievements, and completed an M.A. with first-class honours in mathematics in 1885. In 1886, he was awarded a second Masters in Natural Sciences, making him the first student to be awarded a dual degree from Calcutta University. In the same year he published his third mathematical paper, "*A Note on Elliptic Functions.*" *Mukherjee* was recognised for his achievements by the grant of the Premchand Roychand Fellowship in Mathematics and Physics, Pure and Applied. Still only aged 22, he was further recognised by his election as a *Fellow of the Royal Society of Edinburgh (FRSE)*. By 1888, Mukherjee was a lecturer in mathematics for the recently established *Indian Association for the Cultivation of Science (IACS)*.

*Mukherjee* continued publishing scholarly papers on mathematics and physics into his 30s. By 1893, aged 29, *Mukherjee* had been further elected to the fellowships of *the Physical Society of France* and the *Mathematical Society of Palermo*, and was a member of the *Royal Irish Academy*. He subsequently became a member of the *London Mathematical Society*, the *Paris Mathematical Society* and the *American Mathematical Society* (1900). Although after 1893 he largely abandoned his mathematical pursuits for a legal career, *Mukherjee* has been recognised as the first modern Bhartiya mathematician to enter the field of mathematical research, and founded the Calcutta Mathematical Society in 1908. Among his mathematical contributions, *Mukherjee* determined several crucial derivations of *Gaspare Mainardi*'s answer to determining the oblique trajectory of a system of confocal ellipses. He also made lasting contributions in differential geometry, developing analytical methods of simplifying *Gaspard Monge*'s interpretation of his general differential equation for conics. *Mukherjee* was a *polyglot* learned in *Pali, French* and *Russian*. Apart from his fellowships and memberships in several international academic bodies, he was recognised by an award of the title of *Saraswati* in 1910 from pandits in Nabadwip, followed by that of *Shastravachaspati* in 1912 from the *Dhaka Saraswat Samaj*, Sambudhagama Chakravarty in 1914 and *Bharat Martanda* in 1920. Mukherjee was appointed a *Companion of the Order of the Star of India* (CSI) in June 1909, and *knighted* in December 1911. *Asutosh* played an important role in the early development of mathematics in India by successfully establishing the *Calcutta Mathematical Society (CMS)* on

September 6, 1908 and starting a research journal, *Bulletin of Calcutta Mathematical Society* in 1909.

## 卐 Key Contributions in Mathematics

### 卐 Geometry

*Asutosh* sent 80 problems to the Educational Times, London. Subsequently. he sent the solutions of these questions. He also sent solutions to 155 questions submitted by others (including mathematicians like W K Clifford). Simple proofs and elegance in the derived results, were the traits of *Asutosh*'s mathematics, whether the problem be at an elementary level or of an advanced character. First, consider a simple rider which is one of the 348 problems solved by *Asutosh* in his school days:

*Problem. If two circles cut each other, then the line segment joining their centres bisects the common chord at right angles.*

*Asutosh* solved the problem in two different ways. One of the solutions is given below:

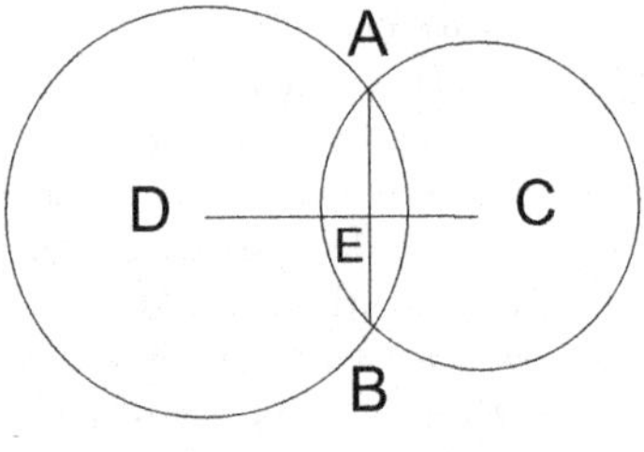


Figure 5.15. Two circles cut each other.

**Asutosh's Solution:**

Let the circles with centres C and D intersect at A and B (see Figure 5.15). It is required to prove that CD bisects AB at right angles. A, B are joined. Then a perpendicular is drawn from C on AB to intersect AB at E. Now, E and D are joined. Hence, AB is a chord to the circle with centre C, and CE is perpendicular to AB; in particular, $\angle CEA = 1$ right angle. So, CE bisects AB at right angles and we have AE = BE, i.e., E is the mid-point of AB.

Again, AB is a chord to the circle with centre D, and E is the mid-point of AB. So, DE is perpendicular to AB and hence, ∠DEA = 1 right angle. So, we have, ∠CEA +∠DEA = 1 right angle + 1 right angle = 2 right angles. Thus, DE and EC are in the same line (CD). This completes the proof.

As an example of his elegant higher-level mathematics, *'On the differential equation of a trajectory'* and *'A general theorem on the differential equations of trajectories'*.

***Asutosh*'s Published Papers**

i. Proof of Euclid I, 25, Messenger of Mathematics, Vol.10, pp.122–123, 1880–81.
ii. Extension of a theorem of Salmon's, Messenger of Mathematics, Vol.13, pp.157–
160, 1883–84.
iii. A note on elliptic functions, Quart. Jour. of Pure and Appl. Maths., Vol.21,
pp.212–217, 1886.
iv. On the differential equations of a trajectory, Jour. of Asiatic Soc., Bengal, Vol.56,
Pt. II, No.1, pp.117–120, 1887.
v. On Monge's differential equation to all conics, Jour. of Asiatic Soc., Bengal, Vol.56,
Pt. II, No.2, pp.134–145, 1887.
vi. A memoir on plane analytic geometry, Jour. of Asiatic Soc., Bengal, Vol.56. Pt. II,
No.3, pp.288–349, 1887.
vii. A general theorem on the differential equations of trajectories, Jour. of Asiatic
Soc., Bengal, Vol.57, Pt. II, No.1, pp.72–99, 1888.
viii. On Poisson's integral, Jour. of Asiatic Soc., Bengal, Vol.57, Pt. II, No.1, pp.100–
106, 1888.
ix. On the differential equation of all parabolas, Jour. of Asiatic Soc., Bengal, Vol.57,
Pt. II, No.4, pp.316–332, 1888.

x. Remarks on Monge's differential equation to all conics, Proc. Asiatic Soc., Bengal,
February 1888.

xi. The geometric interpretation of Monge's differential equation to all conics, Jour.
of Asiatic Soc., Bengal, Vol.58, Pt. II, No.2, pp.181–185, 1889.

xii. Some applications of elliptic functions to problems of mean values, (I), Jour. of
Asiatic Soc., Bengal, Vol.58, Pt. II, No.2, pp.199–213, 1889.

xiii. Some applications of elliptic functions to problems of mean values, (II), Jour. of
Asiatic Soc., Bengal, Vol.58, Pt. II, No.2, pp.213–231, 1889.

xiv. On Clebsch's transformation of the hydrokinetic equations, Jour. of Asiatic Soc.,
Bengal, Vol.59, Pt. II, No.1, pp.56–59, 1890.

xv. Note on Stoke's theorem and hydrokinetic circulation, Jour. of Asiatic Soc., Ben-
gal, Vol.59, Pt. II, No.1, pp.59–61, 1890.

xvi. On a curve of aberrancy, Jour. of Asiatic Soc., Bengal, Vol.59 Pt. II, No.1, pp.61–
63, 1890.

xvii. Mathematical Notes (Questions and Solutions), Educational Times, London, Vols.43
(pp.125–151), 44 (pp.144–182), 45 (pp.146–168), 1890–92.

*Asutosh* had written more papers that remained unpublished. These are: (a) On some definite integrals. (b) On an application of differential equations to the theory of plane cubics. (c) Researches on the number of normals common to two surfaces, two curves or a curve and a surface. (d) Application of Gauss's theory of curvature to the evaluation of double integrals.

The papers discuss the problem of determining the trajectory of a system of confocal ellipses, which was first solved by the Italian mathematician Gaspare Mainardi (1800–1879) in the journal Annli di Scienze-Mathematische e Fisiche, Tome 1, p.251. Mainardi's problem and his own solution are as follows.

*Mainardi's Problem: Determine the oblique trajectory of a system of confocal ellipses.*

**Mainardi's Solution:**

If the distance between the foci is 2h, and the tangent of the angle between any given curve and the trajectory is n, then the equation of the trajectory is given by

$$-2n\, tan^{-1}\sqrt{\left(\frac{h^2}{xM}-1\right)}+\log\frac{1-\sqrt{\left(1-\frac{M}{x}\right)}}{1+\sqrt{\left(1-\frac{M}{x}\right)}}=C$$

where C is the constant of integration, and M is given by the quadratic equation

$$M(x^2+y^2+h^2)=(M^2+h^2)x$$

But *Asutosh* remarked, *"this form of the equation is so complicated that it would be a hopeless task to have to trace the curve from it; indeed, it is so unsymmetrical and inelegant that Professor Forsyth in his splendid work on Differential Equations does not at all give the answer."*

Later, *Asutosh* gave an elegant solution by which the trajectory was *"represented by a pair of remarkably simple equations which admit of an interesting geometrical interpretation"*. Professor Forsyth quoted Asutosh's solution of Mainardi's problem in subsequent editions of his book on differential equations.

*Asutosh*'s simplification of Mainardi's solution is as follows:

*Asutosh*'s Solution: Assume

$$Mx=h^2\cos^2\phi,\ C=2n\lambda$$

Where $\lambda$ is a new constant, substituting it, we get

$$\log\left[\frac{\left\{1-\sqrt{1-\frac{M}{x}}\right\}}{\left\{1+\sqrt{1-\frac{M}{x}}\right\}}\right] = 2n\lambda + 2n\phi$$

$$\Rightarrow \frac{\left\{1-\sqrt{1-\frac{M}{x}}\right\}}{\left\{1+\sqrt{1-\frac{M}{x}}\right\}} = e^{2n(\lambda+\phi)}$$

$$\Rightarrow 1-\frac{M}{x} = \left(\frac{1-e^{2n(\lambda+\phi)}}{1+e^{2n(\lambda+\phi)}}\right)^2$$

$$\Rightarrow \frac{M}{x} = \frac{4e^{2n(\lambda+\phi)}}{\left(1+e^{2n(\lambda+\phi)}\right)^2}$$

But, $M = \left(\frac{h^2}{x}\right)\cos^2\phi$. Substituting this value of M and taking the positive square root

$$\frac{h}{x}\cos\phi = \frac{2e^{n(\lambda+\phi)}}{1+e^{2n(\lambda+\phi)}}$$

So,

$$xch\ cos\phi\ cosh\ n(\lambda+\phi)$$

Again, substituting the value of M and simplifying, we get

$$\frac{x^2}{h^2\cos^2\phi} - \frac{y^2}{h^2\sin^2\phi} = 1$$

pr,

$$y = h\ sin\phi\ \sinh(\lambda+\phi)$$

Therefore, the coordinates of any point on the trajectory may be expressed in a very neat and symmetrical form, in terms of a parameter $\phi$ viz., we have

$$x = h\,cos\phi\,\cosh(\lambda + \phi)\,, y = h\,sin\phi\,\sinh(\lambda + \phi)$$

*Asutosh* also gave a brilliant and intricate geometrical interpretation of the equations above.

The interpretation is as follows (see Figure 5.16):

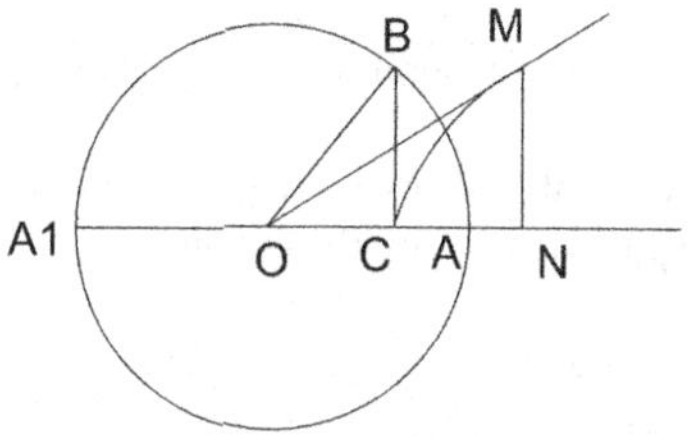


Figure 5.16. Mainardi's solution by Asutosh's trigonometry.

Let A, A1, be the foci of confocal ellipses, such that OA= h and O is the mid-point of AA1. A circle with center O and radius h is drawn. Any point B on the circle is taken such that ∠AOB = $\phi$. Let C be the foot of the perpendicular from B on OA. Then,

$$OC = h\,cos\phi, \quad BC = hsin\,\phi$$

A hyperbola with centre O and transverse and conjugate axes as OC and BC respectively is drawn. Let M be a point on the hyperbola such that the area of the hyperbolic sector OCM equals $n(\lambda + \varphi)$ times the area of the triangle OCB. Asutosh proved that M is a point on the required trajectory, i.e., a point having the coordinates displayed in (4). We sketch his argument here.
Let N be the foot of the perpendicular from M on the line OA; let $\alpha = h\,cos\,\phi$ and $\beta = h\,sin\,\phi$. Thus, the point M on the hyperbola is of the form $(\alpha\,cosh\,\psi, \beta\,sinh\,\psi)$ and the area of the triangle OMN is $\frac{1}{2}\,\alpha\beta\,cosh\,\psi\,sinh\,\psi$. Using integration, the area of the portion CMN can be seen to be $\frac{1}{2}\,\alpha\beta\,cosh\,\psi\,sinh\,\psi\, - \frac{1}{2}\alpha\beta\psi$, showing that the area of OCM is $\frac{1}{2}\,\alpha\beta\psi$. Hence, by hypothesis, $\psi = n(\lambda + \varphi)$. Thus, the coordinates of M satisfy (iv).

In another paper, on a general theorem on the differential equations of trajectories is a development of the earlier paper. Asutosh firmly believed that, "every simple mathematical result could be established by a correspondingly simple process" and re-examined the whole question to see if the artificial process of Mainardi could be simplified. The investigation led him to a very general theorem which shows (roughly) that:

*Whenever the coordinates of any point on a curve can be expressed by means of a single variable parameter, the coordinates of the corresponding point on the trajectory may be similarly expressed.*

The precise statement of the general theorem is given below:

Theorem. Let the coordinates of any point on a plane curve be represented by means of a variable parameter $\theta$, by the equations

$$x = f1(\theta, a), y = f2(\theta, b),$$

where a, b are arbitrary constants.

We consider the system of curves obtained by varying a and b subject to a condition which can be analytically represented by the relations

$$a = f_1(\psi, h), \qquad b = f_2(\psi, h),$$

where $\psi$ is a parameter and h a given constant.

Then, the coordinates of the corresponding point on the oblique trajectory, which intersects any such curve at an angle $\alpha$, is given by a similar expression

$$X = f_1\big(\theta, f_1(\psi, h)\big), \qquad Y = f_2(\theta, f_2(\psi, h)),$$

where $\psi$ is a function of $\theta$ detemined by the differential equation

$$\frac{d\psi}{d\theta} = \frac{nL}{N - nM}$$

where,

$$n = tan\ \alpha$$

$$L = \left(\frac{df_1}{d\theta}\right)^2 + \left(\frac{df_2}{d\theta}\right)^2$$

$$M = \left(\frac{df_1}{d\theta}\right)\left(\frac{df_1}{d\psi}\right) + \left(\frac{df_2}{d\theta}\right)\left(\frac{df_2}{d\psi}\right)$$

$$N = \left(\frac{df_2}{d\theta}\right)\left(\frac{df_1}{d\psi}\right) - \left(\frac{df_1}{d\theta}\right)\left(\frac{df_2}{d\psi}\right)$$

Merely stating and proving the general theorem did not satisfy Asutosh; he also applied it to determine the oblique trajectories of: (i) a system of confocal ellipses (Mainardi's problem), (ii) a system of confocal hyperbolas, (iii) a system of parabolas having a common principal axis and which touch each other at their common vertex, (iv) a pencil of coplanar rays radiating from a point (by two methods), (v) a system of circles which touch a given straight line at a given point, (vi) a system of parabolas which have a common focus and principal axis, vii) the system of curves obtained by varying b in the equation $ex \sin y = ab$, and other problems. He also connected the theorem with the theory of conjugate functions.

**Further Reading**


1. "Asutosh Mukhopadhyay: An eminent educator who made Bengalis proud".
2. "Calcutta Mathematical Society". Calmathsoc.org. Archived from the original on 31 May 2012. Retrieved 12 July 2012.
3. "Grandson of Sir Ashutosh Mukherjee upset at Mamata's silence on letter". The Times of India. 25 March 2013. Retrieved 15 May 2018.
4. "IACS - Annual Report for the Year 1924" (PDF). Archive - IACS. Retrieved 2 October 2017.
5. "No. 28263". The London Gazette (Supplement). 22 June 1909. p. 4855.
6. "No. 28559". The London Gazette. 8 December 1911. p. 9365.
7. "Our History". mslawcollege.org. 2012. Retrieved 28 April 2012. Ashutosh Mukherjee, the then Vice-Chancellor of Calcutta University who was a student of Utkal Gourab Madhusudan Das
8. "Shrimati Jogmaya Devi Passes Away". The Calcutta Municipal Gazette. 68 (13): 301. 19 July 1958. Retrieved 26 December 2021.
9. "Sir Ashutosh Mukherjee: educationist, leader and institution-builder" (PDF). Current Science. Retrieved 29 September 2017.

10. Aparna Basu (1974) The Growth of Education and Political Development in India, 1898–1920, New Delhi: Oxford University Press.
11. Dutta, Ashis (29 June 2013). "First in class". The Hindu.
12. Ghatak, Atulchandra, Ashutosher Chatrajiban Ed. 8th, 1954, p 1, Chakraborty Chatterjee & Co. Ltd.
13. IACS - Annual Report for the Year 1888.
14. Indian Association for the Cultivation of Science: Annual Report for 1983-84. 1984. p. 3.
15. Mukherji, Purabi (11 October 2022). Notable Modern Indian Mathematicians and Statisticians: During the 19th and 20th Centuries of Bengal. Springer Nature. ISBN 978-981-19-6132-8.
16. Patrick Petitjean, Catherine Jami and Anne Marie Moulin eds. (1992) Science and Empires, Boston Study in the Philosophy of Science, Vol. 136, Kluwer Academic Publishers. ISBN 978-94-011-2594-9, doi:10.1007/978-94-011-2594-9
17. Rachana Chakrabarty (2012). "Premchand Roychand Studentship". In Sirajul Islam; Ahmed A. Jamal (eds.). Banglapedia: National Encyclopedia of Bangladesh (Second ed.). Asiatic Society of Bangladesh.
18. Tathagata Roy (2014). The Life & Times of Shyama Prasad Mookerjee. Prabhat Prakashan. p. 11. ISBN 9789350488812.
19. "The mathematician in Asutosh Mukhopadhyay" (PDF). Current Science. Retrieved 29 September 2017.
20. Ashutosh Mukherjee, https://en.wikipedia.org/wiki/Ashutosh_Mukherjee.
21. Sar, Satyabachi (July 2015). "Asutosh Mukhopadhyay and his Mathematical Legacy". Resonance: Journal of Science Education. 20 (7): 575–604. Retrieved 23 June 2019. https://www.ias.ac.in/article/fulltext/reso/020/07/0575-0604.

**Ganesh Prasad (गणेश प्रसाद)**
**(15 November 1876 – 9 March 1935)**

Figure 5.17. Ganesh Prasad (गणेश प्रसाद). (Image Credit: https://en.wikipedia.org/wiki/)

Ganesh Prasad (गणेश प्रसाद) was a mathematician who specialised in the theory of potentials, theory of functions of a real variable, Fourier series and the theory of surfaces (see Figure 5.17). He was trained at the *Universities of Cambridge* and *Göttingen* and on return to Bharat he helped develop the culture of mathematical research in Bharat. The mathematical community of Bharat considers *Ganesh Prasad* as the *'Father of Mathematical Research in Bharat'*. He was also an educator taking special interest in the advancement of primary education in the rural areas of Bharat.

*Ganesh Prasad* was born on 15 November 1876 at Ballia, Uttar Pradesh. He obtained the B.A. degree from *Muir Central College*, Allahabad, M.A. degree from the *Universities in Allahabad* and *Calcutta* and the *D.Sc.* degree from *Allahabad University*. After teaching at the *Kayasth Pathshala*, Allahabad, and at the *Muir Central College*, Allahabad, for about two years, he proceeded to *Cambridge* for higher studies and research.

*Ganesh Prasad* was instrumental in the introduction of compulsory primary education in villages in Uttar Pradesh. He donated from his private savings an amount of Rs. 22,000 for the education of girls in Ballia. He also donated an amount of Rupees two hundred thousand for establishing prizes for the toppers at the M.A. and MSc examinations of the Agra University. He donated large amounts of money to the Allahabad and Banaras Universities also.

Prof. Prasad was well known as a teacher, and his textbooks on the differential and integral calculus is still in use in many Bhartiya universities. His Patna University readership lectures on "*The Place of Partial Differential Equations in Mathematical Physics*" were published in 1924, and since 1928 he had devoted a good deal of his time to the writing of books on higher mathematics and on the history of the subject. He was a man of wonderfully simple habits, and remarkable energy and powers of endurance.

## 卐 Key Contributions in Mathematics

### 卐 Overall

*Prasad* returned to Bharat from Europe in 1904 and was appointed professor of mathematics at the *Muir Central College*, Allahabad. Within a year of his appointment at Allahabad, Prasad was sent to the *Queen's College, Banaras* and he continued there till 1914 when he was invited to head the mathematics department of *Calcutta University*. *Ganesh Prasad* was the *Ras Behari Ghosh Chair* of Applied Mathematics of Calcutta University from 1914 to 1917 and *Hardinge Professor* of Mathematics in the same University from 1923 till his death on 9 March 1935.

In between these two assignments he served *Banaras Hindu University* as professor of mathematics (1917–1923). While at *Banaras*, he helped found the Banaras Mathematical Society. *Ganesh Prasad* was elected President of the *Calcutta Mathematical Society* and the Vice-President of the *Indian Association for Advancement of Science*, Calcutta in 1924 and continued in the same position till his death.

He was a founder member of the *National Institute of Sciences*, India, which has now been rechristened as the *Indian National Science Academy*. *Ganesh Prasad* authored 11 books including "*A Treatise on Spherical Harmonics and the Functions of Bessel and Lame*" and over fifty research papers in mathematics.

*Prof. Prasad* was the first in Bharat to create a school of mathematical research. Many of the papers on mathematics published by young Indian investigators in the last twenty years bear an acknowledgment of indebtedness to him for guidance and help. He founded the *Benares Mathematical Society* in 1918, and was its life-president. He had been for many years the president also of the *Calcutta Mathematical Society*. One of the earliest contributions made by Prof. Prasad was his dissertation entitled "*The Constitution of Matter and Analytical Theories of Heat*" (1903), in which he dealt with the difficulty of interpreting differential coefficients when the molecular constitution of matter is taken into account.

His papers on applied mathematics, published in various journals, dealt likewise with problems in which he skilfully applied his knowledge of the theory of functions of a real variable to potential problems in which the differential coefficients became infinite or did not exist.

His later researches were on the theory of Fourier series and other branches of the theory of functions of a real variable. At the time of his death, he had in hand the completion of a long memoir on "*The Expansion of Zero*".

**Further Reading**


1. Joseph Warren Dauben, Christoph J. Scriba (2002). Writing the history of mathematics: its historical development. Birkhäuser. p. 689. ISBN 9783764361679. (see p.314).
2. Sunil N. Shabde (2012) The Meteoric Life of a Mathematician, A tribute to Dr. N. g. Shabde, page 307, Ganesh Prasad, via Google Books.
3. 1909: A Text-book of Differential Calculus, via Internet Archive, https://archive.org/details/textbookofdiffer00prasrich/page/n5.
4. 1929: "On the differentiability of the integral function", Journal fur die reine und angewandte Mathematik 160: 100–110 doi:10.1515/crll.1929.160.100.
5. 1930: "On the nature of Θ in the mean-value theorem of the differential calculus", Bulletin of the American Mathematical Society 36: 289–91 doi:10.1090/S0002-9904-1930-04937-X.

6. 1933: Some Great Mathematicians of the Nineteenth Century via Internet Archive, https://archive.org/details/in.ernet.dli.2015.80661/page/n9.
7. Dale Hoiberg, Indu Ramchandani (2000). Students' Britannica India: Select essays Volume 6. Popular Prakashan. p. 441. ISBN 978-0-85229-762-9. (see p. 333).
8. Ganesh Prasad, https://en.wikipedia.org/wiki/Ganesh_Prasad.
9. Ganesh Prasad: (1876–1935)". Indian Mathematical Society. Retrieved 19 June 2010.
10. Prof. Ganesh Prasad, https://www.nature.com/articles/135644a0.
11. Sen, S.N. (1992). "Factors in the development of scientific research in India during 1906 – 1930" (PDF). Indian Journal of History of Science. 27 (4): 379–388. Archived from the original (PDF) on 16 March 2012. Retrieved 19 June 2010.
12. Singal, M.K. (1974). "Ganesh Prasad". Bulletin of the Mathematical Association of India. 6: 6–8.

**Bharatikrishna Tirtha (भारतीकृष्ण तीर्थ)**
**(14 March 1884 – 2 February 1960)**

Figure 5.18. Swami Bharatikrishna Tirtha (भारतीकृष्ण तीर्थ).

Jagadguru Śaṅkarācārya Svāmī Bhāratīkṛṣṇa Tīrtha (जगद्गुरु शङ्कराचार्य स्वामि भारतीकृष्ण तीर्थ), born Veṅkatarāmaṇ Śāstrī (वेङ्कतारामण शास्त्री), was a *Hindu monk* (हिन्दू भिक्षु) and *Shankaracharya* (शंकराचार्यः) of *Govardhana Math* in *Puri* in the Bhartiya state of Odisha, from 1925 through 1960 (see Figure 5.18). He is particularly known for his book *Vedic Mathematics,* being the first *Sankaracarya* in history to visit the West, and for his connection with nationalist aspirations, thus earning him the title *'Father of the Vedic Maths'*. *Veṅkatarāmaṇ Śāstrī* was born on 14 March 1884 to a resolute *Tamil Brahmin* family. His father *P. Narasimha Shastri* was a tehsildar at *Tirunelveli* in Madras Presidency, who later became the Deputy Collector of the Presidency.

*Venkataraman* joined the *National College* in Trichinopoly. After this, he moved to the *Church Missionary Society College* and eventually the *Hindu College,* both in Tirunelveli. Venkataraman passed his matriculation examination from *Madras University* in January 1899, where he also finished first. Although *Venkataraman* always performed well in subjects such as mathematics, sciences and humanities, he was also proficient in languages and particularly skilful in Sanskrit. According to his own testimonials, Sanskrit and oratory were his favourite subjects. Due to his knowledge of the language, he was conferred the title "*Saraswati* - सरस्वती" at the age of 16 by the *Madras Sanskrit Association* in July 1899. At about that time, *Venkataraman* was profoundly influenced by his Sanskrit guru *Vedam Venkatrai Shastri* (वेदं वेङ्कटराय सस्त्री).

*Venkataraman* passed the B.A. examination in 1902. He then appeared for the M.A. examination for the *American College of Sciences* in *Rochester*, New York from the Bombay centre in 1903. He also contributed to *W. T. Stead*'s Review of Reviews on diverse topics in religion and science. During his college days, he also wrote extensively on history, sociology, philosophy, politics, and literature.

*Venkataraman* worked under *Gopal Krishna Gokhale* (गोपाल कृष्ण गोखले) in 1905 for the *National Education Movemen*t and the South African Bhartiya problems. However, his inclination towards Hindu studies led him to study the ancient Bhartiya holy scripture *Adhyātma-Vidyā* (अध्यात्म-विद्या). In 1908, he joined the *Sringeri Matha* (शृङ्गेरी मठ्) in Mysore to study under *Svāmī Saccidānandaśivābhinavanṛsiṁha Bhāratī*, the *Sankaracarya* of Sringeri. However, his spiritual practice was interrupted when he was pressured by nationalist leaders to head the newly-started *National College* at Rajamahendry. *Prof. Venkataraman Shastri* taught at the college for three years. But in 1911, he suddenly left the college to go back to *Sringeri Math*.

Returning to Sringeri, *Venkataraman* spent the next eight years studying *Advaita Vedanta* (अद्वैत वेदान्त) and *Sastra (scripture)* – शास्त्र under *Jagadguru Sankaracarya Satcitananda Sivabhinava Nrisimha Bharati* (जगद्गुरु शंकराचार्य सत्चितानन्द शिवभिनव नृसिंह भारती). During those years, the *Jagadguru* initiated *Venkatraman* into *yogic practices* (योगाभ्यासाः). *Venkataraman* also practiced meditation, *Brahma-sadhana* (ब्रह्म साधना) and *Yoga-sādhāna* (योगसाधना), in the nearby forests. He is said to have attained self-realization during his years at *Sringeri Math*. He would leave society and practice meditation in seclusion for many days. During those eight years, he also taught Sanskrit and philosophy at local schools and ashrams. He delivered a series of *sixteen lectures* on *Adi Shankara's* (आदि शङ्कर) philosophy at *Shankar Institute of Philosophy, Amalner* (Khandesh). During that time, he also lectured as a guest professor at institutions in Mumbai, Pune and Khandesh.

**Initiation into Sannyasa**

After *Venkataraman*'s eight-year practice and study of *Vedanta,* he was initiated into sannyasa, in the Tirtha sub-order of the *Dashanami Sampradaya,* in Varanasi by *Jagadguru Śaṅkarācārya Svāmī Trivikrāma Tīrtha* (जगद्गुरु शङ्कराचार्य स्वामि त्रिविक्रम तीर्थ) of *Dwarka Sharada Peetham* in Dwarka on July 4, 1919, receiving the name "*Swami Bharatikrishna Tirtha* (स्वामी भारतीकृष्ण तीर्थ)".

**Shankaracharya of Govardhana Matha**

The *Sankaracarya* of Govardhana Math, *Swami Madhusudhana Tirtha,* was in failing health and was greatly impressed with *Bharatikrishna. Madhusudana* requested *Bharatikrishna* to succeed him at *Govardhana Matha. Bharatikrishna* declined the offer. In 1925, however, *Madhusudhana*'s health worsened and *Bharatikrishna* was compelled to accept the *Govardhana gaddi* (chair). In 1925, *Bharatikrishna* assumed the pontificate of *Govardhana Math.* He installed *Swami Swarupananda Saraswati* as the new *Sankaracarya of Dwarka Sharada Peetham.* As *Shankaracharya* of Govardhana Matha, *Bharatikrishna* toured several countries in thirty-five years to promote *Dharma* (धर्म) and *Bhartiya culture* (भारतीय संस्कृति). He wrote a number of treatises and books on religion, science, mathematics, world peace, and social issues. In 1953, at Nagpur, he founded the *Sri Vishwa Punarnirmana Sangha* (*World Reconstruction Association*). The administrative board initially consisted of *Bharatikrishna*'s disciples and supporters, then later included distinguished personalities.

**卐 Key Contributions in Mathematics**

**卐 Vedic Mathematics**

*Bharatikrishna*'s book, *Vedic Mathematics,* is a list of *sixteen terse sūtras,* or "*aphorisms*", discussing strategies for mental calculation and was first published in 1965. *Bharatikrishna* claimed that he found the sūtras after years of studying the Vedas, a set of sacred ancient Hindu scriptures. It contains a list of mathematical techniques, which were falsely claimed to have been retrieved from the Vedas and to contain advanced mathematical knowledge. *Krishna Tirtha* failed to produce the sources, and scholars unanimously note it to be a mere compendium of tricks for increasing the speed of elementary mathematical calculations sharing no overlap with historical mathematical developments during the *Vedic period.*

The book contains metaphorical aphorisms in the form of *sixteen sutras and thirteen sub-sutras,* which *Krishna Tirtha* states allude to significant mathematical tools. The range of their asserted applications spans from topic as diverse as statics and pneumatics to astronomy and financial domains. Tirtha stated that no part of advanced mathematics lay beyond the realms of his book and propounded that studying it for a couple of hours every day for a year equated to spending about two decades in any standardized education system to become professionally trained in the discipline of mathematics. According to *Krishna Tirtha,* the sutras and other accessory content were found after years of solitary study of the Vedas in a forest. They were supposedly contained in the *pariśiṣṭa* (परिशिष्ट) of the *Atharvaveda* (अथर्ववेदः). He does not provide any more bibliographic clarification on the sourcing. The book's editor, Professor V. S. Agrawala argues that since the Vedas are defined as the traditional repositories of all knowledge, any knowledge can be de facto assumed to be in the Vedas, irrespective of whether it may be physically located in them; he even went to the extent of deeming Krishna Tirtha's work as a pariśiṣṭa in itself.

**Further Reading**


1. "Legitimisation of Vedic mathematics, astrology opposed". The Hindu. 14 August 2001. ISSN 0971-751X. Retrieved 25 November 2019, https://www.thehindu.com/todays-paper/tp-miscellaneous/tp-others/legitimisation-of-vedic-mathematics-astrology-opposed/article27971872.ece.
2. "Neither Vedic Nor Mathematics A statemant signed by SG Dani and other Indian scientists". www.sacw.net. Retrieved 20 October 2019.
3. "Stop this Fraud on our Children". archives.peoplesdemocracy.in. Archived from the original on 14 March 2016. Retrieved 20 October 2019.
4. "tecmath". YouTube. Retrieved 14 March 2020, https://www.youtube.com/channel/UCb7w5aTnt7YeXBcVCY0mgFw
5. "Vedic maths: Not quite adding up - Times of India". The Times of India. 11 January 2015. Retrieved 22 October 2019, https://timesofindia.indiatimes.com/home/education/news/Vedic-maths-Not-quite-adding-up/articleshow/45839520.cms.

6. Agrawala, V. S. (1992). General Editor's note. Vedic Mathematics (pp. v-viii) Motilal Banarsidass Publishers Private Limited.
7. Bal, Hartosh Singh (12 August 2010). "The Fraud of Vedic Maths". The Open. Retrieved 25 November 2019.
8. Behera, Navnita Chadha (1 July 1996). "Perpetuating the divide: Political abuse of history in South Asia". Contemporary South Asia. 5 (2): 191–205. doi:10.1080/09584939608719789. ISSN 0958-4935.
9. Bellos, Alex (2010). "Something about nothing". Alex's Adventures in Numberland. Bloomsbury. ISBN 9781408808863.
10. Biographical sketch by Manjula Trivedi, 1965 in book Vedic Mathematics, pages x, xi.
11. Cooke, Roger L. (2013). "Overview of Mathematics in India". The history of mathematics : a brief course. Hoboken, N.J.: Wiley. p. 212. ISBN 978-1-118-46029-0. OCLC 865012817.
12. Crisman, Karl-Dieter (9 August 2019). "Reviews". The American Mathematical Monthly. 126 (7): 667–672. doi:10.1080/00029890.2019.1606573. ISSN 0002-9890. S2CID 215770851.
13. Dani, SG (22 November 1993). "Myths and reality : On 'Vedic mathematics'" (PDF). TIFR. Archived (PDF) from the original on 6 January 2022. Retrieved 17 November 2022, https://www.tifr.res.in/~vahia/dani-vmsm.pdf
14. Glover, James (17 October 2014). "Everything Vedic in 'Vedic Maths'". The Hindu. Retrieved 4 January 2016.
15. Hasan, Mushirul (1 December 2002). "The BJP's intellectual agenda: Textbooks and imagined history". South Asia: Journal of South Asian Studies. 25 (3): 187–209. doi:10.1080/00856400208723498. ISSN 0085-6401. S2CID 143341141.
16. Hogendijk, Jan (March 2004). "De Veda's en de berekeningen van goeroe Tirthaji" (PDF). Nieuwe Wiskrant. 23 (3): 49–52, http://www.fi.uu.nl/wiskrant/artikelen/233/233maart_hogendijk.pdf .
17. Huddar, S. R.; Rupanagudi, S. R.; Kalpana, M.; Mohan, S. (2013). "Novel high speed vedic mathematics multiplier using compressors". 2013 International Mutli-Conference on Automation, Computing, Communication, Control and Compressed Sensing (IMac4s). IEEE. pp. 465–469. doi:10.1109/iMac4s.2013.6526456. ISBN 978-1-4673-5090-7. S2CID 11124644.
18. Kasliwal, Prabha S.; Patil, B. P.; Gautam, D. K. (1 January 2011). "Performance Evaluation of Squaring Operation by Vedic

Mathematics". IETE Journal of Research. 57 (1): 39–41. doi:10.4103/0377-2063.78327. ISSN 0377-2063. S2CID 120316446.

19. Krishna Tirtha had a Master of Arts in Mathematics.
20. Kunchigi, V.; Kulkarni, L.; Kulkarni, S. (2012). "High speed and area efficient vedic multiplier". 2012 International Conference on Devices, Circuits and Systems (ICDCS). IEEE. pp. 360–364. doi:10.1109/ICDCSyst.2012.6188747. ISBN 978-1-4577-1546-4. S2CID 19077488.
21. Kurien, Prema A. (2007). "Re-visioning Indian History: Internet Hinduism" (PDF). A place at the multicultural table the development of an American Hinduism. Rutgers University Press. p. 164. ISBN 9780813540559. JSTOR j.ctt5hj9tk. OCLC 703221465.
22. Manjula Trivedi, "My Beloved Gurudeva", a dedication to Bharatikrishna, in Bharatikrishna's Vedic Mathematics, page i.

- Manjula Trivedi, page ii.
- Manjula Trivedi, page iii
- Manjula Trivedi, page iii.
- Manjula Trivedi, page iv.
- Manjula Trivedi, page vi.
- Manjula Trivedi, page v-vi.
- Manjula Trivedi, pages i-ii.

23. Mehta, Parth; Gawali, Dhanashri (2009). "Conventional versus Vedic Mathematical Method for Hardware Implementation of a Multiplier". 2009 International Conference on Advances in Computing, Control, and Telecommunication Technologies. IEEE. pp. 640–642. doi:10.1109/ACT.2009.162. ISBN 978-1-4244-5321-4. S2CID 6773150.
24. Myths and reality : On 'Vedic mathematics'. S.G. Dani. School of Mathematics, Tata Institute of Fundamental Research. Frontline, 22 October and 5 November 1993.
25. Nanda, Meera (2000). "The Science Wars in India". The Sokal Hoax: The Sham that Shook the Academy. University of Nebraska Press. pp. 206–213. ISBN 9780803219243.
26. Nanda, Meera (2005). "Postmodernism, Hindu Nationalism, and "Vedic Science"". In Koertge, Noretta (ed.). Scientific values and civic virtues. Oxford University Press. p. 224. doi:10.1093/0195172256.001.0001. ISBN 9780198038467. OCLC 62288153.

27. Narlikar, Jayant V. (4 August 2003). The Scientific Edge: The Indian Scientist from Vedic to Modern Times. Penguin UK. ISBN 9789351189282.
28. Nelson, Dean (7 January 2015). "India's next gift to the world could be Vedic mathematics". Daily Telegraph. ISSN 0307-1235. Retrieved 20 October 2019.
29. Nussbaum, Martha Craven (2008). The Clash Within : Democracy, Religious Violence, and India's Future. Harvard University Press. p. 222. ISBN 9780674030596. OCLC 1006798430.
30. Pandey, Pushp Deep (2003). "Public engagement with mathematics in India" (DjVu). Current Science. 84 (7): 862–863. ISSN 0011-3891. JSTOR 24108037.
31. Pijl, Kees Van Der (2010). "Warrior Heroes in the Indo-European Lineage". The Foreign Encounter in Myth and Religion: Modes of Foreign Relations and Political Economy. Vol. 2. Pluto Press. p. 98. doi:10.2307/j.ctt183h05v.7. ISBN 978-0-7453-2316-9. JSTOR j.ctt183h05v.7.
32. Plofker, Kim (18 January 2009). "Mathematical Thought In Vedic India". Mathematics in India. Princeton University Press. p. 16. ISBN 9780691120676.
33. Prasad, Rajendra (1946). Autobiography (PB). Penguin Books India. p. 136. ISBN 9780143068815.
34. Publication announcement by N.H. Bhagwati, Vice-Chancellor, Banaras Hindu University, March 27, 1965, after title page, Vedic Mathematics
35. S. G. Dani (December 2006). "Myths and reality : On 'Vedic mathematics'"., http://www.tifr.res.in/~vahia/dani-vmsm.pdf
36. Sen, Syamal K.; Agarwal, Ravi P. (1 January 2016), Sen, Syamal K.; Agarwal, Ravi P. (eds.), "5 - Conclusions", Zero, Academic Press, pp. 93–142, ISBN 978-0-08-100774-7, retrieved 23 November 2019
37. Shukla, K.S. (2019). "Vedic Mathematics: The deceptive title of Swamiji's book". In Kolachana, Aditya; Mahesh, K.; Ramasubramanian, K. (eds.). Studies in Indian Mathematics and Astronomy: Selected Articles of Kripa Shankar Shukla. Sources and Studies in the History of Mathematics and Physical Sciences. Singapore: Springer Publishing. doi:10.1007/978-981-13-7326-8. ISBN 9789811373251, https://books.google.com/books?id=k86aDwAAQBAJ
38. Sikand, Yoginder (2009). "Voices for Reform in the Indian Madrasas". In Noor, Farish A.; Sikand, Yoginder; Bruinessen, Martin van (eds.).

The madrasa in Asia : political activism and transnational linkages. ISIM Series on Contemporary Muslim Societies. Amsterdam University Press. p. 61. ISBN 978-81-7304-837-1. JSTOR j.ctt46n10w. OCLC 912632940.

39. South Asian Studies. South Asia Institute, University of Heidelberg, Delhi Branch. 1978. pp. 416, 417.
40. Taylor, McComas (2 October 2014). "Hindu Activism and Academic Censorship in India". South Asia: Journal of South Asian Studies. 37 (4): 717–725. doi:10.1080/00856401.2014.956679. ISSN 0085-6401. S2CID 144408510.
41. The Fraud of Vedic Maths. Hartosh Singh Bal. Open Magazine. 14 August 2010.
42. Trautmann, Thomas R. (2002). Languages and Nations : Conversations in Colonial South India. University of California Press. p. 45. ISBN 9780520931909. OCLC 476020847.
43. W.B. Vasantha Kandasamy; Florentin Smarandache (December 2006). Vedic Mathematics: Vedic Or Mathematics: A Fuzzy and Neutrosophic Analysis (PDF). American Research Press. ISBN 978-1-59973-004-2. Retrieved 23 May 2013.
44. "Understanding ancient Indian mathematics". The Hindu. India. 26 December 2011, http://www.thehindu.com/sci-tech/science/article2747006.ece
45. A biography of Sri Bharati Krishna Tirthaji (1884–1960) at the Wayback Machine (archived 9 April 2005), https://web.archive.org/web/20050409014349/http://www.bespokewear.com/vm/aboutbkt.php
46. A Pictorial Biography of Bharti Krishna Tirthaji Maharaja, includes Historic Pictures and Newspaper Article, https://web.archive.org/web/20070504052911/http://vedicmathsindia.org/father_of_vedic_math.htm
47. An updated version appears in Kandasamy and Smarandache (2006).
48. Bharati Krishna Tirtha, https://en.wikipedia.org/wiki/Bharati_Krishna_Tirtha
49. Dani's efforts to debunk the myth of Vedic Maths have been lauded by fellow mathematicians. Over Bhattacharya, Siddhartha; Das, Tarun; Ghosh, Anish; Shah, Riddhi (26 January 2015). Recent Trends in Ergodic Theory and Dynamical Systems. American Mathematical Society. p. 3. ISBN 9781470409319., M. S. Raghunathan admires his efforts in this regard.
50. Manjula Trivedi, page i.

51. Near-simultaneously, as a 2-part article in Frontline, 22 October and 5 November 1993.
52. Originally published in Dani, S. G. (1993). "'Vedic Mathematics': Myth and Reality". Economic and Political Weekly. 28 (31): 1577–1580. ISSN 0012-9976. JSTOR 4399991.
53. Puroshattam D. Chidgupkar and Mangesh T. Karad (2004). "The Implementation of Vedic Algorithm in Digital Signal Processing" (PDF). Global Journal of Engineering Education, http://www.wiete.com.au/journals/GJEE/Publish/vol8no2/Chidgupkar.pdf
54. S.G. Dani. "Vedic Mathematics": a dubious pursuit, http://www.math.tifr.res.in/~dani/vmss.pdf
55. Svāmī Bhāratī Kṛṣṇa Tīrtha, Vedic Mathematics, Motilal Banarsidass, Delhi (1992). ISBN 81-208-0164-4
56. Trivedi, Manjula, My Beloved Gurudeva, Sri Vishwa Punarnirmana Sangha, Nagpur (1965).
57. Vedic Mathematics, https://en.wikipedia.org/wiki/Vedic_Mathematics.

**Srinivasa Ramanujan (श्रीनिवास रामानुजन)**
**(22 December 1887 – 26 April 1920)**

Figure 5.19. Srinivasa Ramanujan (श्रीनिवास रामानुजन).

Srinivasa Ramanujan (श्रीनिवास रामानुजन) born Srinivasa Ramanujan Aiyangar, was a great mathematician (see Figure 5.19). Though he had almost no formal training in pure mathematics, he made substantial contributions to mathematical analysis, number theory, infinite series, and continued fractions, including solutions to mathematical problems then considered unsolvable. *Ramanujan* initially developed his own mathematical research in isolation. Seeking mathematicians who could better understand his work, in 1913 he began a postal correspondence with the English mathematician *G. H. Hardy* at the *University of Cambridge*, England. Recognising *Ramanujan*'s work as extraordinary, *Hardy* arranged for him to travel to Cambridge. In his notes, Hardy commented that Ramanujan had produced groundbreaking new theorems, including some that "*defeated me completely; I had never seen anything in the least like them before*", and some recently proven but highly advanced results.

During his short life, *Ramanujan* independently compiled nearly 3,900 results (mostly identities and equations). Many were completely *novel*; his original and highly unconventional results, such as the *Ramanujan prime,* the *Ramanujan theta function, partition formulae* and mock *theta functions,* have opened entire new areas of work and inspired a vast amount of further research.

Of his thousands of results, all but a dozen or two have now been proven correct. The *Ramanujan Journal,* a scientific journal, was established to publish work in all areas of mathematics influenced by *Ramanujan,* and his notebooks—containing summaries of his published and unpublished results—have been analysed and studied for decades since his death as a source of new mathematical ideas.

As late as 2012, researchers continued to discover that mere comments in his writings about "*simple properties*" and "*similar outputs*" for certain findings were themselves profound and subtle number theory results that remained unsuspected until nearly a century after his death. He became one of the youngest *Fellows of the Royal Society* and only the second Indian member, and the first Indian to be elected a *Fellow of Trinity College,* Cambridge. Of his original letters, *Hardy* stated that a single look was enough to show they could have been written only by a mathematician of the highest calibre, comparing *Ramanujan* to mathematical geniuses such as *Euler* and *Jacobi.*

In 1919, ill health—now believed to have been hepatic amoebiasis (a complication from episodes of dysentery many years previously)—compelled *Ramanujan*'s return to India, where he died in 1920 at the age of 32. His last letters to Hardy, written in January 1920, show that he was still continuing to produce new mathematical ideas and theorems. His "*lost notebook*", containing discoveries from the last year of his life, caused great excitement among mathematicians when it was rediscovered in 1976.

> *An equation means nothing to me unless it expresses a thought of God.*
>
> —Srinivasa Ramanujan

*Ramanujan*'s father, *Kuppuswamy Srinivasa Iyengar,* originally from *Thanjavur* district, worked as a clerk in a sari shop. His mother, *Komalatammal,* was a housewife and sang at a local *temple* (मंदिरः). They lived in a small traditional home on *Sarangapani Sannidhi* Street in the town of *Kumbakonam*. In December

1889, *Ramanujan* contracted *smallpox*, but recovered, unlike the 4,000 others who died in a bad year in the *Thanjavur* district around this time. He moved with his mother to her parents' house in *Kanchipuram*, near Madras (now Chennai).

On 1 October 1892, *Ramanujan* was enrolled at the local school. After his maternal grandfather lost his job as a court official in *Kanchipuram*, *Ramanujan* and his mother moved back to *Kumbakonam*, and he was enrolled in *Kangayan Primary School*. When his paternal grandfather died, he was sent back to his maternal grandparents, then living in Madras. He did not like school in Madras, and tried to avoid attending. His family enlisted a local constable to make sure he attended school. Within six months, *Ramanujan* was back in *Kumbakonam*. Since *Ramanujan*'s father was at work most of the day, his mother took care of the boy, and they had a close relationship. From her, he learned about tradition and *puranas* (पुराणाः), to sing *religious songs* (धार्मिकगीतानि), to attend pujas at the temple, and to maintain particular eating habits—all part of Brahmin culture. At *Kangayan Primary School*, *Ramanujan* performed well. Just before turning 10, in November 1897, he passed his primary examinations in English, Tamil, geography, and arithmetic with the best scores in the district. That year, *Ramanujan* entered *Town Higher Secondary School*, where he encountered formal mathematics for the first time.

A child prodigy by age 11, he had exhausted the mathematical knowledge of two college students who were lodgers at his home. He was later lent a book written by *S. L. Loney* on advanced trigonometry. He mastered this by the age of 13 while discovering sophisticated theorems on his own. By 14, he received merit certificates and academic awards that continued throughout his school career, and he assisted the school in the logistics of assigning its 1,200 students (each with differing needs) to its approximately 35 teachers. He completed mathematical exams in half the allotted time, and showed a familiarity with geometry and infinite series. *Ramanujan* was shown how to solve cubic equations in 1902. He would later develop his own method to solve the quartic. In 1903, he tried to solve the quintic, not knowing that it was impossible to solve with radicals.

In 1903, when he was 16, *Ramanujan* obtained from a friend a library copy of *A Synopsis of Elementary Results in Pure and Applied Mathematics*, *G. S. Carr*'s collection of 5,000 theorems. *Ramanujan* reportedly studied the contents of the book in detail. The next year, *Ramanujan* independently developed and investigated the *Bernoulli numbers* and calculated the *Euler–Mascheroni constant* up to 15 decimal places. His peers at the time said they "*rarely understood him*" and "*stood in respectful awe*" of him.

When he graduated from *Town Higher Secondary School* in 1904, *Ramanujan* was awarded the *K. Ranganatha Rao prize* for *mathematics* by the school's headmaster, *Krishnaswami Iyer. Iyer* introduced *Ramanujan* as an outstanding student who deserved scores higher than the maximum. He received a scholarship to study at *Government Arts College*, Kumbakonam, but was so intent on mathematics that he could not focus on any other subjects and failed most of them, losing his scholarship in the process.

In August 1905, *Ramanujan* ran away from home, heading towards Visakhapatnam, and stayed in Rajahmundry for about a month. He later enrolled at *Pachaiyappa*'s College in Madras. There, he passed in mathematics, choosing only to attempt questions that appealed to him and leaving the rest unanswered, but performed poorly in other subjects, such as English, physiology, and Sanskrit. In 1910, after a meeting between the 23-year-old *Ramanujan* and the founder of the *Indian Mathematical Society, V. Ramaswamy Aiyer, Ramanujan* began to get recognition in Madras's mathematical circles, leading to his inclusion as a researcher at the University of Madras.

On 14 July 1909, *Ramanujan* married *Janakiammal* (21 March 1899 – 13 April 1994), a girl his mother had selected for him a year earlier and who was ten years old when they married. It was not unusual then for marriages to be arranged with girls at a young age. *Janakiammal* was from *Rajendram*, a village close to *Marudu*r (Karur district) Railway Station. *Ramanujan*'s father did not participate in the marriage ceremony. As was common at that time, *Janakiammal* continued to stay at her maternal home for three years after marriage, until she reached puberty. In 1912, she and *Ramanujan*'s mother joined *Ramanujan* in Madras.

After the marriage, *Ramanujan* developed a *hydrocele testis*. The condition could be treated with a routine surgical operation that would release the blocked fluid in the scrotal sac, but his family could not afford the operation. In January 1910, a doctor volunteered to do the surgery at no cost.

After his successful surgery, *Ramanujan* searched for a job. He stayed at a friend's house while he went from door to door around *Madras* looking for a clerical position. To make money, he tutored students at *Presidency College* who were preparing for their Fellow of Arts exam.

In late 1910, *Ramanujan* was sick again. He feared for his health, and told his friend *R. Radakrishna Iyer* to "*hand [his notebooks] over to Professor Singaravelu Mudaliar [the mathematics professor at Pachaiyappa's College] or to the British professor Edward B. Ross, of the Madras Christian College*."

After Ramanujan recovered and retrieved his notebooks from Iyer, he took a train from Kumbakonam to Villupuram, a city under French control. In 1912, *Ramanujan* moved with his wife and mother to a house in *Saiva Muthaiah Mudali* street, George Town, Madras, where they lived for a few months. In May 1913, upon securing a research position at *Madras University, Ramanujan* moved with his family to Triplicane.

In 1910, Ramanujan met deputy collector *V. Ramaswamy Aiyer, w*ishing for a job at the revenue department where Aiyer worked, Ramanujan showed him his mathematics notebooks.

Aiyer sent *Ramanujan,* with letters of introduction, to his mathematician friends in Madras. Some of them looked at his work and gave him letters of introduction to *R. Ramachandra Rao,* the district collector for Nellore and the secretary of the *Indian Mathematical Society. Rao* was impressed by *Ramanujan*'s research but doubted that it was his own work. *Ramanujan* mentioned a correspondence he had with *Professor Saldhana,* a notable Bombay mathematician, in which *Saldhana* expressed a lack of understanding of his work but concluded that he was not a fraud.

*Ramanujan's* friend *C. V. Rajagopalachari* tried to quell *Rao*'s doubts about *Ramanujan*'s academic integrity. *Rao* agreed to give him another chance, and listened as *Ramanujan* discussed *elliptic integrals, hypergeometric series,* and *his theory of divergent series,* which *Rao* said ultimately convinced him of *Ramanujan's* brilliance. When *Rao* asked him what he wanted, *Ramanujan* replied that he needed work and financial support. *Rao* consented and sent him to Madras. He continued his research with *Rao*'s financial aid. With *Aiyer*'s help, *Ramanujan* had his work published in the *Journal of the Indian Mathematical Society.*

*Ramanujan* departed from Madras aboard the S. S. Nevasa on 17 March 1914. There, *Hardy* and *Littlewood* began to look at *Ramanujan*'s *notebooks. Hardy* had already received 120 theorems from *Ramanujan* in the first two letters, but there were many more results and theorems in the notebooks. Hardy saw that some were wrong, others had already been discovered, and the rest were new breakthroughs. *Ramanujan* spent nearly five years in Cambridge collaborating with *Hardy* and *Littlewood*, and published part of his findings there. *Hardy* and *Ramanujan* had highly contrasting personalities. Their collaboration was a clash of different cultures, beliefs, and working styles. In the previous few decades, the foundations of mathematics had come into question and the need for mathematically rigorous proofs recognised. *Hardy* was an atheist and an apostle of proof and mathematical rigour, whereas *Ramanujan* was a deeply religious man who relied very strongly on his intuition and insights. *Hardy* tried his best to fill the gaps in *Ramanujan*'s education and to mentor him in the need for formal proofs to support his results, without hindering his inspiration—a conflict that neither found easy.

*Ramanujan* was awarded a *Bachelor of Arts by Research* degree (the predecessor of the PhD degree) in March 1916 for his work on highly composite numbers, sections of the first part of which had been published the preceding year in the *Proceedings of the London Mathematical Society*. The paper was more than 50 pages long and proved various properties of such numbers. At age 31, Ramanujan was one of the youngest Fellows in the Royal Society's history. He was elected *"for his investigation in elliptic functions and the Theory of Numbers."*

*Ramanujan* had numerous health problems throughout his life. His health worsened in England; possibly he was also less resilient due to the difficulty of keeping to the strict dietary requirements of his religion there and because of wartime rationing in 1914–18. He was diagnosed with tuberculosis and a severe vitamin deficiency, and confined to a sanatorium. In 1919, he returned to *Kumbakonam*, Madras Presidency, and in 1920 he died at the age of 32. After his death, his brother Tirunarayanan compiled *Ramanujan*'s remaining handwritten notes, consisting of formulae on singular moduli, hypergeometric series and continued fractions.

> *While asleep, I had an unusual experience. There was a red screen formed by flowing blood, as it were. I was observing it. Suddenly a hand began to write on the screen. I became all attention. That hand wrote a number of elliptic integrals. They stuck to my mind. As soon as I woke up, I committed them to writing.*
>
> —Srinivasa Ramanujan

## Key Contributions in Mathematics

### Infinite Series for $\pi$

In mathematics, there is a distinction between insight and formulating or working through a proof. *Ramanujan* proposed an abundance of formulae that could be investigated later in depth. As a byproduct of his work, new directions of research were opened up. Examples of the most intriguing of these formulae include infinite series for $\pi$, one of which is given below:

$$\frac{1}{\pi} = \frac{2\sqrt{2}}{9801} \sum_{k=0}^{\infty} \frac{(4k)!\,(1103 + 26390k)}{(k!)^4 396^{4k}}$$

This result is based on the negative fundamental discriminant $d = -4 \times 58 = -232$ with class number $h(d) = 2$. Further, $26390 = 5 \times 7 \times 13 \times 58$ and $16 \times 9801 = 3962$, which is related to the fact that

$$e^{\pi\sqrt{58}} = 396^4 - 104.000000177 \ldots$$

This might be compared to *Heegner numbers,* which have class number 1 and yield similar formulae.

*Ramanujan*'s series for $\pi$ converges extraordinarily rapidly and forms the basis of some of the fastest algorithms currently used to calculate $\pi$. Truncating the sum to the first term also gives the approximation

$$\frac{9801\sqrt{2}}{4412}$$

for $\pi$, which is correct to six decimal places; truncating it to the first two terms gives a value correct to 14 decimal places. See also the more general *Ramanujan–Sato series.*

His intuition also led him to derive some previously unknown identities, such as:

$$\left(1+2\sum_{n=1}^{\infty}\frac{\cos(n\theta)}{\cosh(n\pi)}\right)^{-2}+\left(1+2\sum_{n=1}^{\infty}\frac{\cos(n\theta)}{\cosh(n\pi)}\right)^{-2}$$

$$=\frac{2\tau^4(\frac{3}{4})}{\pi}=\frac{8\pi^3}{\tau^4(\frac{1}{4})}$$

for all $\theta$ such that $|R(\theta)| < \pi$ and $|\tau(\theta)| < \pi$, ehere $\tau(z)$ s the gamma function, and related to a special value of the *Dedekind eta* function. Expanding into series of powers and equating coefficients of $\theta^0, \theta^4$ and $\theta^8$ gives some deep identities for the *hyperbolic secant*.

In 1918, *Hardy* and *Ramanujan* studied the *partition function* $P(n)$ extensively. They gave a *non-convergent asymptotic series* that permits exact computation of the number of partitions of an integer. *Ramanujan* and *Hardy*'s work in this area gave rise to a powerful new method for finding *asymptotic formulae* called the *circle method*.

In the last year of his life, *Ramanujan* discovered mock *theta functions*. For many years, these functions were a mystery, but they are now known to be the *holomorphic parts of harmonic weak Maass forms*.

One of the theorems had an equation mentioned to be valid for $0 < a < b + \frac{1}{2}$:

$$\int_0^{\infty}\frac{1+\frac{x^2}{(b+1)^2}}{1+\frac{x^2}{a^2}}\times\frac{1+\frac{x^2}{(b+2)^2}}{1+\frac{x^2}{(a+1)^2}}\times\ldots dx=\frac{\sqrt{\pi}}{2}\times\frac{\tau(a+\frac{1}{2})\tau(b+1)\tau(b-a+1)}{\tau(a)\tau(b+\frac{1}{2})\tau(b-a+\frac{1}{2})}$$

Hardy was also impressed by some of *Ramanujan*'s other work relating to infinite series:

$$1-5\left(\frac{1}{2}\right)^3\times 9\left(\frac{1\times 3}{2\times 4}\right)^3-13\left(\frac{1\times 3\times 5}{2\times 4\times 6}\right)^3+\cdots=\frac{2}{\pi}$$

$$1-9\left(\frac{1}{4}\right)^3\times 17\left(\frac{1\times 5}{4\times 8}\right)^3+25\left(\frac{1\times 5\times 9}{4\times 8\times 12}\right)^3+\cdots=\frac{2\sqrt{2}}{\sqrt{\pi}\tau^2(\frac{3}{4})}$$

## The Ramanujan Conjecture

Although there are numerous statements that could have borne the name *Ramanujan* conjecture, one was highly influential on later work. In particular, the connection of this conjecture with conjectures of André Weil in algebraic geometry opened up new areas of research. That *Ramanujan* conjecture is an assertion on the size of the tau-function, which has as generating function the discriminant modular form $\Delta(q)$, a typical cusp form in the theory of modular forms. It was finally proven in 1973, as a consequence of Pierre Deligne's proof of the Weil conjectures. The reduction step involved is complicated. Deligne won a Fields Medal in 1978 for that work.

In his paper "*On certain arithmetical functions*", *Ramanujan* defined the so-called *delta-function,* whose coefficients are called $\tau(n)$ (the *Ramanujan tau* function). He proved many congruences for these numbers, such as

$$\tau(p) \equiv 1 + p11\ mod\ 691$$

for primes $p$.

This congruence (and others like it that *Ramanujan* proved) inspired Jean-Pierre Serre (1954 Fields Medalist) to conjecture that there is a theory of Galois representations that "*explains*" these congruences and more generally all modular forms. $\Delta(z)$ is the first example of a modular form to be studied in this way. Deligne (in his Fields Medal-winning work) proved Serre's conjecture. The proof of *Fermat's Last Theorem* proceeds by first reinterpreting elliptic curves and modular forms in terms of these *Galois representations.* Without this theory, there would be no proof of Fermat's Last Theorem.

## Ramanujan's Notebooks

While still in Madras, *Ramanujan* recorded the bulk of his results in four notebooks of looseleaf paper. They were mostly written up without any derivations. This is probably the origin of the misapprehension that *Ramanujan* was unable to prove his results and simply thought up the final result directly.

This may have been for any number of reasons. Since paper was very expensive, *Ramanujan* did most of his work and perhaps his proofs on slate, after which he transferred the final results to paper. At the time, slates were commonly used by mathematics students in the Madras Presidency.

He was also quite likely to have been influenced by the style of G. S. Carr's book, which stated results without proofs. It is also possible that Ramanujan considered his work to be for his personal interest alone and therefore recorded only the results.

The first notebook has 351 pages with 16 somewhat organised chapters and some unorganised material. The second has 256 pages in 21 chapters and 100 unorganised pages, and the third 33 unorganised pages. The results in his notebooks inspired numerous papers by later mathematicians trying to prove what he had found. In 1976, George Andrews rediscovered a fourth notebook with 87 unorganised pages, the so-called "*lost notebook*".

### Hardy–Ramanujan Number 1729

The number 1729 is known as the *Hardy–Ramanujan* number after a famous visit by *Hardy* to see *Ramanujan* at a hospital.

$$1729 = 1^3 + 12^3 = 9^3 + 10^3$$

Generalisations of this idea have created the notion of "*taxicab numbers*".

### Further Reading


1. A Disappearing Number is a British stage production by the company Complicite that explores the relationship between Hardy and Ramanujan.
2. B Berndt and S Bhargava, Ramanujan - For lowbrows, Amer. Math. Monthly 100 (1993), 644-656.
3. B Berndt, Srinivasa Ramanujan, The American Scholar 58 (1989), 234-244.
4. B Bollobas, Ramanujan - a glimpse of his life and his mathematics, Eureka 48 (1988), 81-98.
5. B Bollobas, Ramanujan - a glimpse of his life and his mathematics, The Cambridge Review (1988), 76-80.
6. B C Berndt and R A Rankin, Ramanujan: Letters and commentary (Providence, Rhode Island, 1995).
7. Berndt, Bruce C. (1985). Ramanujan's Notebooks: Part I. New York: Springer. ISBN 978-0-387-96110-1.

8. Berndt, Bruce C. (1993). Ramanujan's Notebooks: Part IV. New York: Springer. ISBN 978-0-387-94109-7.
9. Berndt, Bruce C. (1998). Butzer, P. L.; Oberschelp, W.; Jongen, H. Th. (eds.). Charlemagne and His Heritage: 1200 Years of Civilization and Science in Europe (PDF). Turnhout, Belgium: Brepols Verlag. pp. 119–146. ISBN 978-2-503-50673-9. Archived (PDF) from the original on 9 September 2004.
10. Berndt, Bruce C. (1999). Ramanujan's Notebooks: Part II. New York: Springer. ISBN 978-0-387-96794-3.
11. Berndt, Bruce C. (2004). Ramanujan's Notebooks: Part III. New York: Springer. ISBN 978-0-387-97503-0.
12. Berndt, Bruce C. (2005). Ramanujan's Notebooks: Part V. New York: Springer. ISBN 978-0-387-94941-3.
13. Berndt, Bruce C. (2006). Number Theory in the Spirit of Ramanujan. Vol. 9. Providence, Rhode Island: American Mathematical Society. ISBN 978-0-8218-4178-5.
14. Berndt, Bruce C.; Rankin, Robert A. (1995). Ramanujan: Letters and Commentary. Vol. 9. Providence, Rhode Island: American Mathematical Society. ISBN 978-0-8218-0287-8.
15. Berndt, Bruce C.; Rankin, Robert A. (2001). Ramanujan: Essays and Surveys. Vol. 22. Providence, Rhode Island: American Mathematical Society. ISBN 978-0-8218-2624-9.
16. Biography in Encyclopaedia Britannica. http://www.britannica.com/biography/Srinivasa-Ramanujan.
17. C T Rajagopal, Stray thoughts on Srinivasa Ramanujan, Math. Teacher (India) 11A (1975), 119-122, and 12 (1976), 138-139.
18. D A B Young, Ramanujan's illness, Notes and Records of the Royal Society of London 48 (1994), 107-119.
19. David Leavitt's novel The Indian Clerk explores the events following Ramanujan's letter to Hardy.
20. E H Neville, Srinivasa Ramanujan, Nature 149 (1942), 292-294.
21. E Shils, Reflections on tradition, centre and periphery and the universal validity of science : the significance of the life of S Ramanujan, Minerva 29 (1991), 393-419.
22. G E Andrews, An introduction to Ramanujan's 'lost' notebook, Amer. Math. Monthly 86 (1979), 89-108.
23. G H Hardy, Ramanujan (Cambridge, 1940).
24. G H Hardy, Srinivasa Ramanujan, Proc. London Math, Soc. 19 (1921), xl-lviii.

25. G H Hardy, The Indian mathematician Ramanujan, Amer. Math. Monthly 44 (3) (1937), 137-155.
26. George E. Andrews and Bruce C. Berndt, Ramanujan's Lost Notebook: Part I (Springer, 2005, ISBN 0-387-25529-X).
27. George E. Andrews and Bruce C. Berndt, Ramanujan's Lost Notebook: Part II, (Springer, 2008, ISBN 978-0-387-77765-8).
28. George E. Andrews and Bruce C. Berndt, Ramanujan's Lost Notebook: Part III, (Springer, 2012, ISBN 978-1-4614-3809-0).
29. George E. Andrews and Bruce C. Berndt, Ramanujan's Lost Notebook: Part IV, (Springer, 2013, ISBN 978-1-4614-4080-2).
30. George E. Andrews and Bruce C. Berndt, Ramanujan's Lost Notebook: Part V, (Springer, 2018, ISBN 978-3-319-77832-7).
31. Google honoured Ramanujan on his 125th birth anniversary by replacing its logo with a doodle on its home page.
32. Hardy, G. H. (1978). Ramanujan. New York: Chelsea Pub. Co. ISBN 978-0-8284-0136-4.
33. Hardy, G. H. (1999). Ramanujan: Twelve Lectures on Subjects Suggested by His Life and Work. Providence, Rhode Island: American Mathematical Society. ISBN 978-0-8218-2023-0.
34. Hardy, G. H. (March 1937). "The Indian Mathematician Ramanujan". The American Mathematical Monthly. 44 (3): 137–155. doi:10.2307/2301659. JSTOR 2301659.
35. Hardy, G. H.; Ramanujan, S. (1917). "Asymptotic Formulae for the Distribution of Integers of Various Types". Proceedings of the London Mathematical Society. 16 (1): 112–132. doi:10.1112/plms/s2-16.1.112.
36. Hardy, G. H.; Ramanujan, Srinivasa (1918). "Asymptotic Formulae in Combinatory Analysis". Proceedings of the London Mathematical Society. 17 (1): 75–115. doi:10.1112/plms/s2-17.1.75.
37. Hardy, G. H.; Ramanujan, Srinivasa (1918). "On the coefficients in the expansions of certain modular functions". Proc. R. Soc. A. 95 (667): 144–155. Bibcode:1918RSPSA..95..144H. doi:10.1098/rspa.1918.0056.
38. Henderson, Harry (1995). Modern Mathematicians. New York: Facts on File Inc. ISBN 978-0-8160-3235-8.
39. J M Borwein and P B Borwein, Ramanujan and pi, Scientific American 258 (2) (1988), 66-73.
40. J N Kapur (ed.), Some eminent Indian mathematicians of the twentieth century (Kapur, 1989).
41. K Ramachandra, Srinivasa Ramanujan (the inventor of the circle method), Hardy-Ramanujan J. 10 (1987), 9-24.

42. K Ramachandra, Srinivasa Ramanujan (the inventor of the circle method), J. Math. Phys. Sci. 21 (1987), 545-564.
43. Kanigel, Robert (1991). The Man Who Knew Infinity: a Life of the Genius Ramanujan. New York: Charles Scribner's Sons. ISBN 978-0-684-19259-8.
44. L Debnath, Srinivasa Ramanujan (1887-1920): a centennial tribute, International journal of mathematical education in science and technology 18 (1987), 821-861.
45. Leavitt, David (2007). The Indian Clerk (paperback ed.). London: Bloomsbury. ISBN 978-0-7475-9370-6.
46. M. N. Krish's thriller novel The Steradian Trail weaves Ramanujan and his accidental discovery into its plot connecting religion, mathematics, finance and economics.
47. M. P. Chaudhary, A simple solution of some integrals given by Srinivasa Ramanujan, (Resonance: J. Sci. Education – publication of Indian Academy of Science, 2008).
48. M.P. Chaudhary, Mock theta functions to mock theta conjectures, SCIENTIA, Series A : Math. Sci., (22)(2012) 33–46.
49. M.P. Chaudhary, On modular relations for the Roger-Ramanujan type identities, Pacific J. Appl. Math., 7(3)(2016) 177–184.
50. Nandan Kudhyadi directed the Indian documentary films The Genius of Srinivasa Ramanujan (2013) and Srinivasa Ramanujan: The Mathematician and His Legacy (2016) about the mathematician.
51. Narlikar, Jayant V. (2003). Scientific Edge: the Indian Scientist From Vedic to Modern Times. New Delhi, India: Penguin Books. ISBN 978-0-14-303028-7.
52. Ore, Biography in Dictionary of Scientific Biography (New York 1970-1990).
53. Ono, Ken; Aczel, Amir D. (13 April 2016). My Search for Ramanujan: How I Learned to Count. Springer. ISBN 978-3319255668.
54. P K Srinivasan, Ramanujan: Am inspiration 2 Vols. (Madras, 1968).
55. P V Seshu Aiyar, The late Mr S Ramanujan, B.A., F.R.S., J. Indian Math. Soc. 12 (1920), 81-86.
56. Partition, a play by Ira Hauptman about Hardy and Ramanujan, was first performed in 2013.
57. Problems posed by Ramanujan, Journal of the Indian Mathematical Society.
58. R A Rankin, Ramanujan as a patient, Proc. Indian Ac. Sci. 93 (1984), 79-100.

59. R A Rankin, Ramanujan's manuscripts and notebooks II, Bull. London Math. Soc. 21 (1989), 351-365.
60. R A Rankin, Ramanujan's manuscripts and notebooks, Bull. London Math. Soc. 14 (1982), 81-97.
61. R A Rankin, Srinivasa Ramanujan (1887- 1920), International journal of mathematical education in science and technology 18 (1987), 861-.
62. R Kanigel, The man who knew infinity: A life of the genius Ramanujan (New York, 1991).
63. R Ramachandra Rao, In memoriam S Ramanujan, B.A., F.R.S., J. Indian Math. Soc. 12 (1920), 87-90.
64. Ramanujan (The Man Who Reshaped 20th Century Mathematics), an Indian docudrama film directed by Akashdeep released in 2018.
65. Ramanujan was mentioned in the 1997 film Good Will Hunting, in a scene where professor Gerald Lambeau (Stellan Skarsgård) explains to Sean Maguire (Robin Williams) the genius of Will Hunting (Matt Damon) by comparing him to Ramanujan.
66. Ramanujan, an Indo-British collaboration film chronicling Ramanujan's life, was released in 2014 by the independent film company Camphor Cinema. The cast and crew include director Gnana Rajasekaran, cinematographer Sunny Joseph and editor B. Lenin. Indian and English stars Abhinay Vaddi, Suhasini Maniratnam, Bhama, Kevin McGowan and Michael Lieber star in pivotal roles.
67. Ramanujan, S. (1914). "Some definite integrals connected with Gauss's sums". Messenger Math. 44: 75–85.
68. Ramanujan, S. (1914). "Some definite integrals". Messenger Math. 44: 10–18.
69. Ramanujan, S. (1915). "Highly Composite Numbers". Proceedings of the London Mathematical Society. 14 (1): 347–409. doi:10.1112/plms/s2_14.1.347.
70. Ramanujan, S. (1915). "On certain infinite series". Messenger Math. 45: 11–15.
71. Ramanujan, S. (1915). "On the number of divisors of a number". The Journal of the Indian Mathematical Society. 7 (4): 131–133.
72. Ramanujan, S. (1915). "Short Note: On the sum of the square roots of the first n natural numbers". The Journal of the Indian Mathematical Society. 7 (5): 173–175.
73. Ramanujan, S. (1916). "A Series for Euler's Constant $\gamma$". Messenger Math. 46: 73–80.

74. Ramanujan, S. (1916). "Some formulae in the analytical theory of numbers". Messenger Math. 45: 81–84.
75. Ramanujan, S. (1917). "On the expression of numbers in the form ax2 + by2 + cz2 + du2". Mathematical Proceedings of the Cambridge Philosophical Society. 19: 11–21.
76. Ramanujan, S. (1919). "A proof of Bertrand's postulate". The Journal of the Indian Mathematical Society. 11 (5): 181–183.
77. Ramanujan, S. (1920). "A class of definite integrals". Quart. J. Pure. Appl. Math. 48: 294–309. hdl:2027/uc1.$b417568.
78. Ramanujan, S. (1921). "Congruence properties of partitions". Math. Z. 9 (1–2): 147–153. doi:10.1007/BF01378341. S2CID 121753215. Posthumously published extract of a longer, unpublished manuscript.
79. Ramanujan, Srinivasa (1919). "Some definite integrals". The Journal of the Indian Mathematical Society. 11 (2): 81–88.
80. Ramanujan, Srinivasa; Hardy, G. H.; Seshu Aiyar, P. V.; Wilson, B. M.; Berndt, Bruce C. (2000). Collected Papers of Srinivasa Ramanujan. AMS. ISBN 978-0-8218-2076-6.
81. S Chandrasekhar, On Ramanujan, in Ramanujan Revisited (Boston, 1988), 1-6.
82. S R Ranganathan, Ramanujan: the man and the mathematician (London, 1967).
83. S Ram, Srinivasa Ramanujan (New Delhi, 1979).
84. S Ramanujan, Collected Papers (Cambridge, 1927).
85. S. Ramanujan (1957). Notebooks (2 Volumes). Bombay: Tata Institute of Fundamental Research.
86. S. Ramanujan (1988). The Lost Notebook and Other Unpublished Papers. New Delhi: Narosa. ISBN 978-3-540-18726-4.
87. S. Ramanujan (2012). Notebooks (2 Volumes). Bombay: Tata Institute of Fundamental Research.
88. Sankaran, T. M. (2005). "Srinivasa Ramanujan- Ganitha lokathile Mahaprathibha" (in Malayalam). Kochi, India: Kerala Sastra Sahithya Parishath.
89. Srinivasa-Ramanujan, https://en.wikipedia.org/wiki/Srinivasa_Ramanujan.
90. Srinivasa-Ramanujan, https://mathshistory.st-andrews.ac.uk/Biographies/Ramanujan/.
91. The Man Who Knew Infinity is a 2015 film based on Kanigel's book of the same name. British actor Dev Patel portrays Ramanujan.

92. The Man Who Loved Numbers is a 1988 PBS NOVA documentary about Ramanujan (S15, E9).
93. The play First Class Man by Alter Ego Productions was based on David Freeman's First Class Man. The play centres around Ramanujan and his complex and dysfunctional relationship with Hardy. On 16 October 2011 it was announced that Roger Spottiswoode, best known for his James Bond film Tomorrow Never Dies, is working on the film version, starring Siddharth.
94. These books contain photocopies of the original notebooks as written by Ramanujan.
95. This book contains photo copies of the pages of the "Lost Notebook".
96. This book was originally published in 1927 after Ramanujan's death. It contains the 37 papers published in professional journals by Ramanujan during his lifetime. The third reprint contains additional commentary by Bruce C. Berndt.
97. This was produced from scanned and microfilmed images of the original manuscripts by expert archivists of Roja Muthiah Research Library, Chennai.

**A. A. Krishnaswami Ayyangar (कृष्णस्वामी अय्यांगर)**
**(1892 – 1953)**

Figure 5.20. A. A. Krishnaswami Ayyangar (कृष्णस्वामी अय्यांगर). (Image Credit: https://en.wikipedia.org/wiki/)

A. A. Krishnaswami Ayyangar (कृष्णस्वामी अय्यांगर) was born on 1 December, 1892. He got his M.A. in Mathematics at the age of 18 and subsequently started teaching Mathematics at his own alma mater, *Pachaiappas College*, Chennai (Madras) (see Figure 5.20). He was married to *Seshammal* and they had four sons, *A. K. Srinivasan, A. K. Ramanujan, A. K. Rajagopal, A. K. Vasudevan*, and two daughters *Mrs. Vedavathi Bhogishayana* and *Mrs. Saroja Krishnamurthi.*

In 1918 he joined the Mathematics Department, *University of Mysore* and retired from there in 1947. He passed away in June 1953. During his tenure at the University, for nearly three decades, he made many contributions to Geometry, Statistics, Astronomy, the History of Bhartiya Mathematics, and other topics.

The originality of *Ayyangar*'s work was brought home to *Rajagopal* on a casual conversation with *Professor R. Sridharan* of the *Tata Institute of Fundamental Research*, Bombay in 1998. *Professor Sridharan* gave him an article he had

written in Science in the West and Bharat, edited by *B. V. Subbarayappa* and *N. Mukunda* (1995, Himalaya Publishing House, Bombay). In this he points out that *Ayyangar*'s article on the *Chakravala method* shows how it differs from the method of continued fractions. He recounted that this point was missed by *A. Weil*, who thought the *Chakravala method* was only an *"experimental fact"* to the Bhartiyas and attributed general proofs to *Fermat* and *Lagrange* much later!
On a different occasion, *Professor Subhash Kak* of *Louisiana State University*, Baton Rouge, stressed to *Rajagopal* that *Ayyangar*'s presentations of Bhartiya works were unique and should be known generally in the community of mathematicians today who are interested in the History of Mathematics. Without the keen interest evinced by *Professor Kak* and his drive to get Bhartiya contributions to the World of Astronomy, Mathematics, Music, and everything else, this collection would have suffered the quiet death of so many things Bhartiya!

## 卐 Key Contributions in Mathematics

### 卐 The Earliest Solution of the Bioquadratic

The principle in *Bhaskara*'s solution of the biquadratic was hidden away in a numerical example. *Bhaskara*'s is the earliest attempt at the *solution of the biquadratic* recognizing only the real positive roots. This article generalizes his principle of his method of solving a biquadratic. He provided details about it in his work *"Current Science, Vol.7, No.4, Oct. 1938"*.

### 卐 Bhaskara and Samclishta Kuttaka

In his work *"Journal of Indian Mathematical Society, 1929-30, Vol.18"*, he worked on *Bhaskara* and *Samclishta Kuttaka*.

*Aryabhata* mentions only two congruences but can be extended to more than two cases by repeating the process. This is done by *Mahaviracharya*. This paper gives an example to solve a set of congruences which is a repetition of a process with no new principle involved. *Thus,* it does not justify an explicit enunciation by *Bhaskara*. If he had done so, he has played a role of a commentator. Therefore, it is likely that a commentator has interpolated it in *Bhaskara*'s text.

*Aryabhata* was the first to hint the general problem of finding a number having given residues with respect to given modulii. *Mahaviracharya* has given a rule different from the *samclishta rule*. *Bhaskara* suggests a novel method (*Samclishta Kuttaka*) different from samclishta rule for solving simultaneous equations in the particular case of the same modulus for all congruences. It furnishes novel method of solving simultaneous congruence equations, with common modulus. *Bhaskara* does not point out the cases of failure, which is a normal custom. This paper indicates the condition in which his rule fails, and gives numerical example.

$$5\,x \equiv 7\ (mod.\,63)$$

$$30\,x \equiv 14\ (mod.\,63)$$

*Bhaskara*'s method will give $35\,x \equiv 49\ (mod.\,63)$ or $5\,x \equiv 49\ (mod.\,9)$ as the representative congruence with $x \equiv 5$ as the least value. This does not satisfy the given congruences and have to choose the next higher value 14.

*Ayyangar* gave new proofs of *Apollonius* and *Brahmagupta*'s theorems.

### 卐 New Light on Bhaskara's Chakravala or Cyclic Method of Solving Indeterminate Equations of the Second Degree in Two Variables

More than thousand years before Euler, Bhartiyas were the first to give a general solution to the equation

$$x^2 - Ny^2 = 1$$

It was based on the principle of composition of quadratic forms.

*Euler's theorem:*

If

$x = a, y = b$ satisfies $\alpha\, x^2 + p = y^2$, and

$x = c,\ y = d$ satisfies $\alpha\, x^2 + q = y^2$, then

$x = bc \pm ad, y = bd \pm \alpha ac$ satisfies $\alpha\, x^2 + pq = y^2$

has been dealt by *Brahmagupta* in his '*Vargaprakriti*' and drawing important corollaries from it.

Special identities are also given by him for deriving integral solutions of $Ny^2 + 1 = x^2$ and of $Ny^2 \pm 4 = x^2$.

Bhaskara gives integral solutions, designating it as *'Chakravala'* after the ancients. Western writers attribute the cyclic method to Brahmagupta but the credit should go to *Bhaskara*. Bhaskara may have given credit to his predecessors for the name *'chakravala'*, as applied to all iterative operations in mathematics.

*Brahmagupta*'s formula for the composition of quadratic forms is also repeated by Bhaskara. Using this both of them derive the solutions of equation (1) from those of

$$N\,y^2 \pm 2 = x^2 \quad or \quad N\,y^2 \pm 4 = x^2$$

*Bhaskara* goes one step further to show that the roots of the equations (1) and (5) can be derived, by successive reduction, from the more general equation

$$N\,y^2 \pm k = x^2$$

where k is an integer. The method of such reduction is the *chakravala method (Cyclic Method)*. The first complete proof of *Bhaskara's Cyclic Method* is also sketched by him.

He also discussed the following:

- *Bhaskara's Condition for Reduction*
- *Reduced Forms*
- *Properties of Bhaskara's Forms*
- *Bhaskara and Fermat*

### 卐 The Bhakshali Manuscript

In his work *"The Mathematics Student, Vol. VII, 1939"*, he performed re-examination and re-assessment of the *Bhakshali Manuscript*. G. R. Kaye has written introduction and notes after the earlier work of Hoernie. G. R. Kaye's biased view of Greek influence of anything Bhartiya prompted others to re-examine his statements.

This article examines the manuscript under the following headings:

- Method of presentation
- Peculiar terminology, abbreviations, and the cross symbol for subtraction
- Decimal Notation and the absence of word numerals
- Symbol for unknown
- Rule of Regula Falsi
- Square-root rule and the process of reconciliation, and
- Series and sequences.

The problems on series and sequences are very original, varied and interesting. In most of the problems, the sum to n terms happens to be proportional to the first term and therefore *Ishtakarma* can apply. This paper discusses eight types of problems using modern notation. The identity

$$T_n = (1 - a_1)(1 - a_2)\ldots.(1 - a_{n-1})a_n$$

is common in Hindu mathematical works from at least the 8th century onwards.

The *Bhakshali* text, apart from peculiarities, is more or less a replica of other Hindu mathematical works such as *Ganitasara Sangraha*. It contains in common with them the following:

- Practical and commercial problems
- Problems of income and expenditure
- Motion-problems
- Profit and loss
- Interest
- Bills of exchange or hundika, and
- Miscellaneous problems involving literary and social references.

### 卐 Fourteen Calendars

In his work on "*The Mathematics Student, Vol. V, 1937*", he described the rule of classifying the yearly calendars into *fourteen distinct types* (seven ordinary and seven leap years) which recur in a given century.

## 卐 Remarks on Bhaskara's Approximation to the Sine of an Angle

In his article *"The Mathematics Student, Vol. 18, 1950"*, he did lots of works on Bhaskara's approximation to sine of an angle. Bhaskara's rational approximation to

$$\sin\left(\frac{\pi}{n}\right)$$

can be written as

$$\frac{16(n-1)}{5n^2 - 4n + 4}$$

which is the best rational approximation.

This can also be expressed as

$$\frac{(n^2 - (n-2)^2)}{n^2 + \left(\frac{1}{4}\right)(n-2)^2}$$

which is *Ganesa*'s variant.

## 卐 The Mathematics of Aryabhata

In his work, *"Quarterly Journal of the Mythic Society, Vol. 16, 1926"*, he did lots of works on Aryabhata's formulas and theories.

*Aryabhata* gives in his *Dasagitika,* a peculiar notation for expressing numbers in terms of the letters of the Sanskrit alphabet. The *twenty-five varga letters* of the alphabet represent the numbers one to twenty-five respectively in the square or odd places while the avarga letters represent the numbers $30, 40, \ldots\ldots, up\ to\ 100$ occupying the even or non-square places. The nine vowels attached to any consonant indicate that the value of the consonant is multiplied by $1, 100, 100^2 \ldots 100^8$ respectively.

This system was used merely for mnemonic purposes and not followed in the *Ganitapada*. His notation and numeration indicate that the Hindus of that age were acquainted with the principle of the position system in the decimal or the centesimal scale. He defines the square and the cube of a number and gives rules for finding the *square-root* and the *cube-root.*

The value of $\pi$ is given by the rule:

*"when the diameter is* 20,000, *the circumference will be* 62,832 *approximately".*

This rule occurs before his rule for the formation of the sine-tables. This leads one to suppose that the above value of $\pi$ was used only for the construction of the sine-tables at intervals of $3.75^0$ and that the less approximate values, such as $\sqrt{10}$, were used elsewhere. He gives the rule for deriving successive sine-differences, in the modern notation,

$$\frac{d^2(Sin\ x)}{dx^2} = -\ Sin\ x$$

The '*Sine*" is equivalent to the modern *sine multiplied* by 3438.

According to the rule, each sine-difference diminished by the quotients of all the previous differences and itself by the first difference gives the next difference. The differences given in Dasagitika are: $225, 224, 222, 219 \ldots\ldots 37, 22, 7$. Same results are also given in Suryasiddhanta with the same rule. *Ayyangar* gives a possible reasoning how the rule may have been developed. As there were some discrepancies in some differences it is possible corrections were made by comparing the results with the actual ones obtained by direct calculation for the common angles $30^0, 45^0, 60^0, 75^0$ *and* $90^0$.

Another topic is on the mathematics of *Sun-dial* and the *Shadows*. In discussing this topic, constructions are given for drawing a circle, a triangle (*equilateral!*) given a side and a rectangle (*square!*) given a diagonal. Directions are given for determining experimentally the horizontal and vertical planes by means of water and plumb-line respectively. *Pythagorean* rule is used for finding the radius of the gnomon-circle given the height of the gnomon and the length of the shadow.

Two rules are given for determining:

- the lengths of the shadow of a gnomon of given height and
- the height of the source of light and its distance from two equal gnomons casting known shadows.

The rules are:

- Shadow $= \frac{(height\ of\ gnomon\ x\ the\ distance\ of\ the\ light\ from\ the\ gnomon)}{the\ difference\ between\ and\ the\ heights\ (the\ gnomon\ \&\ the\ light)}$

- The distance between the end of a shadow and the base of the light,when two gnomons are considered $= \frac{(the\ length\ of\ the\ shadow\ \times the\ distance\ between\ the\ ends\ of\ the\ shadows)}{The\ difference}$

It is not clear which difference is meant in lower equation.

*Aryabhata* in an *Eclipse Problem* a property of the circle is enunciated which is

*'that in a circle, the product of the arrows is the square of the semi-chord of the arc'.*

He gives a theorem, derived from this property, to use in the calculation of eclipses. The next topic in the *Ganitapada* is the arithmetical progression.

A general formula is given for the sum of the terms beginning with the $(p + 1)th$ term.

$$n = [a + (\frac{n-1}{2} + p)\, d]$$

where $a$ is the first term, $d$ the increment.

An alternative form is also given as

*'add the beginning and end terms and multiply by the sum by half the number of terms'.*

These are followed by another formula determining the number of terms. Next follows the contents of a triangular pile and a square pile as:

$$n(n+1)(n+2)/6\ and\ n(n+1)(2n+1)/6$$

The formula for the sum of the cubes is given by

$$\sum n^3 = \left(\sum n\right)^2$$

A pair of semi-geometrical identities to solve simultaneous equations of the types

$$x \pm y = p, \qquad xy = q;$$

$$x \pm y = p, \qquad x^2 + y^2 = q; \qquad xy = p, \qquad x^2 + y^2 = q.$$

Next topic is his rule for finding the interest on the *principle (P),* the interest on the *amount (A)* and the *time (t).*

Solving the quadratic equation

$$P\,r^2\,t^2 + Prt = A$$

determines the *rate of interest.*

After giving this rule he proceeds to enunciate the principle of the rule of three and the usual rules for the division of one fraction by another and for reducing all fractions to a common denominator.

Rules for reversing the steps in a mathematical process are enunciated by *Aryabhata.* This principle is useful for verification purposes and for the solution of equations where one has to clear the variable from all the ramifications in which it is involved. He gives an identity.

In modern notation it is

$$\sum\left(\frac{\Sigma X_r^i - X_r}{n-1}\right) = \sum X_r \ , \qquad i = 1,\ldots,n; \ r = 1,\ldots,n$$

Aryabhata's solution of the linear indeterminate equation is the crown of his mathematics.

He writes the problem as

*"to find a number which leaves residues n, n' with respect to the modulii m, m' respectively".*

If $n›n', m$ is called (*adhikaagrabhaagahaar*) the divisor corresponding to the greater residue and $m'$ is called (*oonaagrabhaaghaar*) the divisor belonging to the lesser residue. While the residue for the modulus $mm'$ is called '*dwichchedagra*'. The rationale and genesis of this method can be explained by an example.

$$\text{Let } m' = 29, n' = 15, m = 45, n = 19$$

We have to find x and y to satisfy the equation:

$$29x + 15 = 45y + 19.$$

The method is to express x in terms of y as

$$x = y + \frac{16y + 4}{29}.$$

Since $\frac{16y+4}{9}$ should be an integer let it be equal to $z$.

Then $y = z + \frac{(13z-4)}{16}$.

Again let $p = \frac{13z-4}{16}$ leading to $z = p + \frac{3p+4}{13}$. At this stage, since the coefficient is small, choose p = 3 so that $3p + 4$ is divisible by 13. To find the value of x, we retrace the steps and get $z = 4, y = 7, x = 1$. Aryabhata calls p as '*mati*'.

Thus, *Aryabhata*'s rule is a method of detached coefficients for carrying out the backward process for evaluating successively $z, y,$ and $x$. This process was called "*Vallikakuttaakaara*" by later mathematicians. *Bhaskara*'s method the '*creeper*' (valli) is extended to its utmost length. *Mahaviracharya* gives exactly *Aryabhata*'s rule using *Aryabhata*'s nomenclature.

## 卐 The Hindu Arabic Numerals

In his work "*Quarterly Journal of the Mythic Society, Vol. 18, 4, 1928*, and *Vol.19, 1, 1929*", he put light on *hindu-arabic numerals.*
There were four different kinds of numerals in use in Bharat from early times. They are *Kharoshti, Brahmi, symbolic word notation,* and *alphabetic notation,* before the decimal came into being with the nine symbols and the zero.
The *Kharoshti script* was written from right to left and the numerals followed this direction. These numerals occur in the Saka inscriptions (first century B.C).

The symbols used are:

- one, two, three vertical strokes for 1, 2, 3 respectively,
- an inclined cross for 4,
- a symbol for 10,
- a cursive combination of two tens for twenty,
- a sign resembling the Brahmi symbol with a vertical strike to its right for 'one hundred'.

The Hindus called the decimal notation *'anka palli'*; the word *'anka'* means a mark or symbol. In the word-numeral notation adapted by *Varahamihira* and others *'anka'* used to represent the numeral 9, probably because nine not ten numerals were in common use. By 9th century, at least, all the ten symbols were in place. In connection with these ten numerals of the new notation (date unknown), a rule came (*ankaana vaamathogathi* - अङ्काण वामथोगति) that is, the order of the numerals is from left to right. One does not know whether this rule refers to the order in writing or to an arrangement in some form of abacus. It is said that Hindu astrologers were using a wooden calculating board called *'paati* - पाति*'*, hence *'paati ganita'* (name for Arithmetic). This paper traces the history of the Hindu numerals. It establishes that Western writers, who had a bias towards Greek mathematicians, are wrong in saying that Hindus did not have their own system of numerals and these numerals were spread by Arab merchants to their country. From there these numerals were adapted by others. Leonardo Fibonacci spread the Hindu numerals in Europe and wrote in 1202 his thesis Liber Abaci.

### 卐 Some Glimpses of Ancient Hindu Mathematics

In his work *"The Mathematics Student, Vol. I, 1933"*, he paved glimpses on some novel works on ancient Bhartiya Hindu mathematics. The early Hindus developed mathematics as a limb of *Vedas (Vedanta)*. The world owes some of the basic ideas such as place value system of notation in Arithmetic, the generalizations of Algebra, the sine function of Trigonometry, and the foundations of Indeterminate Analysis. The early texts contain merely rules, results, and a number of problems but rarely a full worked out mathematical argument.

- *Vedanga Jyotisha* is the oldest work bearing astronomy. The style of the text is characterized by brevity, archaisms, and want of connection between consecutive verses. The text gives rules to calculate relative lengths of day and night. It gives positions of the moon at different parts of the year and the time of the day. The early Hindus, evidently, were well acquainted with arithmetical manipulations including fractions.

- *Sulva Sutra* (about 200 A.D.) deal with the construction of sacrificial altars. It gives construction of squares, rectangles, triangles and parallelograms with given specifications. It also contains the first arithmetical solution of indeterminate equation

$$x^2 + y^2 = z^2$$

and rational approximations to $\sqrt{2}$, $\frac{\sqrt{\pi}}{2}$.

- *Bhakshali* Manuscript has a second approximation for $\sqrt{(a2 + r)}$. This result can also be obtained from the rule found in all the Bhartiya arithmetical works.
- Geometrical problems were first analyzed as problems in arithmetic and then expressed in geometrical language. Area of a *cyclic quadrilateral*, *Pythagorean* Theorem and others were analyzed as arithmetical problems first.
- The next stage after *Sulva Sutra* is the age of *Aryabhata, Varahamihira, Brahmagupta,* and the unknown author of *Suryasiddhanta*. Aryabhata is placed by tradition at the head of the Bhartiya mathematicians but he appears to be the last of the earlier school. His work appears to be the first to give form and an individuality to the mathematical knowledge existed before him. His treatises were translated into Latin, French, and English. He was the first to in Bharat to hint the daily rotation of the earth on its axis. In his work one can find new alphabetic notation adopting both consonants and vowels to denote large numbers in a concise way. His admirer, *Lalla* abandoned it in favor of the more popular and the more ancient word numerals. *Vedanga Jyotisha, Surya Siddhanta,* and Varahamihira use the numeral words. *Brahmagupta* was the first to explicitly enunciate this system.
- A list of important mathematical topics on which there are contributions from the ancient and the medieval Hindus. Among the contributions of ancient Hindus to higher mathematics, the most outstanding, is on Indeterminate Analysis. Aryabhata was the first to indicate a method of getting integral solutions for

$$mx + a = ny + b$$

where $m, n, a, b$ are integers.

He proceeds by successive reductions to simpler equations until one reaches a solution. *Brahmagupta* has given a clearer exposition of the same. *Bhaskara* and others followed with improvements, alterations and extension. *Brahmagupta* was the first to enunciate the principle of composition of roots to solve the indeterminate equations of the second degree. This is similar to the modern principle of compositions of quadratic forms. It may be noted that the so called *Pellian* Equation

$$x^2 - Ny^2 = 1$$

should be called the *Brahmagupta Equation.*

- *Spherical trigonometry* was developed earlier than *Plane trigonometry* to be used in Astronomy. The *semi-chord* and the arrow gave rise to the sine and the versed-sine functions. Hindus divided the circumference of a circle into *21600 minutes*. Sine is considered as a length, and not a ratio, and expressed in terms of the arc. Hindus attributed to the arc all the properties of the angle. They established fundamental formulae of trigonometry and managed their astronomical formulae with the help of three functions: sine, the versed-sine, and the cosine. The works on astronomy from *Suryasiddhanta* to *Siddhanta Siromoni* give interpolation and extrapolation formulas for the tabulation of successive sines.
- *Vedic hymns* contain description of the motions of Sun and the Moon, the seasons, the number of days in a year, the two ayanas (*Devayana* and *Pitriyana*), the intercalary month etc. There are sufficient data in the *Vedas* for postulating the phenomenon of precession of the equinoxes and determining its rate. After the Vedas we have the twenty or so Siddhantas. These were the earliest efforts at crystallizing, in a scientific form, the knowledge of astronomy. Of these twenty, *Suryasiddhanta* and *Brahma Siddhanta* are the most popular and have been revised by other writers starting from *Aryabhata*. *Siddhantas* mention the number of revolutions of the sun, moon, and the planets, and their nodes. They assumed the motions of the planets have mean circular orbits during a *Kalpa* and all of them travel the same number of Yojanas. This constant is described as *Khakaksha* the circumference of the sphere to which solar rays extend. *Bhaskara*'s value for this constant is $187120692 \, x \, 10^8$ while *Suryasiddhanta* gives value as $18712080864 \, x \, 10^6$. Bhartiya astronomers

give the moon's distance from the earth but erred in the distances of the sun and the planets as they were calculated from the assumption of constant velocity. There is enough evidence that the Bhartiya astronomers actually verified the astronomical elements by personal observations of the heavenly bodies.

He provided the names, after Bhaskara, of other astronomers in his writings.

### 卐 A New Continued Fraction

In an article published in "*Current Science, Vol.6, 1937-38*", he proposed *a new Half Regular Continued Fraction* development of quadratic surds, distinct from suggested by *Minnegerod*. Such a development was implied in the method developed by *Bhaskara* for obtaining the integral solutions of the indeterminate equation

$$X^2 - NY^2 = 1.$$

This article rebuts the erroneous view that *Bhaskara*'s method gives rise to the simple continued fraction. The full development of this method is described in the subsequent articles.

### 卐 Theory of the Nearest Square Continued Fraction

In the work published "*The Half-Yearly Journal of the Mysore University, Vol. I, 1940, No.1 and Vol. I, 1941, No. 2*", he did some works on the theory of the *nearest square continued fraction*.

This paper was inspired by the remark of *Sir Thomas Little Heath "Indian Cyclic Method of solving the equation $x^2 - Ny^2 = 1$ in integers due to Bhaskara in 1150 is remarkably enough, the same as that which was rediscovered and expounded by Lagrange in 1768"*. This paper investigates the Bhartiya method of *half-regular continued fraction*. This new continued fraction (the nearest square continued fraction) is a natural sequel to *Bhaskara's cyclic method*. This theory was developed with the help of the simplest mathematics known to Hindus about the fifth century A.D.

It starts with the definition of the quadratic surd of the form $\frac{P+\sqrt{R}}{Q}$ as a standard form if R is a non-square positive integer, and $P\ (\neq\ 0), Q, \frac{R-P2}{Q}$ are integers having no common factor. If $P = 0$ it is sufficient $\frac{R}{Q}$ and $Q$ are relatively prime integers.

### A Peek into Bharat's Mathematical Past

In his work *"The Half-Yearly Journal of the Mysore University, Vol. V, 1945, No. 2"*, he traced the ancient Hindu mathematics from 800 B.C. to 1200 A.D. It has a brief description of the mathematics and astronomy developed by Bhartiya mathematicians of that era. The following are the topics selected:

- Vedanga Jyothisha (1200 B.C.)
- The Sulva – Sutras (between 800 and 500 B.C.)
- Surya Siddhanta (gone through revisions from 500 A.D. to 100 A.D.)
- Aryabhata (Aryabhatiyam: 499 A.D.)
- Varahamihira (6th century)
- Brahmagupta (598 A.D.)
- Sridhara (probably 8th century)
- Mahavira (period between Brahmagupta and Bhaskara)

He also listed Bhartiya Mathematicians in *"The Tamil Encyclopedia, Vol.3, 1st Edition"*. This is written in Tamil language. It covers more or less the topics on mathematics covered in some of the above articles.

**Further Reading**


- "Ambedkar, Gandhi and the Rise of Subcontinental Philosophy". thewire.in. Retrieved 27 January 2022.
- "Google Scholar, publications by Ayyangar". scholar.google.com. Retrieved 5 March 2023.
- Joseph, George Ghverghese (2000). The Crest of the Peacock: Non-European Roots of Mathematics. Princeton University Press.
- Kulshrestha, Chirantan (1981). "A. K. Ramanujan: A Profile". Journal of South Asian Literature. 16 (2): 181–184. ISSN 0091-5637.
- Pearce, Ian G. (2002). Indian Mathematics: Redressing the balance.
- Sridharan, Ramaiyengar (1998). Science in the West and India. Himalaya Publishing House, Bombay.
- "Google Scholar Publications". scholar.google.com. Retrieved 16 March 2023.

- A._A._Krishnaswami_Ayyangar, https://en.wikipedia.org/wiki/A._A._Krishnaswami_Ayyangar.
- Bhaskara and samclishta kuttaka, Ibid, 1929-30.
- Dauben, Joseph W.; Scriba, Christopher J. Writing the History of Mathematics - Its Historical Development. Springer. p. 315.
- Fourteen calendars, Ibid, 1937.
- History of Indian Mathematics (in Tamil), Kalai Kalanjiyam (Tamil Encyclopedia, 1st Edition) Vol. 3.
- Mathematics of Aryabhata, Quarterly Journal of Mythic Society, 1926.
- New light on Bhaskara's Chakravala or cyclic method of solving indeterminate equations of second degree in two variables, Ibid, 1929-30.
- New proofs of old theorems – Apollonius and Brahmagupta, Ibid, 1929-30.
- O'Connor, John J.; Robertson, Edmund F., "A. A. Krishnaswami Ayyangar", MacTutor History of Mathematics Archive, University of St Andrews.
- Peeps into India's mathematical past, Ibid, 1945.
- Remarks on Bhaskara's approximation to the sine of an angle, Ibid, 1950.
- Some glimpses of ancient Hindu mathematics, Mathematics Student, 1933.
- The Bhakshali Manuscript, Ibid, 1939.
- The Earliest Solution of the Biquadratic, Current Science, Vol. 7, 1938.
- The Hindu – Arabic Numerals, Ibid, 1928.
- The New Continued Fraction, Current Science, Vol.6, 1937-38.
- The Training of Indian almanac mathematics, Ibid, 1935.
- Theory of nearest square continued fraction, Journal of Mysore University, 1940.

## Syamadas Mukhopadhyaya (श्यामदास मुखोपाध्याय)
## (22 June 1866 – 8 May 1937)

Figure 5.21. Syamadas Mukhopadhyaya. (Image Credit: https://en.wikipedia.org/wiki/)

Syamadas Mukhopadhyaya (श्यामदास मुखोपाध्याय) was a mathematician who introduced the *four-vertex theorem* and *Mukhopadhyaya's theorem* in plane geometry (see Figure 5.21). *Syamadas Mukhopadhyaya* was born at *Haripal*, Hooghly district, in Bengal Presidency, British Bharat. He graduated from Hooghly College, received his M.A. degree from Presidency College in Calcutta, and his Ph.D. degree from Calcutta University in 1910. He also took classes from the *Indian Association for the Cultivation of Science*.

His father *Babu Ganga Kanta Mukhopadhyay* being in the state judicial service had different places of posting, and as a consequence Syamadas had to study in different institutions at different times. On completion of his graduation from Hooghly College he obtained his M.A. degree in Mathematics from the Presidency College, Calcutta, in 1890.

For his mathematical dissertation entitled '*On the infinitesimal analysis of an arc*', he was awarded the *Griffith*'s Prize of Calcutta University in 1909. In 1910, for his pioneering work on Differential Geometry, he was awarded the Ph. D. degree by the University of Calcutta, and the name of his thesis was

*'Parametric Coefficients in Differential Geometry of Curves'*. *S. D. Mukhopadhyay* was the first Bhartiya recipient of the Ph.D. degree in Mathematics from Calcutta University.

*Mukhopadhyaya* was appointed by *Asustosh Mookerjee* as professor of mathematics in the Rajabazar Science College, University of Calcutta. *Jacques Hadamard* communicated with *Mukhopadyaya* about the latter's work on the geometry of a plane arc and *Wilhelm Blaschke*'s book on geometry had a reference to *Mukhopadhyaya*.

Prior to his appointment in government service as a Professor of Mathematics in the Bethune College, Calcutta, *Syamadas Mukhopadhyay* joined a private college in Calcutta as a Faculty Member soon after completing his M.A. in Mathematics and served there for several years. In Bethune College, the first women's college in Asia, *S. D. Mukhopadhyay* had a very heavy teaching load, and apart from Mathematics, he had to take classes there on English and Philosophy regularly. In 1904, he was transferred to the Presidency College. He served there for eight years till 1912.

In 1912, *Syamadas Mukhopadhyay* was invited to join the newly set up Department of Pure Mathematics in the Calcutta University by Sir Ashutosh Mookerjee, the then Vice-Chancellor of Calcutta University. *S. D. Mukhopadhyay* accepted the offer and joined the department. He served there to the satisfaction of *Sir Ashutosh* and continued his service till 1932, and during the last six years of his tenure, *Syamadas* held the post of a University Professor.

As a talented and eminent teacher of Pure Mathematics, name of *S. D. Mukhopadhyay* spread everywhere. In order to be a direct student of *Professor Mukhopadhyay, R. C. Bose* after completing his M.Sc. in Applied Mathematics from the Delhi University came and joined the Pure Mathematics Department of Calcutta University as a student. He later became a researcher under *S. D. Mukhopadhyay* and earned eminence as a great mathematician. Among many students of *Syamadas Mukhopadhyay*, three deserve special mention. They were *Professor G. Bhar, Professor R. N. Sen* and Professor *M. C. Chaki*. Professor Bhar was a teacher of great repute in Presidency College, *Professor Sen* became the *Hardinge Professor* of Pure Mathe-matics in Calcutta University and also a famous researcher in the field of Differential Geometry, and *Professor Chaki* held the Chair named *'Sir Ashutosh Birth Centenary Professor of Higher*

*Mathematics*' in Calcutta University and also had a number of research publications in many areas of Geometry.

He worked at *Bangabasi College* and then at *Bethune College* in Calcutta, where he lectured in Mathematics, English Literature, and Philosophy. In 1932, he was elected president of the Calcutta Mathematical Society. He served in this capacity until his death from heart failure in 1937. Mukhopadhyaya went to Europe on a Ghose Travelling Fellowship in 1933 to study methods of education. He gave lectures in Paris University.

## 卐 Key Contributions in Mathematics

### 卐 Four-Vertex Theorem

The *four-vertex theorem* of geometry states that the curvature along a simple, closed, smooth plane curve has at least four local extrema (specifically, at least two local maxima and at least two local minima) (see Figure 5.22). The name of the theorem derives from the convention of calling an extreme point of the curvature function a vertex. This theorem has many generalizations, including a version for space curves where a vertex is defined as a *point of vanishing torsion*.

The curvature at any point of a smooth curve in the plane can be defined as the reciprocal of the radius of an osculating circle at that point, or as the norm of the second derivative of a parametric representation of the curve, parameterized consistently with the length along the curve. For the vertices of a curve to be well-defined, the curvature itself should vary continuously, as happens for curves of smoothness $C^2$. A vertex is then a local maximum or local minimum of curvature. If the curvature is constant over an arc of the curve, all points of that arc are considered to be vertices. The four-vertex theorem states that a smooth closed curve always has at least four vertices. An ellipse has exactly four vertices: two local maxima of curvature where it is crossed by the major axis of the ellipse, and two local minima of curvature where it is crossed by the minor axis. In a circle, every point is both a local maximum and a local minimum of curvature, so there are infinitely many vertices. Every curve of constant width has at least six vertices. Although many curves of constant width, such as the Reuleaux triangle, are non-smooth or have circular arcs on their boundaries, there exist smooth curves of constant width that have exactly six vertices.

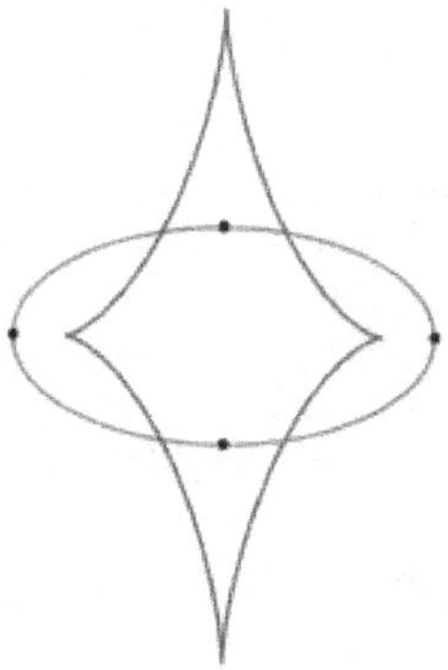

Figure 5.22. An ellipse (red) and its evolute (blue), showing the four vertices of the curve, each vertex corresponding to a cusp on the evolute. (Image Credit: https://en.wikipedia.org/wiki/)

His proof utilizes the fact that a point on the curve is an extremum of the curvature function if and only if the osculating circle at that point has fourth-order contact with the curve; in general, the osculating circle has only third-order contact with the curve.

**Applications**

One corollary of the theorem is that a homogeneous, planar disk rolling on a horizontal surface under gravity has at least 4 balance points. A discrete version of this is that there cannot be a monostatic polygon. However, in three dimensions there do exist monostatic polyhedra, and there also exists a convex, homogeneous object with exactly 2 balance points (one stable, and the other unstable), the *Gömböc*.

There are several discrete versions of the four-vertex theorem, both for convex and non-convex polygons. Here are some of them:

- The sequence of angles of a convex equilateral polygon with at least four vertices has at least four extrema.
- The sequence of side lengths of a convex equiangular polygon with at least four sides has at least four extrema.
- A circle circumscribed around three consecutive vertices of a polygon with at least four vertices is called extremal if it contains all remaining vertices of the polygon, or has none of them in its interior. Such a

convex polygon is generic if it has no four vertices on the same circle. Then every generic convex polygon with at least four vertices has at least four extremal circles.

- Two convex n-gons with equal corresponding side length have either zero or at least 4 sign changes in the cyclic sequence of the corresponding angle differences.
- Two convex n-gons with parallel corresponding sides and equal area have either zero or at least 4 sign changes in the cyclic sequence of the corresponding side lengths differences.

### 卐 Mukhopadhyaya's Theorem

In geometry *Mukhopadhyaya*'s theorem may refer to one of several closely related theorems about the number of vertices of a curve due to Mukhopadhyaya (1909). One version, called the four-vertex theorem, states that a simple convex curve in the plane has at least four vertices, and another version states that a simple convex curve in the affine plane has at least six affine vertices.

### 卐 Synthetic Geometry

*Syamadas Mukhopadhyay's* research work may be broadly classified into 4 parts:

- Use of Synthetic Geometry to solve properties of plane curves
- Non-Euclidean Geometry
- Differential Geometry
- Stereoscopic representations of four-dimensional space.

In the first part, he has dealt with properties of plane curves, especially in their infinitesimal regions using synthetic methods. Here he had developed new method which led to a number of interesting theorems on the existence of minimum number of cyclic and sextactic points between two points of a given curve on a convex oval, etc. In this context two of his theorems deserve special attention. But before that it would be necessary to understand what are cyclic and sextactic points. A cyclic point is a singular point on a plane curve, where the circle of curvature passes through 4 consecutive points instead of 3. On the other hand, a sextactic point is a singular point, where the osculating conic passes through 6 consecutive points instead of 5. At a cyclic point, the circle of curvature may touch the given circle internally or externally. In the former case, the point will be called in-cyclic and in the latter case ex-cyclic. Similarly, at a sextactic point, the osculating conic may touch the given curve internally or externally. In the former case, the point is called in-sextactic, and in the latter case it is called ex-sextactic. With these definitions, S. D. Mukhopadhyay first demonstrated a number of interesting propositions.

Later they led to his famous two theorems stated below:

- Theorem I state that *'the minimum number of cyclic points on a convex oval is 4'*, and Theorem II states that *'the minimum number of sextactic points on a convex oval is 6'*. These two theorems were first published in 1909 in the Bulletin of the Calcutta Mathematical Society. But initially, at that time not much attention was paid to this piece of research.

- Later on, S. D. Mukhopadhyay generalized these two theorems. The generalized Theorem I states that *'If a circle C intersects an oval V in 2n points (n < 2) then there exists at least 2n cyclic points in order on V, of alternately contrary signs, provided the oval has continuity of order 3'*. The generalized Theorem II states that 'If a conic C intersects an oval V in 2n points (n >or = 2), then there exist at least 2n sextactic points in order on V, which are alternatively positive and negative, provided V has continuity of order 5'. Thus S. D. Mukhopadhyay placed his earlier investigations on more rigorous basis.

In the second part of his research work, Syamadas Mukhopadhyay contributed to a type on *Non-Euclidean Geometry* namely, *Hyperbolic Geometry*. Actually, *Hyper-bolic Geometry* is the geometry obtained by assuming all the postulates of Euclid except the fifth one. The fifth postulate of Euclid is replaced by its negation. In the context of *S. D. Mukhopadhyay*'s research in Non-Euclidean Geometry.

S. D. Mukhopadhyay in collaboration with his student G. Bhar published his first research paper on Hyperbolic Geometry in the Bulletin of the Calcutta Mathemat-ical Society in 1920. The paper was titled *'Generalisation of certain theorems in the Hyperbolic Geometry of the triangle'*. This publication was followed by a number of research papers on this kind of Geometry. The investigations carried out by him resulted in the important discovery of the

*'Rectangular Pentagon'* and his beautiful geometrical interpretation of the *Engel-Napier rules.*
To be more specific, *S. D. Mukhopadhyay* extended the well-known concurrency theorems of the angle bisectors and the right bisectors of the sides of an ordinary triangle, to all types of hyperbolic triads of lines and points. It may be noted that by a hyperbolic triad, it is meant a group of three elements (points or lines) lying upon a hyperbolic plane.

In this context, his three research papers titled:

- *'Geometrical investi-gations on the correspondence between a right angled triangle, a three-right-angled quadrilateral and a rectangular pentagon in hyperbolic geometry'*
- *'On general theo-rems of co-intimacy of symmetries and hyperbolic triads'*
- *'Triadic equations in hyperbolic geometry'*

In the third part of his research and investigations, *Professor Syamadas Mukhopadhyay* while dealing with *Differential Geometry of curves introduced for n-dimensional space curves certain differential forms*. He named them parametric coefficients, and using them he expressed many invariant properties of the curves.

The fourth and final part of his research comprises a suggestion of a *stereoscopic device for visualizing figures in four-dimensional space* and his discussion with Bryan in this matter.

In 1912–1913, S. D. Mukhopadhyay had published his note on the stereoscopic representation of four-dimensional space.

Towards the end of his research career, *S. D. Mukhopadhyay* published two important papers entitled *'Lower segments of M-curves'* and *'Cyclic curves of an ellipsoid'* in the *Journal of Indian Mathematical Society*. Mukhopadhyay explained that the name *M-curve* (Monotropic Curve) was due to the German geometrician *Stackel H. Mohrmann* (1907–1934), who in adopting the name gave the following precise definition.

In his investigations, *S. D. Mukhopadhyay* considered M-curves which are not necessarily analytic. He has characterized M-curves in a different way and investigated the associated problem.

**Further Reading**


1. Berger, Marcel (2010). "V.8. The four vertex theorem and its converse; an application to physics". Geometry Revealed. Heidelberg: Springer. pp. 271–278. doi:10.1007/978-3-540-70997-8. ISBN 978-3-540-70996-1. MR 2724440.
2. Calcutta Review 1934.
3. Collected geometrical papers of Syamadas Mukhopadhyaya, Volume 1.
4. Craizer, Marcos; Teixeira, Ralph; Balestro, Vitor (2018). "Closed cycloids in a normed plane". Monatshefte für Mathematik. 185 (1): 43–60. arXiv:1608.01651. doi:10.1007/s00605-017-1030-5. MR 3745700.
5. Dahlberg, Björn (2005). "The converse of the four vertex theorem". Proceedings of the American Mathematical Society. 133 (7): 2131–2135. doi:10.1090/S0002-9939-05-07788-9.
6. DeTurck, D.; Gluck, H.; Pomerleano, D.; Vick, D.S. (2007). "The four vertex theorem and its converse" (PDF). Notices of the American Mathematical Society. 54 (2): 9268. arXiv:math/0609268.
7. Ghomi, Mohammad (2017). "Boundary torsion and convex caps of locally convex surfaces". Journal of Differential Geometry. 105 (3): 427–486. arXiv:1501.07626. doi:10.4310/jdg/1488503004.
8. Gluck, Herman (1971). "The converse to the four-vertex theorem". L'Enseignement mathématique. 17: 295–309. MR 0344998.
9. Graustein, W. C. (1937). "Extensions of the four-vertex theorem". Transactions of the American Mathematical Society. 41 (1): 9–23. doi:10.2307/1989876. MR 1501889.
10. Kneser, Adolf (1912). "Bemerkungen über die Anzahl der Extrema der Krümmung auf geschlossenen Kurven und über verwandte Fragen in einer nicht euklidischen Geometrie". Festschrift Heinrich Weber. Teubner. pp. 170–180.
11. Martinez-Maure, Yves (1996). "A note on the tennis ball theorem". The American Mathematical Monthly. 103 (4): 338–340. doi:10.2307/2975192. JSTOR 2975192. MR 1383672.
12. Mukhopadhyaya, S. (1909). "New methods in the geometry of a plane arc". Bulletin of the Calcutta Mathematical Society. 1: 21–27.
13. Osserman, Robert (1985). "The four-or-more vertex theorem". The American Mathematical Monthly. 92 (5): 332–337. doi:10.2307/2323126. MR 0790188.
14. Pak, I. Lectures on Discrete and Polyhedral Geometry Archived 2009-01-29 at the Wayback Machine, Section 21.

15. Sedykh, V.D. (1994). "Four vertices of a convex space curve". Bulletin of the London Mathematical Society. 26 (2): 177–180. doi:10.1112/blms/26.2.177.
16. Sources in the Development of Mathematics: Series and Products from the Fifteenth to the Twenty-first Century.
17. The Mathematics Student, Volumes 1-4, 1933.
18. "Syamadas Mukhopadhyaya", Bulletin of the Calcutta Mathematical Society, Vol. 29 (1937), pages 115-120.
19. 'A General Theory of Osculating Conics I' [Journal and Proceedings of Asiatic Society of Bengal, (NS), IV, No. 4, (1908), 167–168].
20. 'A General Theory of Osculating Conics II' [Journal and Proceedings of Asiatic Society of Bengal, (NS), IV, No. 10, (1908), 497–509].
21. 'A Note on Current View of Operations Through the Fourth Dimension' [Bulletin of Calcutta Mathematical Society, IX, (1917)].
22. 'A Note on the Stereoscopic Representation of Four-Dimensional Space' [Bulletin of Calcutta Mathematical Society, IV, (1912–13), 15].
23. 'An Exposition of the Axioms of Order of Hilbert' [Indian Physico Mathematical Journal, III, (1932)].
24. 'Circles Incident on an Oval of Undefined Curvature' [Tohuku Journal of Mathematics, Japan, 37, (1931)].
25. 'Cyclic Curves of an Ellipsoid' [Journal of Indian Mathematical Society, XX, (1932), 246–250].
26. 'Extended Minimum Number Theorems of Cyclic and Sextactic on a Plane Convex Oval' [Mathematische Zeitschrift, Band 33, (1931), 648–662].
27. 'Gemattic Extensions of Elementary Chains' [Tohuku Journal of Mathematics, Japan, 38, (1933)].
28. 'Generalied Form of Bohmer's Theorem for Elliptically Curved Non-Analytic Oval' [Mathematische Zeitschrift, Band 30, (1929), 560–571].
29. 'Generalisations of Certain Theorems in the Hyperbolic Geometry of the Triangle' [Bulletin of Calcutta Mathematical Society, XII, (1920–21), 14–28].
30. 'Genesis of an Elementary Arc' [Bulletin of Calcutta Mathematical Society, XVII, No. 4, (1926), 153].
31. 'Geometrical Investigations on the Correspondence Between a Right Angled Triangle, a Three Right Angled Quadrilateral and a Rectangular Pentagon in Hyperbolic Geometry' [Bulletin of Calcutta Mathematical Society, XIII, No. 4, (1922–23), 211–216].
32. 'Geometrical Theory of a Plane Non-Cyclic Arc, Finite as Well as Infinitesimal' [Journal of Asiatic Society of Bengal, (NS), IV, (1908)].

33. 'Intrinsic Parameters in the Differential Geometry of Curves in an N-Space' [Griffiths Memorial Prize Essay, (1910), Calcutta University Publications].
34. 'Lower Segments of M-Curves' [Journal of Indian Mathematical Society, XIX, (1931), 75–80].
35. 'New Methods in the Geometry of a Plane Arc I, Cyclic and Sextactic Points' [Bulletin of Calcutta Mathematical Society, 1, (1909), 31–37].
36. 'New Methods in the Geometry of a Plane Arc II Cyclic Points and Normals' [Bulletin of Calcutta Mathematical Society, X, (1919), 65–72].
37. 'Note on T Hayashi's Paper on the Osculating Ellipses of a Plane Curve' [Circolo Mathematico Palermo, Tomo L 1, (1927)].
38. 'On Rates of Variation of the Osculating Conic' [Bulletin of Calcutta Mathe-matical Society, No. 2, (1909), 125–130].
39. 'Parametric Coefficients in the Differential Geometry of Curves in an N-Space, I, General Conceptions' [Bulletin of Calcutta Mathematical Society, 1, No. 2, (1909), 187–200].
40. 'Parametric Coefficients in the Differential Geometry of Curves in an N-Space, II, Extension of Serret Frenet Formulae to Curves in an N-Dimensional Space' [Bulletin of Calcutta Mathematical Society, 1, No. 4, (1909), 233–234].
41. 'Parametric Coefficients in the Differential Geometry of Curves in an N-Space, III, Fundamental Formulae' [Bulletin of Calcutta Mathematical Society, 11, (1910)].
42. 'Parametric Coefficients in the Differential Geometry of Curves in an N-Space, IV, Expressions of the Coordinates of a Point on a Curve in an N-Space as Power Series in S' [Bulletin of Calcutta Mathematical Society, 111, (1911)].
43. 'Parametric Coefficients in the Differential Geometry of Curves in an N-Space, VI; On Parametric Coefficient and Osculating Spherics to a Curve in an N-Space' [Bulletin of Calcutta Mathematical Society, VIII, (1915)].
44. 'Parametric Coefficients in the Differential Geometry of Curves, V, Principal Directions and Curvatures at a Singular Point of a Curve in an N-Space' [Bulletin of Calcutta Mathematical Society, V, (1913–14), 13–20].
45. 'Reply to Prof Bryan's Criticism' [Bulletin of Calcutta Mathematical Society, VI, (1914–15), 55–56].
46. 'Some General Theorems in the Geometry of Plane Curve' [Sir Astosh Mookerjee Silver Jubilee, II, (1922), Calcutta University Publications].

47. 'Triadic Equations in Hyperbolic Geometry' [Bulletin of Calcutta Mathematical Society, XVIII, No. 4, (1927)].
48. Chaki, M. C. (1990), "Syamadas Mukhopadhyaya (1866–1937)", Journal of Pure Mathematics, 7: 59–65, MR 1306649
49. D. DeTurck, H. Gluck, D. Pomerleano, D.S. Vick, The four vertex theorem and its converse, Notices of the AMS, 54 (2007), no. 2, 192–207.
50. Four-vertex theorem, https://en.wikipedia.org/wiki/Four-vertex_theorem.
51. History of the Calcutta School of Physical Sciences.
52. Mukhopãdhyãya, S. (1929), Collected geometrical papers of Syamadas Mukhopadhyaya, Calcutta University Press.
53. Mukhopadhyaya, Syamadas (1909), "New Methods in the Geometry of a Plane Arc.-I Cyclic and Sextactic Points", Bulletin of Calcutta Mathematical Society, 1 (1): 31–37.
54. Mukhopadhyaya_theorem, https://en.wikipedia.org/wiki/Mukhopadhyaya_theorem
55. Pressley, Andrew (2010). Elementary Differential Geometry. Springer Undergraduate Mathematics Series (2nd ed.). London: Springer-Verlag. Definition 2.1.1, p. 30 and Exercise 2.2.6, p. 44. doi:10.1007/978-1-84882-891-9. ISBN 978-1-84882-890-2. MR 2598317.
56. Purabi Mukherji, Notable Modern Indian Mathematicians and Statisticians: During the 19th and 20th Centuries of Bengal [1st ed. 2022] 981196131X, 9789811961311, Springer.
57. Syamadas_Mukhopadhyaya, https://en.wikipedia.org/wiki/Syamadas_Mukhopadhyaya.
58. With R C Bose 'On General Theorems of Co-intimacy of Symmetries and Hyperbolic Triad' [Bulletin of Calcutta Mathematical Society, XVII, No. 1, (1926), 39–55].

**Prasanta Chandra Mahalanobis (प्रशन्तचन्द्र महालानोबिश)**
**(29 June 1893– 28 June 1972)**

Figure 5.23. Prasanta Chandra Mahalanobis. (Image Credit: https://en.wikipedia.org/wiki/)

Prasanta Chandra Mahalanobis (प्रशन्तचन्द्र महालानोबिश) was a scientist and statistician. He is best remembered for the *Mahalanobis distance*, a statistical measure, and for being one of the members of the first *Planning Commission of free India* (see Figure 5.23). He made pioneering studies in anthropometry in India. He founded the Indian Statistical Institute, and contributed to the design of large-scale sample surveys. For his contributions, *Mahalanobis* has been considered the *Father of statistics in India*.

Father of statistics in India

*Mahalanobis* belonged to a prominent *Bengali Brahmin family* of landed gentry in *Bikrampur*, Dhaka, Bengal Presidency (now in Bangladesh). His grandfather *Gurucharan* (1833–1916) moved to Calcutta in 1854 and built up a business, starting a chemist shop in 1860. *Gurucharan* was influenced by *Debendranath Tagore* (1817–1905), father of the Nobel Prize-winning poet,

*Rabindranath Tagore. Gurucharan* was actively involved in social movements such as the Brahmo Samaj, acting as its treasurer and president. His house on *210 Cornwallis Street* was the centre of the *Brahmo Samaj* (ब्राह्मो समाज). *Gurucharan*'s younger son, *Prabodh Chandra* (1869–1942), was the father of *P. C. Mahalanobis*. Born in the house at 210 Cornwallis Street, *Mahalanobis* grew up in a socially active family surrounded by *intellectuals and reformers.*

*Mahalanobis* received his early schooling at the *Brahmo Boys School* in Calcutta, graduating in 1908. He joined the *Presidency College,* then affiliated with the *University of Calcutta,* where he was taught by teachers who included *Jagadish Chandra Bose,* and *Prafulla Chandra Ray*. Others attending were *Meghnad Saha,* a year junior, and *Subhas Chandra Bose,* two years his junior at college.

*Mahalanobis* received a Bachelor of Science degree with honours in physics in 1912. He left for England in 1913 to join the *University of London*. After missing a train, he stayed with a friend at *King's College, Cambridge*. He was impressed by *King's College Chapel* and his host's friend *M. A. Candeth* suggested that he could try joining there, which he did. He did well in his studies at King's, but also took an interest in cross-country walking and punting on the river. He interacted with the mathematical genius *Srinivasa Ramanujan* during the latter's time at *Cambridge*. After his Tripos in physics, *Mahalanobis* worked with *C. T. R. Wilson* at the *Cavendish Laboratory*. He took a short break and went to Bharat, where he was introduced to the Principal of *Presidency College* and was invited to take classes in physics.

After returning to England, *Mahalanobis* was introduced to the journal *Biometrika*. This interested him so much that he bought a complete set and took them to Bharat. He discovered the utility of statistics to problems in meteorology and anthropology, beginning to work on problems on his journey back to Bharat.

In Calcutta, *Mahalanobis* met *Nirmalkumari (Rani),* daughter of *Heramba Chandra Maitra,* a leading educationist and member of the *Brahmo Samaj*. They married on 27 February 1923, although her father did not completely approve of the union. He was concerned about *Mahalanobis*'s opposition to various clauses in the membership of the student wing of the *Brahmo Samaj,* including prohibitions against members' drinking alcohol and smoking. Sir *Nilratan Sircar, P. C. Mahalanobis*' maternal uncle, took part in the wedding ceremony in place of the father of the bride.

## Key Contributions in Mathematics

### Mahalanobis Distance

*Prof. Mahalanobis*'s most notable contribution to statistics is the creation of the $D^2 - statistic$, commonly known as the *'Mahalanobis Distance'*. His journey began in 1920 when tasked with analyzing anthropometric data of a mixed British- Bhartiya community. An existing interest in anthropometric studies facilitated *Mahalanobis*'s deep and quick engagement in extensive research and analysis.

Through his studies, he innovatively devised a multivariate distance measure, the '$D^2$', for comparing and categorizing populations by the renowned statistician, *Sir R. A. Fisher*. His publication in 1922 marked a seminal contribution in this field, demonstrating his adept application of statistical methods to glean meaningful insights from data. *Mahalanobis*'s pioneering anthropometric statistical analysis laid the groundwork for further research in *'Multivariate Analysis'*, inspiring subsequent exceptional contributions from his students like *R. C. Bose, S. N. Roy*, and *C. R. Rao*. A 1945 collaboration resulted in an exemplary research paper, emblematic of scientific excellence, cementing his legacy as a trailblazer in statistical anthropometric analysis.

Between 1922 and 1936, *Mahalanobis* authored 15 papers, applying statistical techniques to *anthropometry, race-origin, and race-mixture studies*. His research culminated in the formation of the *'Mahalanobis Distance'*, a critical metric in modern statistical literature, especially in taxonomical classification issues. *'Mahalanobis Distance'* is a multi-dimensional generalization of the idea of measuring how many standard deviations away $P$ is from the mean of $D$. This distance is zero for $P$ at the mean of $D$ and grows as $P$ moves away from the mean along each principal component axis. If each of these axes is re-scaled to have unit variance, then the *Mahalanobis distance* corresponds to standard *Euclidean distance* in the transformed space. The *Mahalanobis distance* is thus unitless, scale-invariant, and takes into account the correlations of the data set.

Given a probability distribution $Q$ on $R^N$, with mean $\mu = (\mu_1, \mu_2, \mu_3, \dots \mu_N)^T$ and positive-definitive covariance matrix $S$, the Malalanobis distance of a point $x = (x_1, x_2, x_3, \dots x_N)^T$ from $Q$ is

$$d_M(x, Q) = \sqrt{(x-\mu)^T S^{-1}(x-\mu)}$$

Given two points $x$ and $y$ in $R^N$, the Mahalanobis distance between them with respect to $Q$ is

$$d_M(x, y; Q) = \sqrt{(x-\mu)^T S^{-1}(x-\mu)}$$

Which means that $d_M(x, Q) = d_M(x, \mu; Q)$

Since $S$ is positive-definite, so is $S^{-1}$, thus the square roots are always defined.

We can find useful decompositions of the squared *Mahalanobis distance* that help to explain some reasons for the outlyingness of multivariate observations and also provide a graphical tool for identifying outliers.

By the spectral theorem $S^{-1}$ can be decomposed as $S^{-1} = W^T W$ for some real $N \times N$ matrix, which gives us the equivalent definition

$$d_M(x, y; Q) = ||W(x-y)||$$

Where $||.||$ is the *Euclidean norm*. That is, the *Mahalanobis distance* is the *Euclidean distance* after a whitening transformation. The existence of $W$ is guaranteed by the spectral theorem, but it is not unique. Different choices have different theoretical and practical advantages. In practice, the distribution $Q$ is usually the sample distribution from a set of *independent and identically distributed (IID)* samples from an underlying unknown distribution, so $\mu$ is the sample mean, and $S$ is the covariance matrix of the samples.

When the affine span of the samples is not the entire $R^N$ , the covariance matrix would not be positive-definite, which means the above definition would not work. However, in general, the *Mahalanobis distance* is preserved under any full-rank affine transformation of the affine span of the samples. So, in case the affine span is not the entire

$R^N$, the samples can be first orthogonally projected to $R^N$, where $n$ is the dimension of the affine span of the samples, then the *Mahalanobis distance* can be computed as usual.

## 卐 Statistical Tables

*Prof. P. C. Mahalanobis* played a significant role in the realm of *'Theoretical Statistics'* through his major contributions in the creation and organization of *'Statistical Tables'*. He pioneered in 1932 by cataloging the results derived by *R. A. Fisher* concerning the one percent and five percent points of distribution. A noteworthy historical reference was made by statistician K. *R. Nair*, who mentioned that Mahalanobis' initial tabulations of these points were presented in 1932, now commonly known as Snedecor's F. Snedecor later labeled the variance ratio as *F* in tribute to *Fisher*, unaware of the prior Bhartiya contributions. Additionally, *Fisher*, in his *Statistical Medical Research* (1938), opted not to use the symbol *F* since *Mahalanobis* had already used it in his prior tabulations. *Mahalanobis* further developed three additional tables in 1933, with one titled *'Tables for L-tests'* and another *'Tables for Random Samples from a Normal Population'* getting published in *Sankhya (Volume-I)* in consecutive years, 1933 and 1934 respectively.

## 卐 Field Trails

*Prof. P. C. Mahalanobis* made significant contributions in agronomy research, producing around 50 research papers and reports, either individually or collaboratively. Many of these works were focused on field trials, an endeavor he undertook single-handedly since the 1920s, without much assistance. Simultaneously, *R. A. Fisher*, a prominent statistician from the UK, was engaged in similar research at *Rothamsted station*. Initially, *Mahalanobis* was not aware of *Fisher*'s concurrent work. Fisher, having come across *Mahalanobis*'s publications, reached out by sharing some of his own papers, establishing a professional connection between the two. This relationship proved to be enduring and greatly beneficial for *Professor Mahalanobis* in various aspects of his work.

## 卐 Large Scale Sample Survey

*Professor P. C. Mahalanobis* significantly impacted *'Applied Statistics'* by pioneering the *'Theory of Sample Survey.'* His initial involvement in this field was recorded in 1932, where he elaborated on the fundamentals of sampling at the *Presidency College*, Calcutta. *Mahalanobis*'s understanding of sampling was comprehensive and profound, recognizing the immense potential and the necessity for accurate interpretation of results obtained from sample surveys.

Starting in 1937, he applied the method of random sampling practically, notably in the sample census of the jute crop in Bengal. His work garnered international acclaim, with eminent statisticians like *Professor Harold Hotelling* appreciating its accuracy and economical significance, and also suggesting its superiority to existing methods in the United States.

*Mahalanobis*, along with his team and students, utilized the sample survey methodology across various fields such as agriculture, demography, and socio-economic sectors, marking essential contributions to applied statistics. Their work in the mid-twentieth century laid the foundations for contemporary large-scale sample survey techniques.

Renowned individuals such as *Professors F. Yales* and *Sir R. A. Fisher* praised *Mahalanobis*'s foresight and innovative approaches in the field of sample surveys, acknowledging his role in enhancing the global application of sampling methods, especially in less developed nations.

*Mahalanobis*'s work epitomizes the integration of theory with practical application, emphasizing the utility of statistics in addressing diverse *scientific, sociological, and economic* challenges. His commitment to leveraging statistical methodologies to solve real-world problems remained a steadfast aspect of his research contributions throughout his life.

### 卐 Establishement of Indian Statistical Institute

*Prof. P. C. Mahalanobis* was a monumental figure in Bharat's institutional development, particularly notable for founding the *Indian Statistical Institute (ISI)* in Calcutta in 1931. He aimed to create a space for robust research, teaching, and project execution, ultimately succeeding in establishing ISI as a world-renowned institution. *Mahalanobis* held the position of 'Chief Executive Secretary' at ISI from its inception until his demise, contributing significantly to its global reputation for excellence in interdisciplinary statistical research.

A testament to his leadership, the Bhartiya Parliament recognized ISI as an *'Institute of National Importance'* in 1959. His vision also embraced international collaboration, inviting global experts to contribute, thereby enhancing ISI's international standing. Mahalanobis's role in the inception of the *'National Sample Survey'* is another remarkable achievement, marking a significant stride in statistical application in Bharat.

Driven by a vision of excellence, he influenced the inclusion of statistics as a distinct discipline in the *Indian Science Congress* and was instrumental in the establishment of statistical departments and organizations. His influence extended to the realms of economic planning and policy formulation in post-independence Bharat, marked by his significant contributions to the Planning Commission and various committees, like the *'Indian National Income Committee'*. His profound insights and advisory roles, especially close association with Bharat's first Prime Minister, *Jawaharlal Nehru*, facilitated the infusion of statistical expertise in national economic planning and policy.

*Mahalanobis*'s involvement in various capacities, from committee chairmanships to advisory roles, showcased his multifaceted expertise and influential contributions to areas like industrial quality control, macro-economic policy, and national income study. His writings further underscored his broad vision, addressing crucial subjects such as basic industries, scientific research, economic development, and global peace. *Prof. Mahalanobis's* legacy is emblematic of a transformative era in Bharat's institutional and economic landscape.

**Further Reading**


1. "Fellowship – Mahalanobis, Prasanta Chandra". Indian Academy of Sciences. Retrieved 18 February 2018.
2. "National Sample Survey: How India taught the world the art of collecting data". BBC News. 29 June 2022. Retrieved 30 June 2022.
3. "Portfolio Optimizer". portfoliooptimizer.io/. Retrieved 2022-04-23.
4. "VP Naidu pays homage to Mahalanobis at ISI". The Times of India. 30 June 2018. Retrieved 2 July 2018.
5. "Who was Prasanta Chandra Mahalanobis or 'PCM'?". The Indian Express. 29 June 2018.
6. Adhikari, B. P. (1990). "Social construction of the statistical estimation of crop yield". Paper presented at the XII World Congress of Sociology of the International Sociological Association, Madrid, Spain.

7. Das, Gurucharan. (2000) India Unbound: The Social and Economic Revolution from Independence to the Global Information Age. Anchor Books. p. 432 ISBN 0-375-41164-X.
8. Dasgupta, Somesh (1993). "The evolution of the D2-statistic of Mahalanobis" (PDF). Sankhya. 55: 442–459. Archived from the original (PDF) on 13 November 2013.
9. De Maesschalck, R.; Jouan-Rimbaud, D.; Massart, D. L. (2000). "The Mahalanobis distance". Chemometrics and Intelligent Laboratory Systems. 50 (1): 1–18. doi:10.1016/s0169-7439(99)00047-7.
10. Dronamraju, K. R. (1987). "On Some Aspects of the Life and Work of John Burdon Sanderson Haldane, F.R.S., in India". Notes and Records of the Royal Society. 41 (2): 211–237. doi:10.1098/rsnr.1987.0006. PMID 11622022.
11. Etherington, Thomas R. (2019-04-02). "Mahalanobis distances and ecological niche modelling: correcting a chi-squared probability error". PeerJ. 7: e6678. doi:10.7717/peerj.6678. ISSN 2167-8359. PMC 6450376. PMID 30972255.
12. Etherington, Thomas R. (2021-05-11). "Mahalanobis distances for ecological niche modelling and outlier detection: implications of sample size, error, and bias for selecting and parameterising a multivariate location and scatter method". PeerJ. 9: e11436. doi:10.7717/peerj.11436. ISSN 2167-8359. PMC 8121071. PMID 34026369.
13. Farber, Oren; Kadmon, Ronen (2003). "Assessment of alternative approaches for bioclimatic modeling with special emphasis on the Mahalanobis distance". Ecological Modelling. 160 (1–2): 115–130. doi:10.1016/S0304-3800(02)00327-7.
14. Ghosh, J. K.; Majumder, P. P. (2005). "Mahalanobis, Prasanta Chandra". Encyclopedia of Biostatistics. doi:10.1002/0470011815.b2a17090. ISBN 978-0470849071.
15. Ghosh, J. K.; P. Maiti; T. J. Rao; B. K. Sinha (1999). "Evolution of Statistics in India". Revue Internationale de Statistique. 67 (1): 13–34. doi:10.2307/1403563. JSTOR 1403563.
16. Hagger-Johnson, G. (2005). "Mahalanobis, Prasanta Chandra". Encyclopedia of Statistics in Behavioral Science. doi:10.1002/0470013192.bsa360. ISBN 978-0470860809.
17. Hubert, Mia; Debruyne, Michiel (2010). "Minimum covariance determinant". WIREs Computational Statistics. 2 (1): 36–43. doi:10.1002/wics.61. ISSN 1939-5108. S2CID 123086172.

18. Kessy, Agnan; Lewin, Alex; Strimmer, Korbinian (2018-10-02). "Optimal Whitening and Decorrelation". The American Statistician. 72 (4): 309–314. doi:10.1080/00031305.2016.1277159. ISSN 0003-1305. S2CID 55075085.
19. Kim, M. G. (2000). "Multivariate outliers and decompositions of Mahalanobis distance". Communications in Statistics – Theory and Methods. 29 (7): 1511–1526. doi:10.1080/03610920008832559. S2CID 218567835.
20. Krishnamurthy, Prof. V. "Srinivasa Ramanujan – His life and his genius". www.Krishnamurthy.com. (Expository address delivered on Sep.16, 1987, at Visvesvarayya Auditorium as part of the celebrations of Ramanujan Centenary by the IISC, Bangalore). Retrieved 7 September 2016.
21. Kritzman, M.; Li, Y. (2019-04-02). "Skulls, Financial Turbulence, and Risk Management". Financial Analysts Journal. 66 (5): 30–41. doi:10.2469/faj.v66.n5.3. S2CID 53478656.
22. Mahalanobis, P. C. (1922). "Anthropological observations on the Anglo-Indians of Calcutta. Part I. Analysis of male stature". Records of the Indian Museum. 23: 1–96.
23. Mahalanobis, P. C. (1927). "Analysis of race-mixture in Bengal". J. Proc. Asiatic Soc. Bengal. 23: 301–333. Archived from the original on 13 March 2014.
24. Mahalanobis, Prasanta Chandra (1927); Analysis of race mixture in Bengal, Journal and Proceedings of the Asiatic Society of Bengal, 23:301–333.
25. McLachlan, G. J. "Mahalanobis Distance". ias.ac.in.
26. McLachlan, Geoffrey (4 August 2004). Discriminant Analysis and Statistical Pattern Recognition. John Wiley & Sons. pp. 13–. ISBN 978-0-471-69115-0.
27. Menon, Nikhil (2017). "'Fancy Calculating Machine': Computers and planning in independent India". Modern Asian Studies. 52 (2): 421–457. doi:10.1017/S0026749X16000135. S2CID 148820998.
28. Mohan, Rakesh 2007 Statistical system of India – some reflections. Reserve Bank of India, Department of Statistical Analysis and Computer Services, Mumbai, 29 June 2007. PDF.
29. Mukharji, Projit Bihari (2016). "Profiling the profiloscope: Facialization of race technologies and the rise of biometric nationalism in inter-war British India". History and Technology. 31 (4): 376–396. doi:10.1080/07341512.2015.1127459. S2CID 146570565.

30. O'Connor, John J.; Robertson, Edmund F., "Prasanta Chandra Mahalanobis", MacTutor History of Mathematics Archive, University of St Andrews.
31. Prasanta Chandra Mahalanobis at the Mathematics Genealogy Project.
32. Rao, C. R. (1973). "Prasantha Chandra Mahalanobis 1893-1972". Biographical Memoirs of Fellows of the Royal Society. 19: 455–492. doi:10.1098/rsbm.1973.0017. S2CID 74582754.
33. Rao, C.R. (1972). "Prasanta Chandra Mahalanobis: 1893–1972" (PDF). Biographical Memoirs of Fellows of the Indian National Science Academy. 5: 1–24.
34. Rao, J. N. K. (2006) Interplay Between Sample Survey Theory and Practice: An Appraisal. Survey Methodology Vol. 31, No. 2, pp. 117–138. Statistics Canada, Catalogue No. 12-001 PDF Archived 31 March 2007 at the Wayback Machine
35. Royal Society citation
36. Sunil Khilnani (2016). The Idea of India. p. 93. ISBN 9789351184546. Born in 1893 of a Bengali Brahmin family, Mahalanobis took a First in mathematics and physics from King's College, Cambridge in 1915 and returned to Calcutta to teach physics.
37. The Statesman 25 December 2006 Archived 14 March 2008 at the Wayback Machine
38. Van Aelst, Stefan; Rousseeuw, Peter (2009). "Minimum volume ellipsoid". Wiley Interdisciplinary Reviews: Computational Statistics. 1 (1): 71–82. doi:10.1002/wics.19. ISSN 1939-5108. S2CID 122106661.
39. Venkataraman, G. (1995). Saha and his formula. Hyderabad: Universities Press. p. 3. ISBN 9788173710179.
40. "Mahalanobis distance", Encyclopedia of Mathematics, EMS Press, 2001 [1994]
41. "No. 35399". The London Gazette (Supplement). 30 December 1941. p. 24.
42. Mahalanobis (1983). Prasanta Chandra Mahalanobis. National Book Trust, India.
43. Ashok Rudra (1996). Prasanta Chandra Mahalanobis: A Biography. Oxford University Press. ISBN 978-0-19-563679-6.
44. David Engerman (2018). The Price of Aid: The Economic Cold War in India. Harvard University Press.
45. Mahalanobis distance (Nov-17-2006) – overview of Mahalanobis distance, including MATLAB code.

46. Mahalanobis distance tutorial – interactive online program and spreadsheet computation.
47. Mahalanobis, Prasanta Chandra (1936). "On the generalised distance in statistics" (PDF). Proceedings of the National Institute of Sciences of India. 2 (1): 49–55. Retrieved 2016-09-27.
48. Mahalanobis_distance,
https://en.wikipedia.org/wiki/Mahalanobis_distance.
49. Prasanta_Chandra_Mahalanobis,
https://en.wikipedia.org/wiki/Prasanta_Chandra_Mahalanobis.
50. What is Mahalanobis distance? – intuitive, illustrated explanation, from Rick Wicklin on blogs.sas.com.

# Chapter 6

# List of Astronomers and Mathematicians of the Kerala School

## Background

The region surrounding the south-west coast of the Indian subcontinent, now politically organised as the Kerala State in Bharat, has a long tradition of studies and investigations in all areas related to the branch of *śāstra* (शास्त्रम्)known as *jyotiṣa* (ज्योतिषः). This branch of *śāstra* (शास्त्रम्), in its broadest sense, incorporates several subdisciplines like mathematics, astronomy, astrology, horary astrology, etc. In Indian traditional *jyotiṣa scholarship,* there are no clear-cut boundary lines separating these subdisciplines. Hence the list presented below includes all who would be called a *jyotiṣa-scholar* in the Indian traditional sense. All these persons will be, most likely, well versed in the subdisciplines of mathematics and astronomy as well. The list is an adaptation of the list of mathematicians and astronomers compiled by *K. V. Sarma. Sarma* has referred to all of them as astronomers. *K. V. Sarma* (1919–2005) was an Indian historian of science, particularly the astronomy and mathematics of the Kerala school. He was responsible for bringing to light several of the achievements of the Kerala school. He was editor of the *Vishveshvaranand Indological Research Series,* and published the critical edition of several source works in Sanskrit, including the *Aryabhatiya* of *Aryabhata.* He was recognised as "*the greatest authority on Kerala's astronomical tradition*". Additional information about the persons mentioned in the list are available in books on the history of Malayalam literature and on the history of Sanskrit literature in Kerala. Table 6.1 presents the list of astronomers and mathematicians of the Kerala school.

## The List

Table 6.1 List astronomers and mathematicians of the Kerala school

| S. No. | Name | Period | Works | Known for | Remarks |
|---|---|---|---|---|---|
| 1 | Vararuci I (वररुचि प्रथमः) | 4th century CE | - | *Vararuci vākyas;* introduction of *Kaṭapayādi system* for expressing numbers; father figure in the legend of Parayi petta panthirukulam | - |

| | | | | | |
|---|---|---|---|---|---|
| 2 | Vararuci II (वररुचि द्वितीयः) | - | Kerala-dvādasa-bhāva-vākyani, Vārarucika, Jātakarahasya | - | Might be identical with Vararuci I. |
| 3 | Haridatta (हरिदत्तः) | c.650-700 | *Grahacāraṇiniban dhana, Mahā-mārganibabndhan a* | Promulgation of parahita system in 683. | - |
| 4 | Govindasvāmi n (गोविन्दस्वामिन्) | c. 800 - 850 | *Mahābhāskarīya Bhāshya, Govindakṛiti Govindapaddhati* | - | Court astronomer of King Ravi Varma Kerala; teacher of Śaṅkaranār āyaṇa |
| 5 | Śaṅkaranārāya ṇa (शङ्करणारायणः) | 825 - 900 | *Laghubhāskarīya* | References to the presence of an astronomical observatory at M ahodayapuram ( modern day Kodungallu r). | Student of Govindasv āmin |
| 6 | Udayadivākara (उदयादिवाकारः) | 11th century | A commentary called *Sundari* on *Laghubhāskarīya* | - | Commente d on a work of Jayadeva |
| 7 | Acyuta I (अच्युत प्रथमः) | - | *Devakerala* (also known as *Keralajyotiṣṣa*) | - | - |
| 8 | Keralācarya (केरलाचार्य) | 12th century | *Kerala-saṁhitā, Keralīyapraśnamā rga* | - | - |
| 9 | Vyāghrapāda (व्याघ्रपादः) | - | *Aṅkaṇasāstra* | - | A devotee of Lord Siva enshri ned at Vaikom Temple in |

| | | | | | |
|---|---|---|---|---|---|
| | | | | | Central Kerala. |
| 10 | Kṛṣṇa (कृष्णः) | c. 1200 | *Cintājñāna* | - | |
| 11 | Kṛṣṇa-śiṣya (कृष्णशिष्यः) | c. 1200 | Commentary on *Hora; Prśnaphalaprāptkālanirṇaya* | - | A disciple of Kṛṣṇa |
| 12 | Suryadeva Yajvan (सूर्यदेव यज्वान्) | 1191 - c. 1250 | | - | |
| 13 | Vidyāmādhava (विद्यामाधवः) | - | *Muhūrtadarśana* | - | A member of the Tulu Brahmin family of Nīlamana; six commentaries on *Muhūrtadarśana* have been identified. |
| 14 | Viṣṇu (of Nīlamana) (विष्णुः (नीलामनस्य)) | - | A commentary on *Muhūrtadarśana* | - | Son of Vidyāmādhava |
| 15 | Govinda Bhaṭṭatiri (of Talakkuḷam) (गोविन्द भट्टतिरी) | 1237 - 1295 | *Dasādhyāyi, Muhūrtaratna, Muhūrtapadavi* | Progenitor of the Pazhūr Kaniyār family of astrologers | Belongs to Alathiyur (Malappuram) village. |
| 16 | Tāmaranallūr (तामरनाल्लूर) | 14th century | *Muhūrtavidhi (Tāmaranallūr Bhāāṣā)* | - | The work uses a mixture of Malayalam and Sanskrit languages. |

| | | | | | |
|---|---|---|---|---|---|
| 17 | Nityaprakāśa Yati (नित्यप्रकाश यति) | 14th - 15th century | Commentaries on Hora, both in Malayalam and Sanskrit | - | - |
| 18 | Kumāra Gaṇaka (कुमार गणक) | 14th - 15th century | *Raṇadīpikā* | - | - |
| 19 | Rudra I (रुद्र प्रथमः) | c. 1325 - 1400 | - | - | Teacher of *Parameśvara* |
| 20 | Mādhava of Saṅgamagrāma (संगमग्रामस्य माधवः) | c. 1340 - 1425 | *Golavada, Madhya manayanaprakara, Venvaroha, Chan dravakyani* | Discoverer of the infinite series expansions for sin x and cos x | Greatest mathematic ian-astronomer of the Kerala School |
| 21 | Parameśvara of Vaṭaśśeri (Para meśvara I) (वाट्श्रेरिः परमेश्वरः (परमेश्वरः प्रथमः)) | c. 1360 - 1455 | *Drigganita* (1430) , *Bhatadipika* – Commentary on Āryabhaṭīya of Āryabhaṭa I, *Karmadipika* – Commentary on *Mahabhaskariya* of Bhaskara I, *Paramesvari* – Commentary on *Laghubhaskariya* of Bhaskara I, *Sidhantadipika* – Commentary on *Mahabhaskariyabhashya* of Govi ndasvāmi, *Vivarana* – Commentary on Surya Siddhanta and L ilāvati, *Goladipika* – Spherical geometry and astronomy (composed in | Promulgator of the Drigganita system. | A proponent of observation al astronomy in medieval Bharat who had made a series of eclipse observation s to verify the accuracy of the computatio nal methods then in use |

| | | | | | |
|---|---|---|---|---|---|
| | | | 1443 CE), *Grahanamandana* – Computation of eclipses (Its epoch is 15 July 1411 CE.), *Grahanavyakhyadipika* – On the rationale of the theory of eclipses, *Vakyakarana* – Methods for the derivation of several astronomical tables | | |
| 22 | Dāmodara of Vaṭaśreṇi (Dāmodara I) (वाट्श्रेण्यस्य दामोदरः (दामोदरः प्रथमः)) | c. 1410 - 1510 | - | Teacher of Nilakantha Somayaji | Son of Parameśvara of Vaṭaśśeri |
| 23 | Ravi Nampūtiri (रवि नम्पूतिरि) | c. 1425 - 1500 | *Ācāradīpika'* (a commentary on *Muhūrtadīpika*) | - | - |
| 24 | Nīlakaṇṭha Sōmayāji (नीलकंठ्ह सोमायाजी) | 1444 - 1545 | *Tantrasamgraha*, *Golasara* (Description of basic astronomical elements and procedures), *Sidhhantadarpana*, *Candrachayaganita*, *Aryabhatiya-bhashya* (Elaborate commentary on Aryabhatiya), *Sidhhantadarpana-vyakhya*, *Chandra* | *Tantrasamgraha* ( A comprehensive treatise on astronomy) | - |

| | | | | | |
|---|---|---|---|---|---|
| | | | *chhayaganita-vyakhya, Sundaraja-prasnottara, Grahanadi-grantha, Grahapariksakrama* | | |
| 25 | Śankara of Keḷallūr (केआल्लूरस्य शङ्करः) | c. 1475 - 1575 | - | - | Nīlakaṇṭha Sōmayāji's younger brother. The person entrusted with task of popularizing *Aryabhatiya-bhashya* of Nīlakaṇṭha Sōmayāji. |
| 26 | Citrabhānu (चित्रभानुः) | c. 1475 - 1550 | *Karaṇāmṛta* (a manual on astronomical computations) | - | Pupil of Nīlakaṇṭha Sōmayāji |
| 27 | Citrabhānu Śiṣya (चित्रभानु शिष्या) | c. 1500 - 1575 | *Bhācintāvali* (a work on astrology) | - | - |
| 28 | Nārāyaṇa I (नारायण प्रथमः) | c. 1500 - 1575 | *Laghuvivṛti* on *Pañcabodha IV, Uparāgakriyākrama* (eclipse computations), two commentaries on *Līlavati*, one short and the other five times longer than the short, both called *Kriyākramakarī* and *Karmadīpikā* | - | - |

| | | | | | |
|---|---|---|---|---|---|
| 29 | Śankara Vāriyar (शंकर वरियार्) | c. 1500 - 1560 | *Laghuvivṛti* (commentary on *Tantrasamgraha*) | - | Disciple of Nīlakaṇṭha Sōmayāji and protege of Āzhvāñceri Tamprakkaḷ |
| 30 | Jyeṣṭhadeva (ज्येष्ठदेवः) | c. 1500 - 1610 | *Yuktibhāṣā* (in Malayalam), *Gaṣitayuktibhaṣa* (in Sanskrit), *Dṛkkaraṇa* (in Malayalam) | *Yuktibhāṣā* | The name is probably the Sanskritised form of his personal name in the local Malayalam language. Pupil of Teacher of Dāmodara of Vaṭaśreṇi and teacher of Acyuta Piṣārați. |
| 31 | Jyeṣṭhadeva-Śiṣya (ज्येष्ठदेव-शिष्यः) | c. 1550 - 1625 | A metrical commentary on *Tantrasamgraha* | | Disciple of Jyeṣṭhadeva |
| 32 | Māttūr Nampūtiri-s: Puruṣottama I and Subrahmaṇya I (माट्टूर् नन्पूतिरी-सः पुरुषोत्तम प्रथमः तथा सुब्रह्मण्य प्रथमः) | c. 1475 - 1550 | *Muhūrtapadvī* (a condensed work in about 40 verses dealing with the prescriptions of auspicious times for functions) | The book is highly popular as attested by the availability of several commentaries on it. | Lived in Pāññāḷ village in Thrissur district |
| 33 | Nārāyaṇa of Kaṇvavastu: Nārāyaṇa II | 15th century | *Muhūrtadīpika* (A comprehensive treatise in about | - | - |

| | | | | | |
|---|---|---|---|---|---|
| | (कणववस्तु नारायण: नारायण द्वितीय:) | | 400 verses on the auspicious times for functions) | | |
| 34 | Rudra Vāariyar: Rudra II (रुद्र वारियरः रुद्र द्वितीय:) | c. 1475 - 1550 | Commentary on Varāhamihira's *Hora,* called *Nauka* or *Vivarana,* and *Aṣṭamaṅgala praśna* | - | Belongs to Deśamaṅgalaṃ Vāriyaṃ |
| 35 | Śaṅkara of Mahiṣamaṅgalam : Śaṅkara III (महिषमङ्गलस्य शङ्कर : शङ्कर तृतीय:) | 1494 - 1570 | *Gaṅtasāra, Candragaṅitakrama*and *Ayanacalanādi-gaṅita* (all books on astronomy); *Jatakakrama, Jātakasāra* and *Praśnamāla* ( books on astrology). All these are composed in simple Malayalam poetry or prose. His works on Sanskrit include *Jātakasāra*, commentaries on *Pañcabodha* and *Laghubhāskarīya.* | Popularization of astronomical works among the masses in Kerala. | Hails from Perumanam village near Thrissur. Spent most of his life with his teacher Parameśvaran Poṭṭi of Vāzha-māveli house in Chengannur. |
| 36 | Madhava of Iñcakkāzhvā: Mādhava II (इञ्चक्काइवा का माधव : माधव द्वितीय:) | c. 1500 - 1575 | *Praśnasāra* (incorporates several local practices) | - | Hails from Ramamangalam near Muvvattupuzha |
| 37 | Acyuta Piṣārati : Acyuta II | c. 1550 - 1621 | *Spuṭanirṇaya, Raśigoḷaspuṭānīti, Karaṇottama* (astronomical | Enunciation, for the first time in Bhartiya astronomy, of | Teacher of the poet and grammaria |

| | | | | | |
|---|---|---|---|---|---|
| | (अच्युत पिषारति : अच्युत द्वितीयः) | | computation), *Uparagakriyākrama* (eclipse computations), *Chāyāṣṭaka* (shadow computations); commentaries on *Veṅvāroha* and *Sūryasidhanta* | the correction called "Reduction to the ecliptic". This was introduced in Western astronomy by Tycho Brahe at about the same time. | n Melpattūr Narāyaṇa Bhaṭṭa |
| 38 | Nīlakaṇṭha II (नीलकण्ठ द्वितीयः) | 16th - 17th centuries | *Kanakkusāraṃ* couched in *maṇipravāḷaṃ* style which is a mixture of Malayalam and Sanskrit languages. The work deals also with practices relating to grain transactions, house building, weighing og gold and silver, land tenure, masonry, ground measurement, etc. | - | - |
| 39 | Nārāyaṇa III (नारायण तृतीयः) | date completely unknown | *Laghudarśini* (a short work on astrology) | - | - |
| 40 | Dāmodara of Maṅgalaśreṇi (Dāmodara II) (मङ्गलश्रेण्यस्य दामोदरः (दामोदर द्वितीयः)) | c. 1575 - 1675 | *Praśnarīti, Līlāvati-vyākhyā* | - | Hails from Kaṇṇāṭippa rampu in Chirakkal taluk in North Malabar |

| | | | | | |
|---|---|---|---|---|---|
| 41 | Iṭakramañceri Nampūtiri (इट्क्रमञ्चेरि नम्पूतिरि) | c. 1625 - 1700 | *Bhadradīpa-gaṇita* (compose d in *maṇipravāḷaṃ* style) | - | Pupil of Dāmodara of Maṅgalaśre ṇi |
| 42 | Maṅgalaśreṇivi pra-Śiṣya (मङ्गलश्रेणिविप्र-शिष्यः) | 17th century | *Jyotiṣasaṃgraha* (i n Malayalam) | - | - |
| 43 | Pāṇkkāṭṭu (or, Iṭakkāṭṭu) Namputiri (पङ्क्कट्टु ( नम्पुतिरि)) | c. 1625 - 1725 | *Praśnamārga* | - | *Praśnamārg a* is the most popular and authoritati ve work on *praśna* in Kerala. |
| 44 | Iṭakkāṭṭu (or Eṭakkāṭṭu) Kukkuṇiyāḷ (इट्ककट्टु कुक्कुनियः)) | c. 1675 0 1750 | *Praśnarīti* | - | Pupil of Pāṇkkāṭṭu Namputiri |
| 45 | Rāma-śiṣya (रामशिष्यः) | 17th century | An explanatory rendering in Malayalam verses of the *Laghuhorā* of Varāahamihira | - | - |
| 46 | Puruṣottama II (पुरुषोत्तम द्वितीयः) | c. 1650 - 1725 | *Uparāgaparicched a* (computation of solar and lunar eclipses) | - | - |
| 47 | Putumana Somayāji (पुतुमना सोमयाजी) | c. 1660 - 1740 | *Karaṇapaddhati* (c omprehensive text on astronomical computations), *Nyāyaratna, Veṇv ārohāṣṭaka, Mānas agaṇitam, Jātakāde ś* | Authorship of *Karaṇapaddhati* | A member of the Putuvana family of Śivapuram (Thissur). *K araṇapaddha ti* has been commented |

| | | | | | |
|---|---|---|---|---|---|
| | | | | | in Sanskrit, Malayalam and Tamil. It manuscripts are available in Tamil and Telugu scripts also. |
| 48 | Vāsudevasvami : Vāsudeva I (वासुदेवस्वामी : वासुदेव प्रथमः) | - | *Kāaladīpa* (treatise on natural astrology) | - | The manuscrpt of the book was procured from Punnattūrkoṭṭa Mana, Koṭṭappaṭi, North Malabar |
| 49 | Śyāmaḷavaāraṇa Rāja (श्यामाअवरणं राज०) | - | Commentary on *Kāaladīpa* of Vāsudevasvami | - | Member of Punnattūrkoṭṭa Mana |
| 50 | Dāmodara of Bharadvāja Gotra: Dāmodara III (भरद्वाजगोत्रस्य दामोदर : दामोदर तृतीयः)) | - | *Muhūrtābharaṇa* ( treatise on auspicious times for functions) | - | Hails from Tṛpparaṛṛoṭ village in Malabar |
| 51 | Kṛṣṇa II (कृष्णः द्वितीयः) | - | Commentary on *Āryabhaṭīya* in Malayalam | - | - |
| 52 | Keralīya-dvija (केरली-द्विजः) | - | Malayalam commentary on *Karaṇottama* of Acyuta Piṣāraṭi | - | - |
| 53 | Govinda-śiṣya (गोविन्द-शिष्यः) | - | Commentary called *Bālaprabodhini* on *Jatakapad* | - | Hails from Vaikkam |

| | | | | | |
|---|---|---|---|---|---|
| | | | *dhati* of Parameśvara of Vaṭaśśeṇi | | |
| 54 | Veṇād Brāhmaṇa (वेणाद ब्राह्मणः) | - | *Jātakodaya* (work on astrology in 103 verses) | - | - |
| 55 | Azhvāñceri Taṃprākkaḷ (अझ्वान्सेरि तअंप्राक्कः) | c. 1725 - 1800 | Eleven books in the form of adapted texts and commentaries covering the entire field of astronomy and astrology. *Jyotiśśāstra-saṃgraha, Saṃgraha-sādhana-kṛiyā, Jātaka-sāara-saṃgraha, Jātakānīti-mārga, Phalasāra-samuccaya,* etc. | Efforts to propagate interest in studies on Jyotiṣa among members of the Nampūtiri community | Personal name unknown |
| 56 | Vāsudeva of Vaḷḷimana: Vāsudeva II (वैमनस्य वासुदेवः वासुदेव द्वितीयः) | | A metrical commentary on *Muhūrtapadavi* | - | Hails from Kaṇṇamaṅgalaṃ |
| 57 | Tuppan Nampūtiri of Iṭavaṭṭkkāt (इटवट्टक्कतस्य तुप्पं नम्पूतिरी) | c. 1725 - 1800 | *Muhūrtapadavi* VI | - | Hails from Pāakoḍe in Kunnathunad |
| 58 | Nārāyaṇa of Iṭavaṭṭkkāt : Nārāyaṇa IV (नारायण इटावट्टकत : नारायण चतुर्थ) | c. 1728 - 1800 | *Muhūrtapadavi* VII | - | Younger brother of Tuppan Nampūtiri of Iṭavaṭṭkkāt |

| | | | | | |
|---|---|---|---|---|---|
| 59 | Parameśvara II (परमेश्वर द्वितीयः) | - | Commentaries called *Pārameśvarī* on Varāahamihira's *Horā* | - | - |
| 60 | Parameśvara, pupil of Śankara : Parameśvara III (शङ्करस्य शिष्यः परमेश्वरः : परमेश्वरः तृतीयः) | - | Commentaries called *Jātakacandrikā* on Varāahamihira's *Horā* | - | - |
| 61 | Bhaāradvāja-dvija (भरद्वाजद्विजः) | c, 1750 - 1800 | *Gaṇitayuktayah* (rationales of mathematical and astronomical procedures), *Karaṇadarpaṇa* (advanced manual on astronomical computations) | - | - |
| 62 | Nārāyaṇa of Perumanaṃ (Nārāyaṇa V) (पेरुमाननस्य नारायणः (नारायण पंचमः)) | - | *Tantrasāra* | - | - |
| 63 | Kṛṣṇadāsa (Koccukṛṣṇan Āāśan) (कृष्णदास (कोच्चुकृष्णान् आशान्)) | 1756 - 1812 | *Pañcabodha* VIII (Malayalam verse), *Bhāṣājātakapaddhati* (Malayalam commentary on *Jātakapaddhati* of Parameśvara of Vaṭaśśeṇi, *Kaṇkkuśāstram* (mathematical procedures in | - | Born in the family of Neṭumpayil in Tiruvalla taluk. |

| | | | | | |
|---|---|---|---|---|---|
| | | | Malayalam verse) | | |
| 64 | Śankara of Muktisthala: Śankara IV (मुक्तिस्थलस्य शंकर : शंकर चतुर्थः) | 17th century | *Mantrasāra, Sāmudrasāra, Ārūḍha-praśna, Lāñchana-śāstra* (omens, palmistrym, astrological omens, etc.), *Āyurpraśna* (dealing with life longevity, *Aṣṭamaṅgala*, etc.) | - | Hails from Mukkola |
| 65 | Śaṅkara V (शङ्कर पंचमः) | - | *Jātakasāra* (III), *Praśnasāra* (I) (both in Malayalam verse) | - | - |
| 66 | Bhutanāthapura-Somayāji (भूतनाथपुर-सोमायाजी) | - | *Praśnasāra* (III) | - | - |
| 67 | Śrīkumāra, son of Nīlakaṇṭha (नीलकण्ठसुतः श्रीकुमारः) | - | *Praśnāmṛta* | - | - |
| 68 | Nārāyaṇan Iḷayatu of Maccāṭ: Nārāyaṇa VI (नारायण इ़यातु मक्काट : नारायण षष्ठः) | 1765 - 1843 | *Jyotiṣa-Bhāṣāvalī, Jātakādeśaratna* | - | Patronised by Śaktan Tampurān |
| 69 | Parameśvara of Puradahanapura : Parameśvara IV (पुरादहनपुरस्य परमेश्वरः : परमेश्वरः चतुर्थः) | c. 1775 - 1839 | Commentary called *Varadīpikā* on *Muhūrtapadavi* II of Māttūr Puruṣottama Naṃpūtiri | - | Member of Puradahanapura (Purayannur) family in Vaḷḷuvanāṭ |

| | | | | | |
|---|---|---|---|---|---|
| 70 | Śrīkaṇṭha Vāriyar of Veḷḷrakkaād (वेढृक्कादस्य श्रीकण्ठवरियर्) | | Commentary in Malayalam on *Jātakapaddhati* of Parameśara of Vaṭaśreṇi | - | The commentary is available in print. |
| 71 | Ghaṭigopa (घटिगोपः) | c. 1800 - 1860 | Two commentaries on *Āryabhaṭīya*, one in Malayalam and the other in Sanskrit | - | A devotee of God Padmanabha, the presiding deity of Thiruvananthapuram |
| 72 | Goda Varma, Vidvān Iḷaya Tampurān (गोदा वर्मा, विद्वान इाय ताम्पुरान्) | 1800 - 1851 | Commentaries in Sanskrit on *Gaṇitādhyāya* of *Bhāskarīya-gaṇita* and the *Goladhyāya* of *Siddhāntaśiromaṇi* | - | Member of the royal family of Kodungallur |
| 73 | Śaṅkara Varma of Kaṭattanāṭ (शङ्कर वर्मा कटट्टनाट्) | 1800 - 1838 | *Sadratnamālā* | The computation of the value of the mathematical constant c orrect to 17 decimal places | - |
| 74 | Subrahmaṇya Śāstri (सुब्रह्मण्य शास्त्री) | 1829 - 1888 | *Agaṇita* (procedures for computing the positions of planets for a thousand years | - | Hails from Nalleppalli in Chittur-Cochin |
| 75 | Subrahmaṇya of Kunnattu Mana: Subrahmaṇya II (कुन्नट्टू मन का सुब्रह्मण्य : सुब्रह्मण्य द्वितीयः) | c. 1850 - 1900 | - | Commentary called *Bhāvaprakāśika* on the *Muhūrtadarśana* of Vidyāmādhava | Also known as Subrahmaṇyan Tirumunpu of Kunnattu |

| | | | | | |
|---|---|---|---|---|---|
| | | | | | Mana of Payyanur in North Malabar |
| 76 | Puruṣottaman Mūssatu: Puruṣottama III (पुरुषोत्तमन् मूस्सतु : पुरुषोत्तम तृतीयः) | c. 1850 - 1900 | *Praśnāyana* (a comprehensive work in 1018 verses) | - | - |
| 77 | Rāma Varma Koyittampurān (राम वर्मा कोयितम्पुराण) | 1853 - 1910 | *Jyotiṣapradīpa* | - | - |
| 78 | Rāma Vāriyar of Kaikkuḷaṅra (कैक्कुढरस्य राम वरियरः) | 1833 - 1897 | *Sāmudrikaśāstra, Gauḷiśāstra,* commentaries on *Horā* and *Praś namārga* | - | - |
| 79 | A. R. Rajaraja Varma Koyittampurān (अ० र० राजराज वर्मा कोयत्तमपुराण) | 1853 - 1918 | *Karaṇapariṣkaraṇ a, Pañcāṅgaśuddhi paddhati, Jyotiṣap rakāśna* | - | - |
| 80 | Vāsuṇṇi Mūsstu of Veḷḷānaśśeri (वासुणि मूस्तू वेढानाशरी) | 1855 - 1914 | Commentary in Malayalam on *Pañchabodha* | - | - |
| 81 | Punnaśśeri Nampi Nīlakaṇṭha Śarma (पुन्नाशेरी नम्पी नीलकण्ठ शर्मा) | 1858 - 1935 | *Jyotiśāstrasubodhi ni, Pañcabodhakri yā-Bhāṣa* | - | Works published by Bharata Vilasam Press, Thrissur |

## Further Reading


1. Divakaran, P. P. (8 December 2007), "The First Textbook of Calculus: Yuktibhāṣā", Journal of Indian Philosophy, 35 (5): 417–443, doi:10.1007/s10781-007-9029-1, S2CID 170254981.
2. K. Chandra Hari (25 Oct 2007) Critical Evidence to Fix the Native Place of Āryabhaṭa-I, Current Science 93(8).
3. K. V. Sarma (1972). A History of the Kerala School of Hindu Astronomy. Vishveshvaranand Institute of Sanskrit and Indological Studies, Panjab University, Hoshiarpur. Retrieved 28 January 2023.
4. Mathematician of Kerala School. https://en.wikipedia.org/wiki/List_of_astronomers_and_mathematicians_of_the_Kerala_school.
5. Ulloor S. Parameswara Iyer (1953–1957). Kerala Sahitya Charitram (in five volumes). Thiruvananthapuram: University of Kerala. Retrieved 25 March 2023.
6. Vatakkumkur Raja Raja Varma. Keraleeya Samskrita sahitya Charitram (in six volumes). Kamalalaya Book Depot, Thiruvananthapuram (volume 1) and Magalodayam Press, Thrissur (volumes 2 - 6). (Volume 1 in 1938, volumes 2-6 during 1947 - 1965)

# Chapter 7

# The Hindu Numerals

## 卐 The Hindu Numeral

The *Hindu numeral* (or *Hindu–Arabic numeral system* or *Indo-Arabic numeral system*) is a positional base ten numeral system for representing integers, which can be extended to include non-integers, i.e., the Decimal numeral system, which is the most common system for the symbolic representation of numbers in the world. It was invented between the 1st and 4th centuries by Indian mathematicians. The system was adopted in Arabic mathematics by the 9th century. It became more widely known through the writings of the Persian mathematician *Al-Khwārizmī* (*On the Calculation with Hindu Numerals,* c. 825) and Arab mathematician Al-Kindi (*On the Use of the Hindu Numerals,* c. 830). The system had spread to medieval Europe by the High Middle Ages. The system is based upon *ten (originally nine) glyphs*. The symbols *(glyphs)* used to represent the system are in principle independent of the system itself. The glyphs in actual use are descended from Brahmi numerals and have split into various typographical variants since the Middle Ages.

These symbol sets can be divided into three main families:

- Western Arabic numerals used in the Greater Maghreb and in Europe
- Eastern Arabic numerals used in the Middle East
- Bhartiya numerals in various scripts used in the Bhartiya subcontinent.

The *Hindu–Arabic* or *Indo–Arabic* numerals were invented by mathematicians in Bharat. Persian and Arabic mathematicians called them "*Hindu numerals*". The *Hindu–Arabic* system is designed for positional notation in a decimal system. In a more developed form, positional notation also uses a decimal marker *(at first a mark over the ones digit but now more commonly a decimal point or a decimal comma which separates the ones place from the tenths place*), and also a symbol for "*these digits recur ad infinitum*". In modern usage, this latter symbol is usually a vinculum (*a horizontal line placed over the repeating digits*). In this more developed form, the numeral system can symbolize any rational number using only 13 symbols (*the ten digits, decimal marker, vinculum, and a prepended minus sign to indicate a negative number*). Table 7.1 shows various g lyphs.

## Symbols

Various symbol sets are used to represent numbers in the *Hindu–Arabic numeral system,* most of which developed from the *Brahmi numerals.* The symbols used to represent the system have split into various typographical variants since the Middle Ages, arranged in three main groups:

- The Bhartiya numerals in use with scripts of the Brahmic family in Bharat and Southeast Asia. Each of the roughly dozen major scripts of Bharat has its own numeral glyphs (*as one will note when perusing Unicode character charts*).
- The "*Arabic–Indic*" or "*Eastern Arabic numerals*" used with Arabic script, developed primarily in what is now Iraq. A variant of the Eastern Arabic numerals is used in Persian and Urdu.
- The widespread Western "*Arabic numerals*" used with the Latin, Cyrillic, and Greek alphabets in the table, descended from the "*West Arabic numerals*" which were developed in al-Andalus and the Maghreb (*there are two typographic styles for rendering western Arabic numerals, known as lining figures and text figures*).

Table 7.1 Glyph comparison

| Symbol | | | | | | | | | | Used with scripts | Numerals |
|---|---|---|---|---|---|---|---|---|---|---|---|
| 0 | 1 | 2 | 3 | 4 | 5 | 6 | 7 | 8 | 9 | Arabic, Latin, Cyrillic, and Greek | Arabic numerals |
| ٠ | ١ | ٢ | ٣ | ٤ | ٥ | ٦ | ٧ | ٨ | ٩ | Arabic | |
| ۰ | ۱ | ۲ | ۳ | ۴ | ۵ | ۶ | ۷ | ۸ | ۹ | Persian / Dari / Pashto | Eastern Arabic numerals |
| ۰ | ۱ | ۲ | ۳ | ۴ | ۵ | ۶ | ۷ | ۸ | ۹ | Urdu / Shahmukhi | |
| ⠚ | ⠁ | ⠃ | ⠉ | ⠙ | ⠑ | ⠋ | ⠛ | ⠓ | ⠊ | Braille | Braille numerals |
| 𑁦 | 𑁧 | 𑁨 | 𑁩 | 𑁪 | 𑁫 | 𑁬 | 𑁭 | 𑁮 | 𑁯 | Brahmi | Brahmi numerals |

| | | | | | | | | | | | |
|---|---|---|---|---|---|---|---|---|---|---|---|
| ० | १ | २ | ३ | ४ | ५ | ६ | ७ | ८ | ९ | Devanagari | Devanagari numerals |
| ௦ | ௧ | ௨ | ௩ | ௪ | ௫ | ௬ | ௭ | ௮ | ௯ | Tamil | Tamil numerals |
| ০ | ১ | ২ | ৩ | ৪ | ৫ | ৬ | ৭ | ৮ | ৯ | Eastern Nagari | Bengali numerals |
| ੦ | ੧ | ੨ | ੩ | ੪ | ੫ | ੬ | ੭ | ੮ | ੯ | Gurmukhi | Gurmukhi numerals |
| ૦ | ૧ | ૨ | ૩ | ૪ | ૫ | ૬ | ૭ | ૮ | ૯ | Gujarati | Gujarati numerals |
| ୦ | ୧ | ୨ | ୩ | ୪ | ୫ | ୬ | ୭ | ୮ | ୯ | Odia | Odia numerals |
| ᱐ | ᱑ | ᱒ | ᱓ | ᱔ | ᱕ | ᱖ | ᱗ | ᱘ | ᱙ | Santali | Santali numerals |
| ౦ | ౧ | ౨ | ౩ | ౪ | ౫ | ౬ | ౭ | ౮ | ౯ | Telugu | Telugu script Numerals |
| ೦ | ೧ | ೨ | ೩ | ೪ | ೫ | ೬ | ೭ | ೮ | ೯ | Kannada | Kannada script Numerals |
| ൦ | ൧ | ൨ | ൩ | ൪ | ൫ | ൬ | ൭ | ൮ | ൯ | Malayalam | Malayalam numerals |
| ꯰ | ꯱ | ꯲ | ꯳ | ꯴ | ꯵ | ꯶ | ꯷ | ꯸ | ꯹ | Meitei script | Meitei script Numerals |
| ෦ | ෧ | ෨ | ෩ | ෪ | ෫ | ෬ | ෭ | ෮ | ෯ | Sinhala | Sinhala numerals |
| ༠ | ༡ | ༢ | ༣ | ༤ | ༥ | ༦ | ༧ | ༨ | ༩ | Tibetan | Tibetan numerals |
| ၀ | ၁ | ၂ | ၃ | ၄ | ၅ | ၆ | ၇ | ၈ | ၉ | Burmese | Burmese numerals |
| ᠐ | ᠑ | ᠒ | ᠓ | ᠔ | ᠕ | ᠖ | ᠗ | ᠘ | ᠙ | Mongolian | Mongolian numerals |
| ០ | ១ | ២ | ៣ | ៤ | ៥ | ៦ | ៧ | ៨ | ៩ | Khmer | Khmer numerals |
| ๐ | ๑ | ๒ | ๓ | ๔ | ๕ | ๖ | ๗ | ๘ | ๙ | Thai | Thai numerals |
| ໐ | ໑ | ໒ | ໓ | ໔ | ໕ | ໖ | ໗ | ໘ | ໙ | Lao | Lao script Numerals |
| ꧐ | ꧑ | ꧒ | ꧓ | ꧔ | ꧕ | ꧖ | ꧗ | ꧘ | ꧙ | Javanese | Javanese numerals |

## Origin

The *Brahmi numerals* at the basis of the system predate the Common Era. They replaced the earlier *Kharosthi numerals* used since the 4th century BC. *Brahmi* and *Kharosthi numerals* were used alongside one another in the *Maurya Empire period* (मौर्य साम्राज्य काल), both appearing on the 3rd century BC edicts of *Ashoka* (अशोकः). Buddhist inscriptions from around 300 BC use the symbols that became *1, 4, and 6.* One century later, their use of the symbols that became *2, 4, 6, 7, and 9* was recorded. These *Brahmi numerals* are the ancestors of the *Hindu–Arabic glyphs 1 to 9*, but they were not used as a positional system with a zero, and there were rather separate numerals for each of the tens *(10, 20, 30, etc.).* The actual numeral system, including positional notation and use of *zero*, is in principle independent of the *glyphs used*, and significantly younger than the *Brahmi numerals* (see Figure 7.1.). Some Devanagari numerals are shown in Figure 7.2.

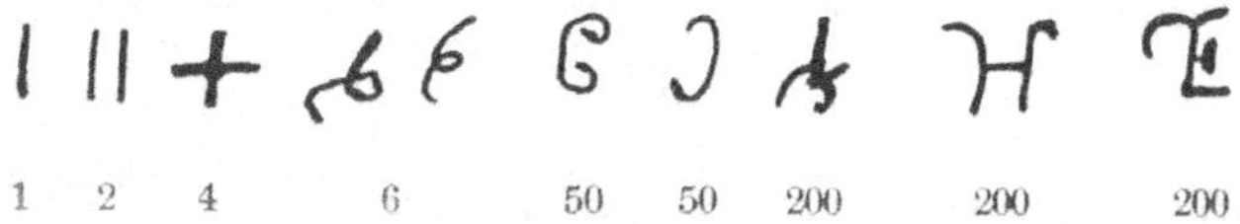


Figure 7.1. The first Brahmi numerals, ancestors of Hindu-Arabic numerals, used by Ashoka in his Edicts of Ashoka c. 250 BC

Figure 7.2. Nagari and Devanagari numerals with handwritten variants

## Development

During the *Vedic period* (वैदिक काल), motivated by geometric construction of the *fire altars* (अग्निवेदिकाः) and *astronomy* (खगोलशास्त्रम्), the use of a numerical system and of basic mathematical operations developed in northern Bharat. *Hindu cosmology* (हिन्दू ब्रह्माण्ड विज्ञान) required the mastery of very large numbers such as the *kalpa* - कल्पः (*the lifetime of the universe*) said to be *4,320,000,000 years* and the "*orbit of the heaven*" said to be *18,712,069,200,000,000 yojanas*.

Numbers were expressed using a "*named place-value notation*", using names for the powers of 10, like *dasa, shatha, sahasra, ayuta, niyuta, prayuta, arbuda, nyarbuda, samudra, madhya, anta, parardha* etc., the last of these being the name for a trillion ($10^{12}$).

For example, the number 26,432 was expressed as "2 *ayuta*, 6 *sahasra*, 4 *shatha*, 3 *dasa*, 2."

In the *Buddhist* text *Lalitavistara* (ललिताविस्तार), the *Buddha* (बुद्धः) is said to have narrated a scheme of numbers up to $10^{53}$. The first Brahmi numerals, ancestors of *Hindu-Arabic numerals*, used by *Ashoka* in his *Edicts of Ashoka* (अशोकस्य आज्ञा) c. 250 BCE. The form of numerals in *Ashoka's inscriptions* (Ashoka's inscriptions) in the *Brahmi script* (*middle of the third century BCE*) involved separate signs for the numbers *1 to 9, 10 to 90, 100 and 1000*. A multiple of 100 or 1000 was represented by a modification (or "*enciphering*") of the sign for the number using the sign for the multiplier number. Such *enciphered numerals* directly represented the named *place-value numerals* used *verbally*. They continued to be used in inscriptions until the end of the 9th century.

In his seminal text of 499 CE, *Aryabhata* devised a *novel positional number system*, using Sanskrit *consonants for small numbers* and *vowels for powers of 10*. Using the *system, numbers up to a billion* could be expressed using *short phrases, e. g., khyu-ghṛ* representing the number *4,320,000*. The system did not catch on because it produced *quite unpronounceable phrases,* but it might have driven home the principle of *positional number system* (called *dasa-gunottara* - दसा-गुणोत्तर, *exponents of 10*) to later mathematicians. A more elegant *katapayadi* (कटपयादी) scheme was devised in later centuries representing a *place-value system including zero*.

## Place-Value Numerals

The place-value system is used in the *Bakhshali* manuscript; the earliest leaves being radiocarbon dated to the period 224–383 AD (see Figure 7.3, Figure 7.4 and Figure 7.5). The development of the positional decimal system takes its origins in Bhartiya mathematics during the Gupta period. Around 500, the astronomer *Aryabhata* uses the word *kha* - ख ("*emptiness*") to mark "*zero*" in tabular arrangements of digits. The 7th century *Brahmasphuta Siddhanta* contains a comparatively advanced understanding of the mathematical role of zero. The Sanskrit translation of the lost 5th century *Prakrit Jaina* cosmological text *Lokavibhaga* (लोकविभागः) may preserve an early instance of positional use of zero.

These Bhartiya developments were taken up in Islamic mathematics in the 8th century, as recorded in *al-Qifti*'s *Chronology of the scholars* (early 13th century). The numeral system came to be known to both the Persian mathematician *Khwarizmi*, who wrote a book, *On the Calculation with Hindu Numerals* in about 825, and the Arab mathematician *Al-Kindi*, who wrote a book, *On the Use of the Hindu Numerals* (كتاب في استعمال العداد الهندي [*kitāb fī isti'māl al-'adād al-hindī*) around 830. Persian scientist *Kushyar Gilani* who wrote *Kitab fi usul hisab al-hind* (*Principles of Hindu Reckoning*) is one of the oldest surviving manuscripts using the Hindu numerals. These books are principally responsible for the diffusion of the Hindu system of numeration throughout the Islamic world and ultimately also to Europe. The first dated and undisputed inscription showing the use of a symbol for zero appears on a stone inscription found at the *Chaturbhuja Temple* (चतुर्भुज मन्दिर) at Gwalior in Bharat, dated 876. In 10th century Islamic mathematics, the system was extended to include fractions, as recorded in a treatise by *Abbasid Caliphate* mathematician *Abu'l-Hasan al-Uqlidisi* in 952–953.

## Place-Value Numerals without Zero

While the numerals in texts and inscriptions used a named place-value notation, a more efficient notation might have been employed in calculations, possibly from the 1st century CE. Computations were carried out on *clay tablets* covered with a *thin layer of sand*, giving rise to the term *dhuli-karana* - धुलि-करणम् (*'sand-work'*) for higher computation. *Karl Menninger* believes that, in such computations, they must have dispensed with the *enciphered numerals* and written down just *sequences of digits to represent the numbers*. A zero would have been represented as a "*missing place*", such as a *dot*. The single manuscript with worked examples available to us, the *Bakhshali* manuscript, uses a place value system with a dot to denote the *zero*. The dot was called the *shunya-sthāna* - शुन्य-स्थानम् *'empty-place'*.

The *Bakhshali* manuscript is a compendium of rules and illustrative examples. Each example is stated as a problem, the solution is described, and it is verified that the problem has been solved. The sample problems are in verse and the commentary is in prose associated with calculations. The problems involve arithmetic, algebra and geometry, including mensuration. The topics covered include fractions, square roots, arithmetic and geometric progressions, solutions of simple equations, simultaneous linear equations, quadratic equations and indeterminate equations of the second degree.

The manuscript is written in an earlier form of *Sharada script* (शारदा लिपि), a script which is known for having been in use mainly from the 8th to the 12th century in the northwestern part of South Asia, such as Kashmir and neighbouring regions. The language of the manuscript, though intended to be Sanskrit, was significantly influenced in its phonetics and morphology by a local artist dialect or dialects, and some of the resultant linguistic peculiarities of the text are shared with *Buddhist Hybrid Sanskrit*. The overlying dialects, though sharing affinities with *Apabhraṃśa* (अपभ्रंशः) and with Old Kashmiri, have not been identified precisely. It is probable that most of the rules and examples had been originally composed in Sanskrit, while one of the sections was written entirely in a dialect. It is possible that the manuscript might be a compilation of fragments from different works composed in a number of language varieties.

A colophon to one of the sections states that it was written by a brahmin identified as "*the son of Chajaka*", a "*king of calculators,*" for the use of *Vasiṣṭha's son Hasika*. The brahmin might have been the author of the commentary as well as the scribe of the manuscript. Near the colophon appears a broken word *rtikāvati* (र्तिकावती), which has been interpreted as the place *Mārtikāvata* mentioned by *Varāhamihira* as being in northwestern Bharat (along with *Takṣaśilā, Gandhāra* etc.), the supposed place where the manuscript might have been written.

The *Bakhshali* manuscript uses numerals with a place-value system, using a dot as a place holder for zero. The dot symbol came to be called the *shunya-bindu* - शुन्य-बिन्दुः (*the dot of the empty place*). References to the concept are found in *Subandhu's Vasavadatta,* which has been dated between 385 and 465 by the scholar *Maan Singh*. Prior to the 2017 carbon dating, a 9th-century inscription of zero on the wall of a temple in Gwalior, Madhya Pradesh, was once thought to be the oldest Bhartiya use of a zero symbol.

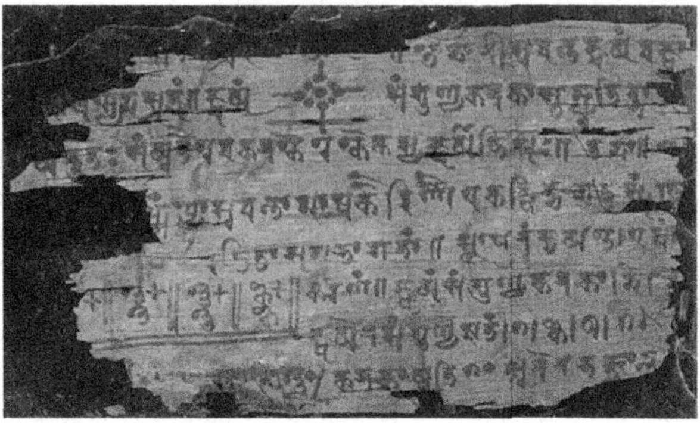

Figure 7.3. One of the *Bakhshali* manuscripts. (Image Credit: https://en.wikipedia.org/wiki/)

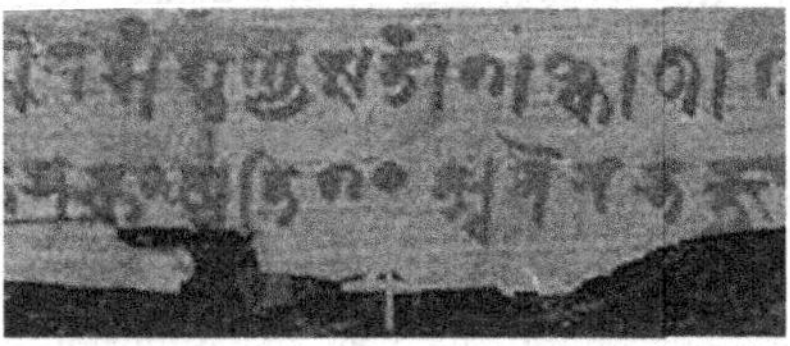

Figure 7.4. *Bakhshali* manuscript, detail of the dot "zero". (Image Credit: https://en.wikipedia.org/wiki/)

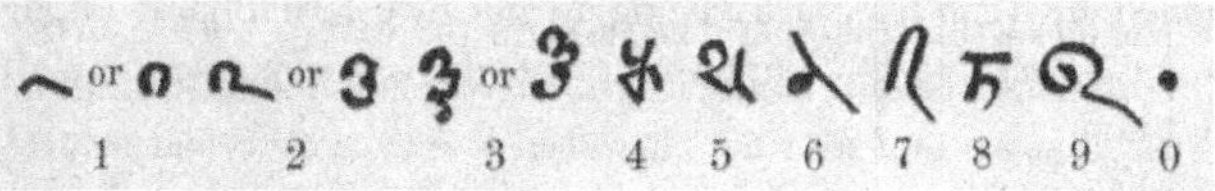


Figure 7.5. The numerals used in the Bakhshali manuscript, dated to sometime between the 3rd and 7th century AD. (Image Credit: https://en.wikipedia.org/wiki/)

Textual references to a place-value system are seen from the 5th century CE onward. A commentary on *Patanjali's* (पतञ्जलि) *Yoga Sutras* (योग सूत्र) from the 5th century reads, "*Just as a line in the hundreds place [means] a hundred, in the tens place ten, and one in the ones place, so one and the same woman is called mother, daughter and sister.*" A system called *bhūta-sankhya* - भूत-सांख्यः (*'object numbers' or 'concrete numbers'*) was employed for representing numerals in Sanskrit verses, by using a concept representing a digit to stand for the digit itself.

The Jain text entitled the *Lokavibhaga*, dated 458 CE, mentions the objectified numeral:

*"panchabhyah khalu shunyebhyah param dve sapta chambaram ekam trini cha rupam cha"*

meaning *'five voids, then two and seven, the sky, one and three and the form'*, i.e., the number 13107200000.

Such objectified numbers were used extensively from the 6th century onward, especially after *Varāhamihira*. Zero is explicitly represented in such numbers as "*the void*" (*sunya*) or the "*heaven-space*" (*ambara akasha*). Correspondingly, the dot used in place of zero in written numerals was referred to as a *sunya-bindu*.

### Place-Value Numerals with Zero

In 628 CE, astronomer-mathematician *Brahmagupta* wrote his text *Brāhmasphuṭasiddhānta* which contained the first mathematical treatment of zero. He defined zero as the result of subtracting a number from itself, postulated negative numbers and discussed their properties under arithmetical operations. His word for zero was *shunya* (void), the same term previously used for the empty spot in 9-digit place-value system. This provided a new perspective on the *shunya-bindu* as a numeral and paved the way for the eventual evolution of a zero digit. The dot continued to be used for at least 100 years afterwards, and transmitted to Southeast Asia and Arabia. Kashmir's Sharada script has retained the dot for zero until this day. Figure 7.6 shows numeral "zero".

Figure 7.6. The numeral "zero" as it appears in two numbers (50 and 270) in an inscription in Gwalior (Dated to the 9th century). (Image Credit: https://en.wikipedia.org/wiki/)

By the end of the 7th century, decimal numbers begin to appear in inscriptions in Southeast Asia as well as in Bharat. Some scholars hold that they appeared even earlier. A 6th century *copper-plate grant* at *Mankani* bearing the numeral 346 (corresponding to 594 CE) is often cited. But its reliability is subject to dispute. The first indisputable occurrence of 0 in an inscription occurs at Gwalior in 876 CE, containing a numeral "*270*" in a notation surprisingly similar to ours. Throughout the 8th and 9th centuries, both the old *Brahmi* numerals and the new decimal numerals were used, sometimes appearing in the same inscriptions. In some documents, a transition is seen to occur around 866 CE.

## Adoption in Arab

Before the rise of the *Caliphate*, the *Hindu–Arabic numeral system* was already moving West and was mentioned in *Syria* in 662 AD by the *Syriac Nestorian* scholar *Severus Sebokht*. Irrespective of whether this is wrong, since all Bhartiya texts after *Aryabhata's Aryabhatiya* used the Bhartiya number system, certainly from this time the Arabs had a translation of a text written in the Bhartiya number system. Figure 7.7 shows first Arabic numerals in Europe appeared.

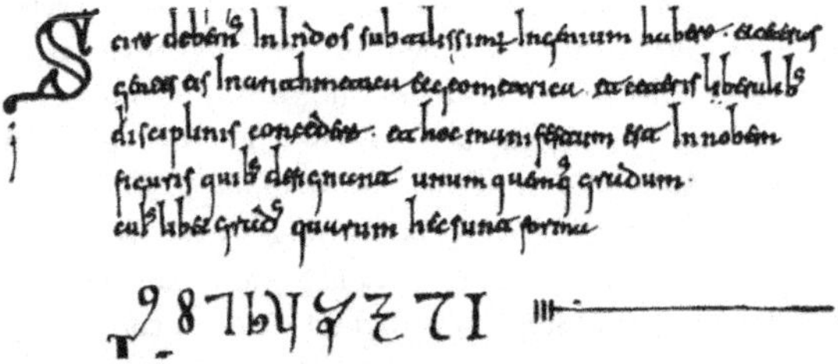

Figure 7.7. The first Arabic numerals in Europe appeared in the Codex Vigilanus in the year 976. (Image Credit: https://en.wikipedia.org/wiki/)

In his text *The Arithmetic of Al-Uqlîdisî (Dordrecht: D. Reidel, 1978), A.S. Saidan's* studies were unable to answer in full how the numerals reached the Arab world. *Al-Uqlidisi* developed a notation to represent idecimal fractions. The numerals came to fame due to their use in the pivotal work of the Persian mathematician *Al-Khwarizmi*, whose book *On the Calculation with Hindu Numerals* was written about 825, and the Arab mathematician *Al-Kindi*, who wrote four volumes "*On the Use of the Indian Numerals*" (*Ketab fi Isti'mal al-'Adad al-Hindi*) about 830. They, amongst other works, contributed to the diffusion of the Bhartiya system of numeration in the Middle East and the West.

## Adoption in Europe

In Christian Europe, the first mention and representation of *Hindu–Arabic numerals (from one to nine, without zero)*, is in the *Codex Vigilanus* (aka *Albeldensis*), an illuminated compilation of various historical documents from the *Visigothic* period in Spain, written in the year 976 by three monks of the *Riojan monastery* of *San Martín de Albelda*. Between 967 and 969, *Gerbert* of *Aurillac* discovered and studied *Arab science* in the *Catalan abbeys*. Later he obtained from these places the book *De multiplicatione et divisione* (*On multiplication and division*). After becoming *Pope Sylvester II* in the year 999, he introduced a new model of *abacus*, the so-called *Abacus of Gerbert*, by adopting tokens representing *Hindu–Arabic numerals*, from *one to nine*.

*Leonardo Fibonacci* brought this system to Europe. His book *Liber Abaci* introduced *Modus Indorum (the method of the Indians)*, today known as *Hindu–Arabic numeral* system or *base-10 positional notation, the use of zero*, and the *decimal place system* to the Latin world. The numeral system came to be called "*Arabic*" by the Europeans. It was used in European mathematics from the 12th century, and entered common use from the 15th century to replace Roman numerals. Figure 7.8 shows the medieval Arabic numbers.

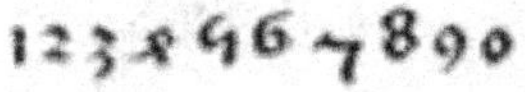

Figure 7.8. Medieval Arabic numbers at World map from *Ptolemy, Cosmographia*. Ulm: Lienhart Holle, 1482. (Image Credit: https://en.wikipedia.org/wiki/)

The system did not come into wide use in Europe, however, until the invention of printing. The numbers appear in their modern form on the titlepage of the "*Conpusicion de la arte de la arismetica y juntamente de geometría*" written by *Juan de Ortega*. The familiar shape of the *Western Arabic glyphs* as now used with the Latin alphabet (*0, 1, 2, 3, 4, 5, 6, 7, 8, 9*) are the product of the late 15th to early 16th century, when they entered early *typesetting*. *Muslim scientists* used the *Babylonian numeral system*, and *merchants* used the *Abjad numerals*, a system similar to the *Greek numeral system* and the *Hebrew numeral system*. Similarly, *Fibonacci*'s introduction of the system to Europe was restricted to learned circles. The credit for first establishing widespread understanding and usage of the *decimal positional notation* among the general population goes to *Adam Ries*, an author of the *German Renaissance*, whose

*1522 Rechenung auff der linihen und federn* (*Calculating on the Lines and with a Quill*) was targeted at the apprentices of *businessmen* and *craftsmen*.

In the last few centuries, the *European variety of Arabic numbers* was spread around the world and gradually became the most commonly used numeral system in the world. Even in many countries in languages which have their own numeral systems, the *European Arabic numerals* are widely used in commerce and mathematics. Figure 7.9 shows the Libro Intitulado Arithmetica Practica, 1549.

Libro inti-
tulado Arithmetica
practica/ muy vtil y proue
choso para toda persona
que quisiere exercitar se en
aprender a contar/ agora
nueuamente hecho por
Juan de Yciar
Viscayno.

1 2 3 4 5 6 7 8 9 0

Cæsaraugustæ. M.D.XLIX.

Figure 7.9. Libro Intitulado Arithmetica Practica, 1549. (Image Credit: https://en.wikipedia.org/wiki/)

### Adoption in East Asia

In 690 AD, *Empress Wu* promulgated *Zetian characters*, one of which was "〇". The word is now used as a synonym for the number zero. In China, *Gautama Siddha* (गौतम सिद्ध) introduced Hindu numerals with zero in 718, but Chinese mathematicians did not find them useful, as they had already had the decimal positional counting rods. In Chinese numerals, a circle (〇) is used to write zero in *Suzhou numerals*. Many historians think it was imported from Bhartiya numerals by *Gautama Siddha* in 718, but some Chinese scholars think it was created from the Chinese text space filler "□". Chinese and Japanese finally adopted the *Hindu–Arabic numerals* in the 19th century, abandoning counting rods.

## Spread of the Western Arabic variant

The *"Western Arabic"* numerals as they were in common use in Europe since the *Baroque period* have secondarily found worldwide use together with the *Latin alphabet*, and even significantly beyond the contemporary spread of the Latin alphabet, intruding into the writing systems in regions where other variants of the *Hindu–Arabic numerals* had been in use, but also in conjunction with Chinese and Japanese writing. Figure 7.10 shows the development of Hindu–Arabic numerals.

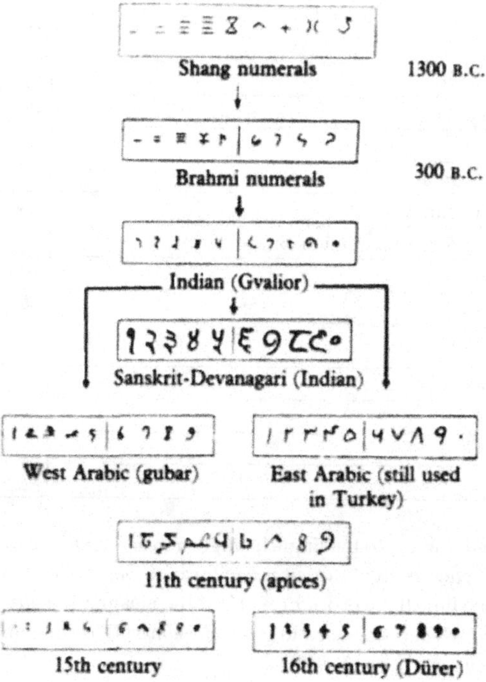


Figure 7.10. Development of Hindu–Arabic numerals. (Image Credit: https://en.wikipedia.org/wiki/)

## 0 (Zero)

0 (zero) is a number representing an empty quantity. As a number, 0 fulfills a central role in mathematics as the additive identity of the integers, real numbers, and other algebraic structures. In *place-value notation* such as decimal, 0 also serves as a numerical digit to indicate that that position's *power of 10* is not multiplied by anything or added to the resulting number. This concept appears to have been difficult to discover. Common names for the number 0 in English are *zero, nought, naught (/nɔːt/), nil*. In contexts where at least one adjacent digit distinguishes it from the letter O, the number is sometimes pronounced as *oh* or *o (/oʊ/)*. Informal or slang terms for 0 include *zilch* and *zip*. Historically, *ought, aught (/ɔːt/)*, and cipher have also been used. Table 7.2 shows varieties of 0 in various formats.

Table 7.2. Varieties of 0.

| | |
|---|---|
| Binary | $0_2$ |
| Ternary | $0_3$ |
| Senary | $0_6$ |
| Octal | $0_8$ |
| Duodecimal | $0_{12}$ |
| Hexadecimal | $0_{16}$ |
| Arabic, Kurdish, Persian, Sindhi, Urdu | ٠ |
| Hindu Numerals | ० |
| Chinese | 零, 〇 |
| Burmese | ၀ |
| Khmer | ០ |
| Thai | ๐ |
| Bangla | ০ |
| Maya |  |

The word zero came into the English language via *French zéro* from the *Italian zero*, a contraction of the *Venetian zevero* form of *Italian zefiro* via *ṣafira* or *ṣifr*. In pre-Islamic time the word *ṣifr* (صفر) had the meaning "*empty*". *Sifr* evolved to mean zero when it was used to translate *śūnya* (शून्य) from Bharat. *The first known English use of zero was in 1598.*

The Italian mathematician Fibonacci is credited with introducing the decimal system to Europe, used the term *zephyrum*. This became *zefiro* in Italian, and was then contracted to zero in *Venetian*. The Italian word *zefiro* was already in existence (meaning "*west wind*" from Latin and Greek *Zephyrus)* and may have influenced the spelling when transcribing *Arabic ṣifr*.

It is often called "*oh*" in the context of reading out a string of digits, such as *telephone numbers, street addresses, credit card numbers, military time, or years* (e.g., the area code *201* would be pronounced "*two oh one*"; a year such as *1907* is often pronounced "*nineteen oh seven*"). The presence of other digits, indicating that the string contains only numbers, avoids confusion with the letter O. For this reason, systems that include strings with both letters and numbers (e.g., Canadian postal codes) may exclude the use of the letter O. Figure 7.11 shows the use of 0 in papyrus.

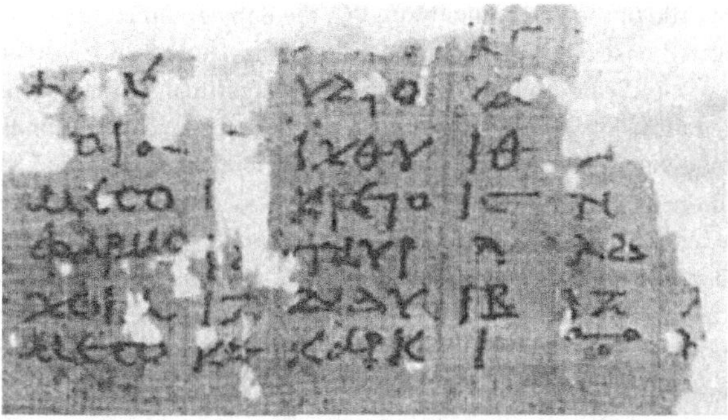

Figure 7.11. Example of the early Greek symbol for zero (lower right corner) from a 2nd-century papyrus. (Image Credit: https://en.wikipedia.org/wiki/)

Pingala used binary numbers in the form of short and long syllables (the latter equal in length to two short syllables), a notation similar to Morse code. Pingala used the Sanskrit word śūnya explicitly to refer to zero. The *Aryabhatiya* (c. 500), states sthānāt sthānaṁ daśaguṇaṁ syāt "from place to place each is ten times the preceding".

> The concept of zero as a written digit in the decimal place value notation was developed in Bharat.

## Ancient use

### Egypt

Ancient Egyptian numerals were of base 10. They used hieroglyphs for the digits and were not positional. By 1770 BC, the Egyptians had a symbol for zero in accounting texts. The symbol *nfr* - 𓄤, meaning beautiful, was also used to indicate the base level in drawings of tombs and pyramids, and distances were measured relative to the base line as being above or below this line.

### Babylon

By the middle of the 2nd millennium BC, the Babylonian mathematics had a sophisticated base 60 positional numeral system. The *lack* of a *positional value (or zero)* was indicated by a space between sexagesimal numerals. In a tablet unearthed at *Kish* (dating to as early as 700 BC), the scribe *Bêl-bân-aplu* used three hooks as a placeholder in the same Babylonian system. By 300 BC, a punctuation symbol (two slanted wedges) was co-opted to serve as this placeholder. The Babylonian placeholder was *not* a *true zero* because it was not used alone, nor was it used at the end of a number. Thus, numbers like 2 *and 120 (2×60), 3 and 180 (3×60), 4 and 240 (4×60)* looked the same, because the *larger numbers lacked a final sexagesimal placeholder*. Only context could differentiate them.

### Greece

The ancient Greeks had *no* symbol for *zero* (*μηδέν,* pronounced '*midén*'), and *did not use a digit* placeholder for it. According to mathematician *Charles Seife,* the ancient Greeks did begin to adopt the Babylonian placeholder zero for their work in astronomy after 500 BC, representing it with the lowercase Greek letter *ό (όμικρον) or omicron.* However, after using the Babylonian placeholder zero for astronomical calculations they would typically convert the numbers back into Greek numerals. Greeks seemed to have a philosophical opposition to using zero as a number. Other scholars give the Greek partial adoption of the Babylonian zero a later date, with the scientist Andreas Nieder giving a date of after 400 BC and the mathematician Robert *Kaplan* dating it after the conquests of *Alexander*. Greeks seemed unsure about the status of zero as a number. Some of them asked themselves, *"How can not being be?"*, leading to philosophical and, by the medieval period, religious arguments about the nature and existence of zero and the vacuum. The paradoxes of Zeno of Elea depend in large part on the uncertain interpretation of zero. Fragment of papyrus with clear Greek script, lower-right corner suggests a tiny zero with a double-headed arrow shape above it. Example of the early Greek symbol for zero (lower right corner) from a 2nd-century papyrus

By AD 150, *Ptolemy*, influenced by *Hipparchus* and the Babylonians, was using a symbol for zero ($\bar{\text{o}}$) in his work on mathematical astronomy called the *Syntaxis Mathematica*, also known as the *Almagest*. This *Hellenistic zero* was perhaps the earliest documented use of a numeral representing zero in the Old World. *Ptolemy* used it many times in his *Almagest (VI.8)* for the *magnitude of solar and lunar eclipses*. It represented the value of both digits and minutes of immersion at first and last contact. Digits varied continuously from 0 to 12 to 0 as the Moon passed over the Sun (a triangular pulse), where twelve digits was the angular diameter of the Sun. Minutes of immersion was tabulated from 0′0″ to 31′20″ to 0′0″, where 0′0″ used the symbol as a placeholder in two positions of his sexagesimal positional numeral system, while the combination meant a zero angle. *Minutes of immersion* was also a continuous function

$$\frac{1}{12}31'20''\sqrt{d(24-d)}$$

(a triangular pulse with convex sides), where d was the digit function and 31′20″ was the sum of the radii of the Sun's and Moon's discs. Ptolemy's symbol was a placeholder as well as a number used by two continuous mathematical functions, one within another, so it meant zero, not none.

**China**

The *Sūnzĭ Suànjīng*, of unknown date but estimated to be dated from the 1st to 5th centuries AD, and Japanese records dated from the 18th century, describe how the 4th century BC Chinese counting rods system enabled one to perform decimal calculations. As noted in *Xiahou Yang's Suanjing* (425–468 AD) that states that to multiply or divide a number by 10, 100, 1000, or 10000, all one needs to do, with rods on the counting board, is to move them forwards, or back, by 1, 2, 3, or 4 places, According to *A History of Mathematics*, the rods "*gave the decimal representation of a number, with an empty space denoting zero*". The counting rod system is considered a positional notation system.

Zero was not treated as a number at that time, but as a "*vacant position*". *Qín Jiǔsháo*'s 1247 Mathematical Treatise in *Nine Sections* is the oldest surviving Chinese mathematical text using a round symbol for zero. Chinese authors had been familiar with the idea of negative numbers by the Han dynasty (2nd century AD), as seen in The Nine Chapters on the *Mathematical Art*.

## 卐 Further Reading


1. "Abu Yusuf Yaqub ibn Ishaq al-Sabbah Al-Kindi". Archived from the original on 2007-10-26. Retrieved 2007-01-12.
2. "Arabic numerals". MacTutor History of Mathematics Archive. Retrieved 2021-05-23.
3. "Babylonian mathematics: View as single page". www.open.edu. Archived from the original on 7 September 2021. Retrieved 7 September 2021.
4. "Carbon dating finds Bakhshali manuscript contains oldest recorded origins of the symbol 'zero'". Bodleian Libraries. 14 September 2017. Retrieved 13 January 2023.
5. "Carbon dating finds Bakhshali manuscript contains oldest recorded origins of the symbol 'zero'". Bodleian Library. 14 September 2017. Archived from the original on 14 September 2017. Retrieved 25 October 2017.
6. "Egyptian numerals". mathshistory.st-andrews.ac.uk. Archived from the original on 15 November 2019. Retrieved 21 December 2019.
7. "Fibonacci Numbers". www.halexandria.org.
8. "Font Survey: 42 of the Best Monospaced Programming Fonts". codeproject.com. 18 August 2010. Archived from the original on 24 January 2012. Retrieved 22 July 2021.
9. "London museum showcases India's contribution to science". www.thehindubusinessline.com. Retrieved 3 February 2022.
10. "Math for Poets and Drummers" (PDF). people.sju.edu. Archived from the original (PDF) on 22 January 2019. Retrieved 20 December 2015.
11. "Mathematics in the Near and Far East" (PDF). grmath4.phpnet.us. p. 262. Archived (PDF) from the original on 4 November 2013. Retrieved 7 June 2012.

12. "Much ado about nothing: ancient Indian text contains earliest zero symbol". The Guardian. 13 September 2017. Archived from the original on 20 November 2017. Retrieved 14 September 2017.
13. *"Null values and the nullable type". IBM. 12 December 2018. Archived from the original on 23 November 2021. Retrieved 23 November 2021. In regard to services, sending a null value as an argument in a remote service call means that no data is sent. Because the receiving parameter is nullable, the receiving function creates a new, uninitialized value for the missing data then passes it to the requested service function.*
14. *"ResultSet (Java Platform SE 8 )"*. docs.oracle.com. Archived from the original on 9 May 2022. Retrieved 9 May 2022.
15. "Zero euro" banknote creator Richard FAILLE strikes again!". 25 June 2017. Archived from the original on 25 April 2023. Retrieved 25 April 2023.
16. "Zero". Maths History. Archived from the original on 21 September 2021. Retrieved 7 September 2021.
17. "Conpusicion de la arte de la arismetica y juntamente de geometría" written by Juan de Ortega.
18. Algebra with Arithmetic of Brahmagupta and Bhaskara, translated to English by Henry Thomas Colebrooke (1817) London.
19. All the pages have been photographed which are available in the book by Hayashi
20. al-Qifti's Chronology of the scholars (early 13th century): Al-Uqlidisi biography by J. J. O'Connor and E. F. Robertson.
21. Arnold Robbins; Nelson Beebe. "Classic Shell Scripting" Archived 24 February 2017 at the Wayback Machine. 2005. p. 274.
22. Aryabhatiya of Aryabhata, translated by Walter Eugene Clark.
23. 'Aught' synonyms Archived 23 August 2014 at the Wayback Machine, Thesaurus.com – Retrieved April 2013.
24. Bemer, R. W. (1967). "Towards standards for handwritten zero and oh: much ado about nothing (and a letter), or a partial dossier on distinguishing between handwritten zero and oh". Communications of the ACM. 10 (8): 513–518. doi:10.1145/363534.363563. S2CID 294510.
25. Berggren, J. Lennart (2007). "Mathematics in Medieval Islam". The Mathematics of Egypt, Mesopotamia, China, India, and Islam: A Sourcebook. Princeton University Press. p. 518. ISBN 978-0-691-11485-9.

26. Bibhutibhusan Datta (1929). "Book Review: G. R. Kaye, The Bakhshâlî Manuscript—A Study in Mediaeval Mathematics, 1927". 35 (4). Bull. Amer. Math. Soc.: 579–580.
27. Bill Casselman (February 2007). "All for Nought". Feature Column. AMS.
28. Bourbaki, Nicolas Elements of the History of Mathematics (1998), p. 46
29. Brezina, Corona (2006), Al-Khwarizmi: The Inventor of Algebra, The Rosen Publishing Group, pp. 39–40, ISBN 978-1-4042-0513-0: "*Historians have speculated on al-Khwarizmi's native language. Since he was born in a former Persian province, he may have spoken the Persian language. It is also possible that he spoke Khwarezmian, a language of the region that is now extinct.*"
30. Brezina, Corona (2006). Al-Khwarizmi: The Inventor of Algebra. The Rosen Publishing Group. ISBN 978-1-4042-0513-0. Archived from the original on 29 February 2020. Retrieved 26 September 2016.
31. Bunt, Lucas Nicolaas Hendrik; Jones, Phillip S.; Bedient, Jack D. (1976). The historical roots of elementary mathematics. Courier Dover Publications. pp. 254–255. ISBN 978-0-486-13968-5. Archived from the original on 23 June 2016. Retrieved 5 January 2016., Extract of pp. 254–255 Archived 10 May 2016 at the Wayback Machine.
32. C. W. Jones, ed., Opera Didascalica, vol. 123C in Corpus Christianorum, Series Latina.
33. Campbell, Douglas M.; Higgins, John C. (1984). Mathematics: People, Problems, Results. Taylor & Francis. ISBN 978-0-534-02879-4.
34. Casselman, Bill. "All for Nought". ams.org. University of British Columbia), American Mathematical Society. Archived from the original on 6 December 2015. Retrieved 20 December 2015.
35. Cepelewicz, Jordana Animals Count and Use Zero. How Far Does Their Number Sense Go? Archived 18 August 2021 at the Wayback Machine, Quanta, August 9, 2021.
36. Chris Woodford 2006, p. 9.
37. Cœdès, George, "A propos de l'origine des chiffres arabes," Bulletin of the School of Oriental Studies, University of London, Vol. 6, No. 2, 1931, pp. 323–328. Diller, Anthony, "New Zeros and Old Khmer," The Mon-Khmer Studies Journal, Vol. 25, 1996, pp. 125–132.
38. Crossley, Lun. 1999, p. 12 "the ancient Chinese system is a place notation system".

39. Darren R. Hayes. "A Practical Guide to Computer Forensics Investigations" Archived 24 February 2017 at the Wayback Machine. 2014. p. 399.
40. Deckers, Michael (2003) [525]. "Cyclus Decemnovennalis Dionysii – Nineteen Year Cycle of Dionysius". Archived from the original on 15 January 2019.
41. Devlin, Hannah (13 September 2017). "Much ado about nothing: ancient Indian text contains earliest zero symbol". The Guardian. ISSN 0261-3077. Archived from the original on 20 November 2017. Retrieved 14 September 2017.
42. Devlin, Hannah (13 September 2017). "Much ado about nothing: ancient Indian text contains earliest zero symbol". The Guardian. ISSN 0261-3077. Retrieved 14 September 2017.
43. Diehl, p. 186.
44. Earliest Uses of Symbols for Fractions by Jeff Miller.
45. Flegg (2002), pp. 6ff.
46. For a modern image.
47. Grimm, R.E., "The Autobiography of Leonardo Pisano", Fibonacci Quarterly 11/1 (February 1973), pp. 99–104.
48. Hansen, Alice (9 June 2008). Primary Mathematics: Extending Knowledge in Practice. SAGE. ISBN 978-0-85725-233-3. Archived from the original on 7 March 2021. Retrieved 7 November 2020.
49. Hayashi 1995, p. 26.
50. Hayashi 1995, p. 54.
51. Helaine Selin, ed. (2008). Encyclopaedia of the History of Science, Technology, and Medicine in Non-Western Cultures. Springer Science & Business Media. p. 198]. ISBN 978-1-4020-4559-2.
52. HLeonardo Pisano: "Contributions to number theory". Encyclopædia Britannica Online, 2006. p. 3. Retrieved 18 September 2006.
53. Hodgkin, Luke (2005). A History of Mathematics: From Mesopotamia to Modernity: From Mesopotamia to Modernity. Oxford University Press. p. 85. ISBN 978-0-19-152383-0.
54. Hoernle 1887.
55. Huggett, Nick (2019). "Zeno's Paradoxes". In Zalta, Edward N. (ed.). The Stanford Encyclopedia of Philosophy (Winter 2019 ed.). Metaphysics Research Lab, Stanford University. Archived from the original on 10 January 2021. Retrieved 9 August 2020.
56. Ifrah 1998, p. 416.
57. Ifrah 1998, p. 417.
58. Ifrah 1998, p. 439.

59. Ifrah, G. The Universal History of Numbers: From prehistory to the invention of the computer. John Wiley and Sons Inc., 2000. Translated from the French by David Bellos, E.F. Harding, Sophie Wood and Ian Monk.
60. Ifrah, Georges (2000), p. 400.
61. Ifrah, Georges (2000), p. 416.
62. Ifrah, Georges (2000). The universal history of numbers: from prehistory to the invention of the computer. David Bellos. New York: Wiley. ISBN 0-471-37568-3. OCLC 42291138.
63. Ifrah, Georges (2000). The Universal History of Numbers: From Prehistory to the Invention of the Computer. Wiley. ISBN 978-0-471-39340-5.
64. In German, a number like 21 is said like "one and twenty", as though being read from right to left. In Biblical Hebrew, this is sometimes done even with larger numbers, as in Esther 1:1, which literally says, "*Ahasuerus which reigned from India even unto Ethiopia, over seven and twenty and a hundred provinces*".
65. It has been claimed that a mid-third century CE text Yavana-jataka (on "Greek horoscopy") employed the device of bhūta-sankhyas (Plofker 2009, p. 47). But it is now considered an error of interpretation. (Mak, Bill M. (2013), "The Transmission of Greek Astral Science into India Reconsidered-Critical Remarks on the Contents and the Newly Discovered Manuscript of the Yavanajātaka", History of Science in South Asia, 1: 1–20, doi:10.18732/H2RP4T).
66. Iztok Fajfar. "Start Programming Using HTML, CSS, and JavaScript" Archived 24 February 2017 at the Wayback Machine. 2015. p. 160.
67. J. J. O'Connor; E. F. Robertson (November 2000). "A history of zero". MacTutor History of Mathematics archive. Retrieved 24 July 2022.
68. J. J. O'Connor; E. F. Robertson (November 2000). "The Bakhshali manuscript". MacTutor History of Mathematics archive. Archived from the original on 9 August 2007. Retrieved 24 July 2022.
69. Jan E.M. Houben "Linguistic Paradox and Diglossia: on the emergence of Sanskrit and Sanskritic language in Ancient India." De Gruyter Open Linguistics (Topical Issue on Historical Sociolinguistic Philology, ed. by Chiara Barbati and Christian Gastgeber.) OPLI – Vol. 4, issue 1: 1–18. DOI: https://doi.org/10.1515/opli-2018-0001.
70. John J O'Connor and Edmund F Robertson (November 2000). "Indian numerals". MacTutor History of Mathematics archive. Archived from the original on 2015-07-06. Retrieved 2022-07-24.

71. John Newsome Crossley; Anthony Wah-Cheung Lun; Kangshen Shen; Shen Kangsheng (1999). The Nine Chapters on the Mathematical Art: Companion and Commentary. Oxford University Press. ISBN 0-19-853936-3.
72. Joseph, G. G. (2000), The Crest of the Peacock, non-European roots of Mathematics, Princeton University Press, pp. 215–216.
73. Joseph, George Gheverghese (2011). The Crest of the Peacock: Non-European Roots of Mathematics (Third ed.). Princeton UP. p. 86. ISBN 978-0-691-13526-7.
74. Kang-Shen Shen; John N. Crossley; Anthony W.C. Lun; Hui Liu (1999). The Nine Chapters on the Mathematical Art: Companion and Commentary. Oxford UP. p. 35. ISBN 978-0-19-853936-0. zero was regarded as a number in India ... whereas the Chinese employed a vacant position.
75. Kaplan, Robert (1999). The Nothing That Is: A Natural History of Zero. New York: Oxford University Press. pp. 68–75. ISBN 978-0-19-514237-2.
76. Kaplan, Robert (28 October 1999). The Nothing that Is: A Natural History of Zero. Oxford University Press. p. 17. ISBN 978-0-19-802945-8. OCLC 466431211. Archived from the original on 24 June 2014. Retrieved 30 April 2022.
77. Kaplan, Robert. (2000). The Nothing That Is: A Natural History of Zero. Oxford: Oxford University Press.
78. Klein, Felix (2009). Elementary Mathematics from an Advanced Standpoint: Arithmetic, Algebra, Analysis. Cosimo, Inc. pp. 25, 80. ISBN 978-1605209319 – via Google Books.
79. Kumar, Raj (2003). Essays on Ancient India. Discovery Publishing House. pp. 196–. ISBN 978-81-7141-682-0.
80. Lay-Yong, Lam (1988). "A Chinese Genesis: Rewriting the History of Our Numeral System". Archive for History of Exact Sciences. 38 (2): 101–108. doi:10.1007/BF00348453. ISSN 0003-9519. JSTOR 41133830.
81. Lemma B.2.2, The integer 0 is even and is not odd, in Penner, Robert C. (1999). Discrete Mathematics: Proof Techniques and Mathematical Structures. World Scientific. p. 34. ISBN 978-981-02-4088-2.
82. Martin Levey and Marvin Petruck, Principles of Hindu Reckoning, translation of Kushyar ibn Labban Kitab fi usul hisab al-hind, p. 3, University of Wisconsin Press, 1965.
83. Mason, Robyn (14 September 2017). "Oxford Radiocarbon Accelerator Unit dates the world's oldest recorded origin of the zero

symbol". School of Archaeology, University of Oxford. Archived from the original on 14 September 2017. Retrieved 14 September 2017.

84. Menninger 2013, p. 395.
85. Menninger 2013, p. 397.
86. Menninger 2013, p. 398.
87. Mercier, Raymond. "Consideration of the Greek symbol 'zero'" (PDF). Home of Kairos. Archived (PDF) from the original on 5 November 2020. Retrieved 28 March 2020.
88. Mortaigne, Véronique (28 November 2014). "The golden age of Mayan civilisation – exhibition review". The Guardian. Archived from the original on 28 November 2014. Retrieved 10 October 2015.
89. Neugebauer, Otto (1969) [1957]. The Exact Sciences in Antiquity (2 ed.). Dover Publications. pp. 13–14, plate 2. ISBN 978-0-486-22332-2.
90. Neugebauer, Otto (2016) [1979]. Ethiopic Astronomy and Computus (Red Sea Press ed.). Red Sea Press. pp. 25, 53, 93, 183, Plate I. ISBN 978-1-56902-440-9.. The pages in this edition have numbers six less than the same pages in the original edition.
91. Nieder, Andreas (19 November 2019). A Brain for Numbers: The Biology of the Number Instinct. MIT Press. p. 286. ISBN 978-0-262-35432-5. Archived from the original on 30 April 2022. Retrieved 30 April 2022.
92. O'Connor, J J; Robertson, E F. "A history of Zero". MacTutor History of Mathematics. Archived from the original on 7 April 2020. Retrieved 28 March 2020.
93. O'Connor, J.J. (January 2004). "Chinese numerals". Mac Tutor. School of Mathematics and Statistics University of St Andrews, Scotland. Archived from the original on 14 June 2020. Retrieved 14 June 2020.
94. O'Connor, Robertson, J.J., E.F. "Aryabhata the Elder". School of Mathematics and Statistics University of St Andrews, Scotland. Archived from the original on 11 July 2015. Retrieved 26 May 2013.
95. Pannekoek, A. (1961). A History of Astronomy. George Allen & Unwin. p. 165.
96. Paul DuBois. "MySQL Cookbook: Solutions for Database Developers and Administrators" Archived 24 February 2017 at the Wayback Machine 2014. p. 204.
97. Pearce, Ian (May 2002). "The Bakhshali manuscript". The MacTutor History of Mathematics archive. Retrieved 2007-07-24.
98. Pedersen, Olaf (2010) [1974]. A Survey of the Almagest. Springer. pp. 232–235. ISBN 978-0-387-84825-9.
99. Plofker 2009, Ch. 2.

100. Plofker 2009, p. 14.
101. Plofker 2009, p. 44.
102. Plofker 2009, p. 45.
103. Plofker 2009, p. 46.
104. Plofker 2009, p. 57.
105. Plofker 2009, pp. 45–46; Menninger 2013, pp. 396–397; Ifrah 1998, p. 400
106. Plofker 2009, pp. 68–69.
107. Plofker 2009, pp. 73–75.
108. Plofker 2009, pp. 75–77.
109. Plofker, Kim (2009), Mathematics in India, Princeton University Press, p. 158, ISBN 978-0-691-12067-6
110. Plofker, Kim (2009). Mathematics in India. Princeton University Press. pp. 54–56. ISBN 978-0-691-12067-6. In the Chandah-sutra of Pingala, dating perhaps the third or second century BC, [ ...] Pingala's use of a zero symbol [śūnya] as a marker seems to be the first known explicit reference to zero. ... In the Chandah-sutra of Pingala, dating perhaps the third or second century BC, there are five questions concerning the possible meters for any value "n". [ ...] The answer is (2)7 = 128, as expected, but instead of seven doublings, the process (explained by the sutra) required only three doublings and two squarings – a handy time saver where "n" is large. Pingala's use of a zero symbol as a marker seems to be the first known explicit reference to zero.
111. Plofker, Kim; Keller, Agathe; Hayashi, Takao; Montelle, Clemency; Wujastyk, Dominik (6 October 2017). "The Bakhshālī Manuscript: A Response to the Bodleian Library's Radiocarbon Dating". History of Science in South Asia. 5 (1): 134–150. doi:10.18732/H2XT07. ISSN 2369-775X.
112. Ptolemy (1998) [1984, c.150]. Ptolemy's Almagest. Translated by Toomer, G. J. Princeton University Press. pp. 306–307. ISBN 0-691-00260-6.
113. Qian, Baocong (1964), Zhongguo Shuxue Shi (The history of Chinese mathematics), Beijing: Kexue Chubanshe.
114. Reid, Constance (1992). From zero to infinity: what makes numbers interesting (4th ed.). Mathematical Association of America. p. 23. ISBN 978-0-88385-505-8. zero neither prime nor composite.
115. Revell, Timothy (14 September 2017). "History of zero pushed back 500 years by ancient Indian text". New Scientist. Archived from the original on 25 October 2017. Retrieved 25 October 2017.

116.Rowlett, Russ (2004-07-04), Roman and "Arabic" Numerals, University of North Carolina at Chapel Hill, retrieved 2019-04-12.
117.Salomon, Richard (1998), Indian Epigraphy: A Guide to the Study of Inscriptions in Sanskrit, Prakrit, and the other Indo-Aryan Languages, Oxford University Press, USA, pp. 61–63, ISBN 978-0-19-535666-3
118.Sarasvati & Jyotishmati 1979, pp. 27, 66.
119.Section VII 11, corresponding to folio 46v.(Hayashi 1995, p. 54)
120.Seife, Charles (1 September 2000). Zero: The Biography of a Dangerous Idea. Penguin. p. 39. ISBN 978-0-14-029647-1. OCLC 1005913932. Archived from the original on 30 April 2022. Retrieved 30 April 2022.
121.Shastri, Ajaya Mitra (1998), "Mankaṇi Charter of Taralasvāmin and the Antiquity of the Decimal Notation", Annals of the Bhandarkar Oriental Research Institute, 79 (1/4): 161–170, JSTOR 41694535.
122.Sigler, L., Fibonacci's Liber Abaci. English translation, Springer, 2003.
123.Singh, Maan (1993). Subandhu, New Delhi: Sahitya Akademi, ISBN 81-7201-509-7, pp. 9–11.
124.Smith & Karpinski 2013, Ch. III; Ifrah 1998, pp. 411–418; Menninger 2013, p. 398
125.Smith & Karpinski 2013, p. 15.
126.Smith & Karpinski 2013, pp. 12–15.
127.Smith, David Eugene; Karpinski, Louis Charles (1911). The Hindu-Arabic numerals. Boston, London, Ginn and Company. p. 52.
128.Steel, Duncan (2000). Marking time: the epic quest to invent the perfect calendar. John Wiley & Sons. p. 113. ISBN 978-0-471-29827-4. In the B.C./A.D. scheme there is no year zero. After 31 December 1 BC came 1 January AD 1. ... If you object to that no-year-zero scheme, then don't use it: use the astronomer's counting scheme, with negative year numbers.
129.Struik, Dirk J. (1987). A Concise History of Mathematics. New York: Dover Publications. pp. 32–33. "In these matrices we find negative numbers, which appear here for the first time in history."
130.Takao Hayashi (2008), "Bakhshālī Manuscript", in Helaine Selin (ed.), Encyclopaedia of the History of Science, Technology, and Medicine in Non-Western Cultures, vol. 1, Springer, pp. B1–B3, ISBN 9781402045592.
131.Vaman Shivaram Apte (1970). Sanskrit Prosody and Important Literary and Geographical Names in the Ancient History of India.

Motilal Banarsidass. pp. 648–649. ISBN 978-81-208-0045-8. Archived from the original on 22 April 2017. Retrieved 21 April 2017.

132. W., Weisstein, Eric. "Zero". mathworld.wolfram.com. Archived from the original on 1 June 2013. Retrieved 4 April 2018.

133. Wallin, Nils-Bertil (19 November 2002). "The History of Zero". YaleGlobal online. The Whitney and Betty Macmillan Center for International and Area Studies at Yale. Archived from the original on 25 August 2016. Retrieved 1 September 2016.

134. Wáng, Qīngxiáng (1999), Sangi o koeta otoko (The man who exceeded counting rods), Tokyo: Tōyō Shoten, ISBN 4-88595-226-3.

135. Weil, Andre (6 December 2012). Number Theory for Beginners. Springer Science & Business Media. ISBN 978-1-4612-9957-8. Archived from the original on 14 June 2021. Retrieved 6 April 2021.

136. Weiss, Ittay (20 September 2017). "Nothing matters: How India's invention of zero helped create modern mathematics". The Conversation. Archived from the original on 12 July 2018. Retrieved 12 July 2018.

137. Will Durant (1950), The Story of Civilization, Volume 4, The Age of Faith, Simon & Schuster, ISBN 978-0-9650007-5-8, p. 241, "In 976, Muhammad ibn Ahmad, in his Keys of the Sciences, remarked that if, in a calculation, no number appears in the place of tens, a little circle should be used "to keep the rows". This circle the Mosloems called ṣifr, "empty" whence our cipher."

138. Will Durant (1950), The Story of Civilization, Volume 4, The Age of Faith: Constantine to Dante – A.D. 325–1300, Simon & Schuster, ISBN 978-0-9650007-5-8, p. 241, "*The Arabic inheritance of science was overwhelmingly Greek, but Hindu influences ranked next. In 773, at Mansur's behest, translations were made of the Siddhantas – Indian astronomical treatises dating as far back as 425 BC; these versions may have the vehicle through which the "Arabic" numerals and the zero were brought from India into Islam. In 813, al-Khwarizmi used the Hindu numerals in his astronomical tables.*"

139. William Darrach Halsey, Emanuel Friedman (1983). Collier's Encyclopedia, with bibliography and index. When the Arabian empire was expanding and contact was made with India, the Hindu numeral system and the early algorithms were adopted by the Arabs.

140. William L. Hosch, ed. (15 August 2010). The Britannica Guide to Numbers and Measurement (Math Explained). The Rosen Publishing Group. pp. 97–98. ISBN 978-1-61530-108-9. Archived from the original on 4 August 2016. Retrieved 26 September 2016.

141.Wu, X.; Ichikawa, T.; Cercone, N. (25 October 1996). Knowledge-Base Assisted Database Retrieval Systems. World Scientific. ISBN 978-981-4501-75-0. Archived from the original on 31 March 2022. Retrieved 7 November 2020.
142.Zero Archived 6 December 2017 at the Wayback Machine, Merriam Webster online Dictionary.
143."A Big Zero: Research uncovers the date of the Bakhshali Manuscript", YouTube video, University of Oxford.
144."Hindu–Arabic Numerals". Archived from the original on 2005-12-27. Retrieved 2005-12-13.
145."The Development of Hindu–Arabic and Traditional Chinese Arithmetic" by Professor Lam Lay Yong, member of the International Academy of the History of Science.
146."Zero". Encyclopædia Britannica (11th ed.). 1911.
147."Zero". Encyclopedia Americana. 1920.
148."zero, n." OED Online. Oxford University Press. December 2011. Archived from the original on 7 March 2012. Retrieved 4 March 2012. French zéro (1515 in Hatzfeld & Darmesteter) or its source Italian zero, for *zefiro, < Arabic çifr.
149.0, https://en.wikipedia.org/wiki/0.
150.A History of Zero, https://web.archive.org/web/20081204042054/http://www-gap.dcs.st-and.ac.uk/~history/HistTopics/Zero.html.
151.Aczel, Amir D. (2015). Finding Zero. New York: Palgrave Macmillan. ISBN 978-1-137-27984-2.
152.Arabic numerals by J J O'Connor and E F Robertson https://web.archive.org/web/20160429163458/http://www-gap.dcs.st-and.ac.uk/%7Ehistory/HistTopics/Arabic_numerals.html.
153.Asimov, Isaac (1978). "Nothing Counts". Asimov on Numbers. New York: Pocket Books. ISBN 978-0-671-82134-0. OCLC 1105483009.
154.Audun Holme, Geometry: Our Cultural Heritage, 2000.
155.Bakhshali_manuscript, https://en.wikipedia.org/wiki/Bakhshali_manuscript
156.Barrow, John D. (2001). The Book of Nothing. Vintage. ISBN 0-09-928845-1.
157.Bourbaki, Nicolas (1998). Elements of the History of Mathematics. Berlin, Heidelberg, and New York: Springer-Verlag. ISBN 3-540-64767-8.
158.Ch. 6 – The Bakhshali manuscript (Ian G. Pearce, Indian Mathematics: Redressing the balance).

159. D E Smith and L C Karpinski, The Hindu-Arabic numerals (Boston, 1911).
160. David H. Bailey, Jonathan Borwein (2011). "A Quartically Convergent Square Root Algorithm: An Exercise in Forensic Paleo-Mathematics" (PDF).
161. Diehl, Richard A. (2004). The Olmecs: America's First Civilization. London: Thames & Hudson.
162. Douglas Harper (2011), Zero Archived 3 July 2017 at the Wayback Machine, Etymology Dictionary, Quote="figure which stands for naught in the Arabic notation," also "the absence of all quantity considered as quantity", c. 1600, from French zéro or directly from Italian zero, from Medieval Latin zephirum, from Arabic sifr "cipher", translation of Sanskrit sunya-m "empty place, desert, naught".
163. Edsger W. Dijkstra: Why numbering should start at zero, EWD831 (PDF of a handwritten manuscript), http://www.cs.utexas.edu/users/EWD/ewd08xx/EWD831.PDF.
164. Filliozat, Pierre-Sylvain (2004), "Ancient Sanskrit Mathematics: An Oral Tradition and a Written Literature", in Chemla, Karine; Cohen, Robert S.; Renn, Jürgen; et al. (eds.), History of Science, History of Text (Boston Series in the Philosophy of Science), Dordrecht: Springer Netherlands, 254 pages, pp. 137–157, doi:10.1007/1-4020-2321-9_7, ISBN 978-1-4020-2320-0.
165. Flegg, Graham (2002). Numbers: Their History and Meaning. Courier Dover Publications. ISBN 0-486-42165-1.
166. G G Joseph, The crest of the peacock (London, 1991).
167. G Ifrah, A universal history of numbers: From prehistory to the invention of the computer (London, 1998).
168. Hayashi, Takao (1995). The Bakhshālī manuscript: an ancient Indian mathematical treatise. Groningen Oriental studies. Groningen: Egbert Forsten. ISBN 978-90-6980-087-5.
169. Hindu–Arabic numerals https://web.archive.org/web/20051227042328/http://www.scit.wlv.ac.uk/university/scit/modules/mm2217/han.htm.
170. History of Hindu Arab Numerals, https://en.wikipedia.org/wiki/History_of_the_Hindu%E2%80%93Arabic_numeral_system.
171. Hoernle, Augustus (1887), On the Bakshali manuscript, Vienna: Alfred Hölder (Editor of the Court and of the University).
172. Hoernle: On the Bakhshali Manuscript, 1887, archive.org.

173.Ifrah, Georges (1998) [first published in French in 1981], The Universal History of Numbers: From Prehistory to the Invention of the Computer, Harvill, ISBN 978-1-860-46324-2.
174.Ifrah, Georges (2000). The Universal History of Numbers: From Prehistory to the Invention of the Computer. Wiley. ISBN 0-471-39340-1.
175.Indian numerals by J J O'Connor and E F Robertson https://web.archive.org/web/20150706140353/http://www-gap.dcs.st-and.ac.uk/%7Ehistory/HistTopics/Indian_numerals.html.
176.Jan Gullberg (1997), Mathematics: From the Birth of Numbers, W.W. Norton & Co., ISBN 978-0-393-04002-9, p. 26, Quote = Zero derives from Hindu sunya – meaning void, emptiness – via Arabic sifr, Latin cephirum, Italian zevero.
177.K W Menninger, Number words and number symbols: A cultural history of numbers (Boston, 1969).
178.Kaplan, Robert (2000). The Nothing That Is: A Natural History of Zero. Oxford University Press.
179.Kaye, George Rusby (2004) [1927]. The Bakhshālī manuscripts: a study in medieval mathematics. New Delhi: Aditya Prakashan. ISBN 978-81-7742-058-6.
180.L C Karpinski, The history of arithmetic (New York, 1965).
181.M N Channabasappa (1976). "On the square root formula in the Bakhshali manuscript" (PDF). Indian J. History Sci. 11 (2): 112–124.
182.Menninger, Karl (1992). Number words and number symbols: a cultural history of numbers. Courier Dover Publications. pp. 399–404. ISBN 978-0-486-27096-8. Archived from the original on 25 April 2016. Retrieved 5 January 2016.
183.Menninger, Karl (2013) [first published by MIT Press in 1969], Number Words and Number Symbols: A Cultural History of Numbers, translated by Paul Broneer, Courier Corporation, ISBN 978-0-486-31977-3.
184.Menninger, Karl W. (1969). Number Words and Number Symbols: A Cultural History of Numbers. MIT Press. ISBN 0-262-13040-8.
185.On the genealogy of modern numerals by Edward Clive Bayley.
186.Plofker, Kim (2009), Mathematics in India, Princeton University Press, ISBN 978-0-691-12067-6.
187.Plofker, Kim, Agathe Keller, Takao Hayashi, Clemency Montelle, and Dominik Wujastyk. 2017. "The Bakhshālī Manuscript: A Response to the Bodleian Library's Radiocarbon Dating." History of Science in South Asia, 5.1: 134–150.

https://journals.library.ualberta.ca/hssa/index.php/hssa/article/view/22.

188. Plofker, Kim, Agathe Keller, Takao Hayashi, Clemency Montelle, and Dominik Wujastyk. 2017. "The Bakhshālī Manuscript: A Response to the Bodleian Library's Radiocarbon Dating". History of Science in South Asia 5 (1). 134–50. https://doi.org/10.18732/H2XT07. Challenges the claims made in the YouTube video "A Big Zero."
189. Plofker, Kim; Agathe Keller; Takao Hayashi; Clemency Montelle; and Dominik Wujastyk. "The Bakhshālī Manuscript: A Response to the Bodleian Library's Radiocarbon Dating" History of Science in South Asia, 5.1: 134–150. doi:10.18732/H2XT07.
190. R Kaplan, The nothing that is : a natural history of zero (London, 1999).
191. Robert Logan (2010), The Poetry of Physics and the Physics of Poetry, World Scientific, ISBN 978-981-4295-92-5, p. 38, Quote = The idea of sunya and place numbers was transmitted to the Arabs who translated sunya or "leave a space" into their language as sifr.
192. Sarasvati, Svami Satya Prakash; Jyotishmati, Usha (1979), The Bakhshali Manuscript: An Ancient Treatise of Indian Arithmetic (PDF), Allahabad: Dr. Ratna Kumari Svadhyaya Sansthan, archived from the original (PDF) on 20 June 2014, retrieved 19 January 2016 with complete text in Devanagari, 110 pages.
193. Sarasvati, Svami Satya Prakash; Jyotishmati, Usha (1979), The Bakhshali Manuscript: An Ancient Treatise of Indian Arithmetic (PDF), Allahabad: Dr. Ratna Kumari Svadhyaya Sansthan, archived from the original (PDF) on 2014-06-20, retrieved 2016-01-19.
194. Searching for the World's First Zero, http://www.huffingtonpost.com/amir-aczel/worlds-first-zero_b_3276709.html.
195. Seife, Charles (2000). Zero: The Biography of a Dangerous Idea. Penguin USA. ISBN 0-14-029647-6.
196. Smith, D. E.; Karpinski, L. C. (2013) [first published in Boston, 1911], The Hindu–Arabic Numerals, Dover, ISBN 978-0486155111.
197. Smithsonian Institution, Oriental Elements of Culture in the Occident, p. 518, at Google Books, Annual Report of the Board of Regents of the Smithsonian Institution; Harvard University Archives, Quote="*Sifr occurs in the meaning of "empty" even in the pre-Islamic time. ... Arabic sifr in the meaning of zero is a translation of the corresponding India sunya.*"

198. The Arabic numeral system – MacTutor History of Mathematics http://www-history.mcs.st-and.ac.uk/HistTopics/Arabic_numerals.html.
199. The Arabic numeral system by: J J O'Connor and E F Robertson Archived 2010-02-23 at the Wayback Machine.
200. The Bakhshali manuscript, MacTutor History of Mathematics archive.
201. The History of Algebra, https://web.archive.org/web/20141009100628/http://www.ucs.louisiana.edu/~sxw8045/history.htm.
202. Weisstein, Eric W. "0". MathWorld, https://mathworld.wolfram.com/0.html
203. Woodford, Chris (2006). Digital Technology. Evans Brothers. ISBN 978-0-237-52725-9. Archived from the original on 17 August 2019. Retrieved 24 March 2016.
204. Zero on In Our Time at the BBC, https://www.bbc.co.uk/programmes/p004y254.
205. Zero Saga, http://home.ubalt.edu/ntsbarsh/zero/ZERO.HTM.

# Chapter 8

# Hindu Time and Metrics

## Introduction

Hindu units of time are described in Hindu texts ranging from microseconds to trillions of years, including cycles of cosmic time that repeat general events in *Hindu cosmology* (हिन्दू ब्रह्माण्ड विज्ञान). Time (*kāla* - काल) is described as eternal. Various fragments of time are described in the *Vedas, Manusmriti, Bhagavata Purana, Vishnu Purana, Mahabharata, Surya Siddhanta* etc.

## Sidereal Metrics

*Sidereal astrology* (पार्श्वज्योतिषशास्त्रम्) maintains the alignment between signs and constellations via corrective systems of Hindu (*Vedic*)-origin known as *ayanamsas* (अयनंसा), to allow for the observed precession of equinoxes, whereas tropical astrology ignores precession. This has caused the two systems, which were aligned around 2,000 years ago, to drift apart over the centuries. *Ayanamsa systems* used in *Hindu astrology (Vedic astrology)* include the *Lahiriayanamsa* and the *Raman ayanamsa*. The *Fagan-Bradley ayanamsa* is an example of an *ayanamsa system* used in *Western sidereal astrology*. As of 2020, sun signs calculated using the *Sri Yukteswar ayanamsa* (श्री युक्तेश्वर अयनम्सा) were around 23 degrees behind tropical sun signs. Per these calculations, persons born between *March 12 - April 12*, for instance, would have the sun sign of *Pisces* (मीन राशिः). By contrast, persons born between *March 21 - April 19* would have the sun sign of Aries (मेष राशिः) per tropical calculations. Table 8.1 shows the sidereal units. Figure 8.1. shows the Hindu measurements in logarithmic scale based on seconds. Table 8.2 shows the SI values of some ancient units. Table 8.3 shows the unit definitions. Table 8.4 provides unit definitions.

Table 8.1 Sidereal units

| Unit | Definition | Value in SI units |
|---|---|---|
| *truti* (त्रुटि) | base unit | ≈ 0.30 μs |
| *renu* (रेणु) | 60 *truti* | ≈ 18 μs |
| *lava* (लव) | 60 *renu* | ≈ 1,080 μs |
| *līkṣaka* (लीक्षक) | 60 *lava* | ≈ 64.8 ms |
| *liptā*(लिप्ता) | 6 *līkṣaka* | ≈ 0.3888 s |

| | | |
|---|---|---|
| *vipala* (विपल) | | |
| *prāṇa*(प्राण) | 10 *liptā* | ≈ 3.888 s |
| *pala* (पल) | 60 *liptā* or 6 *prāṇa* | ≈ 24 s |
| *vighaṭi* (विघटि) | | |
| *vināḍī* (विनाडी) | | |
| *ghaṭi* (घटि) | 60 *vighaṭi* | ≈ 1.44 ks (24 min) |
| *nāḍī* (नाडी) | | |
| *danda* (दण्ड) | | |
| *muhūrta* (मुहूर्त) | 2 *ghaṭi* | ≈ 2.88 ks (48 min) |
| *nakṣhatra ahorātram* (sidereal day; नक्षत्र अहोरात्रम्) | 60 *ghaṭī* | ≈ 86.4 ks (24 h) |
| | 30 *muhūrta* | |

Table 8.2 Units and SI values

| Unit | Definition | Value in SI units |
|---|---|---|
| *truti* | base unit | ≈ *29.6 μs* |
| *tatpara* | 100 *truti* | ≈ *2.96 ms* |
| *nimesha* | 30 *tatpara* | ≈ *88.9 ms* |
| *kāṣṭhā* | 18 *nimesha* | ≈ *1.6 s* |
| *kalā* | 30 *kāṣṭhā* | ≈ *48 s* |
| *ghatika* | 30 *kalā* | ≈ *1.44 ks (24 min)* |
| *muhūrta* (*kṣaṇa*) | 2 *ghatika* | ≈ *2.88 ks (48 min)* |
| *Ahorātram* (sidereal day) | 30 *muhūrta* | ≈ *86.4 ks (24 h)* |
| | | |
| *paramāṇu* | base unit | ≈ *26.3 μs* |
| *aṇu* | 2 *paramāṇu* | ≈ *52.67 μs* |
| *trasareṇu* | 3 *aṇu* | ≈ *158 μs* |

| | | |
|---|---|---|
| *vedha* | 100 *truṭi* | ≈ *47.4 ms* |
| *lava* | 3 *vedha* | ≈ *0.14 s* |
| *nimeṣa* | 3 *lava* | ≈ *0.43 s* |
| *kṣaṇa* | 3 *nimeṣa* | ≈ *1.28 s* |
| *kāṣṭhā* | 5 *kṣaṇa* | ≈ *6.4 s* |
| *laghu* | 15 *kāṣṭhā* | ≈ *96 s (1.6 min)* |
| *danda (nadika)* | 15 *laghu* | ≈ *1.44 ks (24 min)* |
| *muhūrta* | 2 *danda* | ≈ *2.88 ks (48 min)* |
| *ahorātram (sidereal day)* | 30 *muhūrta* | ≈ *86.4 ks (24 h)* |
| *masa (month)* | 30 *ahorātram* | ≈ *2,592 ks* |
| *ritu (season)* | 2 *masa* | ≈ *5,184 ks* |
| *ayana* | 3 *ritu* | ≈ *15,552 ks (6 mth)* |
| *samvatsara (year)* | 2 *ayana* | ≈ *31,104 ks* |

Table 8.3 Unit definitions

| Unit | Definition | Equivalence |
|---|---|---|
| *tithi (lunar day)* | *Time for 12° increase of longitudinal angle between Moon and Sun* | ≈ *1 day (varies 19–26 hours)* |
| *pakṣa (lunar fortnight)* | *15 tithis* | ≈ *15 days* |
| *māsa (lunar month)* | *2 pakṣas: gaura (bright) or śukla pakṣa during waxing moon; kṛṣṇa (dark) pakṣa during waning moon* | ≈ *30 days (29.5 days)* |
| *ṛitu (season)* | *2 māsas* | ≈ *60 days* |
| *ayanam* | *3 ṛitus* | ≈ *180 days* |
| *varsha (lunar year)* | *2 ayanams* | ≈ *360 days (354.36707 days)* |

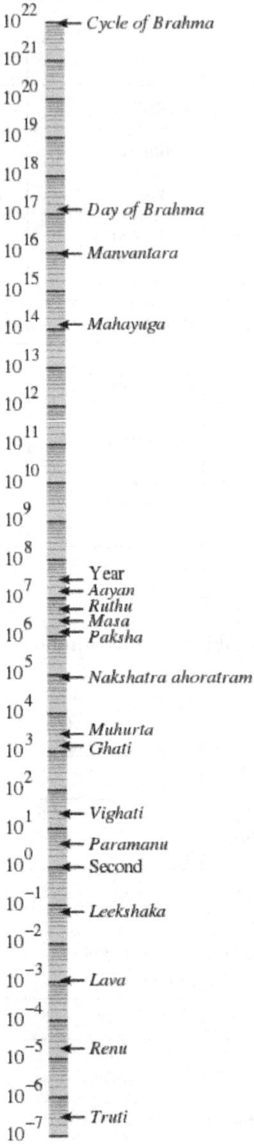


Figure 8.1. Hindu measurements in logarithmic scale based on seconds.

Table 8.4 Cosmic units of time

| Unit | Definition | Human | Pitri | Deva | Manu | Brahma |
|---|---|---|---|---|---|---|
| *mahā-kalpa* | 36,000 *kalpa* & *pralaya* | 311,040,000,000,000 yr | 10,368,000,000,000 yr | 864,000,000,000 yr | ~101,408,450.70 yr | 100 yr |
| *mahā-pralaya* | *mahā-kalpa* length | | | | | |
| *parārdha* | $^1/_2$ *mahā-kalpa* | 155,520,000,000,000 yr | 5,184,000,000,000 yr | 432,000,000,000 yr | ~50,704,225.35 yr | 50 yr |
| *kalpa* | 14 m + 15 ms; 1,000 cy | 4,320,000,000 yr | 144,000,000 yr | 12,000,000 yr | ~1,408.45 yr | 12 h |
| *pralaya* | *kalpa* length | | | | | |
| *manvantara [m]* | 71 *catur-yuga* | 306,720,000 yr | 10,224,000 yr | 852,000 yr | 100 yr | 51.12 min |
| *manvantara-sandhyā [ms]* | *Kṛta-yuga* length | 1,728,000 yr | 57,600 yr | 4,800 yr | ~6.76 mo | 17.28 s |
| *catur-yuga [cy]* | *Kṛta, Tretā, Dvāpara & Kali-yugas* | 4,320,000 yr | 144,000 yr | 12,000 yr | ~1.41 yr | 43.20 s |
| | | | **Kṛta-yuga** | | | |
| | Sum Total | 1,728,000 yr | 57,600 yr | 4,800 yr | ~6.76 mo | 17.28 s |
| *Kṛta-yuga-sandhyā* / *Kṛta-yuga-sandhyāṃśa* | $^1/_{10}$ *Kṛta-yuga* length | 144,000 yr | 4,800 yr | 400 yr | ~16.90 day | 1.44 s |
| *Kṛta-yuga* | 4 *Kali-yuga* lengths | 1,440,000 yr | 48,000 yr | 4,000 yr | ~5.63 mo | 14.40 s |
| | | | **Tretā-yuga** | | | |
| | Sum Total | 1,296,000 yr | 43,200 yr | 3,600 yr | ~5.07 mo | 12.96 s |
| *Tretā-yuga-sandhyā* / *Tretā-yuga-sandhyāṃśa* | $^1/_{10}$ *Tretā-yuga* length | 108,000 yr | 3,600 yr | 300 yr | ~12.68 day | 1.08 s |

| | | | | | | |
|---|---|---|---|---|---|---|
| *Tretā-yuga* | 3 *Kali-yuga* lengths | 1,080,000 yr | 36,000 yr | 3,000 yr | ~4.23 mo | 10.80 s |
| | | | ***Dvāpara-yuga*** | | | |
| | Sum Total | 864,000 yr | 28,800 yr | 2,400 yr | ~3.38 mo | 8.64 s |
| *Dvāpara-yuga-sandhyā* / *Dvāpara-yuga-sandhyāṃśa* | 1/10 *Dvāpara-yuga* length | *72,000 yr* | *2,400 yr* | *200 yr* | *~8.45 day* | *0.72 s* |
| *Dvāpara-yug* | 2 *Kali-yuga* lengths | *720,000 yr* | *24,000 yr* | *2,000 yr* | *~2.82 mo* | *7.20 s* |
| | | | ***Kali-yuga*** | | | |
| | Sum Total | *432,000 yr* | *14,400 yr* | *1,200 yr* | *~1.69 mo* | *4.32 s* |
| *Kali-yuga-sandhyā* / *Kali-yuga-sandhyāṃśa* | 1/10 *Kali-yuga* length | *36,000 yr* | *1,200 yr* | *100 yr* | *~4.23 day* | *0.36 s* |
| *Kali-yuga* | 1,000 *Deva* years | *360,000 yr* | *12,000 yr* | *1,000 yr* | *~1.41 mo* | *3.60 s* |

Hindu texts define lifespans differently for *humans, Pitris (forefathers), Devas (gods), Manus (progenitors of mankind), and Brahma (creator god)*. The division of a year for each is twelve 30-day months or 360 days, where a day is divided into a 12-hour day proper and 12-hour night. A 30-day month amounts to four 7-day weeks with an extra 8th day every two weeks (48-week year). A traditional human year is measured by the sun's northern (*uttarayana* - उत्तरायणः) and southern (*dakshinayana* - दक्षिणायनम्) movements in the sky, where the new year commences only when the sun returns to the same starting point and a pause on the commencement otherwise. *Ebenezer Burgess* postulates an *intercalary month* (अन्तर्कालरी मास) was inserted every five years to anciently maintain the correspondence of the 360-day years with the true solar years (~365.24-day years). For this reason, a traditional 360-day year is equivalent to a modern ~365.24-day solar or tropical year. Table 8.5 shows the comparison of various years.

Table 8.5 Comparison of various years

| Unit | Human | Pitri | Deva | Manu | Brahma |
|---|---|---|---|---|---|
| *Brahma year* | *3,110,400,000,000 yr* | *103,680,000,000 yr* | *8,640,000,000 yr* | *~1,014,084.51 yr* | *1 yr* |
| *Manu year* | *3,067,200 yr* | *102,240 yr* | *8,520 yr* | *1 yr* | *30.67 s* |
| *Deva year* | *360 yr* | *12 yr* | *1 yr* | *~1.01 h* | *3.60 ms* |
| *Pitri year* | *30 yr* | *1 yr* | *1 mo* | *~5.07 min* | *300 μs* |
| *Human year* | *1 yr* | *12 day* | *1 day* | *~10.14 s* | *10 μs* |

## 卐 Cosmic Date

According to *Puranic sources*, *Krishna*'s departure marks the end of the human age of *Dvapara-yuga* and the start *of Kali-yuga,* which is dated to midnight on 17/18 February 3102 BCE of the proleptic Julian calendar. We are currently halfway through *Brahma*'s life (*maha-kalpa*), whose lifespan is equal to the duration of the manifested material elements, from which *Brahma* manifests his universe in kalpa cycles:

- 51st year of 100 (2nd half or *parardha*)
- 1st month of 12
- 1st *kalpa* (*Shveta-Varaha Kalpa*) of 30
- 7th *manvantara* (*Vaivasvatha Manu*) of 14
- 28th *chatur-yuga* of 71
- 4th yuga (Kali-yuga) of 4

A *maha-kalpa* is followed by a *maha-pralaya* (*full dissolution*) of equal length. Each *kalpa (day of Brahma)* is followed by a *pralaya* (*night of Brahma* or *partial dissolution*) of equal length. Preceding the first and following each *manvantara* is a *manvantara-sandhya* (connection period), each with a length of *Krita-yuga* (*Satya-yuga*). Hindu texts specify that the start and end of each of the yugas are marked by astronomical alignments. This cycle's Treta-yuga began with 5 planets residing in the "*Aries*" constellation. This cycle's *Dvapara-yuga* ended with the "*Saptarshi*" constellation (*Ursa major*) residing in the "*Magha*" constellation. The current *Kali-yuga* will end with the Sun, Moon and Jupiter residing in the "*Pushya*" sector.

The history of humanity is divided up into four *yugas* (dharmic ages or world ages)—*Kṛta-yuga (Krita-yuga or Satya-yuga), Tretā-yuga, Dvāpara-yuga and Kali-yuga*—each with a 25% decline in dharmic practices and length, giving proportions (*caraṇas;* pronounced *charanas*) of *4:3:2:1* (*e.g., Satya: 100% start;*

*Kali: 25% start, 0% end*), indicating a de-evolution in spiritual consciousness and an evolution in material consciousness. *Kali-yuga* is followed by *Satya-yuga* of the next cycle, where a cycle is called a *catur-yuga* (pronounced *chatur-yuga* or *mahā-yuga*). Each yuga is divided into a main period (*yuga* proper) and two *yuga-sandhis* (*yuga-sandhyās*; connecting periods)—*yuga-sandhyā* (dawn) and *yuga-sandhyāṃśa* (*yuga-sandhyānśa*; dusk)—where each *yuga-sandhi* lasts for 10% of the main period. Lengths are given in divine years (celestial or *Deva* years), where a divine year lasts for 360 solar (human) years. A *chatur-yuga* lasts for 4.32 million solar (12,000 divine) years with 1,728,000 years of *Krita-yuga*, 1,296,000 years of *Treta-yuga*, 864,000 years of *Dvapara-yuga*, and 432,000 years of *Kali-yuga*.

### 卐 Current Yuga

*Kali-yuga* lasts for 432,000 years and is the 4th of 4 yugas in a cycle as well as the current yuga, with two sandhyas, each lasting for 36,000 years:

| |
|---|
| Kali-yuga started 3102 BCE in past:<br>= current year + Kali-yuga start year - year zero<br>= 2023 + 3102 - 1<br>= 5,124 years |
| Kali-yuga-sandhya (dawn) ends 32,899 CE in future:<br>= Kali-yuga-sandhya - elapsed Kali-yuga<br>= 36,000 - (2023 + 3102 - 1)<br>= 30,876 years |
| Kali-yuga-sandhyamsa (dusk) starts 392,899 CE in future:<br>= Kali-yuga - Kali-yuga-sandhyamsa - elapsed Kali-yuga<br>= 432,000 - 36,000 - (2023 + 3102 - 1)<br>= 390,876 years |
| Kali-yuga ends 428,899 CE in future:<br>= Kali-yuga - elapsed Kali-yuga<br>= 432,000 - (2023 + 3102 - 1) = 426,876 years |

## 卐 Current Chatur Yuga

A chatur-yuga lasts for 4.32 million years, where the current is the 28th of 71:

| |
|---|
| Started 3,891,102 BCE in past:<br><br>= chatur-yuga - Kali-yuga + elapsed Kali-yuga<br>= 4,320,000 - 432,000 + (2023 + 3102 - 1)<br>= 3,893,124 years<br>≈ 3.89 million years |
| Ends 428,899 CE in future:<br><br>= Kali-yuga - elapsed Kali-yuga<br>= 432,000 - (2023 + 3102 - 1)<br>= 426,876 years |

| Yuga | Start (– End) | Length |
|---|---|---|
| *Krita (Satya)* | *3,891,102 BCE* | *1,728,000 (4,800)* |
| *Treta* | *2,163,102 BCE* | *1,296,000 (3,600)* |
| *Dvapara* | *867,102 BCE* | *864,000 (2,400)* |
| *Kali** | *3102 BCE – 428,899 CE* | *432,000 (1,200)* |
| | | Years: 4,320,000 solar (12,000 divine) |

* *Current yuga*

## 卐 Pitri

The lifespan of the Pitris (forefathers) lasts for 100 of their years.

- 24 hours (1 day & night) of Pitris = 1 solar month (masa; Moon's two 15-day fortnights: bright and dark)
- 30 days (1 month) of Pitris = 30 solar months (2.5 solar years)
- 12 months (1 year) of Pitris = 30 solar years (1 month of Devas)
- 100 years (lifespan) of Pitris = 3,000 solar years

## 卐 Deva

The lifespan of the Devas (gods) lasts for 100 of their years.

- 24 hours (1 day & night) of Devas = 1 solar year (Sun's two 180-day motions: northern and southern)
- 30 days (1 month) of Devas = 30 solar years (1 year of Pitris)
- 12 months (1 year) of Devas = 360 solar years
- 100 years (lifespan) of Devas = 36,000 solar years

## 卐 Manu

The lifespan of the Manus (progenitors of mankind) lasts for 100 of their years. Each Manu reigns over a period called a manvantara, each lasting for 71 chatur-yugas (306.72 million years). A total of 14 Manus reign successively in one kalpa (day of Brahma). Preceding the first and following each manvantara is a sandhyā (connection period), each lasting the duration of Satya-yuga (1.728 million years). During each manvantara-sandhyā (a.k.a. manvantara-sandhi), Earth (Bhu-loka) is submerged in water.

- 24 hours (1 day & night) of Manu = 8,520 solar years
- 30 days (1 month) of Manu = 255,600 solar years
- 12 months (1 year) of Manu = 3,067,200 solar years
- 100 years (lifespan) of Manu = 306,720,000 solar years (71 chatur-yugas)

## 卐 Current Manvantara

A manvantara lasts for 306.72 million years, where the current (ruled by Vaivasvatha Manu) is the 7th of 14:

Started in past:

= elapsed 28th chatur-yuga + 27 chatur-yugas
= chatur-yuga - Kali-yuga + elapsed Kali-yuga + 27 chatur-yugas
= (4,320,000 - 432,000 + (2023 + 3102 - 1)) + 4,320,000 * 27
= 120,533,124 years

| ≈ 120.53 million years |
|---|
| Ends in future:<br><br>= remaining 28th chatur-yuga + 43 chatur-yugas<br>= Kali-yuga - elapsed Kali-yuga + 43 chatur-yugas<br>= (432,000 - (2023 + 3102 - 1)) + 4,320,000 * 43<br>= 186,186,876 years<br>≈ 186.19 million years |

### 卐 Brahma

The lifespan of *Brahma* (creator god) lasts for 100 of his years. His 12-hour day or kalpa (a.k.a. day of *Brahma* - ब्रह्मा) is followed by a 12-hour night or *pralaya* (night of *Brahma*) of equal length, each lasting for 4.32 billion years. A kalpa lasts for 1,000 *chatur-yugas* and has 14 *manvantaras* (मन्वन्तराः) and 15 *manvantara-sandhyas* (मन्वन्तर-सन्ध्याः) occurring in it. At the start of *Brahma*'s days, he is re-born and creates the planets and the first living entities. At the end of his days, he and his creations are unmanifest (partial dissolution). His 100-year life (311.04 trillion years) is called a *mahā-kalpa*, which is followed by a *mahā-pralaya* (full dissolution) of equal length, where the bases of the universe, *prakriti*, is manifest at the start and unmanifest at the end of a maha-kalpa. His 100-year life is divided into two 50-year periods, each called a parārdha. In 100 360-day years (*maha-kalpa*), there are a total of 36,000 full days: 36,000 kalpas (days proper) and 36,000 *pralayas* (nights).

- 12 hours (1 day proper: kalpa) of Brahma = 4.32 billion solar years (1,000 chatur-yugas; 14 manvantaras + 15 manvantara-sandhyas)
- 24 hours (1 day & night: kalpa + pralaya) of Brahma = 8.64 billion solar years
- 30 days (1 month) of Brahma = 259.2 billion solar years
- 12 months (1 year) of Brahma = 3.1104 trillion solar years
- 50 years (parardha) of Brahma = 155.52 trillion solar years
- 100 years (lifespan: 2 parardhas) of Brahma = 311.04 trillion solar years

## 卐 Current Kalpa

A kalpa (day of *Brahma*, 12 hours) lasts for 4.32 billion years, where the current (*Shveta-Varaha Kalpa*) is the 1st of 30 in his 1st month of his 51st year:

| |
|---|
| Started in past:<br><br>= elapsed 7th manvantara + 7 manvantara-sandhyas + 6 manvantaras<br>= elapsed 28th chatur-yuga + 27 chatur-yugas + 7 manvantara-sandhyas + 6 manvantaras<br>= chatur-yuga - Kali-yuga + elapsed Kali-yuga + 27 chatur-yugas + 7 manvantara-sandhyas + 6 manvantaras<br>= ((4,320,000 - 432,000 + (2023 + 3102 - 1)) + 4,320,000 * 27) + 1,728,000 * 7 + 306,720,000 * 6<br>= 1,972,949,124 years<br>≈ 1.97 billion years |
| Ends in future:<br><br>= remaining 7th manvantara + 8 manvantara-sandhyas + 7 manvantaras<br>= remaining 28th chatur-yuga + 43 chatur-yugas + 8 manvantara-sandhyas + 7 manvantaras<br>= Kali-yuga - elapsed Kali-yuga + 43 chatur-yugas + 8 manvantara-sandhyas + 7 manvantaras<br>= ((432,000 - (2023 + 3102 - 1)) + 4,320,000 * 43) + 1,728,000 * 8 + 306,720,000 * 7<br>= 2,347,050,876 years<br>≈ 2.35 billion years |

## 卐 Current Maha Kalpa

A *maha-kalpa* (life of Brahma) lasts for 311.04 trillion years:

| |
|---|
| Started in past:<br><br>= elapsed 18,001st kalpa + 18,000 kalpas + 18,000 pralayas |

| |
|---|
| = elapsed 7th manvantara + 7 manvantara-sandhyas + 6 manvantaras + 36,000 kalpas/pralayas<br>= elapsed 28th chatur-yuga + 27 chatur-yugas + 7 manvantara-sandhyas + 6 manvantaras + 36,000 kalpas/pralayas<br>= chatur-yuga - Kali-yuga + elapsed Kali-yuga + 27 chatur-yugas + 7 manvantara-sandhyas + 6 manvantaras + 36,000 kalpas/pralayas<br>= (((4,320,000 - 432,000 + (2023 + 3102 - 1)) + 4,320,000 * 27) + 1,728,000 * 7 + 306,720,000 * 6) + 4,320,000,000 * 36,000<br>= 155,521,972,949,124 years<br>≈ 155.52 trillion years |
| Ends in future:<br><br>= remaining 18,001st kalpa + 17,999 kalpas + 18,000 pralayas<br>= remaining 7th manvantara + 8 manvantara-sandhyas + 7 manvantaras + 35,999 kalpas/pralayas<br>= remaining 28th chatur-yuga + 43 chatur-yugas + 8 manvantara-sandhyas + 7 manvantaras + 35,999 kalpas/pralayas<br>= Kali-yuga - elapsed Kali-yuga + 43 chatur-yugas + 8 manvantara-sandhyas + 7 manvantaras + 35,999 kalpas/pralayas<br>= (((432,000 - (2023 + 3102 - 1)) + 4,320,000 * 43) + 1,728,000 * 8 + 306,720,000 * 7) + 4,320,000,000 * 35,999<br>= 155,518,027,050,876 years<br>≈ 155.52 trillion years |

### Yuga Cycle

A Yuga Cycle (*chatur yuga, maha yuga,* etc.) is a cyclic age (epoch) in Hindu cosmology. Each cycle lasts for 4,320,000 years (12,000 divine years) and repeats four yugas (world ages): *(i) Krita (Satya) Yuga, (ii) Treta Yuga, (iii) Dvapara Yuga, and (iv) Kali Yuga*. As a Yuga Cycle progresses through the four yugas, each yuga's length and humanity's general moral and physical state within each yuga decrease by one-fourth. *Kali Yuga,* which lasts for 432,000 years, is believed to have started in 3102 BCE. Near the end of *Kali Yuga,* when virtues are at their worst, a cataclysm and a re-establishment of dharma occur to usher in the next cycle's *Krita (Satya) Yuga,* prophesied to occur by the *God*

*Kalki*. There are 71 Yuga Cycles in a *manvantara* (age of *Manu*) and 1,000 Yuga Cycles in a *kalpa* (day of Brahma). A Yuga Cycle has several names.

"Age" and "*Yuga*", sometimes with reverential capitalization, commonly denote a "*catur-yuga*", a cycle of four world ages, unless expressly limited by the name of one of its minor ages (e.g., *Kali Yuga*). Its archaic spelling is *yug* (युग), with other forms of *yugam* (युगं), *yugānāṃ* (युगानाम्), and *yuge* (युगे) derived from *yuj* (युज्, 'to join or yoke'), believed derived from **yeug*.

- *Chatur Yuga* (चतुर्युग or *caturyuga, catur-yuga, chaturyuga*, or *chatur-yuga*) 'catur means four; a set of the four ages').

- A cyclic age encompassing the four yuga ages as defined in Hindu texts: *Surya Siddhanta, Manusmriti,* and *Bhagavata Purana.*

- *Daiva Yuga* (Sanskrit: दैवयुग, 'a divine or celestial age; an age of the gods').

- *Deva Yuga* (देवयुग, 'an age of the gods').

- *Divya Yuga* (दिव्य युग, 'a divine or celestial age').

- A cyclic age of the divine, celestrial, or gods (*Devas*) encompassing the four yuga ages ("human ages" or "world ages"). The Hindu texts give a length of 12,000 divine years, where a divine year lasts for 360 solar (human) years.

- *Maha Yuga* (महायुग or 'a great age'): A greater cyclic age encompassing the smaller four yuga ages.

It is theorized that the concept of the four yugas originated sometime after the compilation of the four *Vedas*, but prior to the rest of the Hindu texts, based on the concept's absence in the former writings. It is believed that the four yugas—*Krita (Satya), Treta, Dvapara, and Kali*—are named after throws of an Bhartiya game of long dice, marked with 4-3-2-1 respectively. A dice game is described in the *Rigveda* (ऋग्वेद), *Atharvaveda* (अथर्ववेद), *Upanishads* (उपनिषद), *Ramayana* (रामायण), *Mahabharata* (महाभारत), and *Puranas* (पुराण), while the four yugas are described after the four *Vedas* with no mention of a correlation to dice. A complete description of the four *yugas* and their characteristics are in the *Vishnu Smriti (ch. 20)* - विष्णुस्मृति (अध्याय २०),

*Mahabharata (e.g., Vanaparva 149, 183)* - महाभारत (उदा., वनपर्व १४९, १८३), *Manusmriti (I.81–86)* - मनुस्मृति *(I.८१– ८६), and Puranas (e.g., Brahma, ch. 122–123; Matsya, ch. 142–143; Naradiya, Purvardha, ch. 41)* - पुराणम् (उदा. ब्रह्म, अध्याय १२२-१२३; मत्स्य, अध्याय १४२-१४३; नारदिया, पूर्ववर्धा, अध्याय ४१). The four yugas are also described in the *Bhagavata Purana (3.11.18–20)* - भागवत पुराण *(३.११.१८ - २०).*

## Duration and Structure

Hindu texts describe four *yugas* in a *Yuga Cycle—Krita (Satya) Yuga, Treta Yuga, Dvapara Yuga,* and *Kali Yuga*—where, starting in order from the first age, each *yuga*'s length decreases by one-fourth (25%), giving proportions of *4:3:2:1*. Each yuga is described as having a main period (yuga proper) preceded by its *yuga-sandhyā* (dawn) and followed by its *yuga-sandhyāṃśa* (dusk), where each twilight (dawn/dusk) lasts for one-tenth (10%) of its main period. Lengths are given in divine years (years of the gods), each lasting for 360 solar (human) years.

Each Yuga Cycle lasts for 4,320,000 years (12,000 divine years) with its four yugas and their parts occurring in the following order:

| *Krita* (Satya) *Yuga* |
|---|
| • *Krita* (Satya) Yuga: 1,728,000 (4,800 divine) years |
| • *Krita-yuga-sandhya* (dawn): 144,000 (400 divine) |
| • *Krita-yuga* (proper): 1,440,000 (4,000 divine) |
| • *Krita-yuga-sandhyamsa* (dusk): 144,000 (400 divine) |
| *Treta Yuga* |
| • *Treta Yuga:* 1,296,000 (3,600 divine) years |
| • *Treta-yuga-sandhya* (dawn): 108,000 (300 divine) |
| • *Treta-yuga* (proper): 1,080,000 (3,000 divine) |
| • *Treta-yuga-sandhyamsa* (dusk): 108,000 (300 divine) |
| *Dvapara Yuga* |
| • *Dvapara Yuga*: 864,000 (2,400 divine) years |
| • *Dvapara-yuga-sandhya* (dawn): 72,000 (200 divine) |
| • *Dvapara-yuga* (proper): 720,000 (2,000 divine) |
| • *Dvapara-yuga-sandhyamsa* (dusk): 72,000 (200 divine) |

| *Kali Yuga* |
|---|
| • *Kali Yuga*: 432,000 (1,200 divine) years<br>• *Kali-yuga-sandhya* (dawn): 36,000 (100 divine)<br>• *Kali-yuga* (proper): 360,000 (1,000 divine)<br>• *Kali-yuga-sandhyamsa* (dusk): 36,000 (100 divine) |

### 卐 Yuga Cycle

Table 8.6 shows the structure of current Kalpa.

Table 8.6 Structure of current Kalpa

| Period | Divine Years | Solar Years |
|---|---|---|
| Kalpa | 12,000,000 | 4,320,000,000 |
| *Aadhi Sandhya* - आधि सन्ध्या | 4,800 | 1,728,000 |
| *Swayambhhu Manvantara* - स्वयंभू मन्वन्तर | 852,000 | 306,720,000 |
| *Manvantara Sandhya* - मन्वन्तर सन्ध्या | 4,800 | 1,728,000 |
| *Swarochisha Manvantara* - स्वारोचिषा मन्वन्तर | 852,000 | 306,720,000 |
| *Manvantara Sandhya* - मन्वन्तर सन्ध्या | 4,800 | 1,728,000 |
| *Uttama Manvantara* - *उत्तम मन्वन्तर* | 852,000 | 306,720,000 |
| *Manvantara Sandhya*- मन्वन्तर सन्ध्या | 4,800 | 1,728,000 |
| *Tapas/ Tamas Manvantara* - तपः/ तामस मन्वन्तर | 852,000 | 306,720,000 |
| *Manvantara Sandhya* - मन्वन्तर सन्ध्या | 4,800 | 1,728,000 |
| *Raivata Manvantara* - *रैवत मन्वन्तर* | 852,000 | 306,720,000 |
| *Manvantara Sandhya* - मन्वन्तर सन्ध्या | 4,800 | 1,728,000 |
| *Chakshusha Manvantara* - चक्षुष मन्वन्तर | 852,000 | 306,720,000 |
| *Manvantara Sandhya* - मन्वन्तर सन्ध्या | 4,800 | 1,728,000 |
| *Vaivasvata Manvantara* - वैवस्वतः मन्वन्तर | 852,000 | 306,720,000 |
| *Chatur Yuga 1 to 27* - चतुर्युग १ तः २७ पर्यन्तम् | 324,000 | 116,640,000 |
| | 12,000 | 4,320,000 |
| *Chatur Yuga 28* - चतुर्युग 28 | 4800 | 1,728,000 |
| • *Krita Yuga* - कृतयुग | 3600 | 1,296,000 |
| • *Treta Yuga* - त्रेता युग | 2400 | 864,000 |
| • *Dvapara Yuga* - द्वापर युग | 1200 | 432,000 |
| • ***Kali Yuga*** **- कलियुग** | 516,000 | 185,760,000 |

| | | |
|---|---|---|
| *Chatur Yuga 29 to 71* - चतुर्युग २९ तः ७१ पर्यन्तम् | | |
| *Manvantara Sandhya* - मन्वन्तर सन्ध्या | 4,800 | 1,728,000 |
| *Savarni Manvantara* - सवर्णि मन्वन्तर | 852,000 | 306,720,000 |
| *Manvantara Sandhya* - मन्वन्तर सन्ध्या | 4,800 | 1,728,000 |
| *Daksa Savarni Manvantara* - दक्ष सवर्णि मन्वन्तर | 852,000 | 306,720,000 |
| *Manvantara Sandhya* - मन्वन्तर सन्ध्या | 4,800 | 1,728,000 |
| *Brahma Savarni Manvantara* - ब्रह्म सावर्णि मन्वन्तर | 852,000 | 306,720,000 |
| *Manvantara Sandhya* - मन्वन्तर सन्ध्या | 4,800 | 1,728,000 |
| *Dharma Savarni Manvantara* - धर्म सावर्णि मन्वन्तर | 852,000 | 306,720,000 |
| *Manvantara Sandhya* - मन्वन्तर सन्ध्या | 4,800 | 1,728,000 |
| *Rudra Manvantara* - रुद्र मन्वन्तर | 852,000 | 306,720,000 |
| *Manvantara Sandhya* - मन्वन्तर सन्ध्या | 4,800 | 1,728,000 |
| *Raucya or Deva Savarni Manvantara Sandhya* -रौच्य वा देव सवर्णि मन्वन्तर सन्ध्या | 852,000 | 306,720,000 |
| *Manvantara Sandhya* - मन्वन्तर सन्ध्या | 4,800 | 1,728,000 |
| *Indra Savarni Manvantara* - इन्द्र सावर्णि मन्वन्तर | 852,000 | 306,720,000 |
| *Manvantara Sandhya* - मन्वन्तर सन्ध्या | 4,800 | 1,728,000 |

## 卐 Mahabharata

The *Mahabharata* (12.231.12–31) describes units of time from a wink of the eye (*nimesha* - निमेषः) up to the days (*kalpa*) and nights (*pralaya*) of Brahma.

| |
|---|
| *(12–13) The Rishis, measuring time, have given particular names to particular portions. Five and ten winks of the eye make what is called a Kashtha. Thirty Kashthas make what is called a Kala. Thirty Kalas, with the tenth part of a Kala, make a Muhurta. Thirty Muhurtas make up one day and night. Thirty days and nights form a month, and twelve months form a year.* |
| *(14) Persons well-read in mathematical science say that a year is made up of two solar motions, viz, the northern and the southern.* |

| |
|---|
| *(15) The sun makes the day and the night for men. The night is for the sleep of all living creatures and the day is for work.* |
| *(16) A month of human beings is equal to a day and night of the departed manes. That division consists in this the light half of the month is their day which is for work, and the dark fortnight is their night for sleep.* |
| *(17) A year (of men) is equal to a day and night of the gods. This division consists in this the half year for which the sun travels from the vernal to the autumnal equinox is the day of the gods, and the half year for which the sun moves from the latter to the former is their night.* |
| *(18) Calculating by the days and nights of human beings about which I have told you I shall speak of the day and night of Brahman [(Brahma)] and his years also.* |
| *(19) I shall, in their order, tell you the number of years that are for different purposes calculated differently, in the Krita, the Treta, the Dwapara, and the Kali yugas.* |
| *(20) Four thousand celestial years is the duration of the first or Krita age. The morning of that cycle consists of four hundred years and its evening is of four hundred years.* |
| *(21) Regarding the other cycles, the duration of each gradually decreases by a quarter in respect of both the principal period with the minor portion and the conjoining portion itself.* |
| *(29) The learned say that these twelve thousand celestial years form what is called a cycle. A thousand such cycles form a single day of Brahman [(Brahma)].* |
| *(30) The same is the duration of Brahman's [(Brahma's)] night. With the beginning of Brahman's [(Brahma's)] day the universe begins to come into being. During the period of universal dissolution the Creator sleeps in Yoga meditation. When the period of sleep expires, He awakes.* |

*(31) What is Brahman's [(Brahma's)] day covers a thousand such cycles. His night also covers a thousand similar cycles. They who know this are said to know the day and the night.*

*— Mahabharata 12 (Shanti Parva), Ch. 231[42][i]*

### 卐 Manusmriti

The *Manusmriti* - मनुस्मृति (Laws of Manu; 1.64–80) describes units of time from a twinkling of the eye (nimesha) up to the days (kalpa) and nights (pralaya) of Brahma.

निमेषा दश चाष्टौ च काष्ठा त्रिंशत् तु ताः कला ।
त्रिंशत् कला मुहूर्तः स्यादहोरात्रं तु तावतः ॥ ६४ ॥

nimeṣā daśa cāṣṭau ca kāṣṭhā triṃśat tu tāḥ kalā I
triṃśat kalā muhūrtaḥ syādahorātraṃ tu tāvataḥ II 64 II

(64) Eighteen nimeshas (twinklings of the eye, are one kashtha), thirty kashthas one kala, thirty kalas one muhurta, and as many (muhurtas) one day and night.

अहोरात्रे विभजते सूर्यो मानुषदैविके ।
रात्रिः स्वप्नाय भूतानां चेष्टायै कर्मणामहः ॥ ६५ ॥

ahorātre vibhajate sūryo mānuṣadaivike I
rātriḥ svapnāya bhūtānāṃ ceṣṭāyai karmaṇāmahaḥ II 65 II

(65) The sun divides days and nights, both human and divine, the night (being intended) for the repose of created beings and the day for exertion.

पित्र्ये रात्र्यहनी मासः प्रविभागस्तु पक्षयोः ।
कर्मचेष्टास्वहः कृष्णः शुक्लः स्वप्नाय शर्वरी ॥ ६६ ॥

pitrye rātryahanī māsaḥ pravibhāgastu pakṣayoḥ ।
karmaceṣṭāsvahaḥ kṛṣṇaḥ śuklaḥ svapnāya śarvarī ॥ 66 ॥

(66) A month is a day and night of the manes, but the division is according to fortnights. The dark (fortnight) is their day for active exertion, the bright (fortnight) their night for sleep.

दैवे रात्र्यहनी वर्षं प्रविभागस्तयोः पुनः ।
अहस्तत्रोदगयनं रात्रिः स्याद् दक्षिणायनम् ॥ ६७ ॥

daive rātryahanī varṣaṃ pravibhāgastayoḥ punaḥ ।
ahastatrodagayanaṃ rātriḥ syād dakṣiṇāyanam ॥ 67 ॥

(67) A year is a day and a night of the gods; their division is (as follows): the half year during which the sun progresses to the north will be the day, that during which it goes southwards the night.

ब्राह्मस्य तु क्षपाहस्य यत् प्रमाणं समासतः ।
एकैकशो युगानां तु क्रमशस्तन्निबोधत ॥ ६८ ॥

brāhmasya tu kṣapāhasya yat pramāṇaṃ samāsataḥ ।
ekaikaśo yugānāṃ tu kramaśastannibodhata ॥ 68 ॥

(68) But hear now the brief (description of) the duration of a night and a day of Brahman [(Brahma)] and of the several ages (of the world, yuga) according to their order.

चत्वार्याहुः सहस्राणि वर्षाणां तत् कृतं युगम् ।
तस्य तावत्शती सन्ध्या सन्ध्यांशश्च तथाविधः ॥ ६९ ॥

catvāryāhuḥ sahasrāṇi varṣāṇāṃ tat kṛtaṃ yugam I
tasya tāvatśatī sandhyā sandhyāṃśaśca tathāvidhaḥ II 69 II

(69) They declare that the Krita age (consists of) four thousand years (of the gods); the twilight preceding it consists of as many hundreds, and the twilight following it of the same number.

इतरेषु ससन्ध्येषु ससन्ध्यांशेषु च त्रिषु ।
एकापायेन वर्तन्ते सहस्राणि शतानि च ॥ ७० ॥

itareṣu sasandhyeṣu sasandhyāṃśeṣu ca triṣu I
ekāpāyena vartante sahasrāṇi śatāni ca II 70 II

(70) In the other three ages with their twilights preceding and following, the thousands and hundreds are diminished by one (in each).

यदेतत् परिसङ्ख्यातमादावेव चतुर्युगम् ।
एतद् द्वादशसाहस्रं देवानां युगमुच्यते ॥ ७१ ॥

yadetat parisaṅkhyātamādāveva caturyugam I
etad dvādaśasāhasraṃ devānāṃ yugamucyate II 71 II

(71) These twelve thousand (years) which thus have been just mentioned as the total of four (human) ages, are called one age of the gods.

तद् वै युगसहस्रान्तं ब्राह्मं पुण्यमहर्विदुः ।
रात्रिं च तावतीमेव तेऽहोरात्रविदो जनाः ॥ ७३ ॥

tad vai yugasahasrāntaṃ brāhmaṃ puṇyamaharviduḥ I
rātriṃ ca tāvatīmeva te'horātravido janāḥ II 73 II

(73) Those (only, who) know that the holy day of Brahman [(Brahma)], indeed, ends after (the completion of) one thousand ages (of the gods) and that his night lasts as long, (are really) men acquainted with (the length of) days and nights.

यद् प्राग् द्वादशसाहस्रमुदितं दैविकं युगम् ।
तदेकसप्ततिगुणं मन्वन्तरमिहोच्यते ॥ ७९ ॥

yad prāg dvādaśasāhasramuditaṃ daivikaṃ yugam ।
tadekasaptatiguṇaṃ manvantaramihocyate ॥ 79 ॥

(79) The before-mentioned age of the gods, (or) twelve thousand (of their years), being multiplied by seventy-one, (constitutes what) is here named the period of a Manu (Manvantara).

मन्वन्तराण्यसङ्ख्यानि सर्गः संहार एव च ।
क्रीडन्निवैतत् कुरुते परमेष्ठी पुनः पुनः ॥ ८० ॥

manvantarāṇyasaṅkhyāni sargaḥ saṃhāra eva ca ।
krīḍannivaitat kurute parameṣṭhī punaḥ punaḥ ॥ 80 ॥

(80) The Manvantaras, the creations and destructions (of the world, are) numberless; sporting, as it were, Brahman [(Brahma)] repeats this again and again.

— Manusmriti, Ch. 1

### 卐 Surya Siddhanta

The *Surya Siddhanta (1.10–21)* describes units of time from a respiration (*prana*) up to the 100-year lifespan of *Brahma (maha-kalpa).*

| |
|---|
| lokānām antakṛt kālaḥ kālo 'nyaḥ kalanātmakaḥ ǀ<br>sa dvidhā sthūlasūkṣmatvān mūrtaś cāmūrta ucyate ǁ 10 ǁ<br><br>(10) Time is the destroyer of the worlds; another Time has for its nature to bring to pass. This latter, according as it is gross or minute, is called by two names, real (murta) and unreal (amurta). |
| prāṇādiḥ kathito mūrtas truṭyādyo 'mūrtasaṃjñakaḥ ǀ<br>ṣaḍbhiḥ prāṇair vināḍī syāt tatṣaṣṭyā nāḍikā smṛtā ǁ 11 ǁ<br><br>(11) That which begins with respirations (prana) is called real; that which begins with atoms (truti) is called unreal. Six respirations make a vinadi, sixty of these a nadi; |
| nāḍīṣaṣṭyā tu nākṣatram ahorātraṃ prakīrtitam ǀ<br>tattriṃśatā bhaven māsaḥ sāvano 'rkodayais tathā ǁ 12 ǁ<br><br>(12) And sixty nadis make a sidereal day and night. Of thirty of these sidereal days is composed a month; a civil (savana) month consists of as many sunrises; |
| aindavas tithibhis tadvat saṃkrāntyā saura ucyate ǀ<br>māsair dvādaśabhir varṣaṃ divyaṃ tad aha ucyate ǁ 13 ǁ<br><br>(13) A lunar month, of as many lunar days (tithi); a solar (saura) month is determined by the entrance of the sun into a sign of the zodiac: twelve months make a year. This is called a day of the gods. |

surāsurāṇām anyonyam ahorātraṃ viparyayāt ।
tatṣaṣṭiḥ ṣaḍguṇā divyaṃ varṣam āsuram eva ca ॥ 14 ॥

(14) The day and night of the gods and of the demons are mutually opposed to one another. Six times sixty [360] of them are a year of the gods, and likewise of the demons.

taddvādaśasahasrāṇi caturyugam udāhṛtam ।
sūryābdasaṃkhyayā dvitrisāgarair ayutāhataiḥ ॥ 15 ॥

(15) Twelve thousands of these divine years are denominated a Quadruple Age (caturyuga); of ten thousand times four hundred and thirty-two [4,320,000] solar years

saṃdhyāsaṃdhyāṃśasahitaṃ vijñeyaṃ tac caturyugam ।
kṛtādīnāṃ vyavastheyaṃ dharmapādavyavasthayā ॥ 16 ॥

(16) Is composed that Quadruple Age, with its dawn and twilight. The difference of the Golden and the other Ages, as measured by the difference in the number of the feet of Virtue in each, is as follows:

yugasya daśamo bhāgaś catustridvyekasaṃguṇaḥ ।
kramāt kṛtayugādīnāṃ ṣaṣṭhāṃśaḥ saṃdhyayoḥ svakaḥ ॥ 17 ॥

(17) The tenth part of an Age, multiplied successively by four, three, two, and one, gives the length of the Golden and the other Ages, in order: the sixth part of each belongs to its dawn and twilight.

yugānāṃ saptatiḥ saikā manvantaram ihocyate |
kṛtābdasaṃkhyās tasyānte saṃdhiḥ prokto jalaplavaḥ || 18 ||

(18) One and seventy Ages are styled here a Patriarchate (manvantara); at its end is said to be a twilight which has the number of years of a Golden Age, and which is a deluge.

sasaṃdhayas te manavaḥ kalpe jñeyāś caturdaśa |
kṛtapramāṇaḥ kalpādau saṃdhiḥ pañcadaśaḥ smṛtaḥ || 19 ||

(19) In an Æon (kalpa) are reckoned fourteen such Patriarchs (manu) with their respective twilights; at the commencement of the Æon is a fifteenth dawn, having the length of a Golden Age.

ittham yugasahasreṇa bhūtasaṃhārakārakaḥ |
kalpo brāhmam ahaḥ proktaṃ śarvarī tasya tāvatī || 20 ||

(20) The Æon, thus composed of a thousand Ages, and which brings about the destruction of all that exists, is styled a day of Brahma; his night is of the same length.

paramāyuḥ śataṃ tasya tayāhorātrasaṃkhyayā |
āyuṣo 'rdhamitaṃ tasya śeṣakalpo 'yam ādimaḥ || 21 ||

(21) His extreme age is a hundred, according to this valuation of a day and a night ...

— *Surya Siddhanta, Ch. 1[49]*

## 卐 Greater Cycles

There are 71 Yuga Cycles (306,720,000 years) in a *manvantara*, a period ruled by *Manu*, who is the *progenitor of mankind*. There are 1,000 Yuga Cycles (4,320,000,000 years) in a *kalpa*, a period that is a day (12-hour day proper) of

*Brahma,* who is the creator of the planets and first living entities. There are 14 *manvantaras* (4,294,080,000 years) in a *kalpa* with a remainder of 25,920,000 years assigned to 15 *manvantara-sandhyas* (junctures), each the length of a *Satya Yuga* (1,728,000 years). A *kalpa* is followed by a *pralaya* (night or partial dissolution) of equal length forming a full day (24-hour day). A *maha-kalpa* (life of *Brahma*) lasts for 100 360-day years of *Brahma,* which lasts for 72,000,000 Yuga Cycles (311.04 trillion years) and is followed by a *maha-pralaya* (full dissolution) of equal length.

We are currently halfway through Brahma's life (maha-kalpa):

- 51st year of 100 (2nd half or parardha)
- 1st month of 12
- 1st kalpa (Shveta-Varaha Kalpa) of 30
- 7th manvantara (Vaivasvatha Manu) of 14
- 28th chatur-yuga (a.k.a. Yuga Cycle) of 71
- 4th yuga (Kali Yuga) of 4

*Yuga* dates are used in an ashloka, which is read out at the beginning of Hindu rites to specify the elapsed time in Brahma's life:

*5121 of Kaliyuga year (for 2020 CE) of the 28th Caturyuga of the 7th Manvantra on the first day of the 51st year of Brahma.*

### 卐 Modern Theories

Breaking from the long duration of a Yuga Cycle, new theories have emerged regarding the length, number, and order of the yugas.

#### Sri Yukteswar Giri

Swami Sri Yukteswar Giri - स्वामी श्री युक्तेश्वर गिरि (1855–1936) proposed a Yuga Cycle of 24,000 years in the introduction of his book *The Holy Science* (1894). He claimed the understanding that *Kali Yuga* lasts for 432,000 years was a mistake, which he traced back to *Raja Parikshit* (राजा परीक्षित), just after the descending *Dvapara Yuga* ended (c. 3101 BCE) and all the wise men of his court retired to the *Himalaya Mountains* (हिमालय पर्वत). With no one left to correctly calculate the ages, *Kali Yuga* never officially started. After 499 CE, in ascending *Dvapara Yuga*, when the intellect of men began to develop, but not fully, they noticed mistakes and attempted to correct them by converting what they thought to be *divine years to human years (1:360 ratio). Yukteswar*'s yuga lengths for *Satya, Treta, Dvapara,* and *Kali* are respectively *4,800, 3,600, 2,400, and 1,200* "human" years (12,000 years total). He accepted the four yugas and their *4:3:2:1* length and dharma proportions, but his Yuga Cycle contained *eight yugas,* the original descending set of the four yugas followed by an ascending (reversed) set, where he called each set a "*Daiva Yuga*" or "Electric Couple". His Yuga Cycle lasts for 24,000 years, which he believed equals one precession of the equinoxes (traditionally 25,920 years; 1,920 years difference). He states that the world entered the *Pisces-Virgo* Age in 499 CE ("cycle bottom"), and that the current age of ascending *Dvapara Yuga* started in 1699 CE around the time of scientific discoveries and advancements such as electricity. He explained that in a 24,000-year Yuga Cycle, the Sun completes one orbit around some dual star, becoming nearer and farther to a galactic center, which the pair orbit in a longer period. He called this galactic center Vishnunabhi - विष्णुनाभि (Vishnu's Navel), where *Brahma* regulates *dharma* (धर्मः) or, as *Yukteswar* defined it, *mental virtue. Dharma* is lowest when farthest from *Brahma* at the *descending-ascending intersection* ("*cycle-bottom*"), where the opposite occurs at the "*cycle-top*" when nearest. At *dharma*'s lowest (499 CE), human intellect cannot comprehend anything beyond the gross material world. Comparison of his yuga lengths is shown in Table 8.7.

Table 8.7 Yuga lengths

| Yuga | Start (– End) | Length |
|---|---|---|
| | Descending (12,000 years) | |
| *Krita (Satya)* | *11,501 BCE* | 4,800 |
| *Treta* | *6701 BCE* | 3,600 |
| *Dvapara* | *3101 BCE* | 2,400 |
| *Kali* | *701 BCE* | 1,200 |
| | Ascending (12,000 years) | |
| *Kali* | *499 CE* | 1,200 |
| *Dvapara** | *1699 CE* | 2,400 |
| *Treta* | *4099 CE* | 3,600 |
| *Krita (Satya)* | *7699 CE – 12,499 CE* | 4,800 |
| | | **Years: 24,000** |

## 卐 Manvantara

A *manvantara* (मन्वन्तर), in Hindu cosmology, is a cyclic period of time identifying the duration, reign, or age of a *Manu* - मनुः, the progenitor of mankind. In each manvantara, *seven Rishis* - सप्तर्षि, certain deities, an *Indra* - इन्द्र, a *Manu*, and *kings (sons of Manu)* are created and perish. Each *manvantara* is distinguished by the *Manu* who *rules/reigns* over it, of which we are currently in the seventh *manvantara* of fourteen, which is ruled by *Vaivasvata Manu*.

*Manvantara* sometimes spelled *manwantara or manuantara*, is a compound of *manu* and *antara* (अन्तर, 'interval, period, or term'), creating *manu-antara* (मनु-अन्तर) or *manvantara "the duration of a Manu"*, or his lifespan, with synonym meanings of "*the interval, reign, period, or age of a Manu*".

Sandhya (सन्ध्या, *'union, juncture, or twilight i.e., dawn/morning or dusk/night'*) or sandhi (सन्धि, *'union or juncture'*), sometimes with a compound of *kala* (काल, '*time*'), have been used to represent "*the juncture before or after a manvantara*", a period of universal deluge (flood):

- *manvantara-sandhyā* (मन्वन्तर सन्ध्या)
- *manvantara sandhi* (मन्वन्तर सन्धि)
- *sandhyā-kāla* (सन्ध्या काल) when describing a manvantara
- *sandhikāla* (सन्धि काल) when describing a manvantara

Each *manvantara* lasts for 306,720,000 years (852,000 divine years; 1 divine year = 360 solar years) and repeats seventy-one Yuga Cycles (world ages). In a *kalpa* (day of *Brahma*), which lasts for 4.32 billion years (12 million divine years or 1,000 Yuga Cycles), there are a total of *fourteen manvantaras* (14 x 71 = 994 Yuga Cycles), where each is followed by and the first preceded by a *manvantara-sandhya* (*fifteen sandhyas*) with each *sandhya* lasting for 1,728,000 years (4,800 divine years; the duration of Satya Yuga). During each *manvantara-sandhya*, the earth (*Bhu-loka*) is submerged in water.

Each *kalpa* has 14 *manvantaras* and 15 *manvantara-sandhyas* in the following order:

- *1st manvantara-sandhya (adi sandhya)*
- *1st manvantara*

- *2nd manvantara-sandhya*
- *2nd manvantara*
- ...
- *14th manvantara-sandhya*
- *14th manvantara*
- *15th manvantara-sandhya*

Some Examples form Hindu texts are mentioned below:

> (67) A year is a day and a night of the gods ... (79) The before-mentioned age of the gods, (or) twelve thousand (of their years), being multiplied by seventy-one, (constitutes what) is here named the period of a Manu (Manvantara). (80) The Manvantaras, the creations and destructions (of the world, are) numberless; sporting, as it were, Brahman [(Brahma)] repeats this again and again.
>
> — *Manusmriti, Ch. 1*

> (13) ... twelve months make a year. This is called a day of the gods. (14) ... Six times sixty [360] of them are a year of the gods ... (15) Twelve thousand of these divine years are denominated a Quadruple Age (caturyuga); of ten thousand times four hundred and thirty-two [4,320,000] solar years (18) One and seventy [71] Ages are styled here a Patriarchate (manvantara); at its end is said to be a twilight which has the number of years of a Golden Age, and which is a deluge. (19) In an Æon (kalpa) are reckoned fourteen such Patriarchs (manu) with their respective twilights; at the commencement of the Æon is a fifteenth dawn, having the length of a Golden Age.
>
> — *Surya Siddhanta, Ch. 1*

> Twelve thousand divine years, each composed of (three hundred and sixty) such days, constitute the period of the four Yugas, or ages ... a thousand such aggregates are a day of Brahma, and fourteen Manus reign within that term ... Seven Rishis, certain (secondary) divinities, Indra, Manu, and the kings his sons, are created and perish at one period; and the interval, called a Manwantara, is equal to seventy-one times the number of years contained in the four Yugas, with some additional years: this is the duration of the Manu, the (attendant) divinities, and the rest, which is equal to 852,000 divine years, or to 306,720,000 years of mortals, independent of the additional period. Fourteen times this period constitutes a Brahma day, that is, a day of Brahma ...
>
> — *Vishnu Purana, Part 1, Ch. 3*

### Manus

In our current kalpa (day of *Brahma*), these fourteen *Manu*'s reign in succession:

| | |
|---|---|
| 1. Swayambhu Manu | 8. Savarni Manu |
| 2. Swarochisha Manu | 9. Daksa Savarni Manu |
| 3. Uttama Manu | 10. Brahma Savarni Manu |
| 4. Tapasa/Tamasa Manu | 11. Dharma Savarni Manu |
| 5. Raivata Manu | 12. Rudra Savarni Manu |
| 6. Chakshusha Manu | 13. Raucya or Deva Savarni Manu |
| 7. Vaivasvata Manu (current) | 14. Indra Savarni Manu |

### Hindu Astronomy

In the early texts of Hindu astronomy such as *Surya Siddhanta,* the length of a yuga cycle is used to specify the orbital period of heavenly bodies. Instead of specifying the period of a single orbit of a heavenly body around the Earth, the number of orbits of a heavenly body in a yuga cycle is specified.

**Surya Siddhanta, Ch. 1:[64]**

*(29) In an Age (yuga), the revolutions of the sun, Mercury, and Venus, and of the conjunctions (shighra) of Mars, Saturn, and Jupiter, moving eastward, are four million, three hundred and twenty thousand; (30) Of the moon, fifty-seven million, seven hundred and fifty-three thousand, three hundred and thirty-six; of Mars, two million, two hundred and ninety-six thousand, eight hundred and thirty-two; (31) Of Mercury's conjunction (shighra), seventeen million, nine hundred and thirty-seven thousand, and sixty; of Jupiter, three hundred and sixty-four thousand, two hundred and twenty; (32) Of Venus's conjunction (shigra), seven million, twenty-two thousand, three hundred and seventy-six; of Saturn, one hundred and forty-six thousand, five hundred and sixty-eight; (33) Of the moon's apsis (ucca), in an Age, four hundred and eighty-eight thousand, two hundred and three; of its node (pata), in the contrary direction, two hundred and thirty-two thousand, two hundred and thirty-eight; (34) Of the asterisms, one billion, five hundred and eighty-two million, two hundred and thirty-seven thousand, eight hundred and twenty-eight....*

The orbital period of heavenly bodies can be derived from the above numbers provided the starting point of a yuga cycle is known. According to Burgess, the Surya Siddhanta fixes the starting point of Kali Yuga as:

*The instant at which the Age is made to commence is midnight on the meridian of Ujjayini, at the end of the 588,465th and beginning of the 588,466th day (civil reckoning) of the Julian Period, or between the 17th and 18th of February 1612 J.P., or 3102 B.C.[65]*

Based on this starting point, Ebenezer Burgess calculates the following planetary orbital periods (see Table 8.8):

Table 8.8 Comparison of sidereal revolutions of the planets (geocentric)

| Planet | Surya Siddhanta | | Modern |
|---|---|---|---|
| | Revolutions in a yuga cycle | Revolution length (day hr min sec) | Orbital period (day hr min sec) |
| *Sun* | 4,320,000 | *365 6 12 36.6* | *365 6 9 10.8* |
| *Mercury* | *17,937,060* | *87 23 16 22.3* | *87 23 15 43.9* |
| *Venus* | *7,022,376* | 224 16 45 56.2 | 224 16 49 *8.0* |
| *Mars* | 2,296,832 | 686 23 56 23.5 | 686 23 30 41.4 |

| | | | |
|---|---|---|---|
| *Jupiter* | *364,220* | *4,332 7 41 44.4* | *4,332 14 2 8.6* |
| *Saturn* | *146,568* | *10,765 18 33 13.6* | *10,759 5 16 32.2* |
| *Moon (sidereal)* | *57,753,336* | *27 7 43 12.6* | *27 7 43 11.4* |
| *Moon (synodic)* | *53,433,336* | *29 12 44 2.8* | *29 12 44 2.9* |

## 卐 Kalpa

A *kalpa* is a long period of time (*aeon*) generally between the creation and recreation of a world or universe. Each *kalpa* is divided into 14 manvantara periods, each lasting 71 Yuga Cycles (306,720,000 years). Preceding the first and following each manvantara period is a juncture (sandhya) equal to the length of a *Satya Yuga* (1,728,000 years). A *kalpa* is followed by a *pralaya* (dissolution) of equal length, which together constitute a day and night of *Brahma*. A month of *Brahma* contains thirty such days and nights, or 259.2 billion years. According to the *Mahabharata,* 12 months of *Brahma* (= 360 days) constitute his year, and 100 such years his life called a maha-kalpa (311.04 trillion years or *36,000 kalpa + 36,000 pralaya*). Fifty years of *Brahma* are supposed to have elapsed, and we are now in the *Shveta-Varaha Kalpa* (श्वेत-वराह कल्प) or the first day of his fifty-first year. At the end of a *kalpa,* the world is annihilated by fire. The definition of a *kalpa* equaling 4.32 billion years is found in the *Puranas*—specifically *Vishnu Purana* and *Bhagavata Purana.*

> The duration of the material universe is limited. It is manifested in cycles of kalpas. A kalpa is a day of Brahmā, and one day of Brahmā consists of a thousand cycles of four yugas, or ages: Satya Yuga, Treta Yuga, Dvapara Yuga and Kali Yuga. ... These four yugas, rotating a thousand times, comprise one day of Brahmā, and the same number comprise one night. Brahmā lives one hundred of such "years" and then dies. These "hundred years" total 311 trillion 40 billion (311,040,000,000,000) earth years. By these calculations the life of Brahmā seems fantastic and interminable, but from the viewpoint of eternity it is as brief as a lightning flash. In the Causal Ocean there are innumerable Brahmās rising and disappearing like bubbles. Brahmā and his creation are all part of the material universe, and therefore they are in constant flux.
>
> — *Brihat Swasthani Brata Katha*

The *Matsya Purana* - मत्स्य पुराण *(290.3–12)* lists the names of 30 *kalpas*, each named by *Brahma* based on a significant event in the *kalpa* and the most glorious person in the beginning of the *kalpa*. These 30 *kalpas* or days (along with 30 pralayas or nights) form a 30-day month of *Brahma*.

| | | | | |
|---|---|---|---|---|
| Śveta (current) | Vṛhat | Udāna | Soma | Aghora |
| Nīlalohita | Kandarpa | Gāruda | Mānava | Varāha |
| Vāmadeva | Sadya | Kaurma | Tatpumān | Vairaja |
| Rathantara | Iśāna | Nārasiṁha | Vaikuṇṭha | Gaurī |
| Raurava | Tamah | Samāna | Lakṣmī | Māheśvara |
| Deva | Sārasvata | Āgneya | Sāvitrī | Pitṛ |

The *Vayu Purana* has a different list of names for 33 *kalpas*.

## Number and Distances in Hindu Scriptures

The Hindu scriptures contain many numerical descriptions concerning distances, durations and numbers of items in the universe as seen from the perspective of Hindu cosmology. Table 8.9 and Table 8.10 show the distances and corresponding *yojanas*.

Table 8.9 Various distances

| Property | Number or Measurement |
|---|---|
| Distance from Satyaloka to Vishnuloka (Brahmaloka-sanatana, abode of Brahma) | 26,200,000 yojanas |
| Distance from Dhruvaloka (the pole star) to the Sun | 3,800,000 yojanas |
| Distance from Dhruvaloka to Maharloka | 10,000,000 yojanas |
| Distance from Maharloka to Janaloka | 20,000,000 yojanas |
| Distance from Janaloka to Tapoloka | 80,000,000 yojanas |
| Distance from Tapoloka to Satyaloka | 120,000,000 yojanas |
| Distance from the Sun to Satyaloka | 233,800,000 yojanas |
| Covering of the Universe | 260,000,000 yojanas (2,080,000,000 miles) away from the Sun, each covering layer is 10 times bigger than previous |

| | |
|---|---|
| Distance from lower planets to Śesha Nāga lying on the Garbhodaka Ocean | 30,000 yojanas |
| Depth of Garbhodaka ocean | 249,800,000 yojanas deep |
| Total diameter of the Universe | approximately 500,000,000 yojanas |
| Distance from planet Rāhu to the Sun | 10,000 yojanas |
| Distance from Rāhu to planets of the Siddhas, Cāranas and Vidyādharas | 1,000,000 yojanas |
| Distance from the Earth to the lower planetary systems — Atala, Vitala, Sutala, Talātala, Mahātala, Rasātala and Pātāla | 70,000 yojanas |
| Number of kinds of material bodies for soul, jivaatma (or species of life) in material world | 8,400,000 |
| Number of species living in the water | 900,000 |
| Number of kinds of sthāvara (non-moving living entities such as trees and plants) | 2,000,000 |
| Number of species of insects and reptiles | 1,100,000 |
| Number of species of birds | 1,000,000 |
| Number of varieties of quadrupeds | 3,000,000 |
| Number of human species | 400,000 |
| Number of Manus who manifest in one kalpa (one day of Brahmā) | fourteen Manus |
| Duration of one day of Brahmā | 1,000 chatur-yugas (4.32 billion years) |
| Number of Manus who manifest during one month of Brahmā | 420 Manus |
| Number of Manus who manifest during one year of Brahmā | 5,040 Manus |
| Number of Manus who manifest during the lifetime of one Brahmā | 504,000 Manus |
| Total duration of four yugas (Satya+Treta+Dvapara+Kali) | 4,320,000 years |
| Number of the present Manu (Vaivasvata Manu) | 7th Manu |
| Age of our (Vaivasvata) Manu | 27 chatur-yugas [27 x 4,320,000 solar years] |
| Number of planetary systems in this Brahmanda (material universe) | 14 |
| Duration of 100 years of Brahma (his lifetime duration) | 311,040,000,000,000 years |

| | |
|---|---|
| Duration of one manvantara, the lifespan of one Manu | seventy-one chatur-yugas (=71*4,320,000 years) |
| Duration of Satya yuga | 1,728,000 years |
| Duration of Tretā-yuga | 1,296,000 years |
| Duration of Dvāpara-yuga | 864,000 years |
| Duration of Kali-yuga | 432,000 years |
| Duration of lifetime of human in Satya yuga | around 100,000 years |
| Duration of lifetime of human in Tretā-yuga | around 10,000 years |
| Duration of lifetime of human in Dvāpara-yuga | around 1,000 years |
| Duration of lifetime of human in Kali-yuga | around 100 years which will be reduced gradually to 20 years) |
| Time passed after beginning of Kali-yuga (Battle of Kurukṣetra) | 5,124 years as of 2023 CE (See Kali Yuga) |
| Steeds of Indra | 1,100 |
| Spiritual benefits of following Janmashtami Vows (fasting etc.) | One ... becomes freed from the sins of ten million births. |
| Frequency with which Krishna descends to this planet | every 8,640,000,000 years (once in a day of Brahma) |
| Happiness of liberation into impersonal Brahman (brahmajyoti) compared with happiness of devotional service to Bhagavan | Bhakti is trillion-fold better than impersonalism |

*Note: Where distances are given in yojana a rough equivalent is sometimes given in miles calculated at 8 miles to the yojana. The actual length of the yojana varied throughout its period of use between 4 and 9 miles.*

Table 8.10 Yojana distances

| | |
|---|---|
| yojana-śata | one hundred yojanas (eight hundred miles); SB 4.6.32 |
| niyuta-yojana-viśālaḥ | one million yojanas (eight million miles) wide; SB 5.16.5 |
| nava-yojana-sahasra | 72,000 miles in length; SB 5.16.6 |
| yojana | yojanas (8 mile each SB 5.16.7) |
| ayuta-yojana-utsedhāḥ | ten thousand yojanas high; SB 5.16.9 |
| ayuta-yojana-vistāra-unnāhāḥ | which measure ten thousand yojanas high and wide; SB 5.16.11 |
| sahasra-yojana-unnāhāḥ | one thousand yojanas high; SB 5.16.12 |
| śata-yojana | one hundred yojanas; SB 5.16.12 |

| | |
|---|---|
| ekādaśa-śata-yojana-uttuṅga | 1,100 yojanas high; SB 5.16.16 |
| aṣṭādaśa-yojana-sahasram | eighteen thousand yojanas; SB 5.16.27 |
| ayuta-yojana | ten thousand yojanas; SB 5.16.28 |
| yojana | yojanas; SB 5.20.24 |
| yojana | eight miles; SB 5.20.30 |
| ṣaṭ-triṃśat-lakṣa-yojana-āyataḥ | 3,600,000 yojanas long; SB 5.21.15 |
| nava-koṭi-yojana | of 90,000,000 yojanas; SB 5.21.19 |
| yojana-lakṣa-dvitaye | at a distance of 1,600,000 miles; SB 5.22.14 |
| dvi-lakṣa-yojana-antara-gatāḥ | situated at a distance of 1,600,000 miles; SB 5.22.15 |
| yojana-lakṣa-dvayāt | by a distance of 1,600,000 miles; SB 5.22.16 |
| ekādaśa-lakṣa-yojana-antare | at a distance of 1,100,000 yojanas; SB 5.22.17 |
| trayodaśa-lakṣa-yojana-antarataḥ | another 1,300,000 yojanas; SB 5.23.1 |
| yojana | a measurement equal to eight miles; SB 5.24.1 |
| yojana | a distance of eight miles; SB 5.24.2 |
| śata-yojana | of one hundred yojanas; SB 5.24.6 |
| yojana-ayuta-antareṇa | with an interval of ten thousand yojanas (eighty thousand miles); SB 5.24.7 |
| yojana-ayuta-ayutam | a distance of eighty thousand miles; SB 5.24.27 |
| yojana | eight-mile units of measurement; SB 5.25.1 |
| ayuta-yojana-parimaṇḍale | having a circumference of eighty thousand miles; SB 5.26.14 |
| yojana-śata | of eight hundred miles; SB 5.26.28 |
| yojana | a measurement of eight miles; SB 8.2.1 |
| lakṣa-yojana | one hundred thousand yojanas (eight hundred thousand miles); SB 8.7.9 |
| yojana-śatam | hundreds of miles; SB 8.24.26 |
| yojana-āyāma | occupying eight miles of land; SB 10.12.16 |
| yojana | yojanas (one yojana measures about eight miles); SB 10.57.18 |
| yojana | lengths of eight miles; CC Adi 5.97 |
| yojana | eight miles; CC Adi 5.119 |
| sahasra yojana | thousands of miles; CC Madhya 5.146 |
| karila yojana | harnessed; CC Madhya 14.50 |
| pra-yojana | goal; CC Madhya 20.143 |
| yojana | a distance of eight miles; CC Madhya 21.4 |
| pañcāśat koṭi yojana | four billion miles; CC Madhya 21.84 |

*SB: Śrīmad Bhāgavatam, CC Madhya: Śrī Caitanya Caritāmṛta Madhya, CC Adi: Śrī Caitanya Caritāmṛta Ādi*

## 卐 Lokas

- **Loka-Traya** - the 3 planetary systems

> *"There are three planetary systems, namely the lower worlds, the intermediate worlds and the upper worlds. The human beings on earth are situated at the beginning of the intermediate worlds, but living beings like Brahmā and his contemporaries live in the upper worlds, of which the topmost is Satyaloka. In Satyaloka the inhabitants are fully cognizant of Vedic wisdom, and thus the mystic cloud of material energy is cleared."*
>
> \- *Śrīmad Bhāgavatam 1.19.23*
>
> \-

As stated in the *Bhagavad-gītā,* there are three divisions of material spheres in this universe, namely the *ūrdhva-loka* (topmost planets), *madhya-loka* (midway planets) and *adhaloka* (downward planets). Beyond the *ūrdhva-loka* planets, that is to say above the *Brahmaloka,* are the material coverings of the universes, and above that is the spiritual sky, which is unlimited in expansion, containing unlimited self-illuminated *Vaikuṇṭha* planets inhabited by God Himself along with His associates, who are all eternally liberated living entities.

- *Ūrdhvaloka* (ऊर्ध्वलोक): The "upper world" or ūrdhvaloka is above mount Meru.
  The celestial world, the realms of the gods or heavens, the svarga-lokas - heavenly planets.
- *Madhya-loka*: the middle planetary system, world of mortals, the realms of the humans, animals and plants.
- *Adhaloka* (अधलोक): The adhaloka is the lower world, the subterranean heavens, *Bila-Swarga,* the realms of beings other than humans, like *Daityas, Dānavas and Nāgas* etc.

According to the *purāṇas* this universe is called *loka-traya* - the three lokas (*Bhūrloka, Bhuvarloka and Svarloka*), or three spheres (or worlds) consisting of 14 planetary systems. In the material world, the planetary systems are arranged in three spheres, called triloka, or *Svarga, Martya* and *Pātāla.* The three worlds are the upper, lower and middle intermediate planets of the universe. The human beings on earth are situated at the beginning of the intermediate worlds, but living beings like Brahmā and his contemporaries live in the upper worlds, of which the topmost is Satyaloka.

> *When we speak of loka-traya, we refer to the three primary planetary systems—Bhūḥ, Bhuvaḥ and Svaḥ—into which the universe is divided. Surrounding these planetary systems are the eight directions, namely east, west, north, south, northeast, southeast, northwest and southwest. Lokāloka Mountain has been established as the outer boundary of all the lokas to distribute the rays of the sun and other luminaries equally throughout the universe.*
>
> *- Śrīmad Bhāgavatam 5,20,37*

*Lord Brahmā* is the creator of the three worlds, namely the lower planets, called the *Pātālalokas*, the middle planets, called the *Bhūrlokas*, and the upper planets, called the *Svarlokas*. Still higher planets, such as *Maharloka, Tapoloka, Satyaloka and Brahmaloka*. The upper three planetary systems are called *sāttvika* planets because they provide facilities for a long duration of life and relative freedom from disease and old age, as well as a sense of fearlessness. The great sages and saints are promoted beyond the heavenly planets to Maharloka. In the creation, during *Brahmā*'s day, the three planetary systems—*Svarga, Martya and Pātāla*—revolve, and the inhabitants, including the lower animals, human beings, demigods and *Pitās*, appear and disappear in terms of their fruitive activities.

The cosmic manifestation is grossly divided into three worlds, the upper, lower and middle planetary systems, and then it broadens into the cosmos of fourteen planetary systems. There are fourteen divisions, *Caturdaśa-bhuvana*. There are seven down and seven up. We are in the middle. This is called *Bhūrloka*. Above this, there is *Bhuvarloka*. Above that, there is Janaloka. Above that, there is *Maharloka*. Above that, there is *Satyaloka*. Above that, there is Brahmaloka, the highest planetary system. Similarly, down also, we have got *Tala, Atala, Talātala, Vitala, Pātāla, Rasātala*. This information we get from Vedic literature, fourteen worlds. Each and every universe is consisting of these fourteen planetary systems, and there are innumerable universes.

The fourteen lokas or worlds are generally called as *Caturdaśa-bhuvanas*. Of these, six lokas (worlds) are above the earth (*Bhuloka*) and seven below it.

The vedic concept of the 14 lokas state that they are divided into 7 upper worlds (Svargaloka) and the 7 lower ones, generally known as the Pātālas. The fourteen worlds are enumerated in *Śrīmad-Bhāgavatam,* Second Canto, Fifth Chapter. The upper planetary systems are *(1) Bhū, (2) Bhuvar, (3) Svar, (4) Mahar, (5) Janas, (6) Tapas and (7) Satya. The seven lower planetary systems are (1) Tala, (2) Atala, (3) Vitala, (4) Nitala, (5) Talātala, (6) Mahātala and (7) Sutala*. The lower planets, as a whole, are called Pātāla. Among the upper planetary systems, *Bhū, Bhuvar* and Svar constitute *Svargaloka,* and the rest are called Martya. The entire universe is thus known as *Triloka.*

The universe is divided into fourteen planetary systems. Seven planetary systems, called *Bhūr, Bhuvar, Svar, Mahar, Janas, Tapas and Satya,* are upward planetary systems, one above the other. There are also seven planetary systems downward, known as *Atala, Vitala, Sutala, Talātala, Mahātala, Rasātala and Pātāla,* gradually, one below the other. In this verse, the description begins from the bottom because it is in the line of devotion that the Lord's bodily description should begin from His feet.

We are wandering in these different planetary systems up- and down, life after life. If we are pious, then we are promoted to the higher planetary system, *Svargaloka, Janaloka, Maharloka, Tapoloka, Brahmaloka, Satyaloka*. There are so many... Siddhaloka. And if we are impious, then we are downtrodden. We go down more and more. *Ūrdhvaṁ gacchanti sattva-sthāḥ* (BG 14.18). If you are situated in sattva-guṇa, then you are promoted to the higher planetary system: adho gacchanti tāmasāḥ. Jaghanya-guṇa-vṛtti-sthā adho gacchanti tāmasāḥ. If you are infected with tamo-guṇa, then you go down, down, down, down. Madhye tiṣṭhanti rājasāḥ. And if we are infected by the modes of passion, then we keep ourselves in the middle. But Kṛṣṇa says in the *Bhagavad-gītā, ābrahma-bhuvanāl lokāḥ punar āvartino 'rjuna* (*Bhagavad-gītā 8.16*). Even if you go to the highest planetary system—it is called *Brahmaloka*—where the duration of life is very, very long... That is described in the Bhagavad-gītā: *sahasra-yuga-paryantam ahar yad brahmaṇo viduḥ (Bhagavad-gītā 8.17*).

The planetary systems are situated within the stem of the lotus flower growing from the navel of *lord Viṣṇu.*

From *Viṣṇu*'s navel, the stem of a lotus flower grows, and on that lotus flower the first creature, *Brahmā,* is born. Within the stem of that lotus flower are fourteen divisions of planetary systems, which are created by *Brahmā.*

In the planet Atala, the yawning of a demon has produced three kinds of women, called *svairiṇī* (independent), *kāmiṇī* (lusty) and *puṁścalī* (very easily subdued by men). Below Atala is the planet Vitala, wherein Lord Śiva and his wife Gaurī reside. Because of their presence, a kind of gold is produced called *hāṭaka*. Below Vitala is the planet Sutala, the abode of *Bali Mahārāja*, the most fortunate king. *Bali Mahārāja* was favored by the Supreme Personality of Godhead, Vāmanadeva, because of his intense devotional service. The Lord went to the sacrificial arena of *Bali Mahārāja* and begged him for three paces of land, and on this plea the Lord took from him all his possessions. When *Bali Mahārāja* agreed to all this, the Lord was very pleased, and therefore the Lord serves as his doorkeeper. The description of *Bali Mahārāja* appears in the Eighth Canto of *Śrīmad-Bhāgavatam*.

This idea of 'seven lower planetary systems' may create the impression in a person's mind of seven layers of many globe-like planets orbiting in space somewhere below the assumed Earth globe. The term *'planetary systems'* was *Śrīla Prabhupāda's* choice of words, and it is a term that he used consistently, but our modern conception of what might constitute 'planetary systems' is wholly different from what is being described by *Śukadeva Goswami* as seven huge realms within the one vast Earth circle. From the very beginning of *Śrīmad Bhāgavatam*, *Śrīla Prabhupāda* was translating verses relating to the *Bhū-maṇḍala* without even have a map or image of *Bhū-maṇḍala* to help him conceptualize where each place is situated, and how each place is related to the other.

The Fifth Canto especially presents a detailed description of the universal form that is very difficult to follow without a map, or depiction, or explanation of any kind. Without a map, or image, or explanation, the description of the *virāṭ-rūpa* (universal form) is very difficult to visualize. *Śrīla Prabhupāda* invariably used the word 'planet' to cover the multitude of places mentioned by *Śukadeva Goswami*. However, one should be aware that many of the places that *Śrīla Prabhupāda* refers to as 'planets,' are actually references to places upon the *Bhū-maṇḍala's* vast landscape, or within the depths of the *Bhū-maṇḍala* itself (the subterranean realms). They are not 'other planets' as we conceive them, but part of the Earth itself. As we shall see presently, the conception of seven layers of globe-like planets floating below the Earth is definitely not what is being described by *Śukadeva Goswami*. (Please refer to: Mayesvara Das, *What's Below the Earth?*, Oct 2017) Figure 8.2 shows the fourteen lokas.

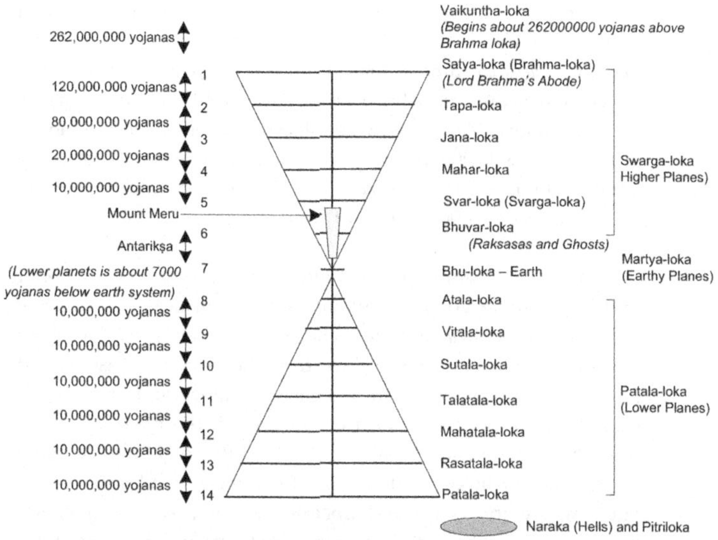


Figure 8.2. The fourteen Lokas (Universes) (Seven Lower and Seven Upper Lokas from *Bhoo-loka* to *Brahma-loka*. (1 yojana = 8 miles = 1.60934 x 8 = 12.875 KM)

- **Satyaloka:** The highest planet is *Satyaloka,* or *Brahmaloka,* where the prime person of this universe, Lord *Brahmā,* resides. We have seen already that we can hardly calculate the wondrous condition of life in *Brahmaloka,* but the highest condition of life, the mode of goodness, can bring us to this. The human beings on earth are situated at the beginning of the intermediate worlds, but living beings like *Brahmā* and his contemporaries live in the upper worlds, of which the topmost *iTapa-lokas Satyaloka.* In *Satyaloka* the inhabitants are fully cognizant of Vedic wisdom, and thus the mystic cloud of material energy is cleared.

- **Tapa-loka:** Tapoloka means those who have performed here severe penances, and they are transferred there in Tapoloka. In Tapoloka, when there is destruction, annihilation of this material world, in the Tapoloka there is no destruction. The higher planetary systems, beginning with *Maharloka, Janaloka* and *Tapoloka,* are not inundated at the time of this devastation. In the upper planetary systems

(*Svargaloka, Janaloka, Tapoloka, Maharloka, Brahmaloka,* etc.) the living entities enjoy a longer duration of life, an increased amount of knowledge, and a generally more blissful existence. *Viśvanātha Cakravartī Ṭhākura* estimates that *Dhruvaloka,* the polestar, is 3,800,000 yojanas above the sun. Above *Dhruvaloka* by 10,000,000 yojanas is *Maharloka,* above *Maharloka* by 20,000,000 yojanas is *Janaloka,* above *Janaloka* by 80,000,000 yojanas is *Tapoloka,* and above *Tapoloka* by 120,000,000 yojanas is *Satyaloka.* Thus, the distance from the sun to *Satyaloka* is 233,800,000 yojanas, or 1,870,400,000 miles.

- **Jana-Loka:** Great saintly persons, after death, they go to *Janaloka* or *Maharlok*a. One in goodness attains higher planetary systems, like *Brahmaloka* or *Janaloka,* and there enjoys godly happiness. The word *amalān* is significant; it means "free from the modes of passion and ignorance." There are impurities in the material world, but the mode of goodness is the purest form of existence in the material world. There are different kinds of planets for different kinds of living entities. Those who die in the mode of goodness are elevated to the planets where great sages and great devotees live. At that time, you happened to be visiting the Lord on *Śvetadvīpa*—that Supreme Lord in whom the Vedas lie down to rest during the period of universal annihilation. A lively discussion arose among the sages on *Janaloka* as to the nature of the Supreme Absolute Truth. Indeed, the same question arose then that you are asking Me now.

- **Mahar-Loka:** The *yogī* alone goes beyond this circle of *Śiśumāra* and attains the planet [Maharloka] where purified saints like *Bhṛgu* and others enjoy a duration of life of 4,300,000,000 solar years. This planet is worshipable even for the saints who are transcendentally situated. The great sages and saints are promoted beyond the heavenly planets to *Maharloka,* but that also is not the place of complete fearlessness because at the end of one *kalpa* the *Maharloka* is annihilated and the inhabitants have to transport themselves to still higher planets. In the *Maharloka* one can live even at time of partial annihilation of the universe. The annihilation begins with volcanic flames of fire emanating from the mouth of Shesha Ananta Deva from the lowest portion of the universe. By such big blazing fire, the heat reaches up to *Maharloka and* the residents thereof pass t*o Brahmaloka* which exists up to twice the duration of *parardha(?)* time.

- **Svar-Loka:** The *karmīs,* or fruitive workers, can elevate themselves to the *Svargaloka* planets, which include the sun and the moon. The moon, the sun and higher planets are called *svargaloka*. There are three different statuses of planets: higher, middle and lower planetary systems. The earth belongs to the middle planetary system. In the word *svārājyasya, svar* refers to *Svargaloka,* the heavenly planet, and *svārājya* refers to the ruler of the heavenly planet, Indra. The denizens of heaven, who live in the planetary systems beginning from *Svarloka,* cannot even see *Lord Viṣṇu* in Śvetadvīpa. Unable to reach the island, they can simply approach the beach of the milk ocean to offer transcendental prayers to the Lord, appealing to Him on special occasions to appear as an incarnation. At the end of Lord *Brahmā's* day, everything up to *Svargaloka* is inundated with water, and the next morning, when there is darkness in the universe, *Brahmā* again brings the phenomenal manifestation into existence. Therefore, he is described here as he who manifests this universe.

- **Bhuvar-Loka:** The demigods, semi-demigods, *Gandharvas, Cāraṇas, Vidyādharas,* etc., are all highly intelligent persons in the upper planets, the human beings are inhabitants of the intermediate planets, and the asuras are inhabitants of the lower planets. While He was passing in the northern direction, all the celestial denizens known as *Cāraṇas* and *Gandharvas,* as well as the *munis and* the damsels of the heavenly planets, prayed and offered Him all respects. The ocean offered Him oblations and a place of residence. The lower systems are called Bhūrloka, the middle systems are called Bhuvarloka, and the higher planetary systems, up to *Brahmaloka,* the highest planetary system of the universe, are called *Svarloka.* This middle portion of the sky is called Bhuvarloka, as confirmed by both *Śrīla Viśvanātha Cakravartī* and *Śrīla Jīva Gosvāmī. Ūrdhvam* means higher planetary system. Here it is said that api trailokya-rājyasya hetoḥ. Trailokya. *Oṁ bhūr bhuvaḥ svaḥ.* This is bhūr-loka. Above this, bhuvar-loka, and above that, *svar-loka,* heavenly planets. *Oṁ bhūr bhuvaḥ svaḥ tat savitur vareṇyam.* This is *Gāyatrī-mantra* (गायत्री-मन्त्रः).

- **Bhūr-Loka:** The planetary system in which we are living is called Bhūrloka, so this is called bhūrloka, earthly planet. If you keep yourself in the rajo-guṇa, then you will keep within this Bhurloka.

And if you are infected with the tamo-guṇa, then you again go to lower grade life, again animal life, plant life. This is the nature's process. You cannot avoid it. Śukadeva Gosvāmī says he will explain the Bhūrloka. In the fifth verse he starts to be more precise about the Bhū-maṇḍala planetary system. It says, "*It resembles a lotus flower.*" It has the shape of a lotus flower. "And the seven islands of Jambūdvīpa resemble the whorl of that flower." Then "*The length and breadth of that island known as Jambūdvīpa...*" The planetary system known as Bhū-maṇḍala resembles a lotus flower, and its seven islands resemble the whorl of that flower. The length and breadth of the island known as Jambūdvīpa, which is situated in the middle of the whorl, are one million yojanas [eight million miles]. Jambūdvīpa is round like the leaf of a lotus flower.

- **Antarikṣa:** Antarikṣa, is the space between Bhūrloka and B*huvarloka*. The sun is situated [vertically] in the middle of the universe, in the area between Bhūrloka and *Bhuvarloka,* which is called *antarikṣa,* outer space. The distance between the sun and the circumference of the universe is *twenty-five koṭi yojanas* [two billion miles]. There are seven lower planetary systems, known as *Atala, Vitala, Sutala, Talātala, Mahātala, Rasātala* and *Pātāla lokas*. All these planetary systems are scattered over the complete universe, which occupies an area of two billion times two billion square miles. The width and length of the seven lower planetary systems are calculated to be exactly the same as those of earth. Demons and *Rakṣasas* live in these lower planetary systems with their wives and children, always engaged in sense gratification and not fearing their next births. Since there is no sunshine in those subterranean planets, time is not divided into days and nights, and consequently fear produced by time does not exist. The sunshine does not reach these planets, but they are illuminated by jewels fixed upon the hoods of snakes. Many great serpents reside there with gems on their hoods, and the effulgence of these gems dissipates the darkness in all directions. Because of these shining gems there is practically no darkness. Those living in these planets do not become old or diseased, and they are not afraid of death from any cause but the time factor, the Supreme Personality of Godhead.

- **Atala:** In the planet Atala, the yawning of a demon has produced three kinds of women, called *svairiṇī* (independent), *kāmiṇī* (lusty) and *puṁścalī* (very easily subdued by men). In Atala there is a demon, the son of *Maya Dānava* named Bala, who created ninety-six kinds of mystic power. Some so-called *yogīs* and *svāmīs* take advantage of this mystic power to cheat people even today. Simply by yawning, the demon Bala created three kinds of women, known as *svairiṇī, kāmiṇī* and *puṁścalī*. The *svairiṇīs* (स्वैरिणी:) like to marry men from their own group, the *kāmiṇīs* marry men from any group, and the *puṁścalīs* (पुंश्चली:) change husbands one after another. If a man enters the planet of Atala, these women immediately capture him and induce him to drink an intoxicating beverage made with a drug known as *hāṭaka* - हाटक [cannabis indica]. This intoxicant endows the man with great sexual prowess, of which the women take advantage for enjoyment. A woman will enchant him with attractive glances, intimate words, smiles of love and then embraces. In this way she induces him to enjoy sex with her to her full satisfaction. Because of his increased sexual power, the man thinks himself stronger than ten thousand elephants and considers himself most perfect. Indeed, illusioned and intoxicated by false pride, he thinks himself God, ignoring impending death.

- **Vitala:** Below Atala is the planet Vitala, wherein Lord Śiva and his wife Gaurī reside. Because of their presence, a kind of gold is produced called hāṭaka. The next planet below Atala is Vitala, wherein Lord Śiva, who is known as the master of gold mines, lives with his personal associates, the ghosts and similar living entities. Lord Śiva, as the progenitor, engages in sex with Bhavānī, the progenitress, to produce living entities, and from the mixture of their vital fluid the river named Hāṭakī is generated. When fire, being made to blaze by the wind, drinks of this river and then sizzles and spits it out, it produces gold called Hāṭaka. The demons who live on that planet with their wives decorate themselves with various ornaments made from that gold, and thus they live there very happily.

- **Sutala:** Below the planet Vitala is another planet, known as Sutala, where the great son of Mahārāja Virocana, Bali Mahārāja, who is celebrated as the most pious king, resides even now. Bali Mahārāja was favored by the Supreme Personality of Godhead, Vāmanadeva, because of his intense devotional service. The Lord went to the sacrificial arena of Bali Mahārāja and begged him for three paces of land, and on this plea the Lord took from him all his possessions.

When Bali Mahārāja agreed to all this, the Lord was very pleased, and therefore the Lord serves as his doorkeeper. The description of Bali Mahārāja appears in the Eighth Canto of Śrīmad-Bhāgavatam. For the welfare of Indra, the King of heaven, Lord Viṣṇu appeared in the form of a dwarf brahmacārī as the son of Aditi and tricked Bali Mahārāja by begging for only three paces of land but taking all the three worlds. Being very pleased with Bali Mahārāja for giving all his possessions, the Lord returned his kingdom and made him richer than the opulent King Indra. Even now, Bali Mahārāja engages in devotional service by worshiping the Supreme Personality of Godhead in the planet of Sutala.

- **Talātala:** Below Sutala is the planet Talātala, the abode of the demon Maya. This demon is always materially happy because he is favored by Lord Śiva, but he cannot achieve spiritual happiness at any time. Talātala is ruled by the Dānava demon named Maya. Maya is known as the ācārya [master] of all the māyāvīs, who can invoke the powers of sorcery. For the benefit of the three worlds, Lord Śiva, who is known as Tripurāri, once set fire to the three kingdoms of Maya, but later, being pleased with him, he returned his kingdom. Since that time, Maya Dānava has been protected by Lord Śiva, and therefore he falsely thinks that he need not fear the Sudarśana cakra of the Supreme Personality of Godhead.

- **Mahātala:** Below Talātala is the planet Mahātala, where there are many snakes with hundreds and thousands of hoods. Mahātala is the abode of many-hooded snakes, descendants of Kadrū, who are always very angry. The great snakes who are prominent are Kuhaka, Takṣaka, Kāliya and Suṣeṇa. The snakes in Mahātala are always disturbed by fear of Garuḍa, the carrier of Lord Viṣṇu, but although they are full of anxiety, some of them nevertheless sport with their wives, children, friends and relatives.

- **Rasātala:** Beneath Mahātala is Rasātala, which is the abode of the demoniac sons of Diti and Danu. They are called Paṇis, Nivāta-kavacas, Kāleyas and Hiraṇya-puravāsīs [those living in Hiraṇya-pura]. They are all enemies of the demigods, and they reside in holes like snakes. From birth they are extremely powerful and cruel, and although they are proud of their strength, they are always defeated by the Sudarśana cakra of the Supreme Personality of Godhead, who rules all the planetary systems. When a female messenger from Indra named Saramā chants a particular curse, the serpentine demons of Mahātala become very afraid of Indra.

- **Pātāla:** Below Rasātala is Pātāla, where the serpent Vasukī lives with his associates. Pātāla is also known as Nāgaloka, where there are many demoniac serpents, the masters of Nāgaloka, such as Śaṅkha, Kulika, Mahāśaṅkha, Śveta, Dhanañjaya, Dhṛtarāṣṭra, Śaṅkhacūḍa, Kambala, Aśvatara and Devadatta. The chief among them is Vāsuki. They are all extremely angry, and they have many, many hoods – some snakes five hoods, some seven, some ten, others a hundred and others a thousand. These hoods are bedecked with valuable gems, and the light emanating from the gems illuminates the entire planetary system of bila-svarga.

The *Bhāgavata Purāṇa* (भागवत पुराण) calls the seven lower regions *bila-svargas* ("subterranean heavens") and they are NOT regarded as planets or planetary systems but they are considered as realms or dwelling places below the earth. These regions are described as being more opulent than the upper regions of the universe, which include heaven. The life here is of pleasure, wealth and luxury, with no distress. The demon architect *Māyā* (माया) has constructed palaces, temples, houses, yards and hotels for foreigners, with jewels. The natural beauty of *Pātāla* is said to surpass that of the upper realms. There is no sunlight in the lower realms, but the darkness is dissipated by the shining of the jewels that the residents of Pātāla wear. There is no old age, no sweat, no disease in Pātāla.

❖ **Naraka - 28 Hells:** The word naraka means "hell." The Twenty-sixth Chapter of Śrīmad-Bhāgavatam describes how a sinful man goes to different hells, where he is punished in various ways by the assistants of Yamarāja. Śukadeva Gosvāmī affirms the existence of the Naraka planets, which lie between the Garbhodaka Ocean and Pātālaloka. Yamarāja is appointed by the Supreme Personality of Godhead to see that the human beings do not violate His rules and regulations. in the fifth Canto, Śrīmad-Bhāgavatam, Śukadeva Gosvāmī has given a vivid description of the naraka planets, or the hellish planets. All the hellish planets are situated in the intermediate space between the three worlds and the Garbhodaka Ocean. They lie on the southern side of the universe, beneath Bhū-maṇḍala, and slightly above the water of the Garbhodaka Ocean. Pitṛloka is also located in this region between the Garbhodaka Ocean and the lower planetary systems. All the residents of Pitṛloka, headed by Agniṣvāttā, meditate in great samādhi on the Supreme Personality of Godhead and always wish their families well. The King of the pitās is Yamarāja, the very powerful son of the sun-god. He resides in Pitṛloka with his personal assistants and, while abiding by the rules and regulations set down by the Supreme Lord, has his agents, the Yamadūtas, bring all the sinful men to him immediately upon their death. After bringing them within his jurisdiction, he properly judges them according to their specific sinful activities and sends them to one of the many hellish planets for suitable punishments. So Yamarāja, his kingdom, his jurisdiction, is within these hellish planets.

Śukadeva Gosvāmī describes the following twenty-eight hells: Tāmisra, Andhatāmisra, Raurava, Mahāraurava, Kumbhīpāka, Kālasūtra, Asi-patravana, Sūkaramukha, Andhakūpa, Kṛmibhojana, Sandaṁśa, Taptasūrmi, Vajrakaṇṭaka-śālmalī, Vaitaraṇī, Pūyoda, Prāṇarodha, Viśasana, Lālābhakṣa, Sārameyādana, Avīci, Ayaḥpāna, Kṣārakardama, Rakṣogaṇa-bhojana, Śūlaprota, Dandaśūka, Avaṭa-nirodhana, Paryāvartana and Sūcīmukha.

## 卐 Bhartiya Units of Measurement

Before the introduction of the Metric system, one may divide the history of Bhartiya systems of measurement into three main periods: the pre-Akbar's period, the period of the Akbar system, and the British colonial period. During pre-Akbar period, weights and measure system varied from region to region, commodity to commodity, and rural to urban areas. The weights were based on the weight of various seeds (specially the wheat berry and Ratti) and lengths were based on the length of arms and width of fingers. Akbar realized a need for a uniform system. He elected the barley corn. This did not replace the existing system. Instead, it just added another system. British entered Bharat as traders. They accepted barley corn as a unit 'grain' for weighing gold. Eventually, British introduced their own system for weighing gold (Troy ounce). In 1956, Government of Bharat passed the Standards of Weights Act, that came into effect in 1958.The metric weight mandatory by October 1960, and the metric measures mandatory by April 1962. In 1956, for metrication, the Bhartiya government defined the Standards of Measurements Act (No. 89 of 1956, amended 1960, 1964) as follows:

$$1 \text{ Tola} = 11.6638038\text{gm} \approx 11.64\text{gm}$$

**Ancient System**

These are the weights and measures popular in North Bharat before the adoption of the metric system. There were different systems in Bengal, the Presidency of Madras, and Bombay. The following nomenclature was prevalent in North Bharat till the metric system came as shown in Table 8.11 and Table 8.12.

Table 8.11 Ancient measurement and conversion

| | |
|---|---|
| 4 Chawal (grain of rice) | 1 Dhan (weight of one wheat berry) |
| 4 Dhan | 1 Ratti (Ratti is the seed of the 'Abrus precatorius'. It is a red seed with a black spot at one end. Abul Fazl refers to it as 'Surkh' in Ain-i-Akbari.) |
| 8 Ratti | 1 Masha |
| 12 Masha (96 Ratti) | 1 bhari |
| 24 Ratti (96 Dhan) | 1 Tak |
| **Conversion** | |
| 1 Bhari | 11.66375 gram |
| 3.75 Troy ounce | 10 bhari |
| Weight of 64 Dhan (Wheat berries) | Weight of 45 Jau (Barley corns) |
| Weight of 1 Barley corn | 64.79891 milligrams |

Table 8.12 Commodity weight system

| | |
|---|---|
| 1 bhari | 4 Siki |
| 1 Kancha | 5 Siki |
| 1 Chhataank | 4 Kancha |
| 1 Chhataank | 5 bhari |
| 1 Adh-pav | 2 Chhatank |
| 2 Chhatank | 1/8 Seer |

| | |
|---|---|
| 1 Pav | 2 Adh-pav |
| 2 Adh-pav | ¼ Seer (Pav means ¼) |
| 1 Adher | 2 Pav |
| 2 Pav | ½ Seer |
| In Hindi ½ Seer | Adha (½) Seer, or Adher |
| 1 Ser | 2 Adher |
| 2 Adher | 4 Pav |
| 4 Pav | 16 Chattank |
| 16 Chattank | 80 Tola |
| 80 Tola | 933.1 grams |
| 1 Savaser | 1 Ser + 1 Pav (1¼ Seer) |
| 1 Savaser weighed | 100 Imperial rupees |
| In Hindi 1¼ Seer | Sava (1¼) Seer, or Savaser |
| 1 Dhaser | 2 Savaser |
| 2 Savaser | 2½ Seer |
| In Hindi 2½ Seer | Dhai (2½) Seer, or Dhaser |
| 1 Paseri | 2 Adisari |
| 2 Adisari | 5 Seer |
| In Hindi 5 Seer | Panch (5) Seer, or Paseri for short |
| 1 Daseri | 2 Pasri |
| 2 Pasri | 10 Seer |
| In Hindi 10 Seer | Das (10) Seer, or Daseri for short |
| 1 Maund (maan or man[मण]) | 4 Daseri |
| 4 Daseri | 8 Pasri |
| 8 Pasri | 40 Seer |

**Rice and Grains Volume Measures**

Grains were not weighed. Special hour-glass shaped measure was used to determine the volume. Table 8.13 shows the rice and grain measure system.

Table 8.13 Rice and Grain Measure System

| | |
|---|---|
| Smallest unit | 1 Nilve |
| 2 Nilve | 1 Kolve |
| 2 Kolve | 1 Chipte (about quarter litre) |
| 2 Chipte | 1 Mapte (about half litre) |
| 2 Mapte | 1 Ser (about one litre) |

**Liquid Volume Measures**

These were hour glass shaped measure used for Milk, Ghee, Oils. The bottom was round like an inverted dome, the top was like flared rim. This shape helped in pouring the liquids (see Table 8.14).

Table 8.14 Liquid Volume Measure System

| | |
|---|---|
| 4 Chhataank | 1 Pav |
| 4 Pav | 1 Seer |
| 40 Seer | 1 Maund |

**Length Measures**

Measure of length is Gaz. To interpret Gaz, depends on what one is measuring and where they are. Bengal: 36", Bombay: 27", Madras: 33", Government Average: 33". The hand measurements were used. Table 8.15 shows the length measure system.

Table 8.15 Length measure system

| | |
|---|---|
| Anguli (width of 3 fingers) | 1 Girah |
| 8 Girah | 1 Hath (*elbow to the end of the middle finger, approximately 18"*) |
| $5\frac{5}{6}$Hath | One Kathi |
| 20 Kathi | One Pand |
| 1 Pand | 1 Beesa |
| 20 Pand | One Begah |
| 2 Hath | 1 Gaz |
| 3 Gaz | Two Karam |
| 3 Karams | 1 Kan |
| 3 Square Kans | 1 Marla |
| 20 Marlas | 1 Kanal |
| 8 Kanals | 1 Ghamaon |
| 9 Kanals 12 Marlas | 1 Acre |
| 4 Kanals | 1 Begah |

*Note: 1 coss= 2000 yards*

## Further Reading


1. "About ISRO – ISRO". Archived from the original on 28 March 2019. Retrieved 28 March 2019.
2. "Aryabhata – ISRO". www.isro.gov.in. Archived from the original on 15 August 2018. Retrieved 15 August 2018.
3. "Ayanamshas in Sidereal Astrology". www.astro.com. Retrieved 29 July 2020.
4. "Balloon X-ray astronomy experiments from India". Archived from the original on 28 May 2002. Retrieved 17 March 2009.
5. "Bhagavad-gītā As It Is 8.17". Bhaktivedanta Vedabase (in Sanskrit and English). Translated by A. C. Bhaktivedanta Swami Prabhupada. Bhaktivedanta Book Trust. 1968. LCCN 68008322. Wikidata Q854700. Retrieved 2020-05-10.
6. "Ch. 103, Episode of Krsna concluded, Sloka 8". Brahma Purana. Vol. Part II. Motilal Banarsidass. 1955. p. 515. It was on the day on which Krishna left the Earth and went to heaven that the Kali age, with time for its body set in.
7. "Ch. 37, Royal Dynasties, Sloka 422". The Vayu Purana. Vol. Part II. Motilal Banarsidass. 1988. p. 824. ISBN 81-208-0455-4. Kali Yuga had started on the very day when Krsna passed away.
8. "Ch. 74, Royal Dynasties, Sloka 241". The Brahmanda Purana. Vol. Part III. Motilal Banarsidass. 1958. p. 950. Kali Yuga began on the day when Krsna passed on to heaven. Understand how it is calculated.
9. "Department of Space and ISRO HQ – ISRO". Archived from the original on 28 March 2019. Retrieved 28 March 2019.
10. "How did the world come into being according to Hinduism?".
11. "Indian ambitions in space go sky-high". New Scientist. 22 January 1981. p. 215. Archived from the original on 13 April 2021. Retrieved 14 March 2021.
12. "India's 'Mr. Missile': A man of the people". 30 July 2015. Archived from the original on 25 September 2015. Retrieved 30 July 2015.
13. "Jai Prakash". Jantar Mantar.
14. "kalpa". Wiktionary. Retrieved 2021-03-25.
15. "Mahattattva, Mahat-tattva: 5 definitions". Wisdom Library. 27 January 2019. Retrieved 10 February 2021. Mahattattva (महत्तत्त्व) or simply Mahat refers to a primordial principle of the nature of both pradhāna and puruṣa, according to the 10th century Saurapurāṇa: one of the various Upapurāṇas depicting Śaivism.—[...] From the disturbed prakṛti and the puruṣa sprang up the seed of mahat, which

is of the nature of both pradhāna and puruṣa. The mahattattva is then covered by the pradhāna and being so covered it differentiates itself as the sāttvika, rājasa and tāmasa-mahat. The pradhāna covers the mahat just as a seed is covered by the skin. Being so covered there spring from the three fold mahat the threefold ahaṃkāra called vaikārika, taijasa and bhūtādi or tāmasa.
16. "Nadivalaya Yantra". Jantar Mantar.
17. "Note on Kali and Dvāpara and their connection with the dice". Wisdom Library. 29 June 2019. Retrieved 2021-03-04. They are in order Kṛta, Tretā, Dvāpara and Kali, and correspond roughly to the Gold, Silver, Brass and Iron Ages of the classics. The Sanskrit names are called after the sides of a die in descending order of their value in play. Thus Kṛta is the side with four dots, while Kali, being the side with only one dot, is always a certain loser.
18. "Rama Yantra". Jantar Mantar.
19. "Rig Veda: Rig-Veda, Book 10: HYMN XC. Puruṣa". Retrieved 26 March 2015.
20. "Samrat Yantra". Jantar Mantar.
21. "sandhyA". Sanskrit Dictionary for Spoken Sanskrit. Retrieved 2021-02-28.
22. "Skanda I, Ch. 18: Curse of the Brahmana, Sloka 6". Bhagavata Purana. Vol. Part I. Motilal Banarsidass Publishers Private Limited. 1950. p. 137. On the very day, and at the very moment the Lord [Krishna] left the earth, on that very day this Kali, the source of irreligiousness, (in this world), entered here.
23. "Śrīmad-Bhāgavatam (Bhāgavata Purāṇa) 12.2.39". Bhaktivedanta Vedabase. Retrieved 2020-05-10.
24. "Śrīmad-Bhāgavatam (Bhāgavata Purāṇa) 12.4.2". Bhaktivedanta Vedabase. Retrieved 2020-05-10.
25. "Śrīmad-Bhāgavatam (Bhāgavata Purāṇa) 3.11.19". Bhaktivedanta Vedabase. Retrieved 10 July 2020.
26. "Śrīmad-Bhāgavatam (Bhāgavata Purāṇa) 9.10.51". Bhaktivedanta Vedabase. Retrieved 2020-05-18. Lord Rāmacandra became King during Tretā-yuga, but because of His good government, the age was like Satya-yuga. Everyone was religious and completely happy.
27. "Star Maps: History, Artistry, and Cartography", p. 17, by Nick Kanas, 2012
28. "Story of Pralaya". Wisdom Library. Retrieved 9 November 2021.
29. "Stratospheric balloon launch bases and sites". StratoCat. Archived from the original on 3 March 2016. Retrieved 4 November 2015.

30. "SUPARCO".
31. "The Saga of Indian Remote Sensing Satellite System – ISRO". www.isro.gov.in. Archived from the original on 27 June 2019. Retrieved 16 March 2021.
32. "Three years of AstroSat – ISRO". www.isro.gov.in. Archived from the original on 30 August 2019. Retrieved 28 September 2018.
33. "Time – Vyasa Mahabharata". Retrieved 31 December 2021.
34. "Timeline of Cycles by René Guénon and Gaston Georgel". Sufi Path of Love. 20 April 2019. Retrieved 13 November 2020.
35. "Unnatamsha Yantra". Jantar Mantar.
36. "Varahamihira". Encyclopædia Britannica. Varāhamihira's knowledge of Western astronomy was thorough. In five sections, his monumental work progresses through native Indian astronomy and culminates in two treatises on Western astronomy, showing calculations based on Greek and Alexandrian reckoning and even giving complete Ptolemaic mathematical charts and tables.
37. "Vedic Time System - वेद Veda". veda.wikidot.com. Retrieved 4 December 2019.
38. "What are Kalpas?". Lion's Roar. December 14, 2016. Retrieved August 29, 2019.
39. "What is a seer?". www.sizes.com. Retrieved 2019-06-20.
40. "Worlds Together Worlds Apart", Fourth Edition, Beginnings Through the 15th century, Tignor, 2014, pg. 5.
41. "yuga". Dictionary.com Unabridged (Online). n.d. Retrieved 27 February 2021.
42. "Zodiac Sign Dates: What Are The Dates for Every Star Sign?". Astrostyle: Astrology and Daily, Weekly, Monthly Horoscopes by The AstroTwins. 28 September 2016. Retrieved 29 July 2020.
43. "चतुर् (catur)". Wiktionary. Retrieved 2021-02-27.
44. "दिव्य (divya)". Wiktionary. Retrieved 2021-02-27.
45. "देव (deva)". Wiktionary. Retrieved 2021-02-27.
46. "दैव (daiva)". Wiktionary. Retrieved 2021-02-27.
47. "मनु (manu)". Wiktionary. Retrieved 2021-02-28.
48. "महा (mahā)". Wiktionary. Retrieved 2021-02-27.
49. "युग (yuga)". Wiktionary. Retrieved 2021-02-27.
50. "বাংলাদেশ মহাকাশ গবেষণা ও দূর অনুধাবন".
51. A human year is divided into twelve equal months, measured by the sun's six month movements in the north BG 8.24 and south BG 8.25, as indicated in Bhagavad-gita.
52. A Mountain Pabbata Sutta (SN 15:5)

53. A survey of Hinduism, Klaus K. Klostermaier, 2007, pp. 495-496
54. Abraham (2008).
55. Account of the several Manus and Manwantaras Vishnu Purana, translated by Horace Hayman Wilson, 1840, Book III: Chapter I. p. 259, The first Manu was Swáyambhuva, then came Swárochisha, then Auttami, then Támasa, then Raivata, then Chákshusha: these six Manus have passed away. The Manu who presides over the seventh Manwantara, which is the present period, is Vaivaswata, the son of the sun...
56. Alban Widgery (1930), The principles of Hindu Ethics, International Journal of Ethics, Vol. 40, No. 2, pages 234–237.
57. Almeida etc. (2001).
58. Amir Muzur, Hans-Martin Sass (2012). Fritz Jahr and the Foundations of Global Bioethics: The Future of Integrative Bioethics. LIT Verlag Münster. p. 348. ISBN 978-3-643-90112-5.
59. Andrew Zimmerman Jones (2009). String Theory For Dummies. John Wiley & Sons. p. 262. ISBN 978-0-470-59584-8.
60. Annette Wilke; Oliver Moebus (2011). Sound and Communication: An Aesthetic Cultural History of Sanskrit Hinduism. Walter de Gruyter. pp. 259–262. ISBN 978-3-11-024003-0.
61. Ariel Glucklich (2008). The Strides of Vishnu: Hindu Culture in Historical Perspective. Oxford University Press. pp. 151–155 (Matsya Purana and other examples). ISBN 978-0-19-971825-2.
62. Ashfaque, Syed Mohammad (1977). "Astronomy in the Indus Valley Civilization A Survey of the Problems and Possibilities of the Ancient Indian Astronomy and Cosmology in the Light of Indus Script Decipherment by the Finnish Scholars". Centaurus. 21 (2): 149–193. Bibcode:1977Cent...21..149A. doi:10.1111/j.1600-0498.1977.tb00351.x.
63. Axel Michaels (2003), Notions of Nature in Traditional Hinduism, Environment across Cultures, Springer, ISBN 978-3-642-07324-3, pages 111–121.
64. Baber, 82–89.
65. Baber, 89–90.
66. Babylon to Voyager and Beyond: A History of Planetary Astronomy. David Leverington. Cambridge University Press, 29 May 2010 – Science – 568 pages. page 41.
67. Bailey, Greg (2008). Gaṇeśapurāṇa – Part II: Krīḍākhaṇḍa. Wiesbaden: Harrassowitz Verlag. p. 5–8. ISBN 978-3-447-05472-0.

68. Barbara A. Holdrege (2015). Bhakti and Embodiment: Fashioning Divine Bodies and Devotional Bodies in Krsna Bhakti. Routledge. p. 334, note 62. ISBN 978-1-317-66910-4.
69. Basu, Major B. D. (1917). "CCLXXXX". The Matsya Puranam. Vol. XVII part II. Sudhindra Natha Vasu, At The Indian Press Allahabad. p. 368.
70. Bhagavad-gita 7.3, *manusyanam sahasresu kascid yatati siddhaye.*
71. Bharatbarsha – A Living Legend.
72. Bhargava & Chakrabarti 2003.
73. Bolton 2001, pp. 64–65.
74. Bolton 2001, pp. 65–68.
75. Bolton, Robert (2001). The Order of the Ages: World History in the Light of a Universal Cosmogony. Sophia Perennis. p. 64. ISBN 978-0-900588-31-0.
76. Brahmagupta, Brahmasphutasiddhanta (628) (cf. Al-Biruni (1030), Indica).
77. Brodd, Jeffrey (2003). World Religions. Winona, MN: Saint Mary's Press. ISBN 978-0-88489-725-5.
78. Brown, W. Norman (1942). "The Creation Myth of the Rig Veda". Journal of the American Oriental Society. 62 (2): 85–98. doi:10.2307/594460. JSTOR 594460.
79. Brown, W. Norman (1964). "The Indian Games of Pachisi, Chaupar, and Chausar". Expedition magazine. Vol. 6, no. 2. Penn Museum. p. 34. ISSN 0014-4738. The Rig-Veda, which we may reasonably consider to have been in its present form before 1000 B.C., has references to the use of dice, and one of its hymns (Book 10, 34) is a charm to cure an inveterate and unsuccessful gambler of the compulsion to gamble that has ruined him. In the Atharva Veda, also, gambling with dice is mentioned (2.3; 4.38; 6.118; 7.52; 7.109). The Aryans of Rig-Vedic times made their dice of the vibhidaka-tree nuts, and we do not know how they used them. Evidently dicing was considered a fitting vice of kings, and in the ritualistic literature of the centuries following the Rig-Veda, say at around 800 B.C., the consecration ceremonies for a king included a game of dice–which the new king must always win–and there was a special officer to take charge of the dice. In the great epic known as the Mahabharata there are two famous instances of kings ruined by gambling.
80. Bryan E. Penprase (5 May 2017). The Power of Stars. Springer. p. 137. ISBN 978-3-319-52597-6.
81. Bryant (2001), 253

82. Bühler 1886, p. 20.
83. Bühler, G. (1886). "Ch. 1, The Creation". In Müller, F. Max (ed.). The Laws of Manu: translated with extracts from seven commentaries. Sacred Books of the East. Vol. XXV. Oxford University Press. p. 20 (1.67–71).
84. Bühler, G. (1886). "Ch. 1, The Creation". In Müller, F. Max (ed.). The Laws of Manu: translated with extracts from seven commentaries. Sacred Books of the East. Vol. XXV. Oxford University Press. pp. 19–20 (1.64–73), 22 (1.79–80).
85. Bühler, G. (1886). "Ch. 1, The Creation". In Müller, F. Max (ed.). The Laws of Manu: translated with extracts from seven commentaries. Sacred Books of the East. Vol. XXV. Oxford University Press. pp. 20 (1.67), 22 (1.79–80).
86. Burgess 1935, p. 17 (1.29–34).
87. Burgess 1935, p. 19.
88. Burgess 1935, p. 41.
89. Burgess 1935, p. 5: The manuscripts without commentary insert, as the first half of v. 11, the usual definition of the length of a respiration: "the time occupied in pronouncing ten long syllables is called a respiration." 10 long syllable (gurvakshara) = 1 respiration (prana, period of four seconds).
90. Burgess 1935, p. 6.
91. Burgess 1935, p. 9: The period of 4,320,000 years is ordinarily styled Great Age (mahayuga), or, as above in two instances [1.15–16], Quadruple Age (caturyuga). In the Surya-Siddhanta, however, the former term is not once found, and the latter occurs only in these verses; elsewhere, Age (yuga) alone is employed to denote it, and always denotes it, unless expressly limited by the name of the Golden (krta) Age.
92. Burgess 1935, p. ix (Introduction): Calculated date of 2163102 B.C. for "the end of the Golden Age (Krta yuga)", the start of Treta yuga, mentioned in Surya Siddhanta 1.57.
93. Burgess 1935, p. ix (Introduction): Calculated date of 2163102 B.C. for "the end of the Golden Age (Krta yuga)", the start of Treta yuga, mentioned in Surya Siddhanta 1.57.
94. Burgess 1935, pp. 12–13 (1.21–24), 19.
95. Burgess 1935, pp. 19, 27.
96. Burgess 1935, pp. 288, 289, chapter XII verse 45-51: The Surya Siddhanta identifies the vernal equinox with the First Point of Aries

and hence does not distinguish between the sidereal and tropical year.

97. Burgess 1935, pp. 8, 9, chapter I verse 13,14.
98. Burgess, Ebenezer (1860). "Ch. I, Of the Mean Motions of the Planets". Translation of the Sûrya-Siddhânta: A text-book of Hindu astronomy, with notes and an appendix. Journal of the American Oriental Society. pp. 10–12 (1.21-24), 17.
99. Burgess, Rev. Ebenezer (1935) [1860]. "Ch. 1: Of the Mean Motions of the Planets.". In Gangooly, Phanindralal (ed.). Translation of the Surya-Siddhanta, A Text-Book of Hindu Astronomy; With notes and an appendix. University of Calcutta. pp. 5–12 (1.10–21).
100. Burgess, Rev. Ebenezer (1935) [1860]. "Ch. 1: Of the Mean Motions of the Planets.". In Gangooly, Phanindralal (ed.). Translation of the Surya-Siddhanta, A Text-Book of Hindu Astronomy; With notes and an appendix. University of Calcutta. pp. 7–9 (1.13–15), 11 (1.18–19).
101. Burgess, Rev. Ebenezer (1935) [1860]. "Ch. 1: Of the Mean Motions of the Planets.". In Gangooly, Phanindralal (ed.). Translation of the Sûrya-Siddhânta: A text-book of Hindu astronomy, with notes and an appendix. University of Calcutta. pp. 7–9 (1.13–17).
102. Buswell, RE Jr.; Lopez, DS Jr. (2014). The Princeton Dictionary of Buddhism (1st ed.). Princeton, New Jersey: Princeton University Press. p. 106. ISBN 978-0-691-15786-3.
103. Calculations as of midnight on 17/18 February 2023 CE. Note, the number of years from 1 BCE to 1 CE is 1 year and not 2 years since there is no year zero.
104. Calculations use a traditional 360-day year (twelve 30-day months) and a standard 24-hour day for all entities:
105. Chapter 224 (CCXXIV) in some sources: Mahabharata 12.224.
106. Charles Lanman, Hymns by Women, Book X, Hymn 125, Rigveda, The Sacred Books of the East Volume IX: India and Brahmanism, Editor: Max Muller, Oxford, pages 46–47.
107. Charles Lanman, The Creation Hymn, Book X, Hymn 129, Rigveda, The Sacred Books of the East Volume IX: India and Brahmanism, Editor: Max Muller, Oxford, page 48.
108. Christopher Chapple (1986), Karma and creativity, State University of New York Press, ISBN 0-88706-251-2, pages 60–64.
109. Clark, Walter (1930). Aryabhatiya: An Ancient Indian Work on Mathematics and Astronomy – An English Translation. The University of Chicago Press.

110.Cremo, M.A., 1999. Puranic time and the archaeological record. In T. Murray (ed.), Time and Archaeology 38–48. London: Routledge. http://catalogue.nla.gov.au/Record/379479.
111.D. Pingree: "History of Mathematical Astronomy in India", Dictionary of Scientific Biography, Vol. 15 (1978), pp. 533–633 (533, 554f.)
112.Daksinottara Bhitti at jantarmantar.org.
113.Dallal, 162.
114.Daniel 1992.
115.Daniélou, Alain (1987) [1st ed. 1985]. While the Gods Play: Shaiva Oracles and Predictions on the Cycles of History and the Destiny of Mankind [La Fantaisie des Dieux et L'Aventure Humaine]. Translated by Bailey, Barbara; Baker, Michael. Inner Traditions International. pp. 193–198. ISBN 0892811153.
116.David Bruce Hughes. Sri Vedanta-sutra, Adhyaya 2. David Bruce Hughes. p. 76.
117.David Christian (1 September 2011). Maps of Time: An Introduction to Big History. University of California Press. p. 18. ISBN 978-0-520-95067-2.
118.Deshpande, S; Nagendra, H. R.; Nagarathna, R (2009). "A randomized control trial of the effect of yoga on Gunas (personality) and Self esteem in normal healthy volunteers". International Journal of Yoga. 2 (1): 13–21. doi:10.4103/0973-6131.43287. PMC 3017961. PMID 21234210.
119.Dick Teresi. Lost Discoveries: The Ancient Roots of Modern Science—from the Babylonians to the Maya. Simon & Schuster. p. 174.
120.Dinkar Joshi (1 January 2005). Glimpses Of Indian Culture. Star Publications. p. 32. ISBN 978-81-7650-190-3. Retrieved 15 December 2012.
121.Doniger, Wendy; Hawley, John Stratton, eds. (1999). "Merriam-Webster's Encyclopedia of World Religions". Merriam-Webster. Merriam-Webster, Incorporated. p. 691 (Manu). ISBN 0-87779-044-2. a day in the life of Brahma is divided into 14 periods called manvantaras ("Manu intervals"), each of which lasts for 306,720,000 years. In every second cycle [(new kalpa after pralaya)] the world is recreated, and a new Manu appears to become the father of the next human race. The present age is considered to be the seventh Manu cycle.
122.Doniger, Wendy; Hawley, John Stratton, eds. (1999). "Merriam-Webster's Encyclopedia of World Religions". Merriam-Webster.

Merriam-Webster, Incorporated. p. 691 (Manu). ISBN 0877790442. a day in the life of Brahma is divided into 14 periods called manvantaras ("Manu intervals"), each of which lasts for 306,720,000 years. In every second cycle [(new kalpa after pralaya)] the world is recreated, and a new Manu appears to become the father of the next human race. The present age is considered to be the seventh Manu cycle.

123. Doniger, Wendy; Hawley, John Stratton, eds. (1999). "Merriam-Webster's Encyclopedia of World Religions". Merriam-Webster. Merriam-Webster, Incorporated. p. 691 (Manu). ISBN 0877790442. a day in the life of Brahma is divided into 14 periods called manvantaras ("Manu intervals"), each of which lasts for 306,720,000 years. In every second cycle [(new kalpa after pralaya)] the world is recreated, and a new Manu appears to become the father of the next human race. The present age is considered to be the seventh Manu cycle.
124. Dutt, Manmatha Nath (1903). "Ch. 231 (CCXXXI)". A Prose English Translation of The Mahabharata (Translated Literally from the Original Sanskrit text). Vol. Book 12 (Shanti Parva). Calcutta: Elysium Press. p. 351 (12.231.17, 19–21, 29).
125. Dutt, Manmatha Nath (1903). "Ch. 231 (CCXXXI)". A Prose English Translation of The Mahabharata (Translated Literally from the Original Sanskrit text). Vol. Book 12 (Shanti Parva). Calcutta: Elysium Press. pp. 350–351 (12.231.12–21, 29–31).
126. Each Kali-yuga-sandhi lasts for 36,000 solar (100 divine) years:
127. East-India Register and Directory. W.H. Allen. 1819-01-01.
128. Edward Quinn (1 January 2009). Critical Companion to George Orwell. Infobase Publishing. p. 188. ISBN 978-1-4381-0873-5. Retrieved 15 December 2012.
129. Elankumaran, S (2004). "Personality, organizational climate and job involvement: An empirical study". Journal of Human Values. 10 (2): 117–130. doi:10.1177/097168580401000205. S2CID 145066724.
130. Epstein, Ronald (2003). Buddhism A to Z. Burlingame, California, United States.: The Buddhist Text Translation Society. ISBN 0-88139-353-3.
131. Epstein, Ronald B.(2002). Buddhist Text Translation Society's Buddhism A to Z p. 204. Buddhist Text Translation Society. ISBN 0-88139-353-3, ISBN 978-0-88139-353-8.

132.F. Max Muller (30 June 2004). The Upanishads, Vol I. Kessinger Publishing. p. 228. ISBN 978-1-4191-8641-7. Retrieved 15 December 2012.
133.Flood, Gavin (24 August 2009). "Hindu concepts". BBC Online. BBC. Archived from the original on 11 April 2014. Retrieved 31 July 2015.
134.Foreign Impact on Indian Life and Culture (c. 326 B.C. to C. 300 A.D.). Satyendra Nath Naskar. Abhinav Publications, 1 January 1996 – History – 253 pages. Pages 56–57
135.Fourth Brâhmana in Brihadaranyaka Upanishad, Fourth Brahmana. Translated by Max Müller as The Upanishads, Part 2 (SBE15) [1879].
136.Ganga Ram Garg (1992). Encyclopaedia of the Hindu World. Concept. p. 446. ISBN 978-81-7022-375-7.
137.Garga Samhita (59).
138.George D. Chryssides; Benjamin E. Zeller (2014). The Bloomsbury Companion to New Religious Movements. Bloomsbury Academic. p. 333. ISBN 978-1-4411-9829-7.
139.Georgis, Faris (2010). Alone in Unity: Torments of an Iraqi God-Seeker in North America. Dorrance Publishing. p. 62. ISBN 978-1-4349-0951-0.
140.Godwin 2011, p. 299.
141.Godwin 2011, p. 301(a): Great Age (Mahayuga).
142.Godwin 2011, p. 301(b): Vishnu Purana, translated by the great Sanskritist Horace Hayman Wilson: One Pararddha, or half [Brahma's] existence, has expired, terminating with the Maha Kalpa called Padma. The Kalpa (or day of Brahma) termed Varaha is the first of the second period of Brahma's existence. ... The Hindu astronomers agree that the Kali Yuga began at midnight between February 17 and 18, 3102 BCE. Consequently it is due to end about 427,000 CE, whereupon a new Golden Age will dawn.
143.Godwin 2011, p. 301: The Hindu astronomers agree that the [Dvapara Yuga ended and] Kali Yuga began at midnight between February 17 and 18, 3102 BCE. Consequently [Kali Yuga] is due to end about 427,000 CE, whereupon a new Golden Age will dawn.
144.Godwin 2011, p. 301: Vishnu Purana, translated by the great Sanskritist Horace Hayman Wilson: One Pararddha, or half [Brahma's] existence, has expired, terminating with the Maha Kalpa called Padma. The Kalpa (or day of Brahma) termed Varaha is the first of the second period of Brahma's existence. ... The Hindu astronomers agree that the Kali Yuga began at midnight between

February 17 and 18, 3102 BCE. Consequently it is due to end about 427,000 CE, whereupon a new Golden Age will dawn.

145.Godwin 2011, p. 304.

146.Godwin 2011, p. 306.

147.Godwin 2011, p. 308.

148.Godwin 2011, p. 309: Daniélou said his figures are accurate to within fifty years.

149.Godwin 2011, pp. 298–299.

150.Godwin 2011, pp. 305–306.

151.Godwin 2011, pp. 306–307.

152.Godwin 2011, pp. 307–310.

153.Godwin 2011, pp. 330, 331, 332.

154.Godwin 2011, pp. 330–331.

155.Godwin 2011, pp. 331–332.

156.Godwin 2011, pp. 332–333.

157.Godwin 2011, pp. 333–334.

158.Godwin, Joscelyn (2011). Atlantis and the Cycles of Time: Prophecies, Traditions, and Occult Revelations. Inner Traditions. pp. 300–301. ISBN 9781594778575.

159.Godwin, Joscelyn (2011). Atlantis and the Cycles of Time: Prophecies, Traditions, and Occult Revelations. Inner Traditions. pp. 300–301. ISBN 9781594778575.

160.Gogerly, Rev. Daniel John; Silva, Rev. David de; Scott, Rev. John (1870). "Budhism: A Lecture delivered before the Colombo Young Men's Christian Association". Journal of the Ceylon Branch of the Royal Asiatic Society. Colombo. 1867–70, Part I: 91–92 (f.n. 4).

161.Gogerly, Silva & Scott 1870, p. 110.

162.Gogerly, Silva & Scott 1870, pp. 106–108.

163.Gogerly, Silva & Scott 1870, pp. 96–97.

164.González-Reimann 2018, p. 415 (World Destruction and Re-creation).

165.González-Reimann, Luis (2018). "Cosmic Cycles, Cosmology, and Cosmography". In Basu, Helene; Jacobsen, Knut A.; Malinar, Angelika; Narayanan, Vasudha (eds.). Brill's Encyclopedia of Hinduism. Vol. 2. Leiden: Brill Publishers. p. 415. doi:10.1163/2212-5019_BEH_COM_1020020. ISBN 978-90-04-17641-6. ISSN 2212-5019. The cycle [of creation and destruction] is either called a yuga (MBh. 1.1.28; 12.327.89; 13.135.11), a kalpa, meaning a formation or a creation (MBh. 6.31.7 [= BhG. 9.7]; 12.326.70; 12.327.23), or a day of the brahman, or of Brahmā, the creator god (MBh. 12.224.28–31).

Sometimes, it is simply referred to as the process of creation and destruction (saṃhāravikṣepa; MBh. 12.271.30, 40, 43, 47–49).
166.Gopal, Madan (1990). K.S. Gautam (ed.). India through the ages. Publication Division, Ministry of Information and Broadcasting, Government of India. p. 79.
167.Graham Chapman; Thackwray Driver (2002). Timescales and Environmental Change. Routledge. pp. 7–8. ISBN 978-1-134-78754-8.
168.Griffith, Ralph T.H. (Transl.): Rigveda Hymn CXXIX. Creation in Hymns of the Rgveda, Vol. II, 1889-92. Munshiram Manoharlal Publishers Pvt. Ltd., New Delhi, 1999.
169.Grimes, John A. (1995). Ganapati: Song of the Self. Albany, New York: State University of New York Press. pp. 101–104. ISBN 0-7914-2439-1. In the Gaṇeśa Purāṇa, Gaṇapati is described as taking a different incarnation (avatāra) in each of the four cosmic ages (yugas). In the kṛta yuga, Gaṇeśa incarnates as Vināyaka (or Mahotkaṭa), the son of Kāśyapa and Aditi. ... During the treta yuga, Gaṇapati incarnates as Mayūreśvara, the son of Lord Śiva. ... During the dvapara yuga, Gaṇeśa incarnates as Gajānāna, the son of Lord Śiva. ... During the kali yuga, Gaṇapati incarnates as Dhūmraketu (or Śūrpakarṇa).
170.Guénon, René (2001) [1st ed. 1970]. Fohr, Samuel D. (ed.). Traditional Forms & Cosmic Cycles [Formes Traditionnelles et Cycles Cosmiques]. Translated by Fohr, Henry D. Sophia Perennis. pp. 5–8. ISBN 0900588179.
171.Gupta 2010, p. 3.
172.Gupta 2010, p. 5.
173.Gupta 2010, p. 6.
174.Gupta 2010, p. 7.
175.Gupta 2010, p. 8.
176.Gupta 2010, p. 9: At the beginning of any Hindu rite, an ashloka in Sanskrit is read out, which means that this ritual is being performed for a certain purpose (name of the purpose) by this particular person (name of the person including father's and family names) at this place (full address with country name) at this time of day (Ghadi and Pala or in hours and minutes) in 5109 of Kalyugi year (for AD 2007, or Hindu calendar year 2064 Sambat) of the 28th Chaturyugee of the 7th Manvantara on the first day of the 51st year of the 2nd Brahma [2nd half of Brahma's life]. This is a wonderful example of counting time from the start of the universe to the present time.
177.Gupta 2010, pp. 5–6.
178.Gupta 2010, pp. 7–8.

179. Gupta, Ravi; Valpey, Kenneth (26 March 2013). The Bhagavata Purana: Sacred Text and Living Tradition. Columbia University Press. ISBN 978-0-231-14999-0.
180. Gupta, S. V. (2010). "Ch. 1.2.4 Time Measurements". In Hull, Robert; Osgood, Richard M. Jr.; Parisi, Jurgen; Warlimont, Hans (eds.). Units of Measurement: Past, Present and Future. International System of Units. Springer Series in Materials Science: 122. Springer. pp. 6–8. ISBN 9783642007378. Paraphrased: Deva day equals solar year. Deva lifespan (36,000 solar years) equals 100 360-day years, each 12 months. Mahayuga equals 12,000 Deva (divine) years (4,320,000 solar years), and is divided into 10 charnas consisting of four Yugas: Satya Yuga (4 charnas of 1,728,000 solar years), Treta Yuga (3 charnas of 1,296,000 solar years), Dvapara Yuga (2 charnas of 864,000 solar years), and Kali Yuga (1 charna of 432,000 solar years). Manvantara equals 71 Mahayugas (306,720,000 solar years). Kalpa (day of Brahma) equals an Adi Sandhya, 14 Manvantaras, and 14 Sandhya Kalas, where 1st Manvantara preceded by Adi Sandhya and each Manvantara followed by Sandhya Kala, each Sandhya lasting same duration as Satya yuga (1,728,000 solar years), during which the entire earth is submerged in water. Day of Brahma equals 1,000 Mahayugas, the same length for a night of Brahma (Bhagavad-gita 8.17). Brahma lifespan (311.04 trillion solar years) equals 100 360-day years, each 12 months. Parardha is 50 Brahma years and we are in the 2nd half of his life. After 100 years of Brahma, the universe starts with a new Brahma. We are currently in the 28th Kali yuga of the first day of the 51st year of the second Parardha in the reign of the 7th (Vaivasvata) Manu. This is the 51st year of the present Brahma and so about 155 trillion years have elapsed. The current Kali Yuga (Iron Age) began at midnight on 17/18 February 3102 BC in the proleptic Julian calendar.
181. Gupta, S. V. (2010). "Ch. 1.2.4 Time Measurements". In Hull, Robert; Osgood, Richard M. Jr.; Parisi, Jurgen; Warlimont, Hans (eds.). Units of Measurement: Past, Present and Future. International System of Units. Springer Series in Materials Science: 122. Springer. pp. 7–8. ISBN 978-3-642-00737-8.
182. Gupta, S. V. (2010). "Ch. 1.2.4 Time Measurements". In Hull, Robert; Osgood, Richard M. Jr.; Parisi, Jurgen; Warlimont, Hans (eds.). Units of Measurement: Past, Present and Future. International System of Units. Springer Series in Materials Science: 122. Springer. pp. 7–8. ISBN 9783642007378. Paraphrased: Mahayuga equals 12,000 Deva

(divine) years (4,320,000 solar years). Manvantara equals 71 Mahayugas (306,720,000 solar years). Kalpa (day of Brahma) equals an Adi Sandhya, 14 Manvantaras (14 x 71 = 994 Mahayugas), and 14 Sandhya Kalas, where 1st Manvantara preceded by Adi Sandhya and each Manvantara followed by Sandhya Kala, each Sandhya lasting same duration as Satya yuga (1,728,000 solar years), during which the entire earth is submerged in water. Day of Brahma equals 1,000 Mahayugas, the same length for a night of Brahma (Bhagavad-gita 8.17). Brahma lifespan (311.04 trillion solar years) equals 100 360-day years, each 12 months. Parardha is 50 Brahma years and we are in the 2nd half of his life. After 100 years of Brahma, the universe starts with a new Brahma. We are currently in the 28th Kali yuga of the first day of the 51st year of the second Parardha in the reign of the 7th (Vaivasvata) Manu.

183. Haag, James. HINDU COSMOGONY/ COSMOLOGY. Routledge, 2 Park Square, Milton Park, Abingdon, Oxon, OX14 4RN. p. 115.
184. Hans Kng (31 October 2006). Tracing The Way: Spiritual Dimensions of the World Religions. A&C Black. p. 50. ISBN 9780826494238.
185. Harris, Melanie J.; Wickramasinghe, N.C.; Lloyd, David; et al. (2002). "Detection of living cells in stratospheric samples" (PDF). Proc. SPIE. Instruments, Methods, and Missions for Astrobiology IV. 4495 (Instruments, Methods, and Missions for Astrobiology IV): 192. Bibcode:2002SPIE.4495..192H. doi:10.1117/12.454758. S2CID 129736236. Archived (PDF) from the original on 22 September 2017. Retrieved 21 September 2019.
186. Hayashi (2008), Aryabhata I.
187. Hayashi (2008), Bhaskara I.
188. Hayashi (2008), Bhaskara II.
189. Hayashi (2008), Brahmagupta.
190. Hayashi (2008), Shripati.
191. Henry White Wallis (1887). The Cosmology of the Ṛigveda: An Essay. Williams and Norgate. p. 117.
192. Henry White Wallis (1887). The Cosmology of the Ṛigveda: An Essay. Williams and Norgate. pp. 61–73.
193. Highlights of Astronomy, Volume 11B: As presented at the XXIIIrd General Assembly of the IAU, 1997. Johannes Andersen Springer, 31 January 1999 – Science – 616 pages. page 721.
194. Hobson-Jobson: A Glossary of Colloquial Anglo-Indian Words and Phrases, and Kindred Terms, Etymological, Historical, Geographical

and Discursive by Col. Henry Yule and A.C. Burnell, Originally published 1903, Reprint, 1968.

195.In Search of the Ultimate Goal of Life by Srila A. C. Bhaktivedanta Prabhupada, *brahmananam sahasrebhyah satra-yaji visisyate.*

196.Indian Astronomy. (2013). In D. Leverington, Encyclopedia of the history of Astronomy and Astrophysics. Cambridge, United Kingdom: Cambridge University Press. Retrieved from http://search.credoreference.com/content/entry/cupaaa/indian_astronomy/0.

197.Indrasena, Buddhike Sri Harsha (2021) [2020]. Resolving the Controversies of Astrology and Vedic Astronomy: Thinking Outside the Box (Revised ed.). Lulu Press. pp. 81–82. ISBN 978-1-716-75246-9. 6.2.2: For all human, divine, and Brahma scales, one year equals 360 'days' of respective scale.

198.J.A.B. van Buitenen (2008).

199.James G. Lochtefeld, Guna, in The Illustrated Encyclopedia of Hinduism: A-M, Vol. 1, Rosen Publishing, ISBN 978-0-8239-3179-8, pages 224, 265, 520.

200.Jamison, Stephanie; Brereton, Joel (2014). The Rigveda: The Earliest Religious Poetry of India. Oxford University Press. pp. 1499–1500, 1607–1609. ISBN 978-0-19-937018-4.

201.Jan N. Bremmer (2007). The Strange World of Human Sacrifice. Peeters Publishers. pp. 170–. ISBN 978-90-429-1843-6. Retrieved 15 December 2012.

202.Janmastami vow, One who eats on Lord Krishna's birthday is lowest of mankind. .. His pious credits of ten million births are at once destroyed. ... One who fasts, follows the Janmastami vow, and keeps an all-night vigil on this day becomes freed from the sins of ten million births. Of this there is no doubt.

203.Jeaneane D. Fowler (1997). Hinduism: Beliefs and Practices. Sussex Academic Press. pp. 10–11. ISBN 978-1-898723-60-8.

204.Jha, Ganganath (1999) [1st ed. 1920]. "Discourse I - Origin of the Work—Creation of the World—Summary of Contents of the Book.". Manusmriti with the Commentary of Medhatithi in Ten Volumes. Vol. Adhyāya 1. Motilal Banarsidass Publishers Pvt. Ltd. ISBN 8120811550 – via Wisdom Library.

205.John A. Grimes (1996). A Concise Dictionary of Indian Philosophy: Sanskrit Terms Defined in English. State University of New York Press. p. 95. ISBN 978-0-7914-3067-5.

206.John E. Mitchiner (2000). Traditions of the Seven Rsis. Motilal Banarsidass. pp. 141–144. ISBN 978-81-208-1324-3.
207.Johnson, W. J., 1951- (12 February 2009). A dictionary of Hinduism (First ed.). Oxford [England]. ISBN 978-0-19-861025-0. OCLC 244416793.
208.Johnson, W.J. (2009). A Dictionary of Hinduism. Oxford University Press. p. 165. ISBN 978-0-19-861025-0.
209.Johnson, W.J. (2009). A Dictionary of Hinduism. Oxford University Press. p. 165. ISBN 978-0-19-861025-0.
210.Johnson, W.J. (2009). A Dictionary of Hinduism. Oxford University Press. p. 165. ISBN 978-0-19-861025-0.
211.Jones, Sir William (1807) [1st ed. 1794]. "The Laws of Menu, Son of Brahma - Chapter The First: On the Creation; with a Summary of the Contents". The Works of Sir William Jones in thirteen volumes. Vol. VII. pp. 101–104.
212.Joseph Lewis Henderson, Maud Oakes (4 September 1990). The Wisdom of the Serpent: The Myths of Death, Rebirth, and Resurrection. Princeton University Press. p. 86. ISBN 0-691-02064-7.
213.Joseph, 306.
214.Joseph, 408.
215.Jyoti Bhusan Das Gupta. Science, Technology, Imperialism, and War. Pearson Education India. p. 33.
216.Kak, Shubash (1995). "The Astronomy of the Age of Geometric Altars". Quarterly Journal of the Royal Astronomical Society. 36: 385–395. Bibcode:1995QJRAS..36..385K.
217.Kali-santarana Upanisad, The sixteen words-Hare Krishna, Hare Krishna, Krishna Krishna, Hare Hare; Hare Rama, Hare Rama, Rama Rama, Hare Hare-are especially meant for completely destroying all the contamination of Kali. To save oneself from the contamination of Kali-yuga, there is no alternative in all the Vedas except the chanting of this sixteen-word mantra.
218.Kane, P. V. (September 1936). Sukthankar, V. S.; Fyzee, A. A. A.; Bhagwat, N. K. (eds.). "Kalivarjya (actions forbidden in the Kali Age)". Journal of the Bombay Branch of the Royal Asiatic Society. The Asiatic Society of Bombay. 12 (1–2): 4.
219.Kenneth Kramer (January 1986). World Scriptures: An Introduction to Comparative Religions. Paulist Press. p. 34. ISBN 978-0-8091-2781-8.
220.King, 240.

221. Knapp, Stephen. "Lord Rama: Fact or Fiction". Stephen Knapp and His Books on Vedic Culture, Eastern Philosophy and Spirituality. Retrieved 2020-05-17. In the Vayu Purana (70.47–48) [published by Motilal Banarsidass] there is a description of the length of Ravana's life. It explains that when Ravana's merit of penance began to decline, he met Lord Rama, the son of Dasarath, in a battle wherein Ravana and his followers were killed in the 24th Treta-yuga. ... The Matsya Purana (47/240,243–246) is another source that also gives more detail of various avataras and says Bhagawan Rama appeared at the end of the 24th Treta-yuga.
222. Kolachana, A.; Montelle, C.; Dhammaloka, J.; Melnad, K.; Mahesh, K.; Vyas, P. (2018). "The Candrārkī of Dinakara: A Text Related to Solar and Lunar Tables". Journal for the History of Astronomy. 49 (3): 306–344. Bibcode:2018JHA....49..306K. doi:10.1177/0021828618787556. S2CID 125563931.
223. Kolachana, Aditya; Montelle, Clemency; Dhammaloka, J.; Melnad, K.; Mahesh, K.; Vyas, P.; Ramasubramanian, K.; Sriram, M.S.; Pai, V. (2018). "A Critical Edition of the Candrārkī of Dinakara: A Text Concerning Solar and Lunar Tables". History of Science in South Asia. 6: 127–161. doi:10.18732/hssa.v6i0.35.
224. Krishan, Yuvraj (1999). Gaṇeśa: Unravelling An Enigma. Delhi: Motilal Banarsidass Publishers. pp. 79–80. ISBN 978-81-208-1413-4.
225. Krishna, the Beautiful Legend of God, pages 11–12, and commentary pages 423–424, by Edwin Bryant.
226. Krishnamurthy, Prof. V. (2019). "Ch. 20: The Cosmic Flow of Time as per Scriptures". Meet the Ancient Scriptures of Hinduism. Notion Press. ISBN 9781684669387. According to the traditional time-keeping ... Thus in Brahma's calendar the present time may be coded as his 51st year - first month - first day - 7th manvantara - 28th maha-yuga - 4th yuga or kaliyuga.
227. Krishnamurthy, V. (2019). "Ch. 20: The Cosmic Flow of Time as per Scriptures". Meet the Ancient Scriptures of Hinduism. Notion Press. ISBN 978-1-68466-938-7. Each manvantara is preceded and followed by a period of 1,728,000 (= 4K) years when the entire earthly universe (bhu-loka) will submerge under water. The period of this deluge is known as manvantara-sandhya (sandhya meaning, twilight). ... According to the traditional time-keeping ... Thus in Brahma's calendar the present time may be coded as his 51st year - first month - first day - 7th manvantara - 28th maha-yuga - 4th yuga or kaliyuga.

228. Krishnamurthy, V. (2019). "Ch. 20: The Cosmic Flow of Time as per Scriptures". Meet the Ancient Scriptures of Hinduism. Notion Press. ISBN 9781684669387. Each manvantara is preceded and followed by a period of 1,728,000 (= 4K) years when the entire earthly universe (bhu-loka) will submerge under water. The period of this deluge is known as manvantara-sandhya (sandhya meaning, twilight). ... According to the traditional time-keeping ... Thus in Brahma's calendar the present time may be coded as his 51st year - first month - first day - 7th manvantara - 28th maha-yuga - 4th yuga or kaliyuga.
229. Krishnamurthy, V. (2019). "Ch. 20: The Cosmic Flow of Time as per Scriptures". Meet the Ancient Scriptures of Hinduism. Notion Press. ISBN 9781684669387. Each manvantara is preceded and followed by a period of 1,728,000 (= 4K) years when the entire earthly universe (bhu-loka) will submerge under water. The period of this deluge is known as manvantara-sandhya (sandhya meaning, twilight).
230. Kumar, Ashwini (2005). Vaastu: The Art And Science Of Living. Sterling Publishers Pvt. Ltd. p. 50. ISBN 81-207-2569-7.
231. Laurie L. Patton (2005). Bringing the Gods to Mind: Mantra and Ritual in Early Indian Sacrifice. University of California Press. pp. 113, 216. ISBN 978-0-520-93088-9.
232. Length of breath in Sukhasana of healthy person.
233. Ludo Rocher (1986). The Purāṇas. Otto Harrassowitz Verlag. pp. 123–125, 130–132. ISBN 978-3-447-02522-5.
234. Mahaprabhura Siksa: The teachings of Chaitanya Mahaprabhu by Bhaktivinode Thakura – Chapter 11. Prayojana: the ultimate goal,
235. Maharishi Mahesh Yogi on the Bhagavad-Gita, a New Translation and Commentary, Chapter 1–6. Penguin Books, 1969, p 128 (v 45) and p 269 v.13
236. Main part of yuga (yuga proper); occurs after its sandhyā (dawn) and before its sandhyāṃśa (dusk).
237. Mani, Vettam (1975). "RAKTAJA". A Comprehensive Dictionary with Special Reference to the Epic and Puranic Literature. Puranic Encyclopedia. Motilal Banarsidass. p. 630. ISBN 0842608222. The following story is told in the Padma Purana (Chapter 14) ... Devendra raised a legal objection to the above injunction of Vishnu as follows: "You, who incarnated yourself as Rama in the twentyseventh yuga of the last Manvantara for the purpose of killing Ravana, killed my son Bali. Therefore I do not wish to procreate Nara as my son." To this objection of Indra, Vishnu assured him that as a penalty for the

mistake of killing Bali, he would be a companion of Nara (Arjuna) who would be born as Indra's son.

238. Mark Juergensmeyer; Wade Clark Roof (2011). Encyclopedia of Global Religion. SAGE Publications. p. 272. ISBN 978-1-4522-6656-5.
239. Matchett, Freda; Yano, Michio (2003). "Part II, Ch. 6: The Puranas / Part III, Ch. 18: Calendar, Astrology, and Astronomy". In Flood, Gavin (ed.). The Blackwell Companion to Hinduism. Blackwell Publishing. pp. 139–140, 390 (Kali yuga epoch). ISBN 0631215352.
240. Matchett, Freda; Yano, Michio (2003). "Part II, Ch. 6: The Puranas / Part III, Ch. 18: Calendar, Astrology, and Astronomy". In Flood, Gavin (ed.). The Blackwell Companion to Hinduism. Blackwell Publishing. pp. 139–140, 390. ISBN 0631215352.
241. Merriam-Webster (1999). "Merriam-Webster's Encyclopedia of World Religions". In Doniger, Wendy; Hawley, John Stratton (eds.). Merriam-Webster. Merriam-Webster, Incorporated. pp. 445 (Hinduism), 1159 (Yuga). ISBN 0877790442.
242. Merriam-Webster (1999). "Merriam-Webster's Encyclopedia of World Religions". In Doniger, Wendy; Hawley, John Stratton (eds.). Merriam-Webster. Merriam-Webster, Incorporated. pp. 445 (Hinduism), 1159 (Yuga). ISBN 0877790442.
243. Merriam-Webster 1999, p. 629 (Kalki): At the end of the present Kali age, when virtue and religion have disappeared into CHAOS and the world is ruled by unjust men, Kalki will appear to destroy the wicked and usher in a new age. ... According to some myths, Kalki's horse will stamp the earth with its right foot, causing the tortoise that supports the world to drop into the deep. Then Kalki will restore the earth to its initial purity.
244. Merriam-Webster 1999, p. 691 (Manu): a day in the life of Brahma is divided into 14 periods called manvantaras ("Manu intervals"), each of which lasts for 306,720,000 years. In every second cycle [(new kalpa after pralaya)] the world is recreated, and a new Manu appears to become the father of the next human race. The present age is considered to be the seventh Manu cycle.
245. Merry I. White; Susan Pollak (2 November 2010). The Cultural Transition: Human Experience and Social Transformation in the Third World and Japan. Edited by Merry I White, Susan Pollak. Taylor & Francis. p. 183. ISBN 978-0-415-58826-3. Retrieved 15 December 2012.

246.Mirabello, Mark (15 September 2016). A Traveler's Guide to the Afterlife: Traditions and Beliefs on Death, Dying, and What Lies Beyond. Inner Traditions / Bear & Co. p. 23. ISBN 978-1-62055-598-9.
247.Mircea Eliade; Charles J. Adams (1987). The Encyclopedia of religion. Macmillan. pp. 100–113, 116–117. ISBN 978-0-02-909730-4.
248.Mishra, R.C (2013). Moksha and the Hindu Worldview. New Delhi, India: SAGE Publications. pp. 21–42.
249.Motilal Banarsidass 1960, p. 2 (fn. 1).
250.Muller, F. Max (2001). The Dhammapada and Sutta-Nipata. Taylor & Francis Group. p. 60. ISBN 9781317849179.
251.Murthy, S. S. N. (August 2005). "Number Symbolism in the Vedas" (PDF). Electronic Journal of Vedic Studies (EJVS). 12 (3): 87–99. ISSN 1084-7561. Steeds of Indra [...] 1100.
252.N. P. Subramania Iyer. Kalaprakasika. Asian Educational Services. p. 3.
253.Narada Bhakti Sutra 39, Purport, Lord Caitanya graphically describes the rarity of gaining the association of a pure devotee. The Lord tells Rupa Gosvami that there are unlimited living entities among 8,400,000 species, and all these living entities are wandering from body to body, planet to planet, within this universe. The few living entities in human bodies may be divided into the uncultured and the cultured – those who are ignorant of the Vedic principles and those who know them. Among those who know the Vedic principles, roughly half simply give lip service to these principles while committing all kinds of sins in violation of these principles. Out of those who actually follow the Vedic principles, most seek material rewards like wealth, good birth, or elevation to heaven. Among millions of pious followers of the Vedic injunctions, one may be actually wise (a jnani). Out of many millions of such jnanis, Lord Caitanya says, one may actually become liberated from birth and death, and out of many millions of such liberated persons, a devotee of the Lord is very difficult to find.
254.Nārada-pañcarātra, The essence of all Vedic knowledge — comprehending the three kinds of Vedic activity [karma-kāṇḍa, jñāna-kāṇḍa and upāsanā-kāṇḍa], the chandas, or Vedic hymns, and the processes for satisfying the demigods — is included in the eight syllables Hare Krsna, Hare Krsna. This is the reality of all Vedānta. The chanting of the holy name is the only means to cross the ocean of nescience.
255.Ōhashi (1993).

256.Ōhashi (1997).
257.Ōhashi (2008), Astronomical Instruments in India.
258.Olivelle 2005, pp. 53–54.
259.Olivelle, Patrick (2005). Manu's Code of Law. Oxford University Press. pp. 3, 241 (1.71). ISBN 0-19-517146-2.
260.Otto Neugebauer (1952) "Tamil Astronomy", Osiris 10: 252–76
261.Panda, Sudhira (2019). "The Bhāsvatī astronomical handbook of Śatānanda". Journal of Astronomical History and Heritage. 22 (3): 536–544. doi:10.3724/SP.J.1440-2807.2019.03.12. S2CID 256574558.
262.Parmeshwaranand, Swami (2001). Encyclopaedic Dictionary of Purāṇas. Vol. 1 (A–C). New Delhi: Sarup & Sons. p. 169. ISBN 81-7625-226-3. The Doubt of Vyāsa: According to the Indian tradition, the sage Vyāsa was the compiler of all the Vedas, and the composer of the Mahābhārata and many other works. The [Bhāgavata Purāṇa] repeats this tradition ...
263.Part 1: 50.41 (city of Bali), 55.3, 55.7. The Vāyu Purāna: Part I. Delhi: Motilal Banarsidass. 1960. pp. 377–382.
264.Part 2: 5.133, 35.73, 35.77, 36.74–85, 37.26–32, 38.21–22, 46.29 (Bali as oblation). The Vāyu Purāna: Part II. Delhi: Motilal Banarsidass. 1960.
265.Patrick McNamara; Wesley J. Wildman (19 July 2012). Science and the World's Religions [3 Volumes]. ABC-CLIO. pp. 180–. ISBN 978-0-313-38732-6. Retrieved 15 December 2012.
266.Pawar, Ashwini (29 July 2015). "I'm proud that I recommended him for ISRO: EV Chitnis". DNA India. Archived from the original on 9 July 2021. Retrieved 13 July 2021.
267.Penprase, Bryan E. (2017). The Power of Stars (2nd ed.). Springer. p. 182. ISBN 9783319525976.
268.Penprase, Bryan E. (2017). The Power of Stars (2nd ed.). Springer. p. 182. ISBN 978-3-319-52597-6.
269.Pierre Cambon, Jean-François Jarrige. "Afghanistan, les trésors retrouvés: Collections du Musée national de Kaboul". Éditions de la Réunion des musées nationaux, 2006 – 297 pages. p269 [5]
270.Pierre Cambon, Jean-François Jarrige. "Afghanistan, les trésors retrouvés: Collections du Musée national de Kaboul". Éditions de la Réunion des musées nationaux, 2006 – 297 pages. p269 [6] "Les influences de l'astronomie grecques sur l'astronomie indienne auraient pu commencer de se manifester plus tot qu'on ne le pensait, des l'epoque Hellenistique en fait, par l'intermediaire des colonies grecques des Greco-Bactriens et Indo-Grecs" (French) Afghanistan, les trésors retrouvés", p269. Translation: "The influence of Greek

astronomy on Indian astronomy may have taken place earlier than thought, as soon as the Hellenistic period, through the agency of the Greek colonies of the Greco-Bactrians and the Indo-Greeks.

271. Pingree, David "Astronomy and Astrology in India and Iran" Isis, Vol. 54, No. 2 (Jun. 1963), pp. 229–246.
272. Pingree, David (1976). "The Recovery of early Greek Astronomy from India". Journal for the History of Astronomy. Science History Publications Ltd. 7 (19): 109–123. Bibcode:1976JHA.....7..109P. doi:10.1177/002182867600700202. S2CID 68858864.
273. Potter, Karl. Dharma and Mokṣa from a Conversational Point of View. Honolulu: University of Hawaii Press. pp. 49–63.
274. Prakriti: Indian philosophy, Encyclopædia Britannica
275. Pralaya The Secret Doctrine by H. P. Blavatsky, Vol. 2, p. 307 THE SEVEN AND FOURTEEN MANUS.
276. Prasad Sinha, Harendra (2006). Bharatiya Darshan Ki Rooprekha. Motilal Banarsidass Publisher. p. 86. ISBN 978-81-208-2144-6. Retrieved 18 March 2015.
277. Pruthi, R. K. (2005). "Ch. 4. Missile Man of India". President A.P.J. Abdul Kalam. Anmol Publications. pp. 61–76. ISBN 978-81-261-1344-6.
278. Raju (2001).
279. Raman, Varadaraja (2012). HINDUISM AND SCIENCE: SOME REFLECTIONS. Malden, USA: Blackwell Publishing Inc. pp. 549–574.
280. Ramasubramanian etc. (1994).
281. Ramasubramanian, K.; Srinivas, M. D.; Sriram, M. S. (25 May 1994). "Modification of the earlier Indian planetary theory by the Kerala astronomers (c. 1500 AD) and the implied heliocentric picture of planetary motion". Current Science. 66 (10): 784–790. ISSN 0011-3891. JSTOR 24098820 – via JSTOR.
282. Ramchandani (2000). Dale Hoiberg (ed.). A to C (Abd Allah ibn al-Abbas to Cypress). New Delhi: Encyclopædia Britannica (India). p. 2. ISBN 978-0-85229-760-5. Archived from the original on 29 August 2016. Retrieved 10 August 2019.
283. Ravi M. Gupta, Kenneth R. Valpey (29 November 2016). The Bhagavata Purana: Sacred Text and Living Tradition. Columbia University Press. p. 60. ISBN 978-0-231-54234-0.
284. Richard L. Thompson (2003), Maya: The World as Virtual Reality, ISBN 9780963530905
285. Richard L. Thompson, Vedic cosmography and astronomy.

286.Rig Veda 10.72 translation by R.T.H. Griffith (1896).
287.Robert M. Torrance (1 April 1999). Encompassing Nature: A Sourcebook. Counterpoint Press. pp. 121–122. ISBN 978-1-58243-009-6. Retrieved 15 December 2012.
288.Robert N. Bellah (2011). Religion in Human Evolution. Harvard University Press. pp. 510–511. ISBN 978-0-674-06309-9.
289.Roshen Dalal (2010). Hinduism: An Alphabetical Guide. Penguin Books. p. 83. ISBN 978-0-14-341421-6.
290.Roshen Dalal (5 October 2011). Hinduism: An Alphabetical Guide. Penguin Books India. pp. 318–319. ISBN 978-0-14-341421-6. Retrieved 15 December 2012.
291.S. M. Razaullah Ansari (2002), History of oriental astronomy: proceedings of the joint discussion-17 at the 23rd General Assembly of the International Astronomical Union, organised by the Commission 41 (History of Astronomy), held in Kyoto, August 25–26, 1997, Springer, pp. 133–4, ISBN 978-1-4020-0657-9.
292.S. M. Razaullah Ansari (2002), History of oriental astronomy: proceedings of the joint discussion-17 at the 23rd General Assembly of the International Astronomical Union, organised by the Commission 41 (History of Astronomy), held in Kyoto, August 25–26, 1997, Springer, p. 138, ISBN 978-1-4020-0657-9.
293.S. M. Razaullah Ansari (2002). History of oriental astronomy: proceedings of the joint discussion-17 at the 23rd General Assembly of the International Astronomical Union, organised by the Commission 41 (History of Astronomy), held in Kyoto, August 25–26, 1997. Springer. p. 141. ISBN 978-1-4020-0657-9.
294.S. N. Sadasivan (1 January 2000). A Social History Of India. APH Publishing. pp. 226–227. ISBN 978-81-7648-170-0. Retrieved 15 December 2012.
295.S.K. Paul, A.N. Prasad (1 November 2007). Reassessing British Literature: Pt. 1. Sarup & Sons. p. 91. ISBN 978-81-7625-764-0. Retrieved 15 December 2012.
296.Sadeh 2013.
297.Sanskriti (1 February 2015). "Concept and Measurement of Time in the Vedas". Sanskriti - Hinduism and Indian Culture Website. Retrieved 31 December 2021.
298.Sarma (2008), Acyuta Pisarati.
299.Sarma (2008), Armillary Spheres in India.
300.Sarma (2008), Lalla.
301.Sarma 2008a.

302.Savage-Smith (1985).
303.See A. Cunningham (1883), A Book of Indian Eras.
304.See Ōhashi (2008) in Astronomy: Indian Astronomy in China.
305.Shailaj Kumar Shrivastava (May 2017), MEASUREMENT UNITS OF LENGTH, MASS AND TIME IN INDIA THROUGH THE AGES, International Journal of Physical and Social Science, retrieved 23 December 2020.
306.Sharma (1995), 8–9.
307.Shilpa, S; Venkatesha Murthy, C. G. (2011). "Understanding personality from Ayurvedic perspective for psychological assessment: A case". AYU. 32 (1): 12–19. doi:10.4103/0974-8520.85716. PMC 3215408. PMID 22131752.
308.Shivaji, S.; Chaturvedi, P.; Begum, Z.; et al. (2009). "Janibacter hoylei sp.nov., Bacillus isronensis sp.nov. and Bacillus aryabhatta sp.nov. isolated from cryotubes used for collecting air from the upper atmosphere". International Journal of Systematic and Evolutionary Microbiology. 59 (12): 2977–2986. doi:10.1099/ijs.0.002527-0. PMID 19643890.
309.Sidharth, B.G (1998). "The Calendric Astronomy of the Vedas". Bulletin of the Astronomical Society of India. 26: 108. Bibcode:1998BASI...26..107S – via NASA Astrophysics Data System.
310.Soiver, Deborah A. (November 1991). The Myths of Narasimha and Vamana: Two Avatars in Cosmological Perspective. State University of New York Press. p. 51. ISBN 978-0-7914-0799-8.
311.Sri Caitanya Caritamrita Madhya 19.138, In this universe there are limitless living entities in 8,400,000 species, and all are wandering within this universe.
312.Sri Caitanya Caritamrta Madhya 19.138, There are 900,000 species living in the water. There are also 2,000,000 nonmoving living entities (sthavara), such as trees and plants. There are also 1,100,000 species of insects and reptiles, and there are 1,000,000 species of birds. As far as quadrupeds are concerned, there are 3,000,000 varieties, and there are 400,000 human species.
313.Sri Caitanya Caritamrta Madhya 19.138, There are 900,000 species living in the water. There are also 2,000,000 nonmoving living entities (sthavara), such as trees and plants. There are also 1,100,000 species of insects and reptiles, and there are 1,000,000 species of birds. As far as quadrupeds are concerned, there are 3,000,000 varieties, and there are 400,000 human species.

314. Sri Caitanya Caritamrta Madhya 19.138, There are 900,000 species living in the water. There are also 2,000,000 nonmoving living entities (sthavara), such as trees and plants. There are also 1,100,000 species of insects and reptiles, and there are 1,000,000 species of birds. As far as quadrupeds are concerned, there are 3,000,000 varieties, and there are 400,000 human species.

315. Sri Caitanya Caritamrta Madhya 19.138, There are 900,000 species living in the water. There are also 2,000,000 nonmoving living entities (sthavara), such as trees and plants. There are also 1,100,000 species of insects and reptiles, and there are 1,000,000 species of birds. As far as quadrupeds are concerned, there are 3,000,000 varieties, and there are 400,000 human species.

316. Sri Caitanya Caritamrta Madhya 19.138, There are 900,000 species living in the water. There are also 2,000,000 nonmoving living entities (sthavara), such as trees and plants. There are also 1,100,000 species of insects and reptiles, and there are 1,000,000 species of birds. As far as quadrupeds are concerned, there are 3,000,000 varieties, and there are 400,000 human species.

317. Srila Rupa Gosvami says that if brahmananda, or the happiness of becoming one with the Supreme, is multiplied by one trillionfold, it still cannot compare to an atomic fraction of the happiness derived from the ocean of devotional service. [1]

318. Śrīmad Bhāgavatam 1.1.21, This age of Kali is not at all suitable for self-realization as was Satya-yuga, the golden age, or Tretā- or Dvāpara-yugas, the silver and copper ages. For self-realization, the people in Satya-yuga, living a lifetime of a hundred thousand years, were able to perform prolonged meditation. And in Tretā-yuga, when the duration of life was ten thousand years, self-realization was attained by performance of great sacrifice. And in the Dvāpara-yuga, when the duration of life was one thousand years, self-realization was attained by worship of the Lord. But in the Kali-yuga, the maximum duration of life being one hundred years only and that combined with various difficulties, the recommended process of self-realization is that of hearing and chanting of the holy name, fame, and pastimes of the Lord.

319. Srimad Bhagavatam 2.1.26, The universe is divided into fourteen planetary systems. Seven planetary systems, called Bhur, Bhuvar, Svar, Mahar, Janas, Tapas and Satya, are upward planetary systems, one above the other. There are also seven planetary systems

downward, known as Atala, Vitala, Sutala, Talatala, Mahatala, Rasatala and Patala, gradually, one below the other.

320. Śrīmad Bhāgavatam 2.3.2-7, Śukadeva Gosvāmī said that out of the 8,400,000 species of living entities, the human form of life is rare and valuable, and out of those rare human beings those who are conscious of the material problems are rarer still, and the still more rare persons are those who are conscious of the value of the Śrīmad-Bhāgavatam, which contains the messages of the Lord and His pure devotees.

321. Śrīmad Bhāgavatam 2.5.39, archived from the original on 29 April 2011, retrieved 15 September 2010, brahmalokaḥ — the spiritual planets; sanātanaḥ — eternal. ... From the forefront of the chest up to the neck of the universal form of the Lord are situated the planetary systems named Janaloka and Tapoloka, whereas Satyaloka, the topmost planetary system, is situated on the head of the form. The spiritual planets, however, are eternal. ... Sometimes Satyaloka, the planet in which Brahmā resides, is also called Brahmaloka. But the Brahmaloka mentioned here is not the same as the Satyaloka planetary system. This Brahmaloka is eternal, whereas the Satyaloka planetary system is not eternal. And to distinguish between the two, the adjective sanātana has been used in this case. According to Śrīla Jīva Gosvāmī, this Brahmaloka is the loka or abode of Brahman, or the Supreme Lord. ... Śrīla Śrīdhara Svāmī, therefore, rightly commenting on this verse /Bhagavad-gītā (9.11)/, says that the Brahmaloka mentioned here is Vaikuṇṭha, the kingdom of God, which is sanātana, or eternal, and is therefore not exactly like the material creations described above.

322. Śrīmad Bhāgavatam 3.11.19, Kali millennium is 1,200 years of the demigods... Kali-yuga, is 1,200 x 360, or 432,000 years.

323. Śrīmad Bhāgavatam 3.11.19, the duration of the Dvāpara millennium equals 2,400 years /of the demigods/ ... The duration of the Dvāpara-yuga is 2,400 x 360, or 864,000 years.

324. Śrīmad Bhāgavatam 3.11.19, The duration of the Satya millennium equals 4,800 years of the years of the demigods... As aforementioned, one year of the demigods is equal to 360 years of the human beings. The duration of the Satya-yuga is therefore 4,800 x 360, or 1,728,000 years.

325. Śrīmad Bhāgavatam 3.11.19, The duration of the Tretā-yuga is 3,600 x 360, or 1,296,000 years.

326.Śrīmad Bhāgavatam 3.26.52, This universal egg, or the universe in the shape of an egg, is called the manifestation of material energy. Its layers of water, air, fire, sky, ego and mahat-tattva increase in thickness one after another. Each layer is ten times bigger than the previous one, and the final outside layer is covered by pradhāna. Within this egg is the universal form of Lord Hari, of whose body the fourteen planetary systems are parts.
327.Srimad Bhagavatam 4.31.10, The duration of one day of Brahma is one thousand times greater than the four yugas, aggregating 4,320,000 years.
328.Śrīmad Bhāgavatam 4.9.22, In the Satya-yuga people generally lived for one hundred thousand years.
329.Srimad Bhagavatam 4.9.62, in Dvapara-yuga they /people/ lived for one thousand years
330.Srimad Bhagavatam 4.9.62, in the Treta-yuga people lived for ten thousand years
331.Śrīmad Bhāgavatam 5.23.9, archived from the original on 6 March 2012, retrieved 15 September 2010, The Vaikuntha planets begin 26,200,000 yojanas (209,600,000 miles) above Satyaloka.
332.Śrīmad Bhāgavatam 7.15.72, As far as kirtana is concerned, the sastras say, sravanam kirtanam visnoh: [SB 7.5.23] one should chant the glories of the Supreme Lord and the holy name of the Supreme Lord. This is clearly stated. Sravanam kirtanam visnoh: one should chant about and glorify Lord Visnu, not any demigod. Unfortunately, there are foolish persons who invent some process of kirtana on the basis of a demigod's name. This is an offense. Kirtana means glorifying the Supreme Lord, not any demigod. Sometimes people invent kālī-kīrtana or śiva-kīrtana, and even big sannyāsīs in the Māyāvāda school say that one may chant any name and still get the same result.
333.Subbaarayappa (1989).
334.Subbarayappa, B. V. (14 September 1989). "Indian astronomy: An historical perspective". In Biswas, S. K.; Mallik, D. C. V.; Vishveshwara, C. V. (eds.). Cosmic Perspectives. Cambridge University Press. pp. 25–40. ISBN 978-0-521-34354-1.
335.Subramanian, Vijaya. "Sidereal and Tropical Zodiac – Vijaya Jyoti". Retrieved 29 July 2020.
336.Sundarraj, M. (1997) [1st ed. 1994]. "Ch. 4 Asvins—Time-Keepers". In Mahalingam, N. (ed.). RG Vedic Studies. Coimbatore: Rukmani Offset Press. p. 219. It is quite clear that the smallest unit was the

"nimisah" ['winking of eyes'], and that time in the general sense of past, present and future was indicated by the word "yuga".
337.Sunil Sehgal (1999). Encyclopaedia of Hinduism: C-G. Sarup & Sons. p. 401. ISBN 978-81-7625-064-1. Retrieved 15 December 2012.
338.Sushil Mittal; Gene Thursby (2012). Hindu World. Routledge. p. 284. ISBN 978-1-134-60875-1.
339.Sutton, George Paul (2006). History of Liquid Propellant Rocket Engines. AIAA. p. 799. ISBN 978-1-56347-649-5. Archived from the original on 13 April 2021. Retrieved 14 March 2021.
340.Swami Sivananda - commentator (1999). Brahma Sutras. Islamic Books. pp. 190–196.
341.Tagare, G. V. (1987). The Vayu Purana, Part I. Motilal Banarsidass. pp. 127 (fn 1), 125–129 (21.26–69), 130–132 (22.9, 20), 133–136 (23.1, 20, 33). ISBN 978-8120803329.
342.Teachings of Lord Caitanya, Chapter 8: The Avataras, during one year of Brahma there are 5,040 Manus.
343.Teachings of Lord Caitanya, Chapter 8: The Avataras, In one kalpa, or one day of Brahma, fourteen Manus are manifest.
344.Teachings of Lord Caitanya, Chapter 8: The Avataras, One day of Brahma is calculated at 4 billion 320 million years, and Brahma lives for one hundred years on this scale.
345.Teachings of Lord Caitanya, Chapter 8: The Avataras, Since Brahma lives for one hundred of his years, it is calculated that there are 504,000 Manus manifest during the lifetime of one Brahma.
346.Teachings of Lord Caitanya, Chapter 8: The Avataras, Thus if fourteen Manus appear in one day of Brahma, there are 420 Manus during one month of Brahma
347.Teresi 2002, p. 174.
348.Teun Goudriaan (2008), Maya: Divine And Human, Motilal Banarsidass, ISBN 978-8120823891, pages 4, 167
349.The Bhagavata Purana (1.18.6),[22] Vishnu Purana (5.38.8),[23] Brahmanda Purana (2.3.74.241),[24] Vayu Purana (2.37.422),[25] and Brahma Purana (2.103.8)[26] state that the day Krishna left the earth was the day that the Dvapara Yuga ended and the Kali Yuga began.
350.The Commentary of Buddhavamsa.
351.The History and Practice of Ancient Astronomy. James Evans. Oxford University Press, 1 October 1998 – History – 496 pages. Page 393 [3]
352.The Vedas: An Introduction to Hinduism's Sacred Texts, Roshen Dalal, p.188

353. Theos Bernard (1999), Hindu Philosophy, Motilal Banarsidass, ISBN 978-81-208-1373-1, pages 74–76
354. Thompson 2007, p. 200.
355. Thompson, Richard L. (2007). The Cosmology of the Bhāgavata Purāna: Mysteries of the Sacred Universe. Delhi: Motilal Banarsidass. p. 239. ISBN 978-81-208-1919-1.
356. Tripathi (2008).
357. Vallée, Guylaine (1 March 2020). "What's Your Vedic Astrological Sign?". Birla Vedic Center. Retrieved 29 July 2020.
358. Varāhamihira. Encyclopædia Britannica (2008)
359. Vishnu Purana, Book I, Ch. III.
360. Vishnu Samhita.
361. Warren, John (1825). A Collection of Memoirs on the Various Modes According to which the Nations of the Southern Parts of India Divide Time.
362. Williams, Clemency; Knudsen, Toke (2005). "South-Central Asian Science". In Glick, Tomas F.; Livesey, Steven John; Wallis, Faith (eds.). Medieval Science, Technology, and Medicine: An Encyclopedia. Routledge. p. 463. ISBN 978-0-415-96930-7.
363. Wilson, H. H. (1895). "Book V, Ch. 38: Arjuna burns the dead, etc., Sloka 8". The Vishnu Purana. S.P.C.K. Press. p. 61. The Parijata tree proceeded to heaven, and on the same day that Hari [Krishna] departed from the earth the dark-bodied Kali age descended.
364. Wilson, H. H. (1940). "The Vishnu Purana: A System of Hindu Mythology and Tradition". London: John Murray. p. 272. Twenty-eight times have the Vedas been arranged by the great Rishis in the Vaivaswata Manwantara in the Dwápara age, and consequently eight and twenty Vyásas have passed away; by whom, in their respective periods, the Veda has been divided into four.
365. Wilson, Horace Hayman (1840). "Ch. III". The Vishnu Purana. Vol. Book I. pp. 23–25.
366. Wood, James (1849-01-01). Wood's Royal Southern Kalendar, Tasmanian Register and General Australasian & East Indian Official Directory ... H. Dowling, Jun.
367. www.wisdomlib.org (21 February 2016). "Mundaka Upanishad: Verse 2.1.10". www.wisdomlib.org. Retrieved 30 June 2022.
368. www.wisdomlib.org (21 February 2016). "Mundaka Upanishad: Verse 2.2.11". www.wisdomlib.org. Retrieved 29 June 2022.
369. Yen, Sheng. Orthodox Chinese Buddhism. p. 104. One great kalpa consists of the four medium kalpas of formation, statis, dissolution,

and nothingness. In other words, from the formation of one billion-world universe, through its destruction, until the beginning of the formation of its replacement billion-world universe is a great kalpa.
370. Yoga Vasistha.
371. Yukteswar 1990, pp. 15–17.
372. Yukteswar 1990, pp. 7–8, 10.
373. Yukteswar 1990, pp. 9–13.
374. Yukteswar, Swami Sri (1990) [1st ed. 1894]. The Holy Science [Kaivalya Darsanam]. Self-Realization Fellowship. pp. 7–17. ISBN 0876120516.
375. "*yeug-". Online Etymology Dictionary. Retrieved 2021-02-27.
376. "caturyuga". Sanskrit Dictionary for Spoken Sanskrit. Retrieved 2021-02-27.
377. "Caturyuga, Catur-yuga". Wisdom Library. 23 November 2017. Retrieved 2021-02-27.
378. "Chapter 36: The Buddhas in the three periods of time". Buddhism in a Nutshell Archives. Hong Kong: Buddhistdoor International. Retrieved 2014-12-21.
379. "daivayuga". Sanskrit Dictionary for Spoken Sanskrit. Retrieved 2021-02-27.
380. "Daivayuga, Daiva-yuga". Wisdom Library. 29 January 2019. Retrieved 2021-02-27.
381. "devayuga". Sanskrit Dictionary for Spoken Sanskrit. Retrieved 2021-02-27.
382. "Devayuga, Deva-yuga". Wisdom Library. 18 August 2017. Retrieved 2021-02-27.
383. "Kala-sandhi, Kāla-sandhi". Wisdom Library. 17 December 2019. Retrieved 2021-02-28.
384. "Kalpa, Kalpā, Kālpa". Wisdom Library. June 2008. Retrieved 2021-03-25.
385. "mahAyuga". Sanskrit Dictionary for Spoken Sanskrit. Retrieved 2021-02-27.
386. "Mahayuga, Maha-yuga, Mahāyuga". Wisdom Library. 6 January 2019. Retrieved 2021-02-27.
387. "Manvantara, Manu-antara". Wisdom Library. 24 June 2012. Retrieved 2021-02-28.
388. "Out of many, many thousands of brahmanas, one who is expert in the performance of sacrifices is considered the best. Out of many thousands of such experts in the performance of sacrifices, one who has surpassed all the knowledge in the Vedas is considered the best.

Out of all those persons who have excelled in the studies of the Vedas, a person who is a devotee of Visnu, the Personality of Godhead, is considered the best. And out of many, many thousands of such worshipers of Visnu, one who is a pure Vaisnava or an exclusive devotee of Sri Krsna is the best because only he will attain the highest abode called Param-dhama, which is described in the Bhagavad-gita by Sri Krsna Himself. (Garuda Purana) ...

389. "Out of millions of great sages who are liberated souls, it is very difficult to find a sober devotee of Lord Narayana." (Bhag. 6.14.5)
390. "Samdhyakala, Saṃdhyākāla, Samdhya-kala". Wisdom Library. 22 August 2019. Retrieved 2021-02-28.
391. "Sandhi, Shandhi". Wisdom Library. 3 August 2014. Retrieved 2021-02-28.
392. "Sandhyakala, Sandhyākāla". Wisdom Library. 29 July 2018. Retrieved 2021-02-28.
393. "yug". Wiktionary. Retrieved 2023-09-01.
394. "yuga". Wiktionary. Retrieved 2021-02-27.
395. "Yuga". Wisdom Library. 29 June 2012. Retrieved 2021-02-27.
396. "अन्तर (antara)". Wiktionary. Retrieved 2021-02-28.
397. "कल्प (kalpa)". Wiktionary. Retrieved 2021-03-25.
398. "काल (kāla)". Wiktionary. Retrieved 2021-02-28.
399. "क्लृप् (kḷp)". Wiktionary. Retrieved 2021-03-25.
400. "युज् (yuj)". Wiktionary. Retrieved 2021-02-27.
401. "संधि (samdhi)". Wiktionary. Retrieved 2021-02-28.
402. "संध्या (samdhyā)". Wiktionary. Retrieved 2021-02-28.
403. "सन्धि (sandhi)". Wiktionary. Retrieved 2021-02-28.
404. (17) By human calculation, a thousand ages [yuga] taken together form the duration of Brahmā's one day. And such also is the duration of his night.
405. (19) The duration of the Satya millennium equals 4,800 years of the years of the demigods; the duration of the Tretā millennium equals 3,600 years of the demigods; the duration of the Dvāpara millennium equals 2,400 years; and that of the Kali millennium is 1,200 years of the demigods. PURPORT: As aforementioned, one year of the demigods is equal to 360 years of the human beings. The duration of the Satya-yuga is therefore 4,800 × 360, or 1,728,000 years. The duration of the Tretā-yuga is 3,600 × 360, or 1,296,000 years. The duration of the Dvāpara-yuga is 2,400 × 360, or 864,000 years. And the last, the Kali-yuga, is 1,200 × 360, or 432,000 years.

406.(2) One thousand cycles of four ages [catur-yuga] constitute a single day of Brahmā, known as a kalpa. In that period, O King, fourteen Manus come and go.

407.(39) The cycle of four ages [catur-yugam] – Satya, Tretā, Dvāpara, and Kali – continues perpetually among living beings on this earth, repeating the same general sequence of events.

408.(71) These twelve thousand (years) which thus have been just mentioned as the total of four (human) ages [caturyugam], are called one age of the gods.

409.Brahma: creator god.

410.Devas: gods, celestials or divine.

❖ HINDUISM: Myths of time and eternity: ... Each yuga is preceded by an intermediate "dawn" and "dusk". The Krita yuga lasts 4,000 god-years, with a dawn and dusk of 400 god-years each, or a total of 4,800 god-years; Treta a total of 3,600 god-years; Dvapara 2,400 god-years; and Kali (the current yuga) 1,200 god-years. A mahayuga thus lasts 12,000 god-years ... Since each god-year lasts 360 human years, a mahayuga is 4,320,000 years long in human time. Two thousand mahayugas form one kalpa (eon) [and pralaya], which is itself but one day in the life of Brahma, whose full life lasts 100 years; the present is the midpoint of his life. Each kalpa is followed by an equally long period of abeyance (pralaya), in which the universe is asleep. Seemingly the universe will come to an end at the end of Brahma's life, but Brahmas too are innumerable, and a new universe is reborn with each new Brahma.

❖ HINDUISM: Myths of time and eternity: ... Each yuga is preceded by an intermediate "dawn" and "dusk." The Krita yuga lasts 4,000 god-years, with a dawn and dusk of 400 god-years each, or a total of 4,800 god-years; Treta a total of 3,600 god-years; Dvapara 2,400 god-years; and Kali (the current yuga) 1,200 god-years. A mahayuga thus lasts 12,000 god-years ... Since each god-year lasts 360 human years, a mahayuga is 4,320,000 years long in human time. Two thousand mahayugas form one kalpa (eon) [and pralaya], which is itself but one day in the life of Brahma, whose full life lasts 100 years; the present is the midpoint of his life. Each kalpa is followed by an equally long period of abeyance (pralaya), in which the universe is asleep. Seemingly the universe will come to an end at the end of Brahma's life, but Brahmas too are innumerable, and a new universe is reborn with each new Brahma.

411.Humans: year equals Sun's northern and southern movements in Earth's sky (a.k.a. solar or tropical year).
412.Manu: progenitor of mankind.
413.Pitris: forefathers or ancestors.
414.Sandhya: 3102 BCE – 32,899 CE
415.Sandhyamsa: 392,899 CE – 428,899 CE
416.Abraham, G. (2008), "Gnomon in India", Encyclopaedia of the History of Science, Technology, and Medicine in Non-Western Cultures (2nd edition) edited by Helaine Selin, pp. 1035–1037, Springer, ISBN 978-1-4020-4559-2.
417.Almeida, D. F.; et al. (2001). "Keralese Mathematics: Its Possible Transmission to Europe and the Consequential Educational Implications". Journal of Natural Geometry. 20: 77–104.
418.Ancient Hindu Astronomy, https://web.archive.org/web/20180207233642/http://www.bharatvani.org/books/ait/ch22.htm
419.anena krama-yogena bhuvi prāṇiṣu vartate ॥ 39 ॥.
420.As one waters the bhakti lata bija, the seed sprouts, and the creeper gradually increases until it penetrates the walls of this universe and goes beyond the Viraja river between the spiritual world and the material world. It attains brahma-loka, the Brahman effulgence. Penetrating through that stratum, it reaches the spiritual sky and the spiritual planet Goloka Vrndavana.
421.Baber, Zaheer (1996), The Science of Empire: Scientific Knowledge, Civilization, and Colonial Rule in India, State University of New York Press, ISBN 0-7914-2919-9.
422.Bhagavad Gita, Chapter 8 verse 17, https://web.archive.org/web/20080527205744/http://bhagavadgitaasitis.com/8/17/en1
423.Bhargava, Pushpa M.; Chakrabarti, Chandana (2003). The Saga of Indian Science Since Independence: In a Nutshell. Universities Press. pp. 39–. ISBN 978-81-7371-435-1. Archived from the original on 13 May 2016. Retrieved 15 November 2015.
424.Bodewitz, Henk (2019). "The Hindu Doctrine of Transmigration: Its Origin and Background". In Heilijgers, Dory; Houben, Jan; van Kooij, Karel (eds.). Vedic Cosmology and Ethics: Selected Studies. Gonda Indological Studies. Vol. 19. Leiden and Boston: Brill Publishers. pp. 3–20. doi:10.1163/9789004400139_002. ISBN 978-90-04-39864-1. JSTOR 10.1163/j.ctvrxk42v.6. S2CID 203303894.
425.Brahma Savarni Manu.

426.Brennand, William (1896), Hindu Astronomy, Chas.Straker & Sons, London
427.Burgess 1935, pp. 285, 286, chapter XII verse 34-36.
428.catur-yuga-sahasraṁ tu brahmaṇo dinam ucyate ।
429.catvāri trīṇi dve caikaṁ kṛtādiṣu yathā-kramam ।
430.Chakshusha Manu.
431.Compare Chaudhary, Govind Vishnu (1997) [1968]. Vedic Numerology-A Treatise On Hindu Astronomy (2 ed.). p. 365.
432.Daksa Savarni Manu
433.Dallal, Ahmad (1999), "Science, Medicine and Technology", The Oxford History of Islam edited by John Esposito, Oxford University Press.
434.Daniel, R.R. (1992). "Space Science in India". Indian Journal of History of Science. New Delhi: Indian National Science Academy. 27 (4): 485–499.
435.Dharma Savarni Manu
436.etad dvādaśasāhasraṃ devānāṃ yugamucyate ॥ 71 ॥
437.Flood, Gavin D. (1996), An Introduction to Hinduism, Cambridge University Press, ISBN 978-0-521-43878-0.
438.For distances in yojanas see the yojana page on vedabase.net. https://web.archive.org/web/20110507121600/http://vedabase.net/y/yojana.
439.González-Reimann, Luis (2018). "Cosmic Cycles, Cosmology, and Cosmography". In Basu, Helene; Jacobsen, Knut A.; Malinar, Angelika; Narayanan, Vasudha (eds.). Brill's Encyclopedia of Hinduism. Vol. 2. Leiden: Brill Publishers. doi:10.1163/2212-5019_BEH_COM_1020020. ISBN 978-90-04-17641-6. ISSN 2212-5019.
440.Gupta, S. V. (2010). "Ch. 1.2.4 Time Measurements". In Hull, Robert; Osgood, Richard M. Jr.; Parisi, Jurgen; Warlimont, Hans (eds.). Units of Measurement: Past, Present and Future. International System of Units. Springer Series in Materials Science: 122. Springer. p. 3. ISBN 9783642007378.
441.Haug, Martin (1863). The Aitareya Brahmanam of the Rigveda, Containing the Earliest Speculations of the Brahmans on the Meaning of the Sacrificial Prayers. ISBN 0-404-57848-9.
442.Hayashi, Takao (2008), Aryabhata I, Encyclopædia Britannica.
443.Hayashi, Takao (2008), Bhaskara I, Encyclopædia Britannica.
444.Hayashi, Takao (2008), Brahmagupta, Encyclopædia Britannica.
445.Hayashi, Takao (2008), Shripati, Encyclopædia Britannica.
446.https://en.wikipedia.org/wiki/Hindu_cosmology.

447.https://en.wikipedia.org/wiki/Hindu_units_of_time.
448.https://en.wikipedia.org/wiki/Indian_astronomy.
449.https://en.wikipedia.org/wiki/Indian_units_of_measurement.
450.https://en.wikipedia.org/wiki/Kalpa_(time).
451.https://en.wikipedia.org/wiki/List_of_numbers_in_Hindu_scriptures.
452.https://en.wikipedia.org/wiki/Manvantara.
453.https://en.wikipedia.org/wiki/Puranic_chronology.
454.https://en.wikipedia.org/wiki/Yuga_Cycle.
455.https://www.iskcon-truth.com/bhu-mandala/fourteen-worlds.html.
456.J.A.B. van Buitenen (2008), calendar, Encyclopædia Britannica.
457.Joseph, George G. (2000), The Crest of the Peacock: Non-European Roots of Mathematics, Penguin Books, ISBN 0-691-00659-8.
458.Joseph, George G. (2000). The Crest of the Peacock: Non-European Roots of Mathematics, 2nd edition. Penguin Books, London. ISBN 0-691-00659-8.
459.Kak, S. (2000). The astronomical code of the Ṛgveda. New Delhi: Munshiram Manoharlal Publishers.
460.Kak, S. C. (1 January 1993). Astronomy of the vedic altars. Vistas in Astronomy: Part 1, 36, 117–140.
461.Kak, Subhash C. "Archaeoastronomy and literature." Current Science 73.7 (1997): 624–627.
462.Kak, Subhash C. "Knowledge of planets in the third millennium BC." Quarterly Journal of the Royal Astronomical Society 37 (1996): 709.
463.Kak, Subhash C. "The astronomy of the age of geometric altars." Quarterly Journal of the Royal Astronomical Society 36 (1995): 385.
464.Kak, Subhash C. (2000). "Birth and Early Development of Indian Astronomy". In Selin, Helaine (2000). Astronomy Across Cultures: The History of Non-Western Astronomy (303–340). Boston: Kluwer. ISBN 0-7923-6363-9.
465.Kak, Subhash. Birth and early development of Indian astronomy. Kluwer, 2000.
466.King, David A (2002). "A Vetustissimus Arabic Text on the Quadrans Vetus". Journal for the History of Astronomy. 33 (3): 237–255. Bibcode:2002JHA....33..237K. doi:10.1177/002182860203300302. S2CID 125329755.
467.Klostermaier, Klaus K. (2003), "Hinduism, History of Science and Religion", Encyclopedia of Science and Religion edited by J. Wentzel Vrede van Huyssteen, pp. 405–410, Macmillan Reference USA, ISBN 0-02-865704-7.

468.kotisv api mahamune.
469.kṛtaṁ tretā dvāparaṁ ca kaliś ceti catur-yugam ।
470.Larson, Gerald. HINDU COSMOGONY/ COSMOLOGY. Routledge, 2 Park Square, Milton Park, Abingdon, Oxon, OX14 4RN. p. 113.
471.Levon Khachikian: "The Ledger of the Merchant Hovhannes Joughayetsi". Journal of the Asiatic Society, Vol 8, No. 3, 1966, p. 182.
472.Maunder, E. Walter (1899), The Indian Eclipse 1898, Hazell Watson and Viney Ltd., London.
473.muktanam api siddhanam.
474.narayana-parayanah.
475.Our only purpose is to present this Bhagavad-gita As It Is in order to guide the conditioned student to the same purpose for which Krsna descends to this planet once in a day of Brahma, or every 8,640,000,000 years. This purpose is stated in Bhagavad-gita, and we have to accept it as it is; otherwise there is no point in trying to understand the Bhagavad-gita and its speaker, Lord Krsna. Lord Krsna first spoke Bhagavad-gita to the sun-god some hundreds of millions of years ago. We have to accept this fact and thus understand the historical significance of Bhagavad-gita, without misinterpretation, on the authority of Krsna. To interpret Bhagavad-gita without any reference to the will of Krsna is the greatest offense. In order to save oneself from this offense, one has to understand the Lord as the Supreme Personality of Godhead, as He was directly understood by Arjuna, Lord Krsna's first disciple. Such understanding of Bhagavad-gita is really profitable and authorized for the welfare of human society in fulfilling the mission of life.
476.Out of many thousands among men, one may endeavor for perfection, and of those who have achieved perfection, hardly one knows Me in truth.
477.Pierre-Yves Bely; Carol Christian; Jean-René Roy (11 March 2010). A Question and Answer Guide to Astronomy. Cambridge University Press. p. 197. ISBN 9780521180665.
478.Prinsep, James (editor Edward Thomas): Essays on Indian Antqities, Historic, Numismatic, and Palaeographic, of the late James Prinsep, F.R.S., to which are added his Useful Tables illustrative of Indian History, Chronology, Modern Coinages, Weights, Measures etc. Two Volumes, Reprint, Indological Book House, Delhi and Varanasi, 1971. Originally published in London, 1858.
479.Project of History of Indian Science, Philosophy and culture, Monograph series, Volume 3. Mathematics, Astronomy and Biology

in Indian Tradition edited by D. P. Chattopadhyaya and Ravinder Kumar.

480. Raivata Manu

481. Raju, C. K. (2001). "Computers, Mathematics Education, and the Alternative Epistemology of the Calculus in the Yuktibhasa". Philosophy East and West. 51 (3): 325–362. doi:10.1353/pew.2001.0045. S2CID 170341845.

482. Ramasubramanian (1994). "Modification of the earlier Indian planetary theory by the Kerala astronomers (c. 1500 CE) and the implied heliocentric picture of planetary motion". Current Science. 66: 784–790.

483. rātriṁ yuga-sahasrāntāṁ te 'ho-rātra-vido janāḥ || 17 ||

484. Raucya or Deva Savarni Manu.

485. Rudra Savarni Manu.

486. sa kalpo yatra manavaś caturdaśa viśām-pate || 2 ||

487. Sadeh, Eligar (2013). Space Strategy in the 21st Century: Theory and Policy. Routledge. ISBN 978-1-136-22623-6. Archived from the original on 6 March 2016. Retrieved 19 February 2021.

488. sahasra-yuga-paryantam ahar yad brahmaṇo viduḥ |

489. saṅkhyātāni sahasrāṇi dvi-guṇāni śatāni ca || 19 ||

490. Sarma, K.V. (2008a). Helaine Selin (ed.). "Astronomy in India". Encyclopaedia of the History of Science, Technology, and Medicine in Non-Western Cultures (2nd ed.). Springer: 317–321. ISBN 978-1-4020-4559-2.

491. Sarma, K.V. (2008b). Helaine Selin (ed.). "Lalla". Encyclopaedia of the History of Science, Technology, and Medicine in Non-Western Cultures (2nd ed.). Springer: 1215. ISBN 978-1-4020-4559-2.

492. Sarma, K.V. (2008c). Helaine Selin (ed.). "Acyuta Pisarati". Encyclopaedia of the History of Science, Technology, and Medicine in Non-Western Cultures (2nd ed.). Springer: 19. ISBN 978-1-4020-4559-2.

493. Sarma, K.V. (2008d). Helaine Selin (ed.). "Armillary Spheres in India". Encyclopaedia of the History of Science, Technology, and Medicine in Non-Western Cultures (2nd ed.). Springer: 243. ISBN 978-1-4020-4559-2.

494. Sarma, Nataraja (2000). "Diffusion of Astronomy in the Ancient World". Endeavour. 24 (4): 157–164. doi:10.1016/s0160-9327(00)01327-2. PMID 11196987.

495. sarva-vedanta-vit-kotya visnu-bhakto visisyate.

496. satra-yaji-sahasrebhyah sarva vedanta-paragah.

497.Savage-Smith, Emilie (1985), Islamicate Celestial Globes: Their History, Construction, and Use, Smithsonian Institution Press.
498.Savarni Manu.
499.Scientific Explanation of Hindu Time Units, https://legendofbharatbarsha.home.blog/2019/12/07/detailed-history-of-time/.
500.Sharma, V.N. (1995), Sawai Jai Singh and His Astronomy, Motilal Banarsidass, ISBN 81-208-1256-5.
501.Sharma, V.N. (2008), "Observatories in India", Encyclopaedia of the History of Science, Technology, and Medicine in Non-Western Cultures (2nd edition) edited by Helaine Selin, pp. 1785–1788, Springer, ISBN 978-1-4020-4559-2.
502.Source (Gupta 2010, p. 6) has typo. 4 yāmas is 1/2 day.
503.Subbaarayappa, B.V. (1989), "Indian astronomy: an historical perspective", Cosmic Perspectives edited by Biswas etc., pp. 25–41. Cambridge University Press. ISBN 0-521-34354-2.
504.sudurlabhah prasantatma.
505.Surya Siddhanta, Chapter I with Commentary and Illustrations, https://web.archive.org/web/20050320030038/http://www.thearchimedeandual.com/platonic/Eastern/surya_siddanta_commentary/surya_siddhanta.htm
506.Swarochisha Manu.
507.Swayambhu Manu.
508.Tapasa/Tamasa Manu.
509.Teresi, Dick (2002). Lost Discoveries: The Ancient Roots of Modern Science — from the Babylonians to the Maya. New York: Simon & Schuster Inc. ISBN 0-684-83718-8.
510.The Àryabhatiya of Àryabhata: The oldest exact astronomical constant?, http://www.jqjacobs.net/astro/aryabhata.html.
511.Tripathi, V.N. (2008), "Astrology in India", Encyclopaedia of the History of Science, Technology, and Medicine in Non-Western Cultures (2nd edition) edited by Helaine Selin, pp. 264–267, Springer, ISBN 978-1-4020-4559-2.
512.Uttama Manu.
513.vaisnavanam sahasrebhya ekanty eko visisyate.
514.Vaivasvata Manu (current).
515.Vasu, S.C. & others (1972). The Matsya Puranam, Part II, Delhi: Oriental Publishers, p.366.
516.When a person receives the seed of devotional service he should take care of it by becoming a gardener and sowing the seed in his heart. If

he waters the seed gradually by the process of sravana and kirtana the seed will begin to sprout.

517. Wilkins, William Joseph (1913) [1st ed. 1882]. Hindu Mythology, Vedic and Purānic (3rd ed.). Calcutta: London Missionary Society. p. 365. In each Manvantara (period of a Manu), seven Rishis, certain deities, an Indra and a Manu, and the kings, his sons, are created and perish.

518. yadetat parisaṅkhyātamādāveva caturyugam ।

519. yatatam api siddhanam kascin mam vetti tattvatah ।

# Chapter 9

# Hindu Calendar

## 卐 Introduction

The Hindu calendar, also called Panchanga (पञ्चाङ्ग), is one of various *lunisolar calendars* that are traditionally used in the Bhartiya subcontinent and Southeast Asia, with further regional variations for social and Hindu religious purposes. They adopt a similar underlying concept for timekeeping based on sidereal year for solar cycle and adjustment of lunar cycles in every three years, but differ in their relative emphasis to moon cycle or the sun cycle and the names of months and when they consider the New Year to start. Of the various regional calendars, the most studied and known Hindu calendars are the *Shalivahana Shaka* - शालिवाहन शक (Based on the *King Shalivahana*, also the Bhartiya national calendar) found in the Deccan region of Southern Bharat and the *Vikram Samvat* - विक्रम संवत् (*Bikrami* - बिक्रमी) found in Nepal and the North and Central regions of Bharat – both of which emphasize the lunar cycle. Their new year starts in spring. In regions such as Tamil Nadu and Kerala, the solar cycle is emphasized and this is called the Tamil calendar (though Tamil Calendar uses month names like in Hindu Calendar) and Malayalam calendar and these have origins in the second half of the 1st millennium CE. A Hindu calendar is sometimes referred to as *Panchangam* (पञ्चाङ्गम्), which is also known as *Panjika* (पञ्जिका) in Eastern Bharat.

The ancient Hindu calendar conceptual design is also found in the *Hebrew* calendar, the *Chinese* calendar, and the *Babylonian* calendar, but different from the *Gregorian* calendar. Unlike the *Gregorian* calendar which adds additional days to the month to adjust for the mismatch between twelve lunar cycles (354 lunar days) and nearly 365 solar days, the Hindu calendar maintains the integrity of the lunar month, but inserts an extra full month, once every 32–33 months, to ensure that the festivals and crop-related rituals fall in the appropriate season.

The Hindu calendars have been in use in the Bhartiya subcontinent since *Vedic* times, and remain in use by the Hindus all over the world, particularly to set *Hindu festival* (हिन्दू पर्व) dates. Early Buddhist communities of Bharat adopted the ancient Vedic calendar, later *Vikrami* calendar and then local Buddhist calendars. Buddhist festivals continue to be scheduled according to a lunar system. The Buddhist calendar and the traditional lunisolar calendars of *Cambodia, Laos, Myanmar, Sri Lanka* and *Thailand* are also based on an older version of the Hindu calendar. Similarly, the ancient Jain traditions have followed the same lunisolar system as the Hindu calendar for festivals, texts and inscriptions. However, the Buddhist and Jain timekeeping systems have attempted to use the *Buddha* (बुद्धः) and the *Mahavira* (महावीर)'s lifetimes as their reference points.

The Hindu calendar is also important to the practice of Hindu astrology and zodiac system. It is also employed for observing the auspicious days of deities and occasions of fasting, such as *Ekadashi* (एकादशी). Figure 9.1 shows a page from the Hindu calendar 1871-72.

Figure 9.1. A page from the Hindu calendar 1871-72. (Image Credit: https://en.wikipedia.org/wiki/)

The Vedic culture developed a sophisticated time keeping methodology and calendars for Vedic rituals, and timekeeping as well as the nature of solar and Moon movements are mentioned in Vedic texts. For example, *Kaushitaki Brahmana* (कौशितकी ब्राह्मण) chapter 19.3 mentions the shift in the relative location of the Sun towards north for 6 months, and south for 6 months. Time keeping was important to Vedic rituals, and *Jyotisha* was the Vedic era field of tracking and predicting the movements of astronomical bodies in order to keep time, in order to fix the day and time of these rituals. This study is one of the six ancient *Vedangas*, or ancillary science connected with the Vedas – the scriptures of Vedic Sanatan Sanskriti.

This Vedanga field developed from actual astronomical studies in ancient Vedic Period. The texts of *Vedic Jyotisha* (वैदिक ज्योतिषः) sciences were translated into the *Chinese language* in the 2nd and 3rd centuries CE, and the Rigvedic passages on astronomy are found in the works of Zhu Jiangyan and Zhi Qian. The beginning of the Hindu calendar was much earlier. Greek historians described *Maurya kings* (मौर्यराजाः) referring to a calendar which originated in 6676 BCE known as *Saptarsi calendar* (सप्तर्षि पञ्चाङ्ग). The *Vikrami* calendar is named after *king Vikramaditya* (राजा विक्रमादित्यः) and starts in 57 BCE.

Hindu scholars kept precise time by observing and calculating the *cycles of Surya* - सूर्यस्य चक्राणि (the Sun), Moon and the planets. These calculations about the Sun appear in various astronomical texts in Sanskrit, such as the 5th-century *Aryabhatiya* by *Aryabhata*, the 6th-century *Romaka* by *Latadeva* and *Panca Siddhantika* by *Varahamihira*, the 7th-century *Khandakhadyaka* by *Brahmagupta* and the 8th-century *Sisyadhivrddida* by *Lalla*. These texts present Surya and various planets and estimate the characteristics of the respective planetary motion. Other texts such as *Surya Siddhanta* dated to have been completed sometime between the 5th century and 10th century present their chapters on various deified planets with stories behind them.

The manuscripts of these texts exist in slightly different versions. They present Surya, planet-based calculations and Surya's relative motion to Earth. These vary in their data, suggesting that the text was open and revised over their lives. For example, the 1st millennium CE Hindu scholars calculated the sidereal length of a year as follows, from their astronomical studies, with slightly different results (see Table 9.1).

Table 9.1 Length of year in various Sanskrit texts

| Hindu text | Estimated length of the sidereal year |
|---|---|
| Surya Siddhanta | 365 days, 6 hours, 12 minutes, 36.56 seconds |
| Paulica Siddhanta | 365 days, 6 hours, 12 minutes, 36 seconds |
| Paracara Siddhanta | 365 days, 6 hours, 12 minutes, 31.50 seconds |
| Arya Siddhanta | 365 days, 6 hours, 12 minutes, 30.84 seconds |
| Laghu Arya Siddhanta | 365 days, 6 hours, 12 minutes, 30 seconds |
| Siddhanta Shiromani | 365 days, 6 hours, 12 minutes, 9 seconds |

The Hindu texts used the lunar cycle for setting months and days, but the solar cycle to set the complete year. This system is similar to the Jewish and Babylonian ancient calendars, creating the same challenge of accounting for the mismatch between the nearly 354 lunar days in twelve months, versus over 365 solar days in a year. They tracked the solar year by observing the entrance and departure of Surya (sun, at sunrise and sunset) in the constellation formed by stars in the sky, which they divided into 12 intervals of 30 degrees each. Like other ancient human cultures, Hindus innovated a number of systems of which intercalary months became most used, that is adding another month every 32.5 months on average. As their calendar keeping and astronomical observations became more sophisticated, the Hindu calendar became more sophisticated with complex rules and greater accuracy.

The *Siddhanta tradition* at the foundation of Hindu calendars predate the Christian era, once had 18 texts of which only 5 have survived into the modern era. These texts provide specific information and formulae on motions of Sun, Moon and planets, to predict their *future relative positions* (भविष्यत् सापेक्षिकस्थानानि), *equinoxes* (विषुवः), *rise* (उदयः) *and set* (अस्तं), *with corrections for prograde* (प्रगतिः), *retrograde motions* (प्रतिगामी गतिः), *as well as parallax* (लंबनम्). These ancient scholars attempted to calculate their time to the accuracy of a *truti* - त्रुति (29.63 microseconds). In their pursuit of accurate tracking of relative movements of celestial bodies for their calendar, they had computed the mean diameter of the Earth, which was very close to the actual 12,742 km (7,918 mi).

Hindu calendars were refined during the *Gupta era* (गुप्त युग) astronomy by *Āryabhaṭa* and *Varāhamihira* in the 5th to 6th century. These, in turn, were based in the astronomical tradition of *Vedāṅga Jyotiṣa,* which in the preceding centuries had been standardised in a number of (non-extant) works known as *Sūrya Siddhānta.* Regional diversification took place in the medieval period. The astronomical foundations were further developed in the medieval period, notably by *Bhāskara II* (12th century).

Later, the term Jyotisha evolved to include Hindu astrology. The astrological application of the Hindu calendar was a field that likely developed in the centuries after the arrival of Greek astrology with Alexander the Great, because their zodiac signs are nearly identical.

The ancient Hindu texts on Jyotisha only discuss timekeeping, and never mention astrology or prophecy. These ancient texts predominantly cover astronomy, but at a rudimentary level. Later medieval era texts such as the Yavana-jataka and the Siddhanta texts are more astrology-related.

Hinduism and Buddhism were the prominent religions of southeast Asia in the 1st millennium CE, prior to the Islamic conquest that started in the 14th century. The Hindus prevailed in Bali, Indonesia, and they have two types of Hindu calendar. One is a 210-day based *Pawukon calendar* which likely is a pre-Hindu system, and another is similar to lunisolar calendar system found in South Bharat and it is called the *Balinese saka* calendar which uses Hindu methodology. The names of month and festivals of *Balinese Hindus,* for the most part, are different, though the significance and legends have some overlap.

The Hindu calendar is based on a geocentric model of the Solar System. A large part of this calendar is defined based on the movement of the Sun and the Moon around the Earth (*saura māna* (सौरा माना) and *cāndra māna* (चन्द्र माना), respectively). Furthermore, it includes synodic, sidereal, and tropical elements. Many variants of the Hindu calendar have been created by including and excluding these elements (solar, lunar, lunisolar etc.) and are in use in different parts of Bharat. Table 9.2 shows various elements of the Hindu calendar.

Table 9.2 Elements of the Hindu calendar

| | Synodic Elements | Sidereal Elements | Tropical Elements |
|---|---|---|---|
| *saura māna* | | *rāśi, sauramāsa, varsa* | *uttarāyana, dakṣiṇāyana, devayāna, pitṛyāṇa, ṛtu* |
| *cāndra māna* | *tithi, paksa, candramāsa, varsa* | | |
| *nākṣatra māna* | | *dina, ghatikā (aka nāḍī), vighaṭikā (aka vināḍī), prāṇa (aka asu)* | |
| *sāvana māna* | *dina* | | |
| | synodic elements | sidereal elements | tropical elements |

*Samvat* refers to era of the several Hindu calendar systems in Nepal and Bharat, in a similar manner to the Christian era. There are several samvat found in historic Buddhist, Hindu and Jain texts and epigraphy, of which three are most significant: Vikrama era, Old Shaka era and Shaka era of 78 CE.

## 卐 Samvat (Year)

The Hindu calendar *saka samvat* (शक संवत्) system is found in Indonesian inscriptions, such as the *Kedukan Bukit* (केदुकन बुकिट) inscription (pictured above) dated to 604 *Śaka,* which is equivalent to 682 CE.

## 卐 Vikram Samvat (Bikram Sambat)

A northern Bhartiya almanac which started in 57 BCE, and is also called the *Vikrama Era* (विक्रम युग). It is related to the *Bikrami* calendar, and is apocryphally linked to *Vikramaditya*. The year starts from the month of *Baishakh* (बैशाख) / *Vaishakha* (वैशाख). This system is common in epigraphic evidence from northern, western, central and eastern Bhartiya subcontinent, particularly after the early centuries of the 1st millennium CE.

## 卐 Saka Samvat

There are two Saka era systems in scholarly use, one is called Old Saka Era, whose epoch is uncertain, probably sometime in the 1st millennium BCE because ancient Buddhist, Jain and Hindu inscriptions and texts use it. However, the starting point of Old Shaka Era is a subject of dispute among scholars. The second system is called Saka Era of 78 AD, or simply Saka Era, a system that is common in epigraphic evidence from southern Bharat. Figure 9.2. The Hindu calendar *saka samvat* system is found in *Indonesian inscriptions.*

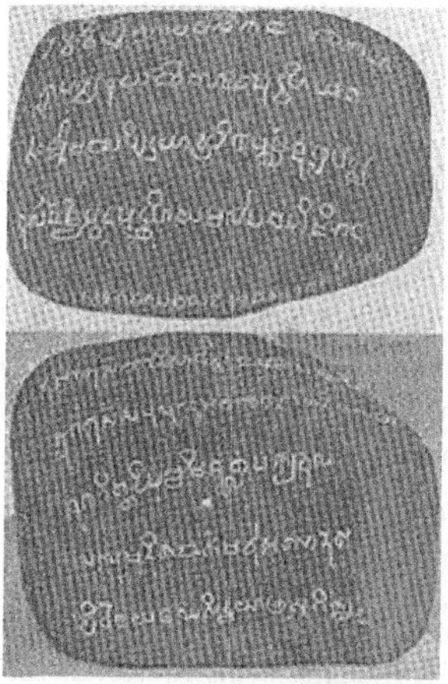

Figure 9.2. The Hindu calendar *saka samvat* system is found in *Indonesian inscriptions*, such as the *Kedukan Bukit* inscription (pictured above) dated to 604 Śaka, which is equivalent to 682 CE. (Image Credit: https://en.wikipedia.org/wiki/)

## 卐 Saka Era of Southeast Asia

The Hindu calendar system in Indonesia is attributed to the legend of Hindus arriving with a sage *Aji Saka* (अजी शक) in 1st-century *Java,* in March 78 CE. Numerous ancient and medieval era texts and inscriptions found in *Indonesian* islands use this reference year. In mainland southeast Asia, the earliest verifiable use of Hindu Saka methodology in inscriptions is marked *Saka* 533 in *Ankor Borei,* which corresponds to 611 CE, while the *Kedukan Bukit* inscription in Sumatra, containing three dates in Saka 604 (682 CE), is the earliest known use of the Shaka era in the Indonesian islands. However, these inscriptions only set the floruit for the use of the Shaka era in these places, and the Hindu calendar likely existed in southeast Asia before these dates to be used in important monuments. Further, the Hindu calendar system remained popular among the Hindus through to the 15th century, and thereafter in Bali.

### 卐 Bhartiya National Calendar (Modern)

Bhartiya calender combines many Hindu calendars into one official standardized one, but old ones remain in use.

### 卐 Months

The Hindu calendar divides the zodiac into twelve division called *rāśi* (राशी). The time taken by the Sun to transit through a *rāśi* is a solar month whose name is identical to the name of the *rāśi*. In practice, solar months are mostly referred as *rāśi* (not months). The solar months are named differently in different regional calendars. While the Malayalam calendar broadly retains the phonetic Sanskrit names, the Bengali and Tamil calendars repurpose the Sanskrit lunar month names (*Chaitra, Vaishaka* etc.) as follows:

- The Tamil calendar replaces *Mesha, Vrisha* etc. with *Chithirai, Vaigasi* etc.
- The Bengali calendar is similar to the Tamil calendar except in that it starts the year with *Boiśākh* (instead of *Choitrô*), followed by *Jyoisthô* etc. The Assamese and Odia calendars too are structured the same way.

The solar months (*rāśi*) along with their equivalent names in the Assamese, Bengali, Malayalam, Odia, Tamil and Gregorian calendar are given in Table 9.3.

Table 9.3 Names of months

| | Vikrami (solar) | Assamese | Bengali | Malayalam | Odia | Tamil | Gregorian |
|---|---|---|---|---|---|---|---|
| 1 | Mina | চ'ত (Söt) | চৈত্র (Choitrô) | മീനം (Mīnaṃ) | ଚୈତ୍ର (Caitra) | பங்குனி (Paṅguṇi) | March –April |

| | | | | | | | | |
|---|---|---|---|---|---|---|---|---|
| 2 | Mēsha | ব'হাগ (Böhag) | বৈশাখ (Boiśākh) | മേടം (Mēḍaṃ) | ବୈଶାଖ (Baiśākha) | சித்திரை (Śittirai) | April – May |
| 3 | Vrisha | জেঠ (Zeth) | জ্যৈষ্ঠ (Jyoisthô) | ഇടവം (Iḍavaṃ) | ଜ୍ୟେଷ୍ଠ (Jyēṣṭha) | வைகாசி (Vaigāsi) | May – June |
| 4 | Mithuna | আহাৰ (Ahar) | আষাঢ় (Āsādh) | മിഥുനം (Mithunaṃ) | ଆଷାଢ଼ (Āṣāḍha) | ஆனி (Āṉi) | June – July |
| 5 | Karka | শাওণ (Xaün) | শ্রাবণ (Śrābôṇ) | കർക്കടകം (Karkkaṭakam) | ଶ୍ରାବଣ (Śrābaṇa) | ஆடி (Āḍi) | July – August |
| 6 | Singa | ভাদ (Bhado) | ভাদ্র (Bhādrô) | ചിങ്ങം (Ciṅṅaṃ) | ଭାଦ୍ରବ (Bhādraba) or ଭାଦ୍ର (Bhādra) | ஆவணி (Āvaṇi) | August – September |
| 7 | Kanya | আহিন (Ahin) | আশ্বিন (Āśhshin) | കന്നി (Kanni) | ଆଶ୍ୱିନ (Āśvina) | புரட்டாசி (Puraṭṭāsi) | September – October |
| 8 | Tula | কাতি (Kati) | কার্তিক (Kārtik) | തുലാം (Tulāṃ) | କାର୍ତ୍ତିକ (Kārttika) | ஐப்பசி (Aippasi) | October – November |
| 9 | Vrischika | আঘোণ (Aghün) | অগ্রহায়ণ (Ôgrôhāyôn) | വൃശ്ചികം (Vr̥ścikaṃ) | ମାର୍ଗଶିର (Mārgaśira) | கார்த்திகை (Kārttigai) | November – December |
| 10 | Dhanus | পোহ (Puh) | পৌষ (Poush) | ധനു (Dhanu) | ପୌଷ (Pauṣa) | மார்கழி (Mārgaḻi) | December – January |
| 11 | Makara | মাঘ (Magh) | মাঘ (Māgh) | മകരം (Makaram) | ମାଘ (Māgha) | தை (Tai) | January – February |

| | | | | | | | |
|---|---|---|---|---|---|---|---|
| 12 | Kumbha | ফাগুন (Phagun) | फाल्गुन (Phālgun) | കുംഭം (Kumbham) | ଫାଲ୍‌ଗୁନ (Phālguna) or ଫଗୁଣ (Phaguṇa) | மாசி (Māsi) | February – March |

The solar months (*rāśi*) along with the approximate correspondence to Hindu seasons and Gregorian months are shown in Table 9.4.

Table 9.4 Seasons in various names

| Rāśi | Gregorian months | Rtu in Devanagari script | Bengali name for Rtu | Malayalam name for Rtu | Tamil name for Rtu |
|---|---|---|---|---|---|
| Mīna Mesh | Mid March–Mid May | वसन्त | বসন্ত (Bôsôntô) | വസന്തം (Vasaṃtam) | இளவேனில் (ilavenil) |
| Vrṣabha Mithuna | Mid May–Mid July | ग्रीष्म | গ্রীষ্ম (Grishsho) | ഗ്രീഷ്മം (Grīṣmam) | முதுவேனில் (mudhuvenil) |
| Karkata Siṃha | Mid July–Mid Sep | वर्षा | বর্ষা (Bôrsha) | വർഷം (Varṣām) | கார் (kaar) |
| Kanyā Tulā | Mid Sep–Mid Nov | शरद् | শরৎ (Shôrôt) | ശരത് (Śarat) | குளிர் (kulir) |
| Vrścika Dhanu | Mid Nov–Mid Jan | हेमन्त | হেমন্ত (Hemôntô) | ഹേമന്തം (Hemantam) | முன்பனி (munpani) |
| Makara Kumbha | Mid Jan–Mid March | शिशिर | শীত (Śeet) | ശിശിരം (Śiśiram) | பின்பனி (pinpani) |

### Lunar months

Lunar months are defined based on lunar cycles, i.e., the regular occurrence of new moon and full moon and the intervening waxing and waning phases of the moon.

### Paksha

A lunar month contains two fortnights called *pakṣa* (पक्ष). One fortnight is the bright, waxing half where the moon size grows and it ends in the full moon. This is called "*Gaura Paksha* (गौरा पक्ष)" or Shukla Paksha. The other half is the darkening, waning fortnight which ends in the new moon. This is called "*Vadhya Paksha* (वध्या पक्ष) " or *Krishna Paksha* (कृष्ण पक्ष). The Hindu festivals typically are either on or the day after the full moon night or the darkest night (*amavasya*, अमावास्या), except for some associated with *Krishna, Durga or Rama*. The lunar months of the hot summer and the busy major cropping-related part of the monsoon season typically do not schedule major festivals.

## 卐 Amanta and Purnimanta Systems

Two traditions have been followed in the Bhartiya subcontinent with respect to lunar months: the *amanta tradition* (अमन्त परम्परा), which ends the lunar month on new moon day (similar to the Islamic calendar) and the *purnimanta tradition* (पूर्णिमन्त परम्परा), which ends it on full moon day. As a consequence, in the *amanta tradition, Shukla paksha* precedes *Krishna paksha* in every lunar month, whereas in the purnimanta tradition, *Krishna paksha* precedes *Shukla paksha* in every lunar month. As a result, a *Shukla paksha* will always belong to the same month in both traditions, whereas a *Krishna paksha* will always be associated with different (but succeeding) months in each tradition. The *amanta* (also known as *Amavasyanta* - अमावस्यान्त or *Mukhyamana* - मुख्यामनः) tradition is followed by most Bhartiya states that have a peninsular coastline (except Assam, West Bengal, Odisha, Tamil Nadu and Kerala, which use their own solar calendars). These states are Gujarat, Maharashtra, Goa, Karnataka, Andhra Pradesh and Telangana. Nepal and most Bhartiya states north of the Vindhya mountains follow the purnimanta (or Gaunamana) tradition. The *purnimanta tradition* was being followed in the Vedic era. It was replaced with the *amanta tradition* as the Hindu calendar system prior to the 1st century BCE, but the *Purnimanta tradition* was restored in 57 BCE by *Vikramaditya*, who wanted to return to the Vedic roots. The presence of this system is one of the factors considered in dating ancient Bhartiya manuscripts and epigraphical evidence that have survived into the modern era.

The two traditions of *Amanta* and *Purnimanta* systems have led to alternate ways of dating any festival or event that occurs in a *Krishna paksha* in the historic Hindu, Buddhist or Jain literature, and contemporary regional literature or festival calendars. For example, the Hindu festival of *Maha Shivaratri* ()महा शिवरात्रि falls on the fourteenth lunar day of *Magha*'s (मघस्य) *Krishna paksha* in the *Amanta system*, while the same exact day is expressed as the fourteenth lunar day of *Phalguna's* (फाल्गुनस्य) *Krishna paksha* in the *Purnimanta system*. Both lunisolar calendar systems are equivalent ways of referring to the same date, and they continue to be in use in different regions, though the *Purnimanta system* is now typically assumed as implied in modern Indology literature if not otherwise specified (see Figure 9.3 and Table 9.3).

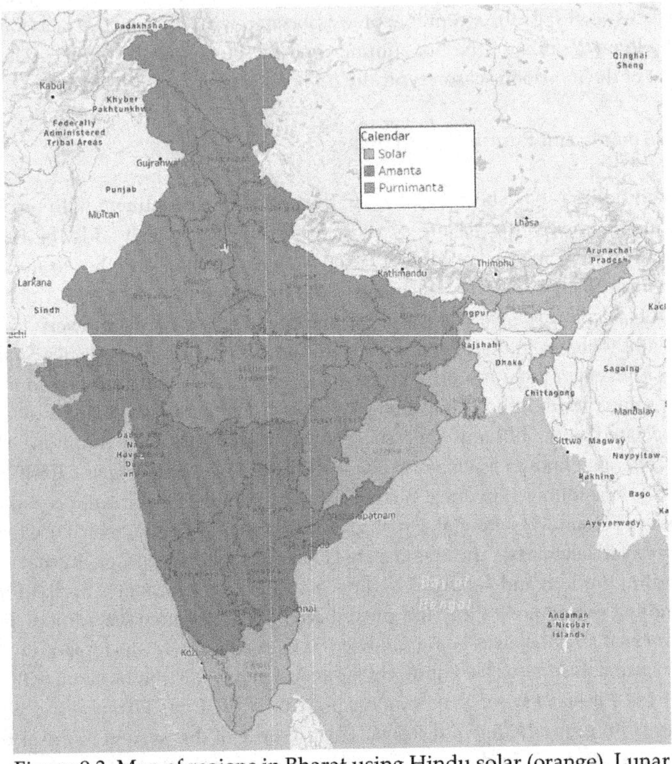


Figure 9.3. Map of regions in Bharat using Hindu solar (orange), Lunar *Amanta* (blue), and Lunar *Purnimanta* (red) Calendars.

Table 9.5. Variations in the naming of lunar months

| | Krishna Paksha | Shukla Paksha | Krishna Paksha |
|---|---|---|---|
| Amanta | Phalguna | Chaitra | |
| Purnimanta | Chaitra | | Amanta |

## 卐 List of Lunar Months

The names of the Hindu months vary by region. Those Hindu calendars which are based on lunar cycle are generally phonetic variants of each other, while the solar cycle are generally variants of each other too, suggesting that the timekeeping knowledge travelled widely across the Bhartiya subcontinent in ancient times. During each lunar month, the Sun transits into a sign of the zodicac (*sankranti* - संक्रान्तिः). The lunar month in which the Sun transits into *Mesha* (मेषः) is named *Chaitra* (चैत्र) and designated as the first month of the lunar year. A few major calendars are summarized in Table 9.6.

Table 9.6. Lunar calendar month names in different Hindu calendars

| # | Vikrami (lunar) | Sankranti | Hindi/ Marathi | Maithili | Meitei (Manipuri) | Nepali | Gregorian |
|---|---|---|---|---|---|---|---|
| 1 | Chaitra | Mēsha | चैत्र | Chait | ꯂꯝꯇꯥ (Lamta) | चैत (Chait) | March – April |
| 2 | Vaisākha | Vrisha | वैशाख | Baishakh | ꯁꯖꯤꯕꯨ (Sajibu) | वैशाख (Baishākh) | April – May |
| 3 | Jyeshta | Mithuna | ज्येष्ठ | Jeth | ꯀꯥꯂꯦꯟ (Kalen) | जेठ (Jēṭh) | May – June |
| 4 | Āshāda | Karka | आषाढ़ / आषाढ | Asadh | ꯏꯉꯥ (Eenga) | असार (Asār) | June – July |
| 5 | Shraavana | Singa | श्रावण | Saon | ꯏꯉꯦꯟ (Eengen) | साउन (Sāun) | July – August |
| 6 | Bhādra | Kanya | भाद्र / भाद्रपद | Bhado | ꯊꯧꯋꯥꯟ (Thouwan) | भदौ (Bhadau) | August – September |
| 7 | Ashwina | Tula | आश्विन | Aasin | ꯂꯥꯡꯕꯟ (Langban) | असोज (Asoj) | September – October |
| 8 | Kartika | Vrischika | कार्तिक | Katik | ꯃꯦꯔꯥ (Mera) | कात्तिक (Kāttik) | October – November |
| 9 | Māargasirsa (Agrahayana) | Dhanus | मार्गशीर्ष | Agahan | ꯍꯤꯌꯥꯡꯀꯩ (Heeyangkei) | मंसिर (Mangsir) | November – December |
| 10 | Pausha | Makara | पौष | Poos | ꯄꯣꯢꯅꯨ (Poinu) | पुष (Puṣ) | December – January |

| | | | | | | | |
|---|---|---|---|---|---|---|---|
| 11 | Māgha | Kumbha | माघ | Magh | ꯋꯥꯛꯆꯤꯡ (Wakching) | माघ (Magh) | January – February |
| 12 | Phālguna | Mina | फाल्गुण / फाल्गुन | Fagun | ꯐꯥꯏꯔꯦꯜ (Fairel) | फागुन (Phagun) | February – March |

## 卐 Corrections Between Lunar and Solar Months

The astronomical basis of the Hindu lunar months. Also illustrates *Adhika Masa* (Year 2-*Bhadrapada* - भद्रपदः) repeats; the first time the Sun moves entirely within *Simha Rashi* (सिंह राशि) thus rendering it an *Adhika Masa*.

Twelve Hindu mas (*māsa,* lunar month) are equal to approximately 354 days, while the length of a sidereal (solar) year is about 365 days. This creates a difference of about eleven days, which is offset every (29.53/10.63) = 2.71 years, or approximately every 32.5 months. *Purushottam Maas* (पुरुषोत्तम मास) or *Adhik Maas* (अधिक मास) is an extra month that is inserted to keep the lunar and solar calendars aligned. The twelve months are subdivided into six lunar seasons timed with the agriculture cycles, blooming of natural flowers, fall of leaves, and weather. To account for the mismatch between lunar and solar calendar, the Hindu scholars adopted intercalary months, where a particular month just repeated. The choice of this month was not random, but timed to sync back the two calendars to the cycle of agriculture and nature. The repetition of a month created the problem of scheduling festivals, weddings and other social events without repetition and confusion. This was resolved by declaring one month as *Shudha* (pure, clean, regular, proper, also called Deva month) and the other *Mala* or *Adhika* (extra, unclean and inauspicious, also called *Asura masa* असुर मास).

The Hindu mathematicians who calculated the best way to adjust the two years, over long periods of a yuga (era, tables calculating 1000 of years), they determined that the best means to intercalate the months is to time the intercalary months on a 19-year cycle, similar to the Metonic cycle used in the Hebrew calendar. This intercalation is generally adopted in the 3rd, 5th, 8th, 11th, 14th, 16th and 19th year of this cycle. Further, the complex rules rule out the repetition of *Mārgasirsa* - मार्गसिरसः (also called *Agrahayana* - अग्रहायणः), *Pausha* (पौषः) and *Maagha* (माघ) lunar months. The historic Hindu texts are not consistent on these rules, with competing ideas flourishing in the Hindu culture.

## 卐 Day

Just like months, the Hindu calendar has two measures of a day, one based on the lunar movement and the other on solar. The solar (*saura* - सौर) day or civil day, called *divasa* (दिवस), has been what most Hindus traditionally use, is easy and empirical to observe, with or without a clock, and it is defined as the period from one sunrise to another. The lunar day is called *tithi* (तिथि), and this is based on complicated measures of lunar movement. A lunar day or *tithi* may, for example, begin in the middle of an afternoon and end next afternoon. Both these days do not directly correspond to a mathematical measure for a day such as equal 24 hours of a solar year, a fact that the Hindu calendar scholars knew, but the system of *divasa* was convenient for the general population. The *tithi* have been the basis for timing rituals and festivals, while *divasa* for everyday use. The Hindu calendars adjust the mismatch in *divasa* and *tithi*, using a methodology similar to the solar and lunar months. In the Bhartiya tradition, the general population's practice has been to treat a tithi as a solar day between one sunrise to next.

A lunar month has 30 *tithi*. The technical standard makes each *tithi* contain different number of hours, but helps the overall integrity of the calendar. Given the variation in the length of a solar day with seasons, and the Moon's relative movements, the start and end time for *tithi* varies over the seasons and over the years, and the *tithi* adjusted to sync with *divasa* periodically with intercalation.

### 卐 Vāsara (Weekday)

Vāsara (वासरः) refers to the weekdays in Sanskrit. Also referred to as Vara and used as a suffix. The correspondence between the names of the week in Hindu and other Indo-European calendars are exact. This alignment of names probably took place sometime during the 3rd century CE. The weekday of a Hindu calendar has been symmetrically divided into 60 *ghatika*, each *ghatika* (24 minutes) is divided into 60 *pala*, each *pala* (24 seconds) is subdivided into 60 *vipala*, and so on (see Table 9.7).

Table 9.7 Vāsara varieties

| Sanskrit | Celestial object | Bengali | Hindi | Latin weekday |
|---|---|---|---|---|
| Ravivāsara<br>रविवासर or<br>Surya vāsara<br>आदित्य वासर | Ravi, Aditya<br>=<br>Sun | Rôbibār<br>রবিবার | Ravivār<br>रविवार | Sunday/<br>dies Solis |
| Somavāsara<br>सोमवासर or<br>Indu vāsara<br>इन्दु वासर | Soma (deity),<br>Chandra<br>=<br>Moon | Śombār<br>সোমবার | Somavār<br>सोमवार | Monday/dies<br>Lunae |
| Maṅgalavāsara<br>मङ्गलवासर or<br>Bhaumavāsara<br>भौम वासर | Maṅgala<br>=<br>Mars | Môṅgôlbār<br>মঙ্গলবার | Maṅgalavār<br>मंगलवार | Tuesday/<br>dies Martis |
| Budhavāsara<br>बुधवासर or<br>Saumya vāsara<br>सौम्य वासर | Budha<br>=<br>Mercury | Budhbār<br>বুধবার | Budhavāra<br>बुधवार | Wednesday/<br>dies<br>Mercuri |
| Guruvāsara<br>गुरुवासर<br>or<br>Brhaspati vāsara<br>बृहस्पतिवासर | Deva-Guru<br>Bṛhaspati<br>= Jupiter | Brihôśpôtibār<br>বৃহস্পতিবার | Guruvār<br>गुरुवार<br>Brihaspativāra<br>बृहस्पतिवार | Thursday/di<br>es Iovis/Jupit<br>er |
| Śukravāsara<br>शुक्रवासर | Śukra<br>=<br>Venus | Śukrôbār<br>শুক্রবার | Śukravār<br>शुक्रवार | Friday/dies<br>Veneris |

| | | | | |
|---|---|---|---|---|
| Śanivāsara<br>शनिवासर<br>or<br>Śaniścaravāsara<br>शनिश्चरवासर | Śani<br>=<br>Saturn | Śônibār | Śanivār | Saturday/<br>dies Saturnis |
| | | শনিবার | शनिवार | |

The term - *vāsara* is often realised as *vāra* or *vaar* in Sanskrit-derived and influenced languages. There are many variations of the names in the regional languages, mostly using alternate names of the celestial bodies involved.

## 卐 Astronomical Basis of the Hindu Calendar

The Hindu calendar is based on a geocentric model of the Solar System. A geocentric model describes the Solar System as seen by an observer on the surface of the Earth. The Hindu calendar defines nine measures of time (मान):

- *brāhma māna*
- *divya māna*
- *pitraya māna*
- *prājāpatya māna*
- *guror māna*
- *saura māna*
- *sāvana māna*
- *candra māna*
- *nākṣatra māna*

Of these, only the last four are in active use and are explained here.

## 卐 Candra māna (चन्द्र माना)

The candra māna (चन्द्र मान) of the Hindu calendar is defined based on the movement of the Moon around the Earth. The new moon (अमावास्य) and full moon (पूर्णिमा) are important markers in this calendar. The *candra māna* of the Hindu calendar defines the following synodic calendar elements:

- Pakṣa: A *pakṣa* (पक्ष) is the time taken by the Moon to move from a new moon to a full moon and vice versa. The waxing phase of the moon is known as the bright side (शुक्ल पक्ष) and the waning phase is known as the dark side (कृष्ण पक्ष). During a *pakṣa*, the Moon advances 180° with respect to the Earth-Sun axis.

- Candramāsa: A *cāndramāsa* (चन्द्रमास) is the time taken by the moon to move from a new moon to the next new moon (अमान्त) or a full moon to the next full moon (as per the *pūrṇimānta* (पूर्णिमान्त). In other words, a cāndramāsa is the synodic period of the Moon, or two *pakṣas*. During a *cāndramāsa*, the Moon advances 360° with respect to the Earth-Sun axis.
- Candra māna varṣa: A *candra māna varṣa* (चन्द्र मान वर्ष) or lunar year is made up of 12 consecutive *candramāsa*. These twelve *candramāsa* are designated by unique names *caitra, vaiśākha*, etc. In some instances, an additional *candramāsa*, known as an *adhikamāsa*, is added to synchronise the *candra māna varṣa* with the solar year or *saura māna varṣa*.
- Tithi: A tithi (तिथि) is the time taken by the Moon to advance 12° with respect to the Earth-Sun axis. In other words, a *tithi* is the time taken for the Moon's elongation (on the ecliptic plane) to increase by 12°. A *tithi* is one fifteenth of a *pakṣa* and one thirtieth of a *cāndramāsa*. A *tithi* corresponds to the concept of a lunar day. Tithi have Sanskrit numbers according by their position in the *pakṣa*, i.e., *prathama* (first), *dvitīya* (second) etc. The fifteenth, that is, the last *tithi* of a *kṛṣṇa pakṣa* is called *amāvāsya* (new moon) and the fifteenth *tithi* of a *śukla pakṣa* is called *pūrṇimā* (full moon).

### 卐 Saura māna

The saura māna (सौर मान) of the Hindu calendar is defined by the movement of the Earth around the Sun. It contains sidereal (निरयन) and tropical (सायन) elements.

- Sidereal elements: Animation that illustrates the twelve *rāśi* (and *sauramāsa*) that make up a *saura māna varṣa*. A *saura māna varṣa* or sidereal year is the time taken by the Sun to orbit the Earth once and return to the starting point with respect to the fixed stars. The starting point is taken to be the position of the Sun when it is in opposition to *Spica* (चित्रा).
- A rāśi (राशि) is a 30° arc of the orbit of the Sun around the Earth (i.e., an arc of the ecliptic). Starting in the vicinity of Zeta Piscium (*revatī*), the twelve (i.e., 360° divided by 30°) *rāśi* are designated *meṣa* (मेष), *vṛṣabha* (वृषभ) etc. A *sauramāsa* (सौरमास) is the time taken by the Sun

to traverse a *rāśi*. *Sauramāsa* get their names from the corresponding *rāśi*. *sauramāsa* corresponds to the concept of a month. The moment in time when the Sun enters a *rāśi* is known as a *saṅkramaṇa* (सङ्क्रमण) or *saṅkrānti* (सङ्क्रान्ति).

## 卐 Tropical Elements

These time periods are defined based on the solstices (*ayana* or अयन) and equinoxes (*viṣuvat* or विषुवत्). The time taken by the Sun to move from the winter solstice to the summer solstice is known as northward movement (*uttarāyaṇa* or उत्तरायण) and time taken by the Sun to move from the summer solstice to the winter solstice is called southward movement (dakṣiṇāyana or दक्षिणायन). Due to the axial tilt of the Earth, the Sun appears to move towards the north from the Tropic of Capricorn to the Tropic of Cancer during *uttarāyaṇa*, and towards the south from the tropic of Cancer to the tropic of Capricorn during *dakṣiṇāyana*. The time taken by the Sun to move from the spring equinox (ecliptic longitude 0°) to the autumnal equinox (ecliptic longitude 180°) is known as *devayāna* (देवयान). The time taken by the Sun to move from the autumnal equinox to the spring equinox is designated as *pitṛyāṇa* (पितृयाण). Due to the axial tilt of the Earth, the Sun appears to be in the north celestial sphere during *devayāna* and the south celestial sphere during *pitṛyāṇa*. In Hindu tradition, the north celestial sphere is consecrated to the gods (deva) and the south celestial sphere is consecrated to the ancestors (*pitṛ*). *Devayāna* and *pitṛyāṇa* are not in active calendric use any longer but do form the basis for *pitṛpakṣa*.

## 卐 A *ṛtu* (ऋतु)

It is the time taken by the Sun to move sixty degrees on its orbit around the Earth. *Ṛtu* corresponds to the concept of a season. The six ṛtu of the year are known as:

1. *Śiśira ṛtu* (winter)
2. *Vasanta ṛtu* (spring)
3. *Grīṣma ṛtu* (summer)
4. *Varṣā ṛtu* the monsoon season, beginning at summer solstice
5. *Śarada ṛtu* (autumn)
6. *Hemanta ṛtu* (pre-winter)

## 卐 Nākṣatra māna

*Nākṣatra māna* (नाक्षत्र मान) is defined with respect to the fixed stars, so all elements are sidereal in nature.

- A dina (दिन) is the time taken by the celestial sphere to complete one sidereal rotation around the Earth. In reality, this movement is caused by the diurnal rotation of the Earth on its axis. This definition is not used in practice but is required for defining the following smaller units of time. Ā *dina* is ~4 minutes short of 24 hours.
- A *ghaṭikā* (घटिका) or *nāḍī* (नाडी) is one sixtieth of a *nakṣatra dina*, or just under 24 minutes.
- A *vighaṭikā* (विघटिका) or vināḍī (विनाडी) is one sixtieth of a *ghaṭikā*, or just under 24 seconds.
- A *prāṇa* (प्राण) or asu (असु) is one sixth of a *vighaṭikā*, or just under four seconds.

## 卐 Sāvana māna

Sāvana māna (सावन मान) of the Hindu calendar defines civil time. A dina (दिन) is the time between two succeeding sunrises. *dina* corresponds to the concept of a solar day. The length of a *dina* varies with daytime length.

## 卐 Nakṣatra

Animation that illustrates the twenty-seven *nakṣatra* (asterism). Apart from the four *māna* explained above, the concept of *nakṣatra* is an important characteristic of the Hindu calendar. This term has multiple meanings:

1. A *nakṣatra* (नक्षत्र) is a star.
2. A *nakṣatra* is an asterism. One of the stars in the asterism is designated as its principal star (*yogatārā* - योगतारा). There are twenty-eight such *nakṣatra* and they are individually named. The name of a *nakṣatra* and its *yogatārā* are identical. For example, *revatī* is an asterism whose principal star is *revatī* (Zeta Piscium).
3. A *nakṣatra* is a 13° 20' arc of the ecliptic. There are twenty-seven such *nakṣatra* (i.e., 360° divided by 13° 20'). Starting in the vicinity of *revatī* (Zeta Piscium), they are named *aśvinī*, *bharaṇī* etc. These names are identical to the names of the asterisms that are located within the respective arc segments. For example, revatī refers to both an asterism and the arc segment within which the asterism is located.

4. In calendric terms, a *nakṣatra* is the time taken by the Moon to traverse a *nakṣatra*. Hence, *nakṣatra* is a sidereal element (unlike the *tithi* which it is similar to) and corresponds to the concept of a day.

## 卐 Combining the Different Measures of Time

The four *māna* explained above are used in combination in the Hindu calendar.

- **adhikamāsa:** As seen above, both the *cāndra māna* and *saura māna* of the calendar define a *varṣa* comprising twelve *māsa*, but the duration of the *varṣa* differ; the *cāndra māna varṣa* is shorter than the *saura māna varṣa* by about eleven *sāvana dina*. As a result, unless explicitly synchronised, these two parts of the calendar will diverge over time, as the *cāndra māna varṣa* will keep "*falling behind*" the *saura māna varṣa*. In order to synchronise these two parts of the calendar, an additional cāndramāsa is introduced into some cāndra māna varṣa. Such a *cāndramāsa* is referred to as *adhikamāsa* (अधिकमास). A *adhikamāsa* takes its name from the name of the *cāndramāsa* which follows, viz. *adhika āśvina* precedes *āśvina*. Most times every *cāndramāsa* witnesses a *saṅkramaṇa*. If a *cāndramāsa* does not witness a *saṅkramaṇa*, that *cāndramāsa* is designated as a *adhikamāsa* thus resulting in the cāndra māna varṣa "*catching up*" with the *saura māna varṣa*. This happens approximately once every two and a half (solar) years.
- **dina and tithi:** As seen above, both the *cāndra māna* and *sāvana māna* of the calendar define the concept of a day as *tithi* and *dina* respectively. *dina* are not named and are not used for calendric purposes. The *tithi* takes precedence instead. Human life is regulated by the rising of the Sun and not by the movement of the Moon through a 12° arc. Hence, the position of the Moon at sunrise is used to determine the *tithi* prevailing at sunrise. This *tithi* is then associated with the entire *sāvana dina*. To illustrate: consider the Gregorian date 18th Sep 2021. Instead of referring to it as "*2nd dina of kanyā masa*" Hindus will refer to it as " *bhādrapada māsa, śukla pakṣa, dvitiyā tithi*", which is the tithi prevailing at sunrise on that sāvana dina. Even though the Moon moves into the trayodaśī arc soon after sunrise (at 6:54AM), that entire *sāvana dina* is considered to be *dvādaśī tithi*.

- **adhika tithi and kṣaya tithi:** It is possible that two consecutive sunrises may have the same *tithi*, i.e., the Moon continues to remain within the same 12° arc across two consecutive sunrises. In such a case, two consecutive *sāvana dina* will be associated with the same *tithi*. The *tithi* associated with the second *sāvana dina* is referred to as a *adhika* (अधिक) (additional) *tithi*. It is also possible that an entire *tithi* elapses between two sunrises, i.e., the Moon traverses a 12° arc in between two sunrises (it enters the arc after one sunrise and exits the arc before the next sunrise). In this such a case, neither *sāvana dina* will be associated with this *tithi*, i.e., this *tithi* will be skipped over in the calendar. Such a *tithi* is referred to as a *kṣaya* (क्षय) (waste) *tithi*.

- **Subdivisions of a sāvana dina:** Above that *a nakṣatra dina* is divided into *ghaṭikā* (of 24 modern minutes each) and v*ighaṭikā* (of 24 modern seconds each). These same units are used to subdivide a *savana dina* using sunrise as the starting point, i.e., the first 24 minutes after sunrise constitute the first *ghaṭikā*, the next 24 minutes the second *ghaṭikā* and so on.
- **pitṛpakṣa:** pitṛpakṣa (पितृपक्ष) is a pakṣa during which the Sun crosses the equator and transitions overhead the southern hemisphere, i.e., the autumnal equinox occurs within *pitṛpakṣa*. *bhādrapada māsa kṛṣṇa pakṣa* is identified with *pitṛpakṣa*. This identification is not always correct. For instance, in the Gregorian year 2020, *bhādrapada māsa kṛṣṇa pakṣa* (भद्रपद मास कृष्णपक्षः) ended with the new moon on 17 September while autumnal equinox occurred five days later, on 22 September.

1. All examples in this article assume the *amānta* tradition.
2. These names are derived from the *nakṣatra* in which the Moon is positioned at the time of full moon.
3. Not everyone is in agreement with this definition. Mercier argues that Ketkar has interpreted a Sanskrit source in a way that is different from other authorities. Yet, this definition is widely used to create Hindu almanacs or *pañcāṅga*).
4. The Surya Siddhantha defines *uttarāyaṇa* and *dakṣiṇāyana* using *rāśi* instead of the equinoxes and solstices. That definition assumed a coincidence of the winter solstice and makara *saṅkramaṇa*. As a result of the precesstion of the equinoxes, that coincidence no longer exists thus making that definition incorrect. To illustrate, as per the Surya Siddhantha definition, the period from winter solstice (Dec 21) to makara *saṅkramaṇa* (Jan 14) is considered part

of *dakṣiṇāyana* but the Sun is moving towards the north during this period.

5. The Surya Siddhanta defines *ṛtu* in terms of various *rāśi* assuming that makara *saṅkramaṇa* coincides with the winter solstice. Due to the precession of the equinoxes, that assumption is no longer true and hence those definitions of ṛtu are no longer accurate.
6. Since a *rāśi* is a 30° arc of the ecliptic, a *ṛtu* can be considered as the time taken by the Sun to transit through two *rāśi.*
7. A sidereal rotation is defined with respect to the fixed stars, i.e., at the end of a sidereal rotation all the fixed stars are back in their starting position.
8. *abhijit* is an asterism for which there is no corresponding arc segment.
9. This is a common calendric technique and is known as intercalation
10. As a result, almost all Hindu festivals are defined in *cāndra māna* terms. Hence these annual festivals do not repeat on the same day on any solar calendar (neither *saurana māna* nor Gregorian).
11. Since the south celestial sphere is consecrated to the ancestors (*pitṛ*), Hindus perform special religious rites in honour of their ancestors during *pitṛpakṣa*.

## 卐 Five Limbs of Time

The complete Vedic calendars contain five *angas* (अङ्गः) or parts of information: (i) lunar day *(tithi)*, (ii) solar day (*diwas*), (iii) asterism *(naksatra)*, (iv) planetary joining (*yoga*) and (v) astronomical period (*karanam).* This structure gives the calendar the name Panchangam. The first two are discussed above.

## 卐 Yoga

The Sanskrit word Yoga means "union, joining, attachment", but in astronomical context, this word means latitudinal and longitudinal information. The longitude of the Sun and the longitude of the Moon are added, and normalised to a value ranging between 0° to 360° (if greater than 360, one subtracts 360). This sum is divided into 27 parts. Each part will now equal 800' (where ' is the symbol of the arcminute which means 1/60 of a degree). These parts are called the yogas. They are labelled:

| | | | |
|---|---|---|---|
| *Viṣkambha* | *Dhṛti* | *Vajra* | *Sādhya* |
| *Prīti* | *Śūla* | *Siddhi* | *Śubha* |
| *Āyuśmān* | *Gaṇḍa* | *Vyatipāta* | *Śukla* |
| *Saubhāgya* | *Vṛddhi* | *Variyas* | *Brahma* |
| *Śobhana* | *Dhruva* | *Parigha* | *Māhendra* |
| *Atigaṇḍa* | *Vyāghatā* | *Śiva* | *Vaidhṛti* |
| *Sukarma* | *Harṣaṇa* | *Siddha* | |

Again, minor variations may exist. The yoga that is active during sunrise of a day is the prevailing yoga for the day.

### 卐 Karaṇa

A *karaṇa* is half of a *tithi*. To be precise, a *karaṇa* is the time required for the angular distance between the Sun and the Moon to increase in steps of 6° starting from 0°. Since the *tithis* are 30 in number, and since 1 *tithi* = 2 *karaṇas,* therefore one would logically expect there to be 60 *karaṇas.* But there are only 11 such *karaṇas* which fill up those slots to accommodate for those 30 *tithis.* There are actually 4 "fixed" *(sthira) karaṇas* and 7 "repeating" *(cara) karaṇas.*

The 4 "fixed" karaṇas are:

1. *Śakuni* (शकुनि)
2. *Catuṣpāda* (चतुष्पाद)
3. *Nāga* (नाग)
4. *Kiṃstughna* (किंस्तुघ्न)

The 7 "repeating" karaṇas are:

1. *Vava or Bava* (बव)
2. *Valava or Bālava* (बालव)
3. *Kaulava* (कौलव)
4. *Taitila or Taitula* (तैतिल)
5. *Gara or Garaja* (गरज)
6. *Vaṇija* (वणिज)
7. *Viṣṭi (Bhadrai* (भद्रा)

Now the first half of the 1st tithi (of *Śukla Pakṣa*) is always *Kiṃtughna karaṇa*. Hence this *karaṇa* is "*fixed*". Next, the 7-repeating *karaṇas* repeat eight times to cover the next 56 half-*tithis*. Thus, these are the "repeating" (cara) *karaṇas*. The 3 remaining half-tithis take the remaining "*fixed*" *karaṇas* in order. Thus, these are also "fixed" *(sthira)*. Thus, one gets 60 *karaṇas* from those 11 *preset karaṇas*. The Vedic day begins at sunrise. The *karaṇa* at sunrise of a particular day shall be the prevailing *karaṇa* for the whole day.

## 卐 Further Reading


1. "Ekadasi: Why Ekadasi is celebrated in Hinduism?-by Dr Bharti Raizada". NewsGram. 22 May 2017.
2. "Maha Shivaratri date". drikpanchang.com.
3. "Satapatha-brahmana Verse 8.7.3.10 [Sanskrit text]". www.wisdomlib.org. 18 September 2021. Retrieved 8 December 2022.
4. "What is Shukla Paksha and Krishna Paksha | Phases of Moon". 22 July 2015. Archived from the original on 7 August 2020. Retrieved 20 August 2018.
5. Andrea Acri (2016). Esoteric Buddhism in Mediaeval Maritime Asia: Networks of Masters, Texts, Icons. ISEAS-Yusof Ishak Institute. pp. 256–258. ISBN 978-981-4695-08-4.
6. Anita Ganeri (2003). Buddhist Festivals Through the Year. BRB. pp. 11–12. ISBN 978-1-58340-375-4.
7. Burgess 1935, p. 104 (II. 64).
8. Burgess 1935, p. 106.
9. Burgess 1935, p. 16 (I. 28).
10. Burgess 1935, p. 207.
11. Burgess 1935, p. 230.
12. Burgess 1935, p. 310 (XIV. 1).
13. Burgess 1935, p. 310 (XIV. 2).
14. Burgess 1935, p. 313 (XIV. 10).
15. Burgess 1935, p. 313 (XIV. 9).
16. Burgess 1935, p. 314 (XIV. 15).
17. Burgess 1935, p. 319 (XIV. 18).
18. Burgess 1935, p. 5 (I. 11).

19. Burgess 1935, p. 7 (I. 13).
20. Burgess 1935, p. 8.
21. Burgess 1935, pp. 202–250.
22. C. K. Raju (2007). Cultural Foundations of Mathematics. Pearson. p. 205. ISBN 978-81-317-0871-2.
23. Christopher John Fuller (2004). The Camphor Flame: Popular Hinduism and Society in India. Princeton University Press. pp. 109–110, 291–293. ISBN 978-0-69112-04-85.
24. Christopher John Fuller (2004). The Camphor Flame: Popular Hinduism and Society in India. Princeton University Press. pp. 109–110. ISBN 978-0-69112-04-85.
25. Christopher John Fuller (2004). The Camphor Flame: Popular Hinduism and Society in India. Princeton University Press. pp. 291–293. ISBN 978-0-69112-04-85.
26. Colette Caillat; J. G. de Casparis (1991). Middle Indo-Aryan and Jain Studies. BRILL. p. 36. ISBN 90-04-09426-1.
27. D. C. Sircar (1965). Indian Epigraphy. Motilal Banarsidass. pp. 304–305 with footnotes. ISBN 978-81-208-1166-9.
28. Duncan Graham (2004). The People Next Door: Understanding Indonesia. University of Western Australia Press. pp. 16–17. ISBN 978-1-920694-09-8.
29. Ebenezer Burgess (1989). P Ganguly, P Sengupta (ed.). Sûrya-Siddhânta: A Text-book of Hindu Astronomy. Motilal Banarsidass (Reprint), Original: Yale University Press, American Oriental Society. pp. 26–27. ISBN 978-81-208-0612-2.
30. Ebenezer Burgess (1989). P Ganguly, P Sengupta (ed.). Sûrya-Siddhânta: A Text-book of Hindu Astronomy. Motilal Banarsidass (Reprint), Original: Yale University Press, American Oriental Society. pp. ix–xi, xxix. ISBN 978-81-208-0612-2.
31. Ebenezer Burgess (1989). P Ganguly, P Sengupta (ed.). Sûrya-Siddhânta: A Text-book of Hindu Astronomy. Motilal Banarsidass (Reprint), Original: Yale University Press, American Oriental Society. pp. vii–xi. ISBN 978-81-208-0612-2.
32. Ebenezer Burgess (1989). Sûrya-Siddhânta: A Text-book of Hindu Astronomy. Motilal Banarsidass. pp. 107–. ISBN 978-81-208-0612-2.
33. Eleanor Nesbitt (2016). Sikhism: a Very Short Introduction. Oxford University Press. pp. 122, 142. ISBN 978-0-19-874557-0.

34. Eleanor Nesbitt (2016). Sikhism: a Very Short Introduction. Oxford University Press. pp. 122–123. ISBN 978-0-19-874557-0.
35. Erik Gregersen (2011). The Britannica Guide to the History of Mathematics. The Rosen Publishing Group. p. 187. ISBN 978-1-61530-127-0.
36. Friedrich Max Müller (1860). A History of Ancient Sanskrit Literature. Williams and Norgate. pp. 210–215.
37. J Fleet (1911). "Arbhatiya". Journal of the Royal Asiatic Society of Great Britain and Ireland. Cambridge University Press for the Royal Asiatic Society: 794–799.
38. J. G. De Casparis (1978). Indonesian Chronology. BRILL Academic. pp. 15–24. ISBN 90-04-05752-8.
39. J. Gordon Melton (2011). Religious Celebrations: An Encyclopedia of Holidays, Festivals, Solemn Observances, and Spiritual Commemorations. ABC-CLIO. pp. 652–653. ISBN 978-1-59884-205-0.
40. James Lochtefeld (2002), "Jyotisha" in The Illustrated Encyclopedia of Hinduism, Vol. 1: A–M, Rosen Publishing, ISBN 0-8239-2287-1, pp. 326–327
41. Jeffery D Long (2013). Jainism: An Introduction. I.B.Tauris. pp. 6–7. ISBN 978-0-85771-392-6.
42. John E. Cort (2001). Jains in the World: Religious Values and Ideology in India. Oxford University Press. p. 228 note 2. ISBN 978-0-19-513234-2.
43. John E. Cort (2001). Jains in the World: Religious Values and Ideology in India. Oxford University Press. pp. 142–146. ISBN 978-0-19-513234-2.
44. Kak, Subhash (2015). "The Mahabharata and the Sindhu-Sarasvati Tradition" (PDF). Sanskrit Magazine. p. 2. Retrieved 22 January 2015.
45. Ketkar 1923. pp. 34–35.
46. Kim Plofker 2009, p. 36.
47. Kim Plofker 2009, pp. 10, 35–36, 67.
48. Kim Plofker 2009, pp. 116–120, 259–261.
49. Kim Plofker 2009, pp. 35–42.
50. Klaus K. Klostermaier (2007). A Survey of Hinduism: Third Edition. State University of New York Press. p. 490. ISBN 978-0-7914-7082-4.

51. Klaus K. Klostermaier (2007). A Survey of Hinduism: Third Edition. State University of New York Press. pp. 490–492. ISBN 978-0-7914-7082-4.
52. Lionel D. Barnett (1994). Antiquities of India: An Account of the History and Culture of Ancient Hindustan. Asian Educational Services. pp. 190–192. ISBN 978-81-206-0530-5.
53. M. C. Ricklefs; P. Voorhoeve; Annabel Teh Gallop (2014). Indonesian Manuscripts in Great Britain: A Catalogue of Manuscripts in Indonesian Languages in British Public Collections. Yayasan Pustaka Obor Indonesia. pp. 49, 69–73, 81. ISBN 978-979-461-883-7.
54. Mercier (2018). pp. 74–75.
55. Monier Monier-Williams (1923). A Sanskrit–English Dictionary. Oxford University Press. p. 353.
56. Monier Monier-Williams, वासर, Sanskrit-English Dictionary, Oxford University Press, page 948.
57. Muriel Marion Underhill (1991). The Hindu Religious Year. Asian Educational Services. pp. 20, 32 note 5. ISBN 978-81-206-0523-7.
58. Muriel Marion Underhill (1991). The Hindu Religious Year. Asian Educational Services. pp. 23, 26–27. ISBN 978-81-206-0523-7.
59. Muriel Marion Underhill (1991). The Hindu Religious Year. Asian Educational Services. pp. 23–28. ISBN 978-81-206-0523-7.
60. Muriel Marion Underhill (1991). The Hindu Religious Year. Asian Educational Services. pp. 24–25. ISBN 978-81-206-0523-7.
61. Muriel Marion Underhill (1991). The Hindu Religious Year. Asian Educational Services. pp. 27–28. ISBN 978-81-206-0523-7.
62. Nachum Dershowitz; Edward M. Reingold (2008). Calendrical Calculations. Cambridge University Press. pp. 123–133, 153–161, 275–311. ISBN 978-0-521-88540-9.
63. Nachum Dershowitz; Edward M. Reingold (2008). Calendrical Calculations. Cambridge University Press. pp. 123–133, 275–311. ISBN 978-0-521-88540-9.
64. Nicholas Campion (2012). Astrology and Cosmology in the World's Religions. New York University Press. pp. 110–111. ISBN 978-0-8147-0842-2.
65. Orazio Marucchi (2011). Christian Epigraphy: An Elementary Treatise with a Collection of Ancient Christian Inscriptions Mainly of

Roman Origin. Cambridge University Press. p. 289. ISBN 978-0-521-23594-5., Quote: "the lunar year consists of 354 days".
66. Peter J. Claus; Sarah Diamond; Margaret Ann Mills (2003). South Asian Folklore: An Encyclopedia : Afghanistan, Bangladesh, India, Nepal, Pakistan, Sri Lanka. Taylor & Francis. pp. 91–93. ISBN 978-0-415-93919-5.
67. Pingree 1973, p. 2.
68. Pingree 1973, pp. 2–3.
69. Richard Salomon (1998). Indian Epigraphy: A Guide to the Study of Inscriptions in Sanskrit, Prakrit, and the other Indo-Aryan Languages. Oxford University Press. pp. 181–183. ISBN 978-0-19-535666-3.
70. Robert E. Buswell Jr.; Donald S. Lopez Jr. (2013). The Princeton Dictionary of Buddhism. Princeton University Press. p. 156. ISBN 978-1-4008-4805-8.
71. Robert Sewell; Śaṅkara Bālakr̥shṇa Dīkshita (1896). The Indian Calendar. S. Sonnenschein. pp. 29–34, 48–56.
72. Robert Sewell; Śaṅkara Bālakr̥shṇa Dīkshita (1896). The Indian Calendar: With Tables for the Conversion of Hindu and Muhammadan Into A.D. Dates, and Vice Versa. S. Sonnenschein. pp. 9–12.
73. Roshen Dalal (2010). Hinduism: An Alphabetical Guide. Penguin Books. p. 89. ISBN 978-0-14-341421-6.
74. Scott L. Montgomery; Alok Kumar (2015). A History of Science in World Cultures: Voices of Knowledge. Routledge. pp. 103–106. ISBN 978-1-317-43906-6.
75. Tej Bhatia (2013). Punjabi. Routledge. pp. 208–209. ISBN 978-1-136-89460-2.
76. The Shalivahan Shaka calendar follows the Amant system. The year begins on the first day of the bright fortnight of the month of Chaitra.Muriel Marion Underhill (1921). The Hindu Religious Year. Association Press. p. 15.
77. Tilak 1955, pp. 20–31
78. Underhill, Muriel Marion (1991). The Hindu Religious Year. Asian Educational Services. pp. 20–21. ISBN 978-81-206-0523-7.
79. V. R. Ramachandra Dikshitar (1993). The Gupta Polity. Motilal Banarsidass. pp. 24–35. ISBN 978-81-208-1024-2.

80. Yukio Ohashi 1993, pp. 185–251.
81. Yukio Ohashi 1999, p. 719.
82. Yukio Ohashi 1999, p. 720.
83. Yukio Ohashi 1999, pp. 719–721.
84. Ahargana - The Astronomy of the Hindu Calendar Explains the various calendric elements of the Hindu calendar by means of astronomical simulations created using Stellarium https://www.youtube.com/@ahargana.
85. Ahargana - The Astronomy of the Hindu Calendar Explains the various calendric elements of the Hindu calendar by means of astronomical simulations created using Stellarium. https://www.youtube.com/@ahargana.
86. B. Richmond (1956). Time Measurement and Calendar Construction. Brill Archive. pp. 80–82. Retrieved 18 September 2011.
87. Burgess 1935, p. 285 (XII. 32).
88. Burgess, Ebenezer (1935). Gangooly, Phanidarlal (ed.). Translation of the Surya Siddhanta – a text-book of Hindu astronomy (PDF). University of Calcutta.
89. Converter: Gregorian and 1957 Normalized Indian Calendar, Shalivahana Hindu calendar, United Kingdom https://web.archive.org/web/20170524154539/http://www.bsswebsite.me.uk/Daysanddates/indiandate.htm.
90. drikPanchang, an online Hindu almanac (IAST: pañcāṅga). https://www.drikpanchang.com/.
91. Hindu Calendar - thedivineindia.com https://www.thedivineindia.com/hindu-calcendar/6931.
92. Hindu Calendar of Nepal The Official Hindu Calendar of Nepal http://nepcal.com/
93. Hindu Calendars Monthly - monthlycalendars.in https://monthlycalendars.in/index.php.
94. Hindu_calendar, https://en.wikipedia.org/wiki/Hindu_calendar.
95. https://en.wikipedia.org/wiki/Astronomical_basis_of_the_Hindu_calendar.
96. Indian panchang and Hindu Calendar Bharat Calendar - Hindu Calendar with in-depth details https://play.google.com/store/apps/details?id=com.india.hindicalender

97. Ketkar, Venkatesh Bapuji (1923). "Indian and Foreign Chronology". Journal of the Asiatic Society of Bombay. British Indian Press. Issue 75, Part 1 of Journal: Extra number.
98. Kim Plofker (2009). Mathematics in India. Princeton University Press. ISBN 978-0-691-12067-6.
99. Kyoto University Gregorian – Saka – Vikrami Calendar Converter Tool, M. YANO and M. FUSHIMI http://www.cc.kyoto-su.ac.jp/~yanom/pancanga/.
100. Maurice Winternitz (1963). History of Indian Literature, Volume 1. Motilal Banarsidass. ISBN 978-81-208-0056-4.
101. Mercier, Raymond (2018). Astronomical Computations for the History of Indian Astronomy. New Delhi: Munshiram Manoharlal Publishers Pvt. Ltd. ISBN 978-81-215-1177-3.
102. Pingree, David (1973). "The Mesopotamian Origin of Early Indian Mathematical Astronomy". Journal for the History of Astronomy. SAGE. 4 (1): 1–12. Bibcode:1973JHA.....4....1P. doi:10.1177/002182867300400102. S2CID 125228353.
103. Pingree, David (1981). Jyotihśāstra : Astral and Mathematical Literature. Otto Harrassowitz. ISBN 978-3447021654.
104. Rai Bahadur Pandit Gaurishankar Hirachand Ojha, The Paleography of India, 2 ed., Ajmer, 1918, reprinted Manshuram Manoharlal publishers, 1993.
105. Reingold and Dershowitz, Calendrical Calculations, Millennium Edition, Cambridge University Press, latest 2nd edition 3rd printing released November 2004. ISBN 0-521-77752-6.
106. S. Balachandra Rao, Indian Astronomy: An Introduction, Universities Press, Hyderabad, 2000.
107. Stellarium, the astronomy software that was used to create the animations featured in this article. https://stellarium.org/.
108. Tilak, Bal Gangadhar (1955). The Orion or Researches into the Antiquity of the Vedas (PDF). Tilak Bros.
109. Yukio Ohashi (1993). "Development of Astronomical Observations in Vedic and post-Vedic India". Indian Journal of History of Science. 28 (3).
110. Yukio Ohashi (1999). Johannes Andersen (ed.). Highlights of Astronomy, Volume 11B. Springer Science. ISBN 978-0-7923-5556-4.

Made in United States
North Haven, CT
04 November 2026

21766912R00378